Lecture Notes in Computer Science 9141

Commenced Publication in 1973
Founding and Former Series Editors:
Gerhard Goos, Juris Hartmanis, and Jan van Leeuwen

More information about this series at http://www.springer.com/series/7407

Ying Tan · Yuhui Shi
Fernando Buarque · Alexander Gelbukh
Swagatam Das · Andries Engelbrecht (Eds.)

Advances in Swarm and Computational Intelligence

6th International Conference, ICSI 2015
held in conjunction with the Second BRICS
Congress, CCI 2015
Beijing, China, June 25–28, 2015
Proceedings, Part II

 Springer

Editors

Ying Tan
Peking University
Beijing
China

Yuhui Shi
Xi'an Jiaotong-Liverpool University
Suzhou
China

Fernando Buarque
Universidade de Pernambuco
Recife
Brazil

Alexander Gelbukh
Instituto Politécnico Nacional
Mexico
Mexico

Swagatam Das
Indian Statistical Institute
Kolkata
India

Andries Engelbrecht
University of Pretoria
Pretoria
South Africa

ISSN 0302-9743   ISSN 1611-3349 (electronic)
Lecture Notes in Computer Science
ISBN 978-3-319-20471-0   ISBN 978-3-319-20472-7 (eBook)
DOI 10.1007/978-3-319-20472-7

Library of Congress Control Number: 2015941139

LNCS Sublibrary: SL1 – Theoretical Computer Science and General Issues

Springer Cham Heidelberg New York Dordrecht London

Springer International Publishing AG Switzerland is part of Springer Science+Business Media
(www.springer.com)

Preface

This book and its companion volumes, LNCS vols. 9140, 9141, and 9142, constitute the proceedings of the 6th International Conference on Swarm Intelligence in conjunction with the Second BRICS Congress on Computational Intelligence (ICSI-CCI 2015) held during June 25–28, 2015, in Beijing, China.

The theme of ICSI-CCI 2015 was "Serving Our Society and Life with Intelligence." With the advent of big data analysis and intelligent computing techniques, we are facing new challenges to make the information transparent and understandable efficiently. ICSI-CCI 2015 provided an excellent opportunity for academics and practitioners to present and discuss the latest scientific results and methods as well as the innovative ideas and advantages in theories, technologies, and applications in both swarm intelligence and computational intelligence. The technical program covered all aspects of swarm intelligence, neural networks, evolutionary computation, and fuzzy systems applied to all fields of computer vision, signal processing, machine learning, data mining, robotics, scheduling, game theory, DB, parallel realization, etc.

The 6th International Conference on Swarm Intelligence (ICSI 2015) was the sixth international gathering for researchers working on all aspects of swarm intelligence, following successful and fruitful events in Hefei (ICSI 2014), Harbin (ICSI 2013), Shenzhen (ICSI 2012), Chongqing (ICSI 2011), and Beijing (ICSI 2010), which provided a high-level academic forum for the participants to disseminate their new research findings and discuss emerging areas of research. It also created a stimulating environment for the participants to interact and exchange information on future challenges and opportunities in the field of swarm intelligence research. The Second BRICS Congress on Computational Intelligence (BRICS-CCI 2015) was the second gathering for BRICS researchers who are interested in computational intelligence after the successful Recife event (BRICS-CCI 2013) in Brazil. These two prestigious conferences were held jointly in Beijing this year so as to share common mutual ideas, promote transverse fusion, and stimulate innovation.

Beijing is the capital of China and is now one of the largest cities in the world. As the cultural, educational, and high-tech center of the nation, Beijing possesses many world-class conference facilities, communication infrastructures, and hotels, and has successfully hosted many important international conferences and events such as the 2008 Beijing Olympic Games and the 2014 Asia-Pacific Economic Cooperation (APEC), among others. In addition, Beijing has rich cultural and historical attractions such as the Great Wall, the Forbidden City, the Summer Palace, and the Temple of Heaven. The participants of ICSI-CCI 2015 had the opportunity to enjoy Peking operas, beautiful landscapes, and the hospitality of the Chinese people, Chinese cuisine, and a modern China.

ICSI-CCI 2015 received 294 submissions from about 816 authors in 52 countries and regions (Algeria, Argentina, Australia, Austria, Bangladesh, Belgium, Brazil, Brunei Darussalam, Canada, Chile, China, Christmas Island, Croatia, Czech Republic,

Egypt, Finland, France, Georgia, Germany, Greece, Hong Kong, India, Ireland, Islamic Republic of Iran, Iraq, Italy, Japan, Republic of Korea, Macao, Malaysia, Mexico, Myanmar, New Zealand, Nigeria, Pakistan, Poland, Romania, Russian Federation, Saudi Arabia, Serbia, Singapore, South Africa, Spain, Sweden, Switzerland, Chinese Taiwan, Thailand, Tunisia, Turkey, UK, USA, Vietnam) across six continents (Asia, Europe, North America, South America, Africa, and Oceania). Each submission was reviewed by at least two reviewers, and on average 2.7 reviewers. Based on rigorous reviews by the Program Committee members and reviewers, 161 high-quality papers were selected for publication in this proceedings volume with an acceptance rate of 54.76 %. The papers are organized in 28 cohesive sections covering all major topics of swarm intelligence and computational intelligence research and development.

As organizers of ICSI-CCI 2015, we would like to express our sincere thanks to Peking University and Xian Jiaotong-Liverpool University for their sponsorship, as well as to the IEEE Computational Intelligence Society, World Federation on Soft Computing, and International Neural Network Society for their technical co-sponsorship. We appreciate the Natural Science Foundation of China and Beijing Xinhui Hitech Company for its financial and logistic support. We would also like to thank the members of the Advisory Committee for their guidance, the members of the international Program Committee and additional reviewers for reviewing the papers, and the members of the Publications Committee for checking the accepted papers in a short period of time. Particularly, we are grateful to Springer for publishing the proceedings in their prestigious series of *Lecture Notes in Computer Science*. Moreover, we wish to express our heartfelt appreciation to the plenary speakers, session chairs, and student helpers. In addition, there are still many more colleagues, associates, friends, and supporters who helped us in immeasurable ways; we express our sincere gratitude to them all. Last but not the least, we would like to thank all the speakers, authors, and participants for their great contributions that made ICSI-CCI 2015 successful and all the hard work worthwhile.

April 2015

Ying Tan
Yuhui Shi
Fernando Buarque
Alexander Gelbukh
Swagatam Das
Andries Engelbrecht

Organization

Honorary Chairs

Xingui He	Peking University, China
Xin Yao	University of Birmingham, UK

Joint General Chair

Ying Tan	Peking University, China

Joint General Co-chairs

Fernando Buarque	University of Pernambuco, Brazil
Alexander Gelbukh	Instituto Politécnico Nacional, Mexico, and Sholokhov
	Moscow State University for the Humanities, Russia
Swagatam Das	Indian Statistical Institute, India
Andries Engelbrecht	University of Pretoria, South Africa

Advisory Committee Chairs

Jun Wang	Chinese University of Hong Kong, HKSAR, China
Derong Liu	University of Chicago, USA, and Institute
	of Automation, Chinese Academy of Science, China

General Program Committee Chair

Yuhui Shi	Xi'an Jiaotong-Liverpool University, China

PC Track Chairs

Shi Cheng	Nottingham University Ningbo, China
Andreas Janecek	University of Vienna, Austria
Antonio de Padua Braga	Federal University of Minas Gerais, Brazil
Zhigang Zeng	Huazhong University of Science and Technology,
	China
Wenjian Luo	University of Science and Technology of China, China

Technical Committee Co-chairs

Kalyanmoy Deb	Indian Institute of Technology, India
Martin Middendorf	University of Leipzig, Germany

Yaochu Jin   University of Surrey, UK
Xiaodong Li   RMIT University, Australia
Gary G. Yen   Oklahoma University, USA
Rachid Chelouah   EISTI, Cergy, France
Kunqing Xie   Peking University, China

Special Sessions Chairs

Benlian Xu   Changshu Institute of Technology, China
Yujun Zheng   Zhejing University of Technology, China
Carmelo Bastos   University of Pernambuco, Brazil

Publications Co-chairs

Radu-Emil Precup   Polytechnic University of Timisoara, Romania
Tom Hankes   Radboud University, Netherlands

Competition Session Chair

Jane Liang   Zhengzhou University, China

Tutorial/Symposia Sessions Chairs

Jose Alfredo Ferreira Costa   Federal University of Rio Grande do Norte, Brazil
Jianhua Liu   Fujian University of Technology, China
Xing Bo   University of Limpopo, South Africa
Chao Zhang   Peking University, China

Publicity Co-chairs

Yew-Soon Ong   Nanyang Technological University, Singapore
Hussein Abbass   The University of New South Wales, ADFA, Australia
Carlos A. Coello Coello   CINVESTAV-IPN, Mexico
Eugene Semenkin   Siberian Aerospace University, Russia
Pramod Kumar Singh   Indian Institute of Information Technology and
                                      Management, India
Komla Folly   University of Cape Town, South Africa
Haibin Duan   Beihang University, China

Finance and Registration Chairs

Chao Deng   Peking University, China
Suicheng Gu   Google Corporation, USA

Conference Secretariat

Weiwei Hu   Peking University, China

ICSI 2015 Program Committee

Hafaifa Ahmed   University of Djelfa, Algeria
Peter Andras   Keele University, UK
Esther Andrés   INTA, Spain
Yukun Bao   Huazhong University of Science and Technology,
                                   China
Helio Barbosa   LNCC, Laboratório Nacional de Computação
                                   Científica, Brazil
Christian Blum   IKERBASQUE, Basque Foundation for Science, Spain
Salim Bouzerdoum   University of Wollongong, Australia
David Camacho   Universidad Autonoma de Madrid, Spain
Bin Cao   Tsinghua University, China
Kit Yan Chan   DEBII, Australia
Rachid Chelouah   EISTI, Cergy-Pontoise, France
Mu-Song Chen   Da-Yeh University, Taiwan
Walter Chen   National Taipei University of Technology, Taiwan
Shi Cheng   The University of Nottingham Ningbo, China
Chaohua Dai   Southwest Jiaotong University, China
Prithviraj Dasgupta   University of Nebraska, Omaha, USA
Mingcong Deng   Tokyo University of Agriculture and Technology,
                                   Japan
Yongsheng Ding   Donghua University, China
Yongsheng Dong   Henan University of Science and Technology, China
Madalina Drugan   Vrije Universiteit Brussel, Belgium
Mark Embrechts   RPI, USA
Andries Engelbrecht   University of Pretoria, South Africa
Zhun Fan   Technical University of Denmark, Denmark
Jianwu Fang   Xi'an Institute of Optics and Precision Mechanics
                                   of CAS, China
Carmelo-Bastos Filho   University of Pernambuco, Brazil
Shangce Gao   University of Toyama, Japan
Ying Gao   Guangzhou University, China
Suicheng Gu   University of Pittsburgh, USA
Ping Guo   Beijing Normal University, China
Fei Han   Jiangsu University, China
Guang-Bin Huang   Nanyang Technological University, Singapore
Amir Hussain   University of Stirling, UK
Changan Jiang   RIKEN-TRI Collaboration Center for
                                   Human- Interactive Robot Research, Japan
Liu Jianhua   Fujian University of Technology, China
Colin Johnson   University of Kent, UK

Chen Junfeng   Hohai University, China
Liangjun Ke   Xian Jiaotong University, China
Farrukh Khan   FAST-NUCES Islamabad, Pakistan
Thanatchai   Suranaree University of Technology, Thailand
  Kulworawanichpong
Germano Lambert-Torres   Itajuba Federal University, Brazil
Xiujuan Lei   Shaanxi Normal University, China
Bin Li   University of Science and Technology of China, China
Xuelong Li   Xi'an Institute of Optics and Precision Mechanics
                                  of Chinese Academy of Sciences, China
Jane-J. Liang   Zhengzhou University, China
Andrei Lihu   Polytechnic University of Timisoara, Romania
Bin Liu   Nanjing University of Post and Telecommunications,
                                  China
Ju Liu   Shandong University, China
Wenlian Lu   Fudan University, China
Wenjian Luo   University of Science and Technology of China, China
Chengying Mao   Jiangxi University of Finance and Economics, China
Bernd Meyer   Monash University, Australia
Martin Middendorf   University of Leipzig, Germany
Hongwei Mo   Harbin University of Engineering, China
Jonathan Mwaura   University of Pretoria, South Africa
Yan Pei   The University of Aizu, Japan
Radu-Emil Precup   Polytechnic University of Timisoara, Romania
Kai Qin   RMIT University, Australia
Quande Qin   Shenzhen University, China
Robert Reynolds   Wayne State University, USA
Guangchen Ruan   Indiana University, USA
Indrajit Saha   University of Warsaw, Poland
Dr. Sudip Kumar Sahana   BITMESRA, India
Yuhui Shi   Xi'an Jiaotong-Liverpool University, China
Zhongzhi Shi   Institute of Computing Technology, Chinese Academy
                                  of Sciences, China
Xiang Su   OULU, France
Ying Tan   Peking University, China
T.O. Ting   Xi'an Jiaotong-Liverpool University, China
Mario Ventresca   Purdue University, USA
Dujuan Wang   Dalian University of Technology, China
Guoyin Wang   Chongqing University of Posts and
                                  Telecommunications, China
Jiahai Wang   Sun Yat-sen University, China
Lei Wang   Tongji University, China
Ling Wang   Tsinghua University, China
Lipo Wang   Nanyang Technological University, Singapore
Qi Wang   Xi'an Institute of Optics and Precision Mechanics
                                  of CAS, China

Zhenzhen Wang	Jinling Institute of Technology, China
Fang Wei	Southern Yangtze University, China
Ka-Chun Wong	University of Toronto, Canada
Zhou Wu	City University of Hong Kong, HKSAR, China
Shunren Xia	Zhejiang University, China
Bo Xing	University of Limpopo, South Africa
Benlian Xu	Changshu Institute of Technology, China
Rui Xu	Hohai University, China
Bing Xue	Victoria University of Wellington, New Zealand
Wu Yali	Xi'an University of Technology, China
Yingjie Yang	De Montfort University, UK
Guo Yi-Nan	China University of Mining and Technology, China
Peng-Yeng Yin	National Chi Nan University, Taiwan
Ling Yu	Jinan University, China
Zhi-Hui Zhan	Sun Yat-sen University, China
Defu Zhang	Xiamen University, China
Jie Zhang	Newcastle University, UK
Jun Zhang	Waseda University, Japan
Junqi Zhang	Tongji University, China
Lifeng Zhang	Renmin University, China
Mengjie Zhang	Victoria University of Wellington, New Zealand
Qieshi Zhang	Waseda University, Japan
Yong Zhang	China University of Mining and Technology, China
Wenming Zheng	Southeast University, China
Yujun Zheng	Zhejiang University of Technology, China
Zhongyang Zheng	Peking University, China
Guokang Zhu	Chinese Academy of Sciences, China
Zexuan Zhu	Shenzhen University, China
Xingquan Zuo	Beijing University of Posts and Telecommunications, China

BRICS CCI 2015 Program Committee

Hussein Abbass	The University of New South Wales, Australia
Mohd Helmy Abd Wahab	Universiti Tun Hussein Onn Malaysia
Aluizio Araujo	Federal University of Pernambuco, Brazil
Rosangela Ballini	State University of Campinas, Brazil
Gang Bao	Huazhong University of Science and Technology, China
Guilherme Barreto	Federal University of Ceará, Brazil
Carmelo J.A. Bastos Filho	University of Pernambuco, Brazil
Antonio De Padua Braga	Federal University of Minas Gerais, Brazil
Felipe Campelo	Federal University of Minas Gerais, Brazil
Cristiano Castro	Universidade Federal de Lavras, Brazil
Wei-Neng Chen	Sun Yat-Sen University, China
Shi Cheng	The University of Nottingham Ningbo, China

Ding Wang	Institute of Automation, Chinese Academy of Sciences, China
Qinglai Wei	Northeastern University, China
Benlian Xu	Changshu Institute of Technology, China
Takashi Yoneyama	Aeronautics Institute of Technology (ITA), Brazil
Yang Yu	Nanjing University, China
Xiao-Jun Zeng	University of Manchester, UK
Zhigang Zeng	Huazhong University of Science and Technology, China
Zhi-Hui Zhan	Sun Yat-sen University, China
Mengjie Zhang	Victoria University of Wellington, New Zealand
Yong Zhang	China University of Mining and Technology, China
Liang Zhao	University of São Paulo, Brazil
Zexuan Zhu	Shenzhen University, China
Xingquan Zuo	Beijing University of Posts and Telecommunications, China

Additional Reviewers for ICSI 2015

Bello Orgaz, Gema	Lin, Ying
Cai, Xinye	Liu, Jing
Chan, Tak Ming	Liu, Zhenbao
Cheng, Shi	Lu, Bingbing
Deanney, Dan	Manolessou, Marietta
Devin, Florent	Marshall, Linda
Ding, Ke	Menéndez, Héctor
Ding, Ming	Oliveira, Sergio Campello
Dong, Xianguang	Peng, Chengbin
Fan, Zhun	Qin, Quande
Geng, Na	Ramírez-Atencia, Cristian
Gonzalez-Pardo, Antonio	Ren, Xiaodong
Han, Fang	Rodríguez Fernández, Víctor
Helbig, Marde	Senoussi, Houcine
Jiang, Yunzhi	Shang, Ke
Jiang, Ziheng	Shen, Zhe-Ping
Jordan, Tobias	Sze-To, Antonio
Jun, Bo	Tao, Fazhan
Junfeng, Chen	Wang, Aihui
Keyu, Yan	Wen, Shengjun
Li, Jinlong	Wu, Yanfeng
Li, Junzhi	Xia, Changhong
Li, Wenye	Xu, Biao
Li, Xin	Xue, Yu
Li, Yanjun	Yan, Jingwei
Li, Yuanlong	Yang, Chun

Yaqi, Wu
Yassa, Sonia
Yu, Chao
Yu, Weijie
Yuan, Bo

Zhang, Jianhua
Zhang, Tong
Zhao, Minru
Zhao, Yunlong
Zong, Yuan

Additional Reviewers for BRICS-CCI 2015

Amaral, Jorge
Bertini, João
Cheng, Shi
Ding, Ke
Dong, Xianguang
Forero, Leonardo
Ge, Jing
Hu, Weiwei
Jayne, Chrisina
Lang, Liu
Li, Junzhi
Lin, Ying

Luo, Wenjian
Ma, Hongwen
Marques Da Silva, Alisson
Mi, Guyue
Panpan, Wang
Rativa Millan, Diego Jose
Xun, Li
Yan, Pengfei
Yang, Qiang
Yu, Chao
Zheng, Shaoqiu

Contents – Part II

Neural Networks and Fuzzy Methods

Data Mining Approaches

Information Security

Automation Control

Combinatorial Optimization Algorithms

Constrained Optimization Algorithms

Scheduling and Path Planning

Machine Learning

Blind Source Separation

Swarm Interaction Behavior

Parameters and System Optimization

Neural Networks and Fuzzy Methods

Apply Stacked Auto-Encoder
to Spam Detection

Guyue Mi[1,2], Yang Gao[1,2], and Ying Tan[1,2(✉)]

[1] Key Laboratory of Machine Perception (MOE), Peking University, Beijing, China
[2] Department of Machine Intelligence, School of Electronics Engineering
and Computer Science, Peking University, Beijing 100871, China
{gymi,gaoyang0115,ytan}@pku.edu.cn

Abstract. In this paper, we apply Stacked Auto-encoder, one of the
main types of deep networks, hot topic of machine learning recently,
to spam detection and comprehensively compare its performance with
other prevalent machine learning techniques those are commonly used in
spam filtering. Experiments were conducted on five benchmark corpora,
namely PU1, PU2, PU3, PUA and Enron-Spam. Accuracy and F_1 measure are selected as the main criteria in analyzing and discussing the
results. Experimental results demonstrate that Stacked Auto-encoder
performs better than Naive Bayes, Support Vector Machine, Decision
Tree, Boosting, Random Forest and traditional Artificial Neural Network both in accuracy and F_1 measure, which endows deep learning
with application in spam filtering in the real world.

Keywords: Spam detection · Machine learning · Artificial neural network · Deep learning · Stacked auto-encoder

1 Introduction

Email has become one of the most commonly used communication tools in our
daily work and life due to its advantages of low cost, high efficiency and good
convenience. However, the above characteristics are also concerned and exploited
by the ones who want to spread advertisement, bad information or even computer
virus to send spam emails. Spam, generally defined as unsolicited bulk email
(UBE) or unsolicited commercial email (UCE) [1], has caused many problems to
our normal email communication. Ferris Research Group [2] has revealed that
large amount of spam not only occupied network bandwidth and server storage,
but also wasted users' time on reading and deleting them, which resulted in loss
of productivity. Moreover, the spam with malware threatens internet safety and
personal privacy.

According to Symantec Internet Security Threat Report 2014 [3], although
the total number of bots (computers that are infected and controlled to send
spam) worldwide has declined from 3.4 million to 2.3 million in 2013 compared
with that of 2012, the overall spam rate only dropped 3%, which is still up to 66%
of the whole email traffic. What's worse, the phishing rate and virus rate both

© Springer International Publishing Switzerland 2015
Y. Tan et al. (Eds.): ICSI-CCI 2015, Part II, LNCS 9141, pp. 3–15, 2015.
DOI: 10.1007/978-3-319-20472-7_1

increased. In 2013, one out of every 196 emails contained virus and one out of every 392 emails was identified as phishing, while the corresponding proportions in 2012 were 1 in 291 and 1 in 414 respectively. In addition, adult, sex and dating related spam dominated in 2013 and made up 70% of the total spam, which was an increase of 15% compared with that of 2012. The statistics from Cyren Internet Threats Trend Report [4] demonstrate that spam made up 68% of all global emails in the third quarter of 2014, with a daily average of 56 billion. Thus, it is still necessary and urgent to take measures to solve the spam problem.

To address this problem, researchers have proposed numbers of anti-spam approaches from different perspectives, including legal means, working out corresponding acts to regulate email sending [5,6]; email protocol methods, improving the control strategies of email protocols [7,8]; simple techniques, such as address protection [9], black/white list [10,11], keywords filtering [12] and so on; and intelligent detection, considering the spam filtering problem as a typical two-class classification problem, which could be solved by the supervised machine learning methods [12–14]. Among all these anti-spam approaches, intelligent detection is the most effective and widely used. On the one hand, intelligent detection is highly automated and do not need much human intervention; On the other hand, intelligent detection has the characteristics of high accuracy, robustness and strong noise tolerance, and it can adapt to the dynamic changes of the emails' content and users' interests.

There are three main related research fields for intelligent spam detection as well as other classification or pattern recognition problems, namely feature selection, feature construction and classifier design, corresponding to the three core steps of intelligent spam detection. The purpose of feature selection lies in reducing the number of features to be further processed and the affect from possible noisy features, so as to reduce the computational complexity and enhance the categorization accuracy respectively. Several feature selection metrics have been proposed and proved to be effective, such as Information Gain (IG) [15], Document Frequency (DF) [16], Term Frequency Variance (TFV) [17], Chi Square (χ^2) [16], Odds Ratio (OR) [18], Term Strength (TS) [16] and so on. Feature construction approaches transform the set of features available into a new set of features by finding relationships between existing features and constructing feature vectors to represent samples. Bag-of-Words (BoW), also known as Space Vector Model, is the most widely used feature construction approach in spam detection [19]. Other feature construction approaches for spam detection have also been studied, like Sparse Binary Polynomial Hashing (SBPH) [20], Orthogonal Sparse Bigrams (OSB) [21], immune concentration based approaches [22–27] and term space partition (TSP) based approach [28] etc. Supervised machine learning methods have been successfully and widely applied for classifier design in spam detection, and the prevalent ones are introduced in Section 2.

This paper applies Stacked Auto-Encoders (SAE), one of the main types of deep neural networks, to intelligent spam detection. And presents a comparative study of SAE with other prevalent supervised machine learning methods to verify the effectiveness of deep learning on spam detection. Experiments were conducted on five benchmark corpora PU1, PU2, PU3, PUA and Enron-Spam to investigate the performance of SAE and other machine learning methods. Accuracy and F_1 measure are selected as the main criteria in analyzing and discussing the results.

The rest of this paper is organized as follows: Section 2 introduces the prevalent machine learning methods that are applied in spam detection. Stacked Auto-Encoders is presented in detail in section 3. Section 4 gives the experimental results and corresponding analysis. Finally, we conclude the paper in Section 5.

2   Prevalent Machine Learning Methods

2.1   Naive Bayes

The Bayes methods compute the probability $P(C = c_k | X = x)$ that the sample x belongs to each category c_k and obtain the final category of sample x according to the maximum value of the probability that has been achieved.

$$P(C = c_k | X = x) = \frac{P(X = x | C = c_k)P(C = c_k)}{P(X = x)} \quad (1)$$

According to the Bayes formula shown in Eq.1, the key part of the Bayes methods is computing the probability $P(X = x | C = c_k)$. Naive Bayes (NB) is the most widely used Bayes method, and it assumes that the sample x is composed of multiple features w_j which are mutually independent in the calculation process, thus $P(X = x | C = c_k)$ could be achieved by computing $P(W = w_j | C = c_k)$. Sahami et al. [29] introduced NB into spam detection, and now it has been widely used in commercial spam filtering system and open source software of spam detection based on its simplicity in implementation and high accuracy.

2.2   Support Vector Machine

The core idea of Support Vector Machine (SVM) is to find the optimal hyperplane and make the classification margin maximized. The targets of training process is maximizing the classification margin and minimizing the structural risk, and obtaining weight vector of the optimal hyperplane by calculation on the training set. For linearly inseparable issues, SVM makes it linearly separable by mapping the training data from the original space to a higher-dimensional space with kernel functions and computes corresponding optimal hyperplane. Drucker et al. [30] applied SVM to spam detection and achieved better performance compared with Ripper and Rocchio. In addition, best performance of SVM was achieved when boolean BoW is employed as feature construction approach other than multi-value BoW.

2.3   Decision Tree

Decision Tree (DT) constructs a tree from top to bottom according to the predefined sequence of attributes, where nodes corresponds to attributes and edges corresponds to attribute values. Each path from the root to the leaves could be seen as a rule. Selecting the sequence of attributes based on IG is one of the commonly used methods in DT. The famous DT algorithms are ID3 and C4.5 etc. Carreras et al. [31] applied DT to spam filtering and adopted RLM distance other than IG for attributes selection. Currently, DT is often used as a weak learner of Boosting methods due to its mediocre performance.

2.4   Artificial Neural Network

Artificial Neural Network (ANN) is proposed by taking inspiration from mechanism of biological neural networks and consists of a large number of interconnected artificial neurons. There are three types of neurons: the input layer neurons, hidden layer neurons and output layer neurons. In the learning (training) process, the connection weights of ANN are dynamically adjusted in accordance with the input and output values of the training data to approximate the mapping function of the input and output values. In the classification process, the input data transfer in the network layer by layer beginning from the input layer. The activation value of each neuron is calculated according to the predefined activation function and effect of each neuron in the classification is determined by the connection weights. Performance of ANN is mainly influenced by three factors: input and the activation function, network structure and connection weights. Clark et al. [32] adopted ANN to classify emails with a fully connected neural network, and used back-propagation (BP) algorithm for training. Experimental results showed that ANN could achieve better performance than NB and k-Nearest Neighbor.

2.5   Boosting

Boosting could be seen as a voting technology based on existing learning methods, other than a particular learning method itself. AdaBoost (Adaptive Boosting) is a typical Boosting method. The core idea of this method is giving more attention to the samples those are difficult to be classified in the learning process [33]. During the training process, weights of samples are dynamically adjusted in accordance with their classification results by the constructed classifiers, and the samples those are difficult to be classified would be selected for learning with greater probability when the new classifiers are built. Finally, new samples are weighted classified according to the performance of each classifier. Carreras et al. [31] applied AdaBoost to spam detection by using DT as base classifier, and AdaBoost achieved better performance than DT and NB in the experiment.

2.6   Random Forest

Random Forest (RF) samples repeatedly from the original sample set by utilizing the re-sampling method bootsrap and constructs Decision Tree model on each bootsrap sampling. Then the DT models constructed are combined to give the prediction by voting. Koprinska et al. [17] applied RF to email classification and compared its performance with that of other methods. Experimental results indicate that RF is promising approach for spam filtering and outperforms DT, SVM and NB, with DT and SVM being also more complex than RF.

3 Stacked Auto-Encoder

Artificial Neural Networks are traditional computational models for machine learning and pattern recognition. Former researches mainly focus on shallow neural networks (i.e. neural networks with one hidden layer or two hidden layers). Since deep neural networks have shown excellent performance in recent years, Deep Learning (DL) has become a hot topic in artificial intelligence. Deep Learning algorithms attempt to learn multiple levels of representation of increasing complexity or abstraction and deep multi-layer neural networks are the basic architectures of DL.

Stacked Auto-Encoder (SAE) is one of the main types of deep networks. It is a stacked ensemble of auto-encoders and has more excellent computational ability [34].

The structure of auto-encoder is show in Fig.1, which consists of three layers, namely input layer, code layer and reconstruction layer. The original input X ontoro at the input layer, and X is encoded to Y through forward-propagation in the neural network. Further, Y is decoded to X'. In auto-encoder, X' has the same dimensionality with X and is seen as a reconstruction of the original input X.

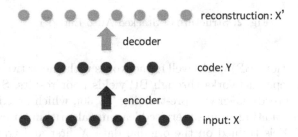

Fig. 1. Structure of Auto-encoder

Simply speaking, the auto-encoder transforms the input vector $X = (x_1, x_2, ..., x_n)$ to vector $Y = (y_1, y_2, ...y_m)$, where n indicates dimensionality of the input vector and m indicates dimensionality of the code vector. Next, the input vector X is reconstructed to X' from the code vector Y with the constraint that $|X - X'|$ is minimized. The training objective of this neural network is minimizing the reconstruction error, and the objective function is defined as follows:

$$J = \sum \|X - X'\| \tag{2}$$

where the sum operation is executed on all input samples. In this paper, the network is trained with the gradient descent BP algorithm, which is widely used in the training of artificial neural networks.

Actually, the code Y is a nonlinear abstract of original input data X and represents some features of X. SAE learns multiple levels of representation of the input vector X, in which the high layer encodes the low layer and each layer represents features of the input with increasing abstraction, and reconstructs X to X' from the last layer with constraint that $|X - X'|$ is minimized, as shown in Fig.2. SAE is a stacked ensemble of multiple auto-encoders, in which X is encoded to Y, Y is encoded to Z and Z is encoded to W successively. Reconstruction is taken in the reverse order. W is decoded to Z', Z' is decoded to Y' and Y' is decoded to X' successively. The training objective of SAE is the same with auto-encoder, which is minimizing the reconstruction error of X.

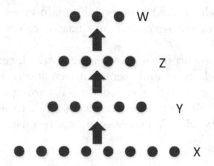

Fig. 2. Structure of Stacked Auto-Encoder

Back Propagation (BP) works well for networks with one or two hidden layers, while training deeper networks through BP yields poor results. SAE is trained by adopting the greedy layerwise pre-training [35,36], which greedily trains one layer at a time, exploiting an unsupervised learning algorithm for each layer. The auto-encoder X-Y is trained on the original data X first and transforms X to code Y. Then the auto-encoder Y-Z is trained the same as above based on Y and encodes Y to Z. Finally, the auto-encoder Z-W is trained on Z. After training of the three individual auto-encoders, the weights got is used to initialize the weight of SAE. The objective function is optimized by using BP algorithm.

Initialization strategy of BP weights is introduced in the training of SAE. The laywise pre-training of each auto-encoder could preliminarily determine the distribution of the initial data and make the wights of the neural network reflect the data characteristics. Initializing SAE with the weights got by pre-training could locate the initial solution close to the optimal solution and effectively reduce the convergence time. SAE possesses greater nonlinear capability than auto-encoder by laywise encoding of the original data. Each layer presents an individual abstract of the original data and higher layer has higher level abstraction. SAE is considered having strong learning ability and able to mine the effective features of original data sufficiently. In this paper, we applied SAE in spam detection to verify the performance of DL in spam filtering.

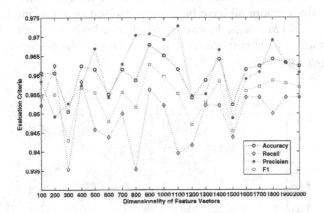

Fig. 3. Performance of AdaBoost with Varied Feature Vector Dimensionality on PU1

4   Experiments

4.1   Experimental Setup

In the experiments, Information Gain (IG) [15] and Bag-of-Words (BoW) [19] are selected as feature selection strategy and feature construction approach respectively for transforming email samples into feature vectors. IG is the most widely employed feature goodness criterion in machine learning area. It measures the number of bits of information obtained for class prediction by knowing the presence or absence of a certain feature in a sample. When applied in spam detection, IG of term t_i is calculated as

$$IG(t_i) = \sum_{c \in (s,h)} \sum_{t \in (t_i, \bar{t}_i)} P(t,c) \log \frac{P(t,c)}{P(t)P(c)} \tag{3}$$

where c denotes the class of an email, s stands for spam, and h stands for ham, t_i and $\bar{t}_i$ denotes the presence and absence of term t_i respectively. BoW, also known as Space Vector Model, is one of the most widely used feature construction approaches in spam detection. It transforms an email m to a n-dimensional feature vector $x = [x_1, x_2, ..., x_n]$ by utilizing a preselected term set $T = [t_1, t_2, ..., t_n]$, where the value x_i is given as a function of the occurrence of t_i in m, depending on the representation of the features adopted. We take binary representation, where x_i is equal to 1 when t_i occurs in m, and 0 otherwise.

Experiments were conducted on PU1, PU2, PU3, PUA [37] and Enron-Spam [38], which are all benchmark corpora widely used for effectiveness evaluation in spam detection. Among them, PU1 contains 1099 emails, 481 of which are spam; PU2 contains 721 emails, and 142 of them are spam; 4139 emails are included in PU3 and 1826 of them are spam; 1142 emails are included in PUA and 572 of them are spam; and Enron-Spam contains 33716 emails, 17171 of which are spam. Emails in the five corpora all have been preprocessed by removing header fields, attachment and HTML tags, leaving subject and body text only. For privacy protection, emails in PU corpora have been encrypted by replacing meaningful terms with specific numbers.

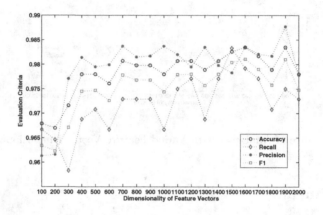

Fig. 4. Performance of Random Forest with Varied Feature Vector Dimensionality on PU1

In addition, SAE was implemented in MATLAB by utilizing the toolbox for deep learning [39], and a six-layer neural network is employed, in which the computational elements of each layer are 2000, 500, 250, 125, 10, 1 respectively. WEKA toolkit [40] was utilized in implementation of the machine learning models selected for comparison, namely Naive Bayes, Support Vector Machine, C4.5, Multilayer Perceptron (MLP), AdaBoost (C4.5 is selected as base learner) and Random Forest. Since determining the number of terms (features) selected for further classification, selection of dimensionality of feature vectors for each machine learning method can not only affect the computational complexity, but also influence the classification performance. Dimensionality of feature vectors for most of the techniques above are set in accordance with the previous researches, where that of MLP is set to 2000, the same as SAE [35] and those of NB, SVM and C4.5 are set to 500 [41]. While the dimensionality of feature vectors for AdaBoost and RF are investigated by conducting experiments on the relatively smaller corpus PU1, and set to 900 and 1600 respectively, as shown in Fig.3 and Fig.4. 10-fold cross validation was utilized on PU corpora and 6-fold cross validation on Enron-Spam according to the number of parts each of the corpora has been already divided into. Accuracy and F_1 measure are the main evaluation criteria, as they can reflect the overall performance of spam detection.

4.2   Performance Comparison

Table 1 to 5 show the performance of different machine learning methods in spam detection when incorporated with IG and BoW. As can be seen, SAE performs the best in most of the cases in terms of both accuracy and F_1 measure, except that it has a similar performance with RF on PU1 (as mentioned above, we take accuracy and F_1 measure as comparison criteria without focusing on precision and recall, which are incorporated into the calculation of F_1 measure and can be reflected by F_1 measure). This indicates that SAE can work well and outperform the current machine learning methods in spam detection, verifying the effectiveness of deep learning in this area and endowing it with application in the real world.

By comparing the performance of the above machine learning methods between different corpus, we can see that SAE can achieve similar and relatively higher accuracy and F_1 measure on all of the corpus selected for the experiments (as well as MLP and SVM), demonstrating that SAE possesses good stability

Table 1. Performance comparison of SAE with other prevalent machine learning techniques on PU1

Method	Precision(%)	Recall(%)	Accuracy(%)	F_1(%)
NB	97.74	82.50	91.47	89.37
C4.5	91.24	89.17	91.28	90.01
SVM	96.19	95.62	96.33	95.77
AdaBoost	97.08	95.62	96.79	96.28
RF	98.36	97.92	**98.35**	**98.11**
MLP	97.99	97.50	97.98	97.68
SAE	98.16	97.71	98.17	97.89

Table 2. Performance comparison of SAE with other prevalent machine learning techniques on PU2

Method	Precision(%)	Recall(%)	Accuracy(%)	F_1(%)
NB	82.67	70.71	90.70	75.34
C4.5	79.26	71.43	89.72	73.96
SVM	90.99	78.57	94.08	83.92
AdaBoost	91.74	75.71	93.66	82.23
RF	97.46	65.00	92.68	76.83
MLP	90.85	89.29	95.91	89.57
SAE	93.29	88.57	**96.34**	**90.37**

Table 3. Performance comparison of SAE with other prevalent machine learning techniques on PU3

Method	Precision(%)	Recall(%)	Accuracy(%)	F_1(%)
NB	92.75	78.63	87.72	84.93
C4.5	91.10	91.76	92.25	91.34
SVM	95.44	93.96	95.33	94.67
AdaBoost	95.54	94.56	95.62	95.02
RF	97.50	95.66	96.97	96.55
MLP	96.72	95.66	96.63	96.17
SAE	96.77	97.14	**97.24**	**96.91**

Table 4. Performance comparison of SAE with other prevalent machine learning techniques on PUA

Method	Precision(%)	Recall(%)	Accuracy(%)	F_1(%)
NB	95.39	94.21	94.65	94.63
C4.5	87.02	92.63	88.68	89.30
SVM	91.84	94.39	92.63	92.87
AdaBoost	91.08	97.02	93.42	93.80
RF	92.15	97.02	94.04	94.36
MLP	94.24	97.02	95.35	95.49
SAE	94.49	97.90	**95.88**	**96.04**

Table 5. Performance comparison of SAE with other prevalent machine learning techniques on Enron-Spam

Method	Precision(%)	Recall(%)	Accuracy(%)	F_1(%)
NB	77.74	98.41	87.45	86.10
C4.5	82.88	97.07	90.33	89.02
SVM	89.64	98.74	94.63	93.86
AdaBoost	89.13	98.78	94.25	93.57
RF	91.46	99.28	96.06	95.11
MLP	92.70	98.71	96.23	95.54
SAE	94.90	98.95	**97.49**	**96.84**

and robustness due to its strong learning ability, which is concerned more in the real world application. While the others usually occur with the phenomenon that the performance declines significantly on some of the corpus selected.

In addition, NB is a simple and efficient method which can recognize majority of the samples, hence it is widely used in the real spam filtering systems. AdaBoost and RF are both ensemble methods based on weak learners and obtain better performance than NB and C4.5 (which are weak learners) in the experiments.

It is worth mentioning that the training of SAE is really time consuming as well as MLP (and AdaBoost on large corpus). However, this could be settled by the offline training of real world spam filters and strong computational ability of modern computers and servers.

5   Conclusion

In this paper, we applied SAE to spam detection and compared its performance with the prevalent machine learning techniques those are commonly used in this area. Comprehensive experiments were conducted on public benchmark corpus and a six-layer neural network was employed to investigate the performance of SAE. The results demonstrate that SAE not only outperforms other machine leaning methods in terms of classification accuracy and F_1 measure, but also possesses stronger stability and robustness. This verifies the effectiveness of deep learning in spam filtering and endows it with application in real world meanwhile. In future work, we intend to further apply the other types of deep networks in spam detection and construct novel feature construction models in accordance with the characteristics of spam problem and advantages of different deep networks.

Acknowledgments. This work was supported by the Natural Science Foundation of China (NSFC) under grant no. 61375119, 61170057 and 60875080, and partially supported by National Key Basic Research Development Plan (973 Plan) Project of China with grant no. 2015CB352300.

References

1. Cranor, L., LaMacchia, B.: Spam!. Communications of the ACM **41**(8), 74–83 (1998)
2. Research, F.: Spam, spammers, and spam control: A white paper by ferris research. Technical Report (2009)
3. Corporation, S.: Internet security threat report: 2014. Technical Report (2014)
4. Cyren: Internet threats trend report: October 2014. Technical Report (2014)
5. Lugaresi, N.: European union vs. spam: a legal response. In: Proceedings of the First Conference on E-mail and Anti-Spam (2004)
6. Moustakas, E., Ranganathan, C., Duquenoy, P.: Combating spam through legislation: a comparative analysis of us and european approaches. In: Proceedings of the Second Conference on Email and Anti-Spam, pp. 1–8 (2005)
7. Marsono, M.N.: Towards improving e-mail content classification for spam control: architecture, abstraction, and strategies. PhD thesis, University of Victoria (2007)
8. Duan, Z., Dong, Y., Gopalan, K.: Dmtp: Controlling spam through message delivery differentiation. Computer Networks **51**(10), 2616–2630 (2007)
9. Hershkop, S.: Behavior-based email analysis with application to spam detection. PhD thesis, Columbia University (2006)
10. Sanz, E., Gomez Hidalgo, J., Cortizo Perez, J.: Email spam filtering. Advances in Computers **74**, 45–114 (2008)

11. Heron, S.: Technologies for spam detection. Network Security **2009**(1), 11–15 (2009)
12. Cormack, G.: Email spam filtering: A systematic review. Foundations and Trends in Information Retrieval **1**(4), 335–455 (2007)
13. Carpinter, J., Hunt, R.: Tightening the net: A review of current and next generation spam filtering tools. Computers & security **25**(8), 566–578 (2006)
14. Kotsiantis, S.: Supervised machine learning: A review of classification techniques. Informatica **31**, 249–268 (2007)
15. Yang, Y.: Noise reduction in a statistical approach to text categorization. In: Proceedings of the 18th Annual International ACM SIGIR Conference on Research and Development in Information Retrieval, pp. 256–263. ACM (1995)
16. Yang, Y., Pedersen, J.: A comparative study on feature selection in text categorization. In: Machine Learning-International Workshop Then Conference-, Morgan Kaufmann Publishers, INC, pp. 412–420 (1997)
17. Koprinska, I., Poon, J., Clark, J., Chan, J.: Learning to classify e-mail. Information Sciences **177**(10), 2167–2187 (2007)
18. Shaw, W.: Term-relevance computations and perfect retrieval performance. Information Processing & Management **31**(4), 491–498 (1995)
19. Guzella, T., Caminhas, W.: A review of machine learning approaches to spam filtering. Expert Systems with Applications **36**(7), 10206–10222 (2009)
20. Yerazunis, W.: Sparse binary polynomial hashing and the crm114 discriminator. the Web (2003). http://crm114.sourceforge.net/CRM114paper.html
21. Siefkes, C., Assis, F., Chhabra, S., Yerazunis, W.S.: Combining winnow and orthogonal sparse bigrams for incremental spam filtering. In: Boulicaut, J.-F., Esposito, F., Giannotti, F., Pedreschi, D. (eds.) PKDD 2004. LNCS (LNAI), vol. 3202, pp. 410–421. Springer, Heidelberg (2004)
22. Tan, Y., Deng, C., Ruan, G.: Concentration based feature construction approach for spam detection. In: Neural Networks, 2009. IJCNN 2009. International Joint Conference on, pp. 3088–3093. IEEE (2009)
23. Ruan, G., Tan, Y.: A three-layer back-propagation neural network for spam detection using artificial immune concentration. Soft Computing-A Fusion of Foundations, Methodologies and Applications **14**(2), 139–150 (2010)
24. Zhu, Y., Tan, Y.: Extracting discriminative information from e-mail for spam detection inspired by immune system. In: 2010 IEEE Congress on Evolutionary Computation (CEC), pp. 1–7. IEEE (2010)
25. Zhu, Y., Tan, Y.: A local-concentration-based feature extraction approach for spam filtering. IEEE Transactions on Information Forensics and Security **6**(2), 486–497 (2011)
26. Mi, G., Zhang, P., Tan, Y.: A multi-resolution-concentration based feature construction approach for spam filtering. In: The 2013 International Joint Conference on Neural Networks (IJCNN), pp. 1–8. IEEE (2013)
27. Gao, Y., Mi, G., Tan, Y.: An adaptive concentration selection model for spam detection. In: Tan, Y., Shi, Y., Coello, C.A.C. (eds.) ICSI 2014, Part I. LNCS, vol. 8794, pp. 223–233. Springer, Heidelberg (2014)
28. Mi, G., Zhang, P., Tan, Y.: Feature construction approach for email categorization based on term space partition. In: The 2013 International Joint Conference on Neural Networks (IJCNN), pp. 1–8. IEEE (2013)
29. Sahami, M., Dumais, S., Heckerman, D., Horvitz, E.: A bayesian approach to filtering junk e-mail. In: Learning for Text Categorization: Papers from the 1998 Workshop, Madison, Wisconsin, vol. 62, pp. 98–105. AAAI Technical Report WS-98-05 (1998)

30. Drucker, H., Wu, D., Vapnik, V.: Support vector machines for spam categorization. IEEE Transactions on Neural Networks **10**(5), 1048–1054 (1999)
31. Carreras, X., Marquez, L.: Boosting trees for anti-spam email filtering. arXiv preprint cs/0109015 (2001)
32. Clark, J., Koprinska, I., Poon, J.: Linger-a smart personal assistant for e-mail classification. In: Proc. of the 13th Intern. Conference on Artificial Neural Networks (ICANN 2003), Istanbul, Turkey, pp. 26–29, June 2003
33. Rokach, L.: Ensemble-based classifiers. Artificial Intelligence Review **33**(1), 1–39 (2010)
34. Vincent, P., Larochelle, H., Lajoie, I., Bengio, Y., Manzagol, P.A.: Stacked denoising autoencoders: Learning useful representations in a deep network with a local denoising criterion. The Journal of Machine Learning Research **11**, 3371–3408 (2010)
35. Hinton, G.E., Salakhutdinov, R.R.: Reducing the dimensionality of data with neural networks. Science **313**(5786), 504–507 (2006)
36. Bengio, Y., Lamblin, P., Popovici, D., Larochelle, H., et al.: Greedy layer-wise training of deep networks. Advances in neural information processing systems **19**, 153 (2007)
37. Androutsopoulos, I., Paliouras, G., Michelakis, E.: Learning to filter unsolicited commercial e-mail. "DEMOKRITOS", National Center for Scientific Research (2004)
38. Metsis, V., Androutsopoulos, I., Paliouras, G.: Spam filtering with naive bayes-which naive bayes. In: Third Conference on Email and Anti-Spam (CEAS), vol. 17, pp. 28–69 (2006)
39. Palm, R.B.: Prediction as a candidate for learning deep hierarchical models of data. Technical University of Denmark, Palm 25 (2012)
40. Hall, M., Frank, E., Holmes, G., Pfahringer, B., Reutemann, P., Witten, I.: The weka data mining software: an update. ACM SIGKDD Explorations Newsletter **11**(1), 10–18 (2009)
41. Zhu, Y., Mi, G., Tan, Y.: Query based hybrid learning models for adaptively adjusting locality. In: The 2012 International Joint Conference on Neural Networks (IJCNN), pp. 1–8. IEEE (2012)

Interval-Valued Intuitionistic Fuzzy Prioritized Ordered Weighted Averaging Operator and Its Application in Threat Assessment

Hua Wu[1,2]([✉]) and Xiuqin Su[1]

[1] Key Laboratory of Ultrafast Photoelectric Diagnostics Technology,
Xi'an Institute of Optics and Precision Mechanics of Chinese,
Academy of Sciences, Xi'an 710119, People's Republic of China
sunshinesmilewh@gmail.com, suxiuqin@opt.ac.cn
[2] University of Chinese Academy of Sciences, Beijing 100049,
People's Republic of China

Abstract. The threat assessment of aerial targets is the basic method for improving the ability of air defense system to deal with multiple objects in complex environment. Considering both the uncertain and imprecise information and prioritization relationship of attributes, we propose an interval-valued intuitionistic fuzzy prioritized ordered weighted averaging (IVIF-POWA) operator-based threat assessment model. The contribution of this paper is twofold: 1) An IVIF-POWA operator is proposed, which expresses the imprecise and uncertain information with more suitable interval-valued intuitionistic fuzzy sets. 2) This work proposes a simple yet effective threat assessment model based on IVIF-POWA operator. Its efficiency and effectiveness are validated by comparing it with some popular operators in a numerical example.

Keywords: Interval-valued intuitionistic fuzzy set · Ordered weighted averaging operator · Prioritization relationship

1 Introduction

Threat assessment of aerial targets is basis of command and decision making in modern air defense systems, which makes the systems more intelligent and precise. In the more complicated environment, the accurate threat assessment is becoming a challenging task.

To solve this problem, vast efforts have been established in many related works. The main alternatives are, Bayes networks [8,18], fuzzy belief reasoning [5], multiple attribute decision making (MADM) [2,11], knowledge reasoning [10], aggregation operators [4,21], TOPSIS [6,7], etc. Among them, because of the simplicity and effectiveness, the MADM and the aggregation operator-based threat assessment are recently becoming the primary choice. For instance, Qu and He [2] proposed a method of threat assessment based on MADM, which is usually applied to tactical command and control systems, such as surface-to-air and air-to-air

© Springer International Publishing Switzerland 2015
Y. Tan et al. (Eds.): ICSI-CCI 2015, Part II, LNCS 9141, pp. 16–24, 2015.
DOI: 10.1007/978-3-319-20472-7_2

defense systems. Wang *et al.* [11] introduced a MADM method with unknown attribute weights and the subjective bias towards projects of the decision-maker. Zhang *et al.* [21] proposed an order weighted aggregation operator-based evaluation model with subjective preference information of decision-making. With the development of aggregation operators, the assessment models based on the suitable aggregation operators play important roles in various threat assessments.

The aggregation operator is an interesting research topic and many important results are received [3,12–14,16,17]. Yager [14] proposed the ordered weighted averaging (OWA) operator, as a parameterized class of mean type aggregation operators. Yager [16,17] introduced prioritized aggregation (PA) operator and ordered weighted averaging (OWA) operator. Xu [12] developed some aggregation operators, including IFWA, IFHWA and IFOWA, and discussed various properties of these operators. Yu and Xu [13] proposed a prioritized intuitionistic fuzzy averaging operator and applied it in multi-attribute decision making.

In the aerial environment, the information of targets is usually imprecise, uncertain and incomplete. It is quite difficult to describe the targets with crisp numbers and even the intuitionistic fuzzy numbers (IFNs) by the experts. Hence, this paper proposes an interval-valued intuitionistic fuzzy prioritized ordered weighted averaging (IVIF-POWA) operator-based threat assessment model.

In summary, this paper proposes an IVIF-POWA operator. It expresses the information of threat assessment with a more suitable interval-valued intuitionistic fuzzy (IVIF) sets. In addition, the IVIF-POWA operator considers both the prioritization relationship and ordered position of attribute simultaneously, which is different from the conventional IVIFN-based operators. Then, a threat assessment model of aerial targets based on interval-valued intuitionistic fuzzy operator is constructed. The superiority of the proposed method is validated by comparing with some popular methods in a numerical example.

2   Preliminary

In this section, some preliminary concepts constituting the basis of this work are briefly introduced as follows.

Definition 1. *[1] Let $X = \{x_1, x_2, \cdots, x_n\}$ be a universe of discourse, then an interval-valued intuitionistic fuzzy set a in X is given*

$$a = (\mu_a(x), v_a(x)), x \in X \qquad (1)$$

where $\mu_a(x) \subseteq [0,1], v_a(x) \subseteq [0,1]$ and $0 \leq (\sup \mu_a(x) + \sup v_a(x)) \leq 1$.

The pair $(\mu_a(x), v_a(x))$ is defined as an interval-valued intuitionistic fuzzy number (IVIFN). For convenience, an IVIFN is denoted as $([\mu_a^-, \mu_a^+], [v_a^-, v_a^+])$ where $[\mu_a^-, \mu_a^+] \subseteq [0,1]$, $[v_a^-, v_a^+] \subseteq [0,1]$ and $0 \leq \mu_a^+ + v_a^+ \leq 1$.

Let A and B be any two IVIFNs. The basic operations are defined in [20], including $A \cap B$, $A \cup B$, $A \oplus B, A \otimes B$, λA and A^λ.

Definition 2. *[19] Let $A = ([\mu_a^-, \mu_a^+], [\upsilon_a^-, \upsilon_a^+])$ be an IVIFN. The score function s is defined as*

$$s(A) = \frac{1}{4}(2 + \mu_a^- - \upsilon_a^- + \mu_a^+ - \upsilon_a^+). \tag{2}$$

Definition 3. *[14] An ordered weighted averaging (OWA) operator of dimension n is a mapping OWA: $R^n \to R$, which has an associated weight vector $w = (w_1, w_2, ..., w_n)^T$, with $w_j \in [0,1]$ and $\sum_{j=1}^{n} w_j = 1$. The OWA is defined as*

$$OWA(a_1, a_2, ..., a_n) = \sum_{j=1}^{n} w_j b_j, \tag{3}$$

where b_j is the jth largest of $a_i (i = 1, 2, ..., n)$.

We assume a collection of attributes $A = \{a_1, a_2, ..., a_n\}$ that are prioritized, such that $a_i > a_j$ if $i < j (i, j = 1, 2, .., n)$. And we assume for a given alternative x, $a_i(x) \in [0,1]$ is the degree of satisfaction to the ith attribute by alternative x.

Definition 4. *[16] A prioritized ordered averaging (POWA) operator of dimension n is a mapping POWA: $R^n \to R$. For any alternative x, there is*

$$POWA(a_1, a_2, ..., a_n) = \sum_{i=1}^{n} w_i a_{\delta(i)}(x), \tag{4}$$

where $\delta(i)$ is the index of the ith satisfied attribute. The weight vector $w = (w_1, w_2, ..., w_n)^T$ is an associated weight vector based on a basic unit monotonic (BUM) function [15].

3   The Proposed IVIF-POWA Operator

In threat assessment, it is quite difficult to describe the aerial circumstance with crisp numbers by the experts. And the information usually changes with time. Therefore, we develop an interval-valued intuitionistic fuzzy prioritized ordered weighted averaging (IVIF-POWA) operator by extending the POWA [16].

Definition 5. *Let $a_i = ([\mu_{a_i}^-, \mu_{a_i}^+], [\upsilon_{a_i}^-, \upsilon_{a_i}^+])$ $(i = 1, 2, ..., n)$ be a collection of IVIFNs. A interval-valued intuitionistic fuzzy prioritized ordered weighted aggregation (IVIF-POWA) operator of dimension n is a mapping $\tilde{F}_{(p,w)}: R^n \to R$. For any alternative x, there is*

$$\begin{aligned}
\tilde{F}_{(p,w)}&(a_1, a_2, ..., a_n) \\
&= \sum_{i=1}^{n} w_i a_{\delta(i)} \\
&= \left(\left[1 - \prod_{i=1}^{n} \left(1 - \mu_{a_{\delta(i)}}^-\right)^{w_i}, 1 - \prod_{i=1}^{n} \left(1 - \mu_{a_{\delta(i)}}^+\right)^{w_i} \right], \right. \\
&\qquad \left. \left[\prod_{i=1}^{n} \left(\upsilon_{a_{\delta(i)}}^-\right)^{w_i}, \prod_{i=1}^{n} \left(\upsilon_{a_{\delta(i)}}^+\right)^{w_i} \right] \right).
\end{aligned} \tag{5}$$

where $\delta(i)$ is the index of the ith most satisfied attribute. The weight vector $w = (w_1, w_2, ..., w_n)^T$ is an associated weight vector based on a BUM function, with $w_i \in [0,1]$ and $\sum_{i=1}^n w_i = 1$.

Theorem 1 (Idempotency). *Let $a_i = ([\mu_{a_i}^-, \mu_{a_i}^+], [\upsilon_{a_i}^-, \upsilon_{a_i}^+])(i = 1, 2, ..., n)$ be a collection of the IVIFNs. Then every IVIFN is equal, i.e., $a_i = a$, for all i. Then there is*

$$\widetilde{F}_{(p,w)}(a_1, a_2, ..., a_n) = a. \tag{6}$$

Proof (of Theorem 1). By Definition 5, we have

$$\widetilde{F}_{(p,w)}(a_1, a_2, ..., a_n) = \sum_{i=1}^n w_i a_{\delta(i)} = \sum_{i=1}^n w_i a = a.$$

Theorem 2 (Monotonicity). *Let $a_i = ([\mu_{a_i}^-, \mu_{a_i}^+], [\upsilon_{a_i}^-, \upsilon_{a_i}^+])$ and $a'_i = ([\mu_{a'_i}^-, \mu_{a'_i}^+], [\upsilon_{a'_i}^-, \upsilon_{a'_i}^+])$ $(i = 1, 2, ..., n)$ be two collections of the IVIFNs. If $\mu_{a_i}^- \leq \mu_{a'_i}^-$, $\mu_{a_i}^+ \leq \mu_{a'_i}^+$, $\upsilon_{a_i}^- \geq \upsilon_{a'_i}^-$ and $\upsilon_{a_i}^+ \geq \upsilon_{a'_i}^+$, for all i, then*

$$\widetilde{F}_{(p,w)}(a_1, a_2, ..., a_n) \leq \widetilde{F}_{(p,w)}(a'_1, a'_2, ..., a'_n). \tag{7}$$

Proof (of Theorem 2). Because $\mu_{a_i}^- \leq \mu_{a'_i}^-$, for all i, we know that

$$1 - \prod_{i=1}^n \left(1 - \mu_{a_{\delta(i)}}^-\right)^{w_i} \leq 1 - \prod_{i=1}^n \left(1 - \mu_{a'_{\delta(i)}}^-\right)^{w_i}.$$

In a similar way, we also have

$$1 - \prod_{i=1}^n \left(1 - \mu_{a_{\delta(i)}}^+\right)^{w_i} \leq 1 - \prod_{i=1}^n \left(1 - \mu_{a'_{\delta(i)}}^+\right)^{w_i}, \quad \prod_{i=1}^n \left(\upsilon_{a_{\delta(i)}}^-\right)^{w_i} \geq \prod_{i=1}^n \left(\upsilon_{a'_{\delta(i)}}^-\right)^{w_i}$$

and $\prod_{i=1}^n \left(\upsilon_{a_{\delta(i)}}^+\right)^{w_i} \geq \prod_{i=1}^n \left(\upsilon_{a'_{\delta(i)}}^+\right)^{w_i}$.

It follows that $\widetilde{F}_{(p,w)}(a_1, a_2, ..., a_n) \leq \widetilde{F}_{(p,w)}(a'_1, a'_2, ..., a'_n)$.

Theorem 3 (Boundary). *Let $a_i = ([\mu_{a_i}^-, \mu_{a_i}^+], [\upsilon_{a_i}^-, \upsilon_{a_i}^+])(i = 1, 2, ..., n)$ be a collection of the IVIFNs. Then*

$$a_* \leq \widetilde{F}_{(p,w)}(a_1, a_2, ..., a_n) \leq a^*, \tag{8}$$

where $a_* = ([min(\mu_{a_i}^-), min(\mu_{a_i}^+)], [max(\upsilon_{a_i}^-), max(\upsilon_{a_i}^+)]), a^* = ([max(\mu_{a_i}^-), max(\mu_{a_i}^+)], [min(\upsilon_{a_i}^-), min(\upsilon_{a_i}^+)])$.

Proof (of Theorem 3). According Theorem 2, we have

$$\widetilde{F}_{(p,w)} \underbrace{(a_*, ..., a_*)}_{n} \leq \widetilde{F}_{(p,w)}(a_1, a_2, ..., a_n) \leq \widetilde{F}_{(p,w)} \underbrace{(a^*, ..., a^*)}_{n}.$$

By Theorem 1, we get $\widetilde{F}_{(p,w)} \underbrace{(a_*, ..., a_*)}_{n} = a_*, \widetilde{F}_{(p,w)} \underbrace{(a^*, ..., a^*)}_{n} = a^*$.

Thus, $a_* \leq \widetilde{F}_{(p,w)}(a_1, a_2, ..., a_n) \leq a^*$.

4   The Threat Assessment Model

In this section, we propose an IVIF-POWA operator-based threat assessment model in complicated aerial environment, as follows.

Step 1: Let $A = \{a_1, a_2, ..., a_n\}$ be a collection of selected attributes, which have priorities such that $a_i > a_j$ if $i < j(i, j = 1, 2, .., n)$. This is a linear ordering with a_1 having the highest priority. We assume that $X = \{x_1, x_2, ..., x_m\}$ is a collection of alternatives.

Step 2: Acquire the score functions s_{a_i} of $a_i(x_k)(i = 1, 2, ..., n)$ using (2).

Step 3: Compute the degree of satisfaction of $a_i(i = 1, 2, ..., n)$.

$$S_i = a_i, (i = 1, 2, ..., n) \tag{9}$$

Step 4: Calculate the importance weights T_i.

$$\begin{cases} T_1 = 1 \\ T_i = \prod_{k=1}^{i-1} S_k, (i = 2, 3, ..., n) \end{cases} \tag{10}$$

Step 5: Compute the normalized priorities based on important weights.

$$r_i = \frac{T_i}{\sum_{i=1}^{n} T_i}. \tag{11}$$

Step 6: According to the score functions s_{a_i}, order the attribute satisfactions and obtain $\delta(i)$ of $a_i(x_k)$.

Step 7: Give the BUM function for the assessment and obtain the weights w_i of the IVIF-POWA operator. The BUM function is

$$g(z) = z^2, \tag{12}$$

and the weights are

$$w_i = g(R_i) - g(R_{i-1}). \tag{13}$$

where $R_0 = 0, R_i = \sum_{k=1}^{i} r_{\delta(i)}, i = 1, 2, ..., n$.

Step 8: According to the results and (5), obtain the aggregated values.

Step 9: Calculate the score of $Z_a(x_k)$ using (2), and rank the results.

5   Practical Example

In this section, a numerical example is given to validate the efficiency of proposed method. Six attributes are considered in threat assessment, which are representative and typical.

5.1   The IVIF-POWA Operator-Based Threat Assessment Model

In this example, there are six different aerial targets $X = \{x_1, x_2, x_3, x_4, x_5, x_6\}$ that are planned to be evaluation. Furthermore, there are six attributes of each

Table 1. The interval-valued fuzzy number of aerial targets

	u_1	u_2	u_3	u_4	u_5	u_6
x_1	([0.90,0.95], [0,0.05])	([0.1,0.15], [0.78,0.83])	([0.928,1], [0,0])	([1,1], [0,0])	([0.734,0.746], [0.21,0.25])	([0.98,1], [0,0])
x_2	([0.50,0.55], [0.38,0.43])	([0.30,0.35], [0.58,0.63])	([0.563,0.662], [0.197,0.261])	([0.922,0.946], [0.03,0.039])	([0.676,0.702], [0.199,0.253])	([0.543,0.588], [0.314,0.336])
x_3	([0.70,0.75], [0.18,0.23])	([0.90,0.95], [0,0.05])	([0.245,0.296], [0.470,0.585])	([0.754,0.789], [0.162,0.196])	([0.333,0.343], [0.482,0.519])	([0.677,0.712], [0.245,0.251])
x_4	([0.50,0.55], [0.38,0.43])	([0.30,0.35], [0.58,0.63])	([0.035,0.04], [0.692,0.783])	([0.882,0.933], [0.03,0.06])	([0.553,0.612], [0.267,0.312])	([0.962,0.969], [0,0.02])
x_5	([0.30,0.35], [0.58,0.63])	([0.50,0.55], [0.38,0.43])	([0.721,0.847], [0.104,0.142])	([0.963,0.970], [0,0.02])	([0.131,0.208], [0.491,0.526])	([0.961,1], [0,0])
x_6	([0.10,0.15], [0.78,0.83])	([0.70,0.75], [0.18,23])	([0.288,0.292], [0.472,0.586])	([0.812,0.843], [0.122,0.131])	([0.299,0.325], [0.487,0.558])	([0.676,0.685], [0.225,0.249])

Table 2. The score functions s_{a_i}

	s_{a_1}	s_{a_2}	s_{a_3}	s_{a_4}	s_{a_5}	s_{a_6}
x_1	0.995	0.950	0.984	0.755	0.982	0.160
x_2	0.620	0.560	0.950	0.732	0.692	0.360
x_3	0.723	0.760	0.796	0.419	0.372	0.950
x_4	0.978	0.560	0.931	0.647	0.150	0.360
x_5	0.990	0.360	0.978	0.331	0.831	0.560
x_6	0.722	0.160	0.851	0.395	0.381	0.760

Table 3. The importance weights $T_i(x_k)$

	T_1	T_2	T_3	T_4	T_5	T_6
x_1	1	0.995	0.945	0.930	0.702	0.689
x_2	1	0.620	0.347	0.330	0.241	0.167
x_3	1	0.723	0.550	0.438	0.183	0.068
x_4	1	0.978	0.548	0.510	0.330	0.049
x_5	1	0.990	0.357	0.349	0.115	0.096
x_6	1	0.722	0.116	0.098	0.039	0.015

target in this threat assessment $U = \{u_1, u_2, u_3, u_4, u_5, u_6\}$, including Target type, Interference ability, Flight time, Flight velocity, Short-cut route and Flight height in Table 1. The prioritization relationship of these attributes is given by experts, $u_6 > u_1 > u_4 > u_5 > u_3 > u_2$.

The threat assessment based on IVIF-POWA operator is given as follows.

Step 1: Let $A = \{a_1, a_2, a_3, a_4, a_5, a_6\}$ be a collection of attributes, and $a_1 > a_2 > a_3 > a_4 > a_5 > a_6$. According to the analysis above, we can get $a_1 = u_6, a_2 = u_1, a_3 = u_4, a_4 = u_5, a_5 = u_3$ and $a_6 = u_2$.

Step 2: Compute the score functions s_{a_i} of $a_i(x_k)$ using (2) in Table 2.

Step 3: According to (9), we get the degree of satisfaction.

$S_0 = 1$, $S_1 = a_1(x_k)$, $S_2 = a_2(x_k)$, $S_3 = a_3(x_k)$, $S_4 = a_4(x_k)$, $S_5 = a_5(x_k)$, $S_6 = a_6(x_k)$, where $k = 1, 2, ..., 6$.

Step 4: Then, the importance weights $T_i(x_k)(i = 1, 2, ..., 6; k = 1, 2, ..., 6)$ are calculated using (10). The calculated results are shown in Table 3.

Step 5: Calculate the normalized priority r_i using (11).

Step 6: Order the attribute satisfactions and obtain $\delta(i)$ of $a_i(x_k)$.

Step 7: Calculate the weights w_i of the IVIF-POWA operator using (12) and (13). The computed results are demonstrated in Table 4.

Step 8: According to the results and (5), we obtain

$Z(x_1) = ([0.815, 1], [0, 0])$, $Z(x_2) = ([0.526, 0.576], [0.384, 0.374])$,
$Z(x_3) = ([0.580, 0.619], [0, 0.326])$, $Z(x_4) = ([0.623, 0.677], [0, 0.280])$,
$Z(x_5) = ([0.628, 1], [0, 0])$ and $Z(x_6) = ([0.367, 0.396], [0.501, 0.544])$.

Step 9: Calculate the scores of $Z_a(x_k)$ using (2), and rank the results.

$s_{Z_a}(x_1) = 0.954$, $s_{Z_a}(x_2) = 0.599$, $s_{Z_a}(x_3) = 0.718$, $s_{Z_a}(x_4) = 0.755$, $s_{Z_a}(x_5) = 0.907$, $s_{Z_a}(x_6) = 0.430$.

Finally, the ranking is $x_1 > x_5 > x_4 > x_3 > x_2 > x_6$.

Table 4. The weight w_i

	w_1	w_2	w_3	w_4	w_5	w_6
x_1	0.036	0.101	0.117	0.226	0.276	0.245
x_2	0.017	0.046	0.053	0.388	0.378	0.120
x_3	0.001	0.043	0.162	0.420	0.255	0.120
x_4	0.086	0.120	0.158	0.427	0.026	0.184
x_5	0.118	0.099	0.039	0.034	0.484	0.226
x_6	0.003	0.001	0.319	0.059	0.025	0.594

5.2   Some Comparative Models

Two comparative models are selected to prove effectiveness the proposed model.

Comparative model 1: Chen [3] proposed a method for decision making based on the IVIF prioritized aggregation operator. And the results calculated by this method are shown as follows,

$IVIFPA_1 = ([1,1],[0,0]), IVIFPA_2 = ([0.820,0.879],[0.045,0.070]),$
$IVIFPA_3 = ([0.914,0.944],[0.020,0.033]), IVIFPA_4 = ([0.991,0.995],[0,0.002])$
$IVIFPA_5 = ([0.983,1],[0,0]), IVIFPA_6 = ([0.724,0.756],[0.132,0.174]).$

Then, we can get the score function of $IVIFPA_i$ using (2),
$E_1 = 1$, $E_2 = 0.896$, $E_3 = 0.951$, $E_4 = 0.996$, $E_5 = 0.996$, $E_6 = 0.7694$.

Finally, the ranking is $x_1 > x_4 = x_5 > x_3 > x_2 > x_6$.

Comparative model 2: Sivaraman et al. [9] proposed a method for complete ranking of incomplete interval information. The attribute weight vector is $\{0.2, 0.2, 0.13, 0.13, 0.16, 0.18\}^T$, and the results calculated by this method are shown as follows,

$L(x_1) = 0.943$, $L(x_2) = 0.557$, $L(x_3) = 0.700$, $L(x_4) = 0.698$,
$L(x_5) = 0.878$, $L(x_6) = 0.412$.

Finally, the ranking is $x_1 > x_5 > x_3 > x_4 > x_2 > x_6$.

5.3 Performance Analysis

According to the rankings of three different models, it can be observed that the proposed model can adequately handle the vague assessment result by the IVIFNs and prioritized relationship. Comparative model 1 is also based on prioritized relationship, but it ignores the position of the given attributes. Hence, Comparative model 1 cannot distinguish the $4th$ target and the $5th$ target, such that $x_4 = x_5$. As for Comparative model 2, it is mainly based on the novel accuracy function and the given weights. And it may have the different ranking results according to the different given weights. In short, from the ranking results above, the proposed method can more efficiently handle the vague assessment and obtain consistent result with majority of the moderators.

6 Conclusion

In this paper, an IVIF-POWA operator is proposed, which considers both the prioritization relationship and ordered position of attribute simultaneously. Then, we address the threat assessment problem based on the proposed operator and multiple attributes. Furthermore, the relationship of attributes is considered in the threat assessment model, including priority and orders of them. Through experimental analysis, the obtained result of threat assessment by our proposed model is more reasonable for the defense system.

References

1. Atanassov, K., Gargov, G.: Interval valued intuitionistic fuzzy sets. Fuzzy sets and systems **31**(3), 343–349 (1989)
2. Qu, C., He, Y.: A method of threat assessment using multiple attribute decision making. In: 6th International Conference on Signal Processing, vol. 2, pp. 1091–1095 (2002)

3. Chen, T.Y.: A prioritized aggregation operator-based approach to multiple criteria decision making using interval-valued intuitionistic fuzzy sets: A comparative perspective. Information Sciences **281**, 97–112 (2014)
4. Du, Y., Wang, Y.: A variable weight model based on IFHPWA operator for air target threat evaluation. Electronics Optics & Control **21**(3), 23–28 (2014)
5. Gao, J., Liang, W., Yang, J.: Application of fuzzy beliefs in threat assessment of aerial targets. In: 5th International Conference on Intelligent Human-Machine Systems and Cybernetics, vol. 2, pp. 87–92 (2013)
6. Guo, H., Xu, H.J., Liu, L.: Threat assessment for air combat target based on interval TOPSIS. Systems Engineering and Electronics **31**(12), 2914–2917 (2009)
7. Gu, H., Song, B.: Study on effectiveness evaluation of weapon systems based on grey relational analysis and TOPSIS. Journal of Systems Engineering and Electronics **20**(1), 106–111 (2009)
8. Lampinen, T., Laitinen, T., Ropponen, J.: Joint threat assessment with asset profiling and entity bayes net. In: 12th International Conference on Information Fusion, pp. 420–427 (2009)
9. Lakshmana Goonathi Nayagan, L., Muralikrishnan, S., Sivaraman, G.: Multicriteria decision-making method based on interval-valued intuitionistic fuzzy sets. Expert Systems with Applications **38**(3), 1464–1467 (2011)
10. Tang, Z.l., Zhang, A.: A method of knowledge reasoning for threat assessment in the battlefield. In: International Conference on Information Technology and Computer Science, vol. 2, pp. 32–35 (2009)
11. Wang, X.Y., Liu, Z.W., Hou, C.Z., Zhang, C., Yuan, J.M.: Method of object threat assessment based on fuzzy MADM. Control and Decision **22**(8), 859–863 (2007)
12. Xu, Z.: Intuitionistic fuzzy aggregation operators. IEEE Trans. Fuzzy Systems **15**(6), 1179–1187 (2007)
13. Yu, X., Xu, Z.: Prioritized intuitionistic fuzzy aggregation operators. Information fusion **14**(1), 108–116 (2013)
14. Yager, R.R.: On ordered weighted averaging aggregation operators in multicriteria decisionmaking. IEEE Trans. Systems, Man and Cybernetics **18**(1), 183–190 (1988)
15. Yager, R.R.: Quantifier guided aggregation using owa operators. International Journal of Intelligent Systems **11**(1), 49–73 (1996)
16. Yager, R.R.: Prioritized aggregation operators. International Journal of Approximate Reasoning **48**(1), 263–274 (2008)
17. Yager, R.R.: Prioritized OWA aggregation. Fuzzy Optimization and Decision Making **8**(3), 245–262 (2009)
18. Hou, Y., Guo, W., Zhu, Z.: Threat assessment based on variable parameter dynamic bayesian network. In: Proc. Chinese Control Conference (CCC), pp. 1230–1235 (2010)
19. Yu, D., Wu, Y., Lu, T.: Interval-valued intuitionistic fuzzy prioritized operators and their application in group decision making. Knowledge-Based Systems **30**, 57–66 (2012)
20. Xu, Z.: Methods for aggregating interval-valued intuitionistic fuzzy information and their application to decision making. Control and Decision **22**(2), 215–219 (2007)
21. Zhang, S., Shen, M.X., Wang, J.J., Zhou, H.: A group decision-making method of the threat evaluation based on OWA operator. Journal of Air Force Engineering University (Natural Science Edition) **8**(5), 60–62 (2007)

Optimizing Production and Distribution Problem for Agricultural Products with Type-2 Fuzzy Variables

Xuejie Bai[1,2](✉) and Lijun Xu[3]

[1] College of Science, Agricultural University of Hebei, Baoding 071001, Hebei, China
[2] College of Management, Hebei University, Baoding 071002, Hebei, China
baixuejie123@sina.com
[3] Bureau of Animal Husbandry and Fisheries of Baoding,
Baoding 071005, Hebei, China

Abstract. This paper focuses on generating the optimal solutions of the production and distribution for agricultural products under fuzzy environment, where the crop's yield is characterized by type-2 fuzzy variable with known type-2 possibility distributions. In order to formulate the problem within the framework of the credibility optimization, we employ the possibility value-at-risk (VaR) reduction method to the type-2 fuzzy yield, and then reformulate the multi-fold fuzzy production and distribution problem as the chance constrained programming model. On the basis of the critical value formula for possibility value-at-risk reduced fuzzy variable, original fuzzy production and distribution model is converted into its equivalent parametric mixed integer programming form, which can be solved by general-purpose software. Numerical experiment is implemented to highlight the application of the fuzzy production and distribution model as well as the effectiveness of the solution method.

Keywords: Agricultural products · Production and distribution · Type-2 fuzzy variable · Reduction method · Chance constrained programming

1 Introduction

As an overall research field comprising of cultivation, harvest, storage, processing and distribution, the production and distribution plan of agricultural products plays an important role in the architecture of advanced planning systems. Because of several reasons, such as the national focus on recent cases of agricultural produce contamination, the changing attitudes of a more health conscious and the preference of the better informed consumers [1], the production and distribution problem has attracted many researchers' attention lately [2,13]. On the other hand, due to the complex environment during the the whole process, some critical parameters in the production and distribution problem are always treated as uncertain variables to meet the practice-oriented situations. Thus,

© Springer International Publishing Switzerland 2015
Y. Tan et al. (Eds.): ICSI-CCI 2015, Part II, LNCS 9141, pp. 25–32, 2015.
DOI: 10.1007/978-3-319-20472-7_3

some authors have modified traditional deterministic models to account for the uncertainty in most farm activities by modeling the unknown parameters as random variables with known probability distribution or fuzzy variables with known possibility distributions. For instance, Ahumada and Villalobos [3] selected a two-stage stochastic program with risk level to build a tactical planning model for the production and distribution. Under the market demand's uncertainty, Cai et al. [8] determined the optimal decisions for a fresh product supply chain which is composed of a producer, a third-party logistics provider and the distributor. Yu and Nagurney [16] developed a network-based food supply chain model under oligopolistic competition and perishability with a focus on fresh produce and investigated a case study focused on the cantaloupe market.

In a fuzzy decision system, fuzziness usually is characterized by fuzzy sets. In general, fuzzy set requires crisp membership function which cannot be obtained in practical problems. To overcome this difficulty, the type-2 fuzzy set as an extension of an ordinary fuzzy set was introduced by Zadeh [17] in 1975. After that, there are a lot of researchers to study, extend and apply type-2 fuzzy sets [4,10–12,14]. Among them, Liu and Liu [11] adopted a variable-based approach to depict type-2 fuzzy phenomenon and presented the fuzzy possibility theory which is a generalization of the usual possibility theory. Bai and Liu [4] proposed the possibility value-at-risk (VaR) reduction method which was employed to the supply chain network design problem [5]. To the best of our knowledge, there is little research for modeling production and distribution problem of agricultural products from type-2 fuzziness standpoint. In the current development, we will formulate a new fuzzy production and distribution model for agricultural products, in which the crops' yields are characterized by type-2 fuzzy variables. More precisely, the fuzzy yield can be represented by parametric possibility distributions, which are obtained by using the possibility VaR reduction method. In order to solve the proposed model, we employ the critical value formula of the reduced fuzzy variables and turn the original model with credibility constraints into its equivalent parametric mixed integer programming which can be solved by general-purpose software. One numerical experiment is performed for the sake of illustration.

The rest of this paper is organized as follows. In Section 2, we presents a detailed description for the production and distribution problem. In Section 3, we reformulate the production and distribution problem with type-2 fuzzy variables as a chance constrained programming, discuss the equivalent parametric representation of credibility constraints and solve the mixed integer linear programming by optimization software. In Section 4, one numerical example is given to highlight the application as well as the effectiveness of the solution method. Section 5 summarizes the main conclusions in our paper.

2   Statements of Production and Distribution Problem

In this section, we will construct a new type of fuzzy programming model for the production and distribution plan of agricultural products. In order to describe conveniently the problem, we display the required parameters in Table 1.

Table 1. List of notations

Notations	Definitions
i	index of crops, $i = 1, 2, \ldots, n$;
M	the maximum possible acres of land
f_i	the nonnegative planting cost for different crop i
s_i	the unit selling price for crop i
p_i	the unit purchase price for crop i
d_i	the demand for crop i
q_i	the yield for every crop i
x_i	acres of land devoted to crop i
y_i	the amount purchased from market for crop i
z_i	the amount sold for crop i

In the following, we discuss the establishment of the objective function and constraints.

Objective Function:
The costs is given by the combination of external conditions, which are out of the producer's control, such as expected market prices, and those determined by the producer himself, such as what and how much to plant in a given season. The aim of the model is to minimize the total costs of a producer. That is to say, the objective function is

$$\min \quad \sum_{i=1}^{n} f_i x_i + \sum_{i=1}^{n} p_i y_i - \sum_{i=1}^{n} s_i z_i.$$

Constraints:
Firstly, the constraint represents the main resources limiting the operations. Usually this resource is the result of strategic decisions, such as land available. The subsequent constraint makes sure that the resource used by a solution does not exceed the total availability of land, i.e.,

$$\sum_{i=1}^{n} x_i \leq M.$$

Secondly, when the values taken by the random fuzzy variables are available, the customers' demands can be satisfied by corrective or recourse actions. This can be expressed as the following inequality,

$$q_i x_i + y_i - z_i \geq d_i, \quad i = 1, 2, \ldots, n.$$

Based on the notations and discussion, the production and distribution model for agricultural product can be established as follows:

$$(1) \quad \begin{cases} \min \sum_{i=1}^{n} f_i x_i + \sum_{i=1}^{n} p_i y_i - \sum_{i=1}^{n} s_i z_i \\ \text{s.t.} \sum_{i=1}^{n} x_i \leq M \\ \quad q_i x_i + y_i - z_i \geq d_i, \quad i = 1, 2, \ldots, n \\ \quad x_i, y_i, z_i \geq 0, \quad i = 1, 2, \ldots, n. \end{cases}$$

The above mathematical model (1) is formulated with certain parameters taking fixed values, which can be effectively solved by standard software. However, in real world, it is difficult to figure out the crisp values of some parameters due to the complexity of the decision environment. In this case, the associated parameters can be endowed with uncertain characteristics, such as randomness or fuzziness. As for stochastic method, it is often required that the historical data are sufficiently large in order to estimate the probability distributions. If the sample size is too small owing to the incompleteness of a prior information, one may resort to the fuzzy theory. In this paper, we concern particularly on the formulation and solution for the production and distribution problem of agricultural products when several parameters are characterized by type-2 fuzzy variables.

3   Reformulation of Production and Distribution Problem with Type-2 Fuzzy Variables

In the production and distribution problem, because of the complexity of the decision environment and the incompleteness of a prior information, it is not easy to predict precisely each crop's yield. So we assume that the crop's yield is type-2 fuzzy variable pre-specified by professional judgments, denoted by $\widetilde{q}_i$. Then we have the subsequent formulation:

$$
\begin{cases}
\min \sum_{i=1}^{n} f_i x_i + \sum_{i=1}^{n} p_i y_i - \sum_{i=1}^{n} s_i z_i \\
\text{s.t. } \sum_{i=1}^{n} x_i \leq M \\
\quad \widetilde{q}_i x_i + y_i - z_i \geq d_i, \quad i = 1, 2, \ldots, n \\
\quad x_i, y_i, z_i \geq 0, \quad i = 1, 2, \ldots, n.
\end{cases}
\tag{2}
$$

Apparently, the second constraint of model (2) include a type-2 fuzzy variable, so it is impossible to judge whether a decision vector is feasible or not. Therefore, the form (2) is not well-defined mathematically. To build a meaningful model, we can employ the possibility VaR reduction method to simplify the type-2 fuzzy variables $\widetilde{q}_i$ so as to generate its reduced fuzzy variable $\overline{q}_i$, and then reformulate the model on the basis on credibility measure.

In literature, the fuzzy optimization have been applied to a variety of decision-making problems up to now (see Bai and Liu [6], Kundu et al. [9], Yang et al [15]). Through minimizing the total cost subject to credibility constraint, we finally reformulate the production and distribution problem into the following chance-constrained programming:

$$
\begin{cases}
\min \sum_{i=1}^{n} f_i x_i + \sum_{i=1}^{n} p_i y_i - \sum_{i=1}^{n} s_i z_i \\
\text{s.t. } \sum_{i=1}^{n} x_i \leq M \\
\quad \mathrm{Cr}\{\overline{q}_i x_i + y_i - z_i \geq d_i\} \geq \beta_i, \quad i = 1, 2, \ldots, n \\
\quad x_i, y_i, z_i \geq 0, \quad i = 1, 2, \ldots, n,
\end{cases}
\tag{3}
$$

where β_i is predetermined credibility confidence level. The chance constraint aims to require that the credibility measure of demand constraint should not be less than the confidence level given in advance.

For solution convenience, we here give an equivalent form to further simplify the proposed chance constrained model. Assume $\tilde{q}_1, \tilde{q}_2, \ldots, \tilde{q}_{n-1}$ and $\tilde{q}_n$ are mutually independent type-2 triangular fuzzy variables such that their elements are defined by $\tilde{q}_i = (r_{1,i}, r_{2,i}, r_{3,i}; \theta_{l,i}, \theta_{r,i})$. Obviously, $\tilde{q}_1, \tilde{q}_2, \ldots, \tilde{q}_{n-1}$ and $\tilde{q}_n$ are mutually independent fuzzy variables. Thus, the credibility constraint $\mathrm{Cr}\{\tilde{q}_i x_i + y_i - z_i \geq d_i\} \geq \beta_i$ has the following equivalent expression:

$$d_i - y_i + z_i \leq (\tilde{q}_i x_i)_{\sup}(\beta_i) = x_i \tilde{q}_{i,\sup}(\beta_i).$$

Based on Theorems 1 and 2 [6], if we denote $\theta = (\theta_{l,i}, \theta_{r,i})$, then the critical values of the upper reduced fuzzy variable $\tilde{q}_i^U$ have the following parametric possibility distributions

$$\tilde{q}_{i,\sup}^U(\beta_i; \theta, \alpha) = \begin{cases} r_{3,i} - \frac{2\beta_i(r_{3,i} - r_{2,i})}{1 + \theta_{r,i} - \alpha_i \theta_{r,i}}, & \text{if } \beta_i \in [0, \frac{1 + \theta_{r,i} - \alpha_i \theta_{r,i}}{4}] \\ r_{2,i} + \frac{(1 - 2\beta_i)(r_{3,i} - r_{2,i})}{1 - \theta_{r,i} + \alpha_i \theta_{r,i}}, & \text{if } \beta_i \in [\frac{1 + \theta_{r,i} - \alpha_i \theta_{r,i}}{4}, \frac{1}{2}] \\ r_{2,i} - \frac{(2\beta_i - 1)(r_{2,i} - r_{1,i})}{1 - \theta_{r,i} + \alpha_i \theta_{r,i}}, & \text{if } \beta_i \in [\frac{1}{2}, \frac{3 - \theta_{r,i} + \alpha_i \theta_{r,i}}{4}] \\ r_{1,i} + \frac{2(1 - \beta_i)(r_{2,i} - r_{1,i})}{1 + \theta_{r,i} - \alpha_i \theta_{r,i}}, & \text{if } \beta_i \in [\frac{3 - \theta_{r,i} + \alpha_i \theta_{r,i}}{4}, 1], \end{cases} \quad (4)$$

while the critical values of the lower reduced fuzzy variable $\tilde{q}_i^L$ have the following parametric possibility distributions

$$\tilde{q}_{i,\sup}^L(\beta_i; \theta, \alpha) = \begin{cases} r_{3,i} - \frac{2\beta_i(r_{3,i} - r_{2,i})}{1 - \theta_{l,i} + \alpha_i \theta_{l,i}}, & \text{if } \beta_i \in [0, \frac{1 - \theta_{l,i} + \alpha_i \theta_{l,i}}{4}] \\ r_{2,i} + \frac{(1 - 2\beta_i)(r_{3,i} - r_{2,i})}{1 + \theta_{l,i} - \alpha_i \theta_{l,i}}, & \text{if } \beta_i \in [\frac{1 - \theta_{l,i} + \alpha_i \theta_{l,i}}{4}, \frac{1}{2}] \\ r_{2,i} - \frac{(2\beta_i - 1)(r_{2,i} - r_{1,i})}{1 + \theta_{l,i} - \alpha_i \theta_{l,i}}, & \text{if } \beta_i \in [\frac{1}{2}, \frac{3 + \theta_{l,i} - \alpha_i \theta_{l,i}}{4}] \\ r_{1,i} + \frac{2(1 - \beta_i)(r_{2,i} - r_{1,i})}{1 - \theta_{l,i} + \alpha_i \theta_{l,i}}, & \text{if } \beta_i \in [\frac{3 + \theta_{l,i} - \alpha_i \theta_{l,i}}{4}, 1]. \end{cases} \quad (5)$$

Through the above discussion, the chance constrained programming (3) can be turned into the following equivalent parametric programming

$$\begin{cases} \min \sum_{i=1}^n f_i x_i + \sum_{i=1}^n p_i y_i - \sum_{i=1}^n s_i z_i \\ \text{s.t. } \sum_{i=1}^n x_i \leq M \\ \quad \tilde{q}_{i,\sup}(\beta_i) x_i + y_i - z_i \geq d_i, \quad i = 1, 2, \ldots, n \\ \quad x_i, y_i, z_i \geq 0, \quad i = 1, 2, \ldots, n, \end{cases} \quad (6)$$

where $\tilde{q}_{i,\sup}(\beta_i)$ is determined by Eqs. (4) or (5). Model (6) is a parametric linear programming, which can be solved by optimization software, such as Lingo.

4   Numerical Example

In this section, we propose an example to demonstrate the idea. The example is described as follows, which was considered in [7] with the assumption that the crops' yields were assumed to be random variables.

Consider a farmer who specializes in raising grain, corn, and sugar beets on his 500 acres of land. During the winter, he wants to decide how much land to devote to each crop.

The farmer knows that at least 200 tons (T) of wheat and 240 T of corn are needed for cattle feed. These amounts can be raised on the farm or bought from a wholesaler. The corresponding planting cost are \$150, \$230 and \$260 per ton of wheat, corn and sugar beet, respectively. Any production in excess of the feeding requirement would be sold. Selling prices are \$170 and \$150 per ton of wheat and corn. The purchase prices are 40% more than this due to the wholesaler's margin and transportation costs, i.e., \$238 and \$210 per ton of wheat and corn. Another profitable crop is sugar beet which sells at \$36/T; however, the European Commission imposes a quota on sugar beet production. Any amount in excess of the quota can be sold only at \$10/T. The farmer's quota for next year is 6000T.

In this paper, we generalize the problem by assuming the yields are characterized by type-2 triangular fuzzy variables. Assume that the yields of three crops $\widetilde{q}_1, \widetilde{q}_2$ and $\widetilde{q}_3$ are the type-2 triangular fuzzy variables defined as

$$\widetilde{q}_1 \sim (1.5, 2.5, 2.75; \theta_{l,1}, \theta_{r,1}), \widetilde{q}_2 \sim (2.4, 3, 3.9; \theta_{l,2}, \theta_{r,2}), \widetilde{q}_3 \sim (16, 20, 28; \theta_{l,3}, \theta_{r,3}).$$

Let x_1, x_2 and x_3 be the acres of land devoted to wheat, corn and sugar beets, y_1 and z_1 be tons of wheat purchased and sold, y_2 and z_2 be tons of corn purchased and sold, and z_{31} and z_{32} be tons of sugar beets sold at the favorable price and lower price, respectively. As a result, we can set up the model as follows:

$$\begin{cases} \min\ 150x_1 + 230x_2 + 260x_3 + 238y_1 - 170z_1 \\ \quad + 210y_2 - 150z_2 - 36z_{31} - 10z_{32} \\ \text{s.t.}\ x_1 + x_2 + x_3 \leq 500 \\ \quad \overline{q}_{1,\sup}(\beta_1)x_1 + y_1 - z_1 \geq 200 \\ \quad \overline{q}_{2,\sup}(\beta_2)x_2 + y_2 - z_2 \geq 240 \\ \quad \overline{q}_{3,\sup}(\beta_3)x_3 - z_{31} - z_{32} \geq 0 \\ \quad z_{31} \leq 6000 \\ \quad x_1, x_2, x_3, y_1, z_1, y_2, z_2, z_{31}, z_{32} \geq 0. \end{cases} \tag{7}$$

For simplicity, we set the parameters $\theta_{l,1} = \theta_{l,2} = \theta_{l,3} = \theta_l, \theta_{r,1} = \theta_{r,2} = \theta_{r,3} = \theta_r$ and $\alpha_i = \alpha, \beta_i = \beta$ for each i in model (7). Given $\alpha = \beta = \theta_l = \theta_r = 0.95$, when making use of the upper possibility VaR reduction method, we have the following optimal solution

$$\begin{aligned} &x_1 = 36.07,\ x_2 = 97.67,\ x_3 = 366.26; \\ &y_1 = 142.44,\ z_1 = 0;\quad y_2 = 0,\quad z_2 = 0;\ z_{31} = 6000,\ z_{32} = 0, \end{aligned} \tag{8}$$

with objective value -58996.50. It is easy to understand the optimal solution. The farmer allocate enough land to sugar beets to reach the quota of 6000 T. Then he devotes enough land to corn to satisfy the minimum requirement. The rest of the land is distributed to wheat production and some wheat need be purchased to meet the basic requirement. In fact, the optimal solution follows

Table 2. Profit for each crop in the computational result (8)

	Wheat	Corn	Sugar Beets
Yield (T/acre)	1.5955	2.4573	16.3819
Planting cost ($/acre)	150	230	260
Selling price ($/T)	170	150	36 under 6000T
			10 above 6000T
Purchase price ($/T)	238	210	
Minimum requirement	200	240	
Profit ($/acre)		121.235 138.595	329.748 under 6000T
			−96.181 above 6000T

a simple heuristic rule: to devote land in order of decreasing profit per acre as shown in Table 2.

In order to identify parameters' influence on solution quality, we compute the optimal values by adjusting the values of parameters α, β and (θ_l, θ_r), and report the computational results in Table 3, where "Reduc." implies the type-2 fuzzy variable's reduction mode.

Table 3. Computational results with various values of parameters

(θ_l, θ_r)	α	β	Reduc.	Value$_{opt}$	x_1	x_2	x_3	y_1	z_1	y_2	z_2	z_{31}	z_{32}
(1, 1)	0.1	0.9	upper	−59618.96	37.18	97.43	365.38	140.31	0	0	0	6000	0
(1, 1)	0.1	0.9	lower	−92462.29	85.06	87.35	327.59	23.17	0	0	0	6000	0
(0.8, 0.2)	0.9	0.9	upper	−65502.19	47.19	95.33	357.48	119.95	0	0	0	6000	0
(0.8, 0.2)	0.9	0.6	upper	−106449.4	103.83	83.40	312.77	0	38.39	0	0	6000	0
(0.3, 0.7)	0.9	0.8	lower	−80336.65	69.39	90.66	339.95	67.29	0	0	0	6000	0
(0.3, 0.7)	0.1	0.7	lower	−100036.1	94.45	85.38	320.17	0	6.38	0	0	6000	0
(0.1, 0.1)	0.7	0.8	upper	−92728.35	85.38	87.29	327.33	22.19	0	0	0	6000	0
(0.1, 0.1)	0.9	0.6	lower	−66286.92	48.47	95.06	356.47	117.21	0	0	0	6000	0

It follows from Table 3 that there are many differences among these solutions. Obviously, the optimal values and the corresponding optimal solutions depend on the values of parameters.

5   Conclusions

In this paper, the proposed fuzzy chance constrained programming is applied to find most optimal schedule in which the crop's yield is expressed as fuzzy variable with parametric possibility distribution generated by VaR reduction method. After that we discuss the equivalent representation of credibility constraints and obtain a parametric programming associated with original model. Finally, one numerical example to demonstrate the effectiveness of the proposed model.

Acknowledgments. This work is supported partially by the National Natural Science Foundation of China (No.61374184) and the Scientific Research and Development Program of Baoding (Nos.14ZN021, 14ZN022, 14ZN023).

References

1. Ahumada, O., Villalobos, J.R.: Application of Planning Models in the Agri-food Supply Chain: A Review. European Journal of Operational Research **196**(1), 1–20 (2009)
2. Ahumada, O., Villalobos, J.R.: A Tactical Model for Planning the Production and Distribution of Fresh Produce. Annals of Operations Research **190**(1), 339–358 (2011)
3. Ahumada, O., Villalobos, J.R., Mason, A.N.: Tactical Planning of the Production and Distribution of Fresh Agricultural Products under Uncertainty. Agricultural Systems **112**, 17–26 (2012)
4. Bai, X., Liu, Y.: Semideviations of Reduced Fuzzy Variables: A Possibility Approach. Fuzzy Optimization and Decision Making **13**(2), 173–196 (2014)
5. Bai, X., Liu, Y.: Robust Optimization of Supply Chain Network Design in Fuzzy Decision System. Journal of Intelligent Manufacturing (2014). doi:10.1007/s10845-014-0939-y
6. Bai, X., Liu, Y.: Minimum Risk Facility Location-allocation Problem with Type-2 Fuzzy Variables. The Scientific World Journal, Article ID: 472623, 9 (2014)
7. Birge, J.R., Louveaux, F.: Introduction to Stochastic Programming. Springer-Verlag, New York (1997)
8. Cai, X., Chen, J., Xiao, Y., Xu, X., Yu, G.: Fresh-product Supply Chain Management with Logistics Outsourcing. Omega **41**, 752–765 (2013)
9. Kundu, P., Kar, S., Marti, M.: Fixed Charge Transportation Problem with Type-2 Fuzzy Variables. Information Sciences **255**, 170–186 (2014)
10. Liu, Y., Bai, X.: Linear Combinations of T2 Fuzzy Variables. Journal of Uncertain Systems **8**(1), 78–80 (2014)
11. Liu, Z., Liu, Y.: Type-2 Fuzzy Variables and Their Arithmetic. Soft Computing **14**(7), 729–747 (2010)
12. Qin, R., Liu, Y., Liu, Z.: Modeling Fuzzy Data Envelopment Analysis by Parametric Programming Method. Expert Systems with Applications **38**(7), 8648–8663 (2011)
13. Shukla, M., Jharkharia, S.: Agri-fresh Produce Supply Chain Management: A State-of-the-art Literature Review. International Journal of Operations & Production Management **33**(2), 114–158 (2013)
14. Wu, X., Liu, Y.: Optimizing Fuzzy Portfolio Selection Problems by Parametric Quadratic Programming. Fuzzy Optimization and Decision Making **11**(4), 411–449 (2012)
15. Yang, K., Liu, Y., Yang, G.: Optimizing Fuzzy p-hub Center Problem with Generalized Value-at-Risk Criterion. Applied Mathematical Modelling **38**, 3987–4005 (2014)
16. Yu, M., Nagurney, A.: Competitive Food Supply Chain Networks with Application to Fresh Produce. European Journal of Operational Research **224**(2), 273–282 (2013)
17. Zadeh, L.A.: The Concept of a Linguistic Variable and Its Application to Approximate Reasoning–I. Information Sciences **8**, 199–249 (1975)

Application of Fuzzy Set FSCOM in the Evaluation of Water Quality Level

Yongxi Lv[(✉)] and Zhenghua Pan

School of Science, Jiangnan University, Wuxi, China
18206181184@163.com

Abstract. Water quality level evaluation problem is an important content in water resource quality research. Based on fuzzy set FSCOM theory with three kinds of negation, water quality level fuzzy comprehensive evaluation is studied. Taking Beijing Beiyunhe Jiuxian bridge section water quality fuzzy comprehensive evaluation as example, according to the given instance and the Surface Water Environmental Quality Standard, analyses three kinds of different negative sets and their relationships and determines all membership functions in water quality evaluation. Finally gives the water quality evaluation and puts forward general steps and method of fuzzy set FSCOM comprehensive evaluation application in practical problems. The paper shows that it's effective to use fuzzy set FSCOM theory to solve practical comprehensive evaluation problems.

Keywords: Water resource · Water environmental quality standard · Three kinds of negative relations · FScom theory comprehensive evaluation

1 Introduction

With the development of social economy and the improvement of people's living standard, water resource pollution problem has attracted more and more attention. To make rational use of different grades of water resource and to have a better control of water pollution, water resource quality level fuzzy comprehensive evaluation method must be established. There are some frequent used water resource quality evaluation methods, such as single factor method, analytic hierarchy process method, matter element analysis method, comprehensive index method, grey system method, fuzzy mathematical method, etc. [1]. But all these methods mentioned above have some limitations. The shortcoming of single factor method is that water quality are judged as bad, no matter how many factors beyond limits, obviously, the poor degree of water quality is different. The single factor method evaluation result is unsatisfactory, it can not be served as a better choice to deal with water quality problem on the basis of paper [2] and [3]. The comprehensive index method relies heavily on weight. Although the traditional fuzzy mathematical method has considered the uncertain nature of water environment system and the connections and ambiguity between water environment pollution degree and water quality level classification, it does not take the internal contact of five kinds of water resource quality into consideration. Traditional fuzzy comprehensive evaluation method takes the five degree of water quality as independent parts, it doesn't distinguish

© Springer International Publishing Switzerland 2015
Y. Tan et al. (Eds.): ICSI-CCI 2015, Part II, LNCS 9141, pp. 33–41, 2015.
DOI: 10.1007/978-3-319-20472-7_4

and analyze the relations and properties among different water quality levels. All the existing method don't have a thorough analysis and estimate of the internal relations among different water quality levels.

In 2011, Pan Zhenghua defined fuzzy set FSCOM, which is a new fuzzy set with contradictory negation, opposite negation and medium negation [4], fuzzy set FSCOM theory clearly describes the different negative relations in fuzzy information. Based on fuzzy set FSCOM theory, the paper analyzes and determines fuzzy set "water quality is better" is the opposite negation of fuzzy set "water quality is worse"; fuzzy set "water quality is good", fuzzy set "water quality is medium" and fuzzy set "water quality is bad" are medium negation of fuzzy set "water quality is better" and fuzzy set "water quality is worse". Taking Beijing Beiyunhe Jiuxian bridge section water quality fuzzy comprehensive evaluation as example, the membership functions of different water quality levels and FSCOM water quality level fuzzy comprehensive evaluation are given. The paper shows it's effective and useful to use fuzzy set FSCOM theory to solve practical comprehensive evaluation problems.

2   Fuzzy Set FSCOM with Three Kinds of Negations

Definition 1[5]. Let fuzzy subset $A \in P(U)$, $\lambda \in (0, 1)$.

(1) If mapping $\Psi^{\daleth}: A(U) \to [0, 1]$ confirmed a fuzzy subset $A^{\daleth}$ on U, $A^{\daleth}(x) = \Psi^{\daleth}(A(x))$. $A^{\daleth}$ is called opposite negation set of A.

(2) If mapping $\Psi^{\sim}: A(U) \to [0, 1]$, $\Psi^{\sim}(A(x))$ satisfies following five kinds conditions:

$$\Psi^{\sim}(A(x)) = 1 - \lambda + \frac{2\lambda - 1}{1 - \lambda}(A(x) - \lambda) \quad \text{if } A(x) \in (\lambda, 1] \text{ and } \lambda \in [\tfrac{1}{2}, 1) \tag{1}$$

$$\Psi^{\sim}(A(x)) = 1 - \lambda + \frac{2\lambda - 1}{1 - \lambda}A(x) \quad \text{if } A(x) \in [0, 1 - \lambda) \text{ and } \lambda \in [\tfrac{1}{2}, 1) \tag{2}$$

$$\Psi^{\sim}(A(x)) = \lambda + \frac{1 - 2\lambda}{\lambda}A(x) \quad \text{if } A(x) \in [0, \lambda) \text{ and } \lambda \in (0, \tfrac{1}{2}] \tag{3}$$

$$\Psi^{\sim}(A(x)) = \lambda + (A(x) + \lambda - 1)\frac{1 - 2\lambda}{\lambda} \quad \text{if } A(x) \in (1 - \lambda, 1] \text{ and } \lambda \in (0, \tfrac{1}{2}] \tag{4}$$

$$\Psi^{\sim}(A(x)) = \text{Max}(A(x), 1 - A(x)) \tag{5}$$

$A^{\sim}(x) = \Psi^{\sim}(A(x))$. $A^{\sim}$ is called medium negation set of A.

(3) If mapping $\Psi^{\neg}: A(U) \to [0, 1]$ satisfies $\Psi^{\neg}(A(x)) = \text{Max}(A^{\daleth}(x), A^{\sim}(x))$, $\Psi^{\neg}$ confirmed a fuzzy subset on U, written as $A^{\neg}$, $A^{\neg}(x) = \Psi^{\neg}(A(x))$. $A^{\neg}$ is called contradictory negation set of A.

This fuzzy set on domain U is defined by the definition 1 and definition 2, which is called 'Fuzzy Set with Contradictory negation, Opposite negation and Medium negation', for short FSCOM.

3   Application of Fuzzy Set FSCOM in the Evaluation of Water Quality Level

According to the surface water environmental function and protection aims of waters, The water quality is divided into I, II, III, IV, V five levels(in Table 1) in the Surface

Water Environmental Quality Standard GB 3838- 2002 [6]. Main factors affecting water quality are: pH value, DO (dissolved oxygen), NH_3-N, COD(chemical oxygen demand), BOD_5(five-day biochemical oxygen demand), T-P(total phosphorus), T-N (total nitrogen), Cu, Zn, Cd, Cr, Pb.

3.1   Example of Water Quality Evaluation

Based on fuzzy set FSCOM theory with contradictory negation, opposite negation and medium negation, how to evaluate water quality classification problem? Let's take a practical problem as example to illustrate it next.

Example. Beijing Beiyunhe Jiuxian bridge section water quality data and corresponding weight value are showed in table 2 and table 3 [3]. According to the relations among concentration of affecting factors and water quality classification (table 1), determine which level does Beiyunhe Jiuxian bridge section water quality belong to.

Table 1. Basic factor limit of surface water environmental quality standard

Water quality index	Classification standard limit				
	I	II	III	IV	V
pH	6-9	6-9	6-9	6-9	6-9
DO	≥7.5	≥6	≥5	≥3	≥2
NH_3-N	≤0.015	≤0.5	≤1.0	≤1.5	≤2.0
COD	≤15	≤15	≤20	≤30	≤40
BOD_5	≤3	≤3	≤4	≤6	≤10
T-P	≤0.02	≤0.1	≤0.2	≤0.3	≤0.4
Cu	≤0.01	≤1.0	≤1.0	≤1.0	≤1.0
Zn	≤0.05	≤1.0	≤1.0	≤2.0	≤2.0
Cd	≤0.001	≤0.005	≤0.005	≤0.005	≤0.01
Cr	≤0.01	≤0.05	≤0.05	≤0.05	≤0.1
Pb	≤0.01	≤0.01	≤0.05	≤0.05	≤0.1

Table 2. Beijing Jiuxian bridge section water quality data

Water quality index	concentration (mg/kg)
DO	4.28
NH_3-N	5.90
COD	82.7
BOD_5	9.54
T-P	0.42
Cu	0.012
Zn	0.44
Cd	0.007
Cr	0.319
Pb	0.056

Table 3. The weight of each water quality index

Water quality index	weight
DO	0.049
NH$_3$-N	0.265
COD	0.155
BOD$_5$	0.083
T-P	0.092
Cu	0.001
Zn	0.016
Cd	0.006
Cr	0.276
Pb	0.057

3.2   Fuzzy Sets and Water Quality Comprehensive Evaluation in the Example

According to GB 3838-2002[6], water quality is divided into five levels (I, II, III, IV, V). Water quality gets worse along with the increase of number (I stands for water quality is better, V stands for water quality is worse). In order to have a thorough description and understanding of different water quality classification, the five kinds of water quality grades (I, II, III, IV, V) stand for "water quality is better", "water quality is good", "water quality is medium", "water quality is bad" and "water quality is worse" respectively. On the basis of FSCOM theory, "water quality is better", "water quality is good", "water quality is medium", "water quality is bad" and "water quality is worse" are different fuzzy sets, and they have following relations:

"water quality is better" is the opposite negation of fuzzy set "water quality is worse"; "water quality is good", "water quality is medium" and "water quality is bad" are different λ-medium negation of fuzzy sets "water quality is better" and "water quality is worse"; "water quality is good" is the opposite negation of fuzzy set "water quality is bad", "water quality is medium" is their medium negation.

By the fuzzy comprehensive evaluation principle in [7], the five fuzzy sets can be expressed as follows:

- h: stands for fuzzy set "water quality is worse" ;
- $h^\daleth$: stands for fuzzy set "water quality is better" ;
- $h_1^\sim$: stands for fuzzy set "water quality is bad" ;
- $h_2^\sim$:stands for fuzzy set "water quality is medium" ;
- $h_3^\sim$:stands for fuzzy set "water quality is good" ;

    Now we can establish factor set and evaluation set:

(1) factor set U={DO, NH3-N, COD, BOD$_5$, T-P, Cu, Zn, Cr, Pb, Cd};
(2) evaluation set T={h, $h_1^\sim$, $h_2^\sim$, $h_3^\sim$, $h^\daleth$ }.

    According to [8, 9], membership functions of different factors are established using one dimensional Euclidean distance formula.

Membership function of DO about "water quality is worse" is:

$$h(x) = \begin{cases} 0 & x \geq 7.5 \\ 1 - \dfrac{d(x,3)}{d(3,7.5)} & 3 < x < 7.5 \\ 1 & x \leq 3 \end{cases} \qquad (6)$$

Membership function of NH$_3$-N about "water quality is worse" is:

$$h(x) = \begin{cases} 0 & x \leq 0.015 \\ 1 - \dfrac{d(x,1.5)}{d(0.015,1.5)} & 0.015 < x < 1.5 \\ 1 & x \geq 1.5 \end{cases} \qquad (7)$$

Membership function of COD about "water quality is worse" is:

$$h(x) = \begin{cases} 0 & x \leq 15 \\ 1 - \dfrac{d(x,30)}{d(15,30)} & 15 < x < 30 \\ 1 & x \geq 30 \end{cases} \qquad (8)$$

Membership function of "water quality is worse" of other factors can be established as well. On the basis of fuzzy set FSCOM theory and λ-medium negation set, "water quality is better" is the opposite negation of fuzzy set "water quality is worse", membership function of "water quality is better" is $h^\neg(x) = 1 - h(x)$; "water quality is good" is the opposite negation of fuzzy set "water quality is bad", membership function of "water quality is good" is $h_3^\sim(x) = 1 - h_1^\sim(x)$. To establish other medium negative sets membership function, parameter λ must be firstly determined. Suppose that the parameter of fuzzy set "water quality is bad" to "water quality is worse" is λ_1, the parameter of fuzzy set "water quality is medium" to "water quality is bad" is λ_2, fuzzy set "water quality is good" is the opposite negation of fuzzy set "water quality is bad", so the parameter of fuzzy set "water quality is good" to "water quality is worse" is $1-\lambda_1$. We know that concentration 6mg/kg of DO is the lower limit of fuzzy set "water quality is good", while 5mg/kg of DO is the upper limit of fuzzy set "water quality is bad" from the surface water environmental quality standard in table 1.

By the membership function (6) of DO, the degree of membership of lower and upper limits can be easily calculated:

$$h(6) = 1 - \frac{d(3,6)}{d(7.5,3)} = \frac{1}{3} = 0.333 , \quad h(5) = 0.556 .$$

The membership degree of DO concentration value 6mg/kg about fuzzy set "water quality is bad" is $h_1^\sim(6)$, and membership degree of 5mg/kg about fuzzy set "water quality is good" is $h_3^\sim(5)$, moreover, they have the relation below:

$$h_1^\sim(6) = h_3^\sim(5) = 1 - h_1^\sim(5) .$$

when $h(6)=0.333$ and $h(5)=0.556$, let's discuss the value of parameter λ :

(a) if $\lambda_1 >= 0.5$, according to the second and the first situation in definition 1,

$$h_1^\sim(6) = 1 - \lambda_1 + 0.333 \times \frac{2\lambda_1 - 1}{1 - \lambda_1} , \; h_1^\sim(5) = 1 - \lambda_1 + (0.556 - \lambda_1)\frac{2\lambda_1 - 1}{1 - \lambda_1} ,$$ we can get $\lambda_1 = 0.5$,

but this means to every concentration x, $h_1^\sim(x) \equiv 0.5$, obviously unreasonable.

(b) if $\lambda_1 < 0.5$, according to the third situation in definition 1, $h_1^\sim(6) = \lambda_1$ $+0.333(1\text{-}2\lambda_1)/\lambda_1$, $\lambda_1 = 0.5$ does not satisfy the hypothesis.

Because of the situations considered above, $h_1^\sim(x) = \max(h(6),1\text{-}h(6))=0.667$ can only be calculated reasonably by the fifth possibility in definition 1, and $h_1^\sim(5) = 0.333$, $\lambda_1 = 0.623$. Any concentration value $x \in [3,7.5]$, the membership degree about "water quality is bad" is $\dfrac{d(5,3)}{d(7.5,3)} = 0.444$, and the membership degree about "water quality is medium" is $\dfrac{d(6,5)}{d(7.5,3)} = 0.222$. Obviously the membership degree about "water quality is medium" is 0.5 times as much as the membership degree about "water quality is bad". So $\lambda_2 = \lambda_1 - 0.5\lambda_1 = 0.312$.

In conclusion, all parameters of DO are:

The parameter of fuzzy set "water quality is bad" to fuzzy set "water quality is worse" is $\lambda_1 = 0.623$; the parameter of fuzzy set "water quality is medium" to fuzzy set "water quality is bad" is $\lambda_2 = 0.25$ and the third parameter $\lambda_3 = 1 - \lambda_1 = 0.377$.

Similarly, the parameters of other factors can be calculated as well, the results are showed in table 4:

Table 4. Other factors affecting water quality parameter value

Water quality index	parameters		
	λ_1	λ_2	λ_3
NH$_3$-H	0.655	0.655	0.345
COD	0.335	0.168	0.665
BOD$_5$	0.339	0.170	0.661
T-P	0.323	0.323	0.677
Cu	0.97	0.97	0.03
Zn	0.49	0.49	0.51
Cd	0.993	0.993	0.007
Cr	0.993	0.993	0.007
Pb	0.998	0.333	0.002

According to the data in table 2, concentration of DO is 4.28 mg/kg, membership degree about fuzzy set "water quality is worse" is $h(4.28) = 1 - \dfrac{d(4.28,3)}{d(7.5,3)} = 0.72$. The parameter of fuzzy set "water quality is bad" to "water quality is worse" is $\lambda_1 = 0.623$, the parameter of fuzzy set "water quality is medium" to fuzzy set "water quality is bad" is $\lambda_2 = 0.25$; "water quality is good" to "water quality is worse" is $\lambda_3 = 1 - \lambda_1 = 0.377$, consider definition 2 in this paper, we can get

$$h_1^\sim(4.28) = 0.961, \ h_2^\sim(4.28) = 0.72, \ h_3^\sim(4.28) = 1 - h_1^\sim(4.28) = 0.039.$$

While fuzzy set "water quality is better" is the opposite negation of fuzzy set "water quality is worse", membership of concentration 4.28 mg/kg about fuzzy set "water quality is better" is $h^\daleth(4.28)=1-h(4.28)=0.28$. For NH$_3$-N, we can get $h(5.90)=1, h^\daleth(5.90)=1-h(5.90)=0$, $h_1^\sim(5.90)=h_2^\sim(5.90)=0.665$, $h_3^\sim(5.90)=0.345$.

Membership degree of other factors can be determined in the same way. The membership matrix about water quality level of all factors are given below, after normalized the evaluation results based on [10], the membership matrix is

$$R' = \begin{pmatrix}
0.2646 & 0.3532 & 0.2646 & 0.0147 & 0.1029 \\
0.3766 & 0.2467 & 0.2467 & 0.1299 & 0 \\
0.3163 & 0.2103 & 0.2631 & 0.2103 & 0 \\
0.3173 & 0.2097 & 0.2633 & 0.2097 & 0 \\
0.3332 & 0.2223 & 0.2223 & 0.2223 & 0 \\
0.0016 & 0.0539 & 0.0539 & 0.0759 & 0.8147 \\
0.0806 & 0.1997 & 0.1997 & 0.1997 & 0.3204 \\
0.9225 & 0.0065 & 0.0065 & 0.0646 & 0 \\
0.9225 & 0.0065 & 0.0065 & 0.0646 & 0 \\
0.5984 & 0.0012 & 0.3992 & 0.0012 & 0
\end{pmatrix}$$

According to the weight value in table 3, the weight set is

$$V = \{0.049, 0.265, 0.155, 0.083, 0.092, 0.001, 0.016, 0.006, 0.276, 0.057\},$$

Using compositional operation of membership matrix and weight set

$$P = V \circ R' = \{0.5143, 0.1583, 0.1893, 0.1272, 0.0110\}.$$

Elements in P represent the membership degree of every water quality level of Jiuxian bridge under the consideration of weights: the membership of water quality belongs to V is 0.5143, belongs to IV is 0.1583, etc. By the principle of maximum degree of membership, we have the following evaluation conclusion for Beijing Beiyunhe Jiuxian bridge section water quality:

Water quality of Beijing Beiyunhe Jiuxian bridge section belongs to "water quality is worse" (V level). Factors like COD, BOD5 and Cd concentration values exceed the standard, heavy metals such as Pb and Cr also exceed IV water quality limits, the water has been polluted by heavy metals seriously.

Paper [3] also gave the fuzzy comprehensive evaluation of Beijing Beiyunhe Jiuxian bridge water quality and $P = \{0.092, 0.016, 0.049, 0.057, 0.265\}$, water quality belongs to V level, which is consistent with the final result in this paper. But the difference is the basis of fuzzy comprehensive evaluation method in paper [3] is tradition fuzzy set theory, it doesn't analyze the internal relations among different water quality levels, the five levels of water quality are treated as independent fuzzy sets when establish membership function, the whole evaluation process is done by the establishment of fifty membership functions. This paper gives a new fuzzy set FSCOM fuzzy comprehensive evaluation, after analyzing the internal relations among negative fuzzy sets, fuzzy comprehensive evaluation method is established according to different water levels. Evaluation method is better than traditional fuzzy comprehensive evaluation method.

3.3   General Steps and Method of Fuzzy Set FSCOM in Practical Problem Evaluation Application

According to paper [8, 9, 11, 12] and the water quality evaluation method mentioned above, the general steps and method of fuzzy set FSCOM in practical problem evaluation application are as follows:

(1) Analyze and extract useful fuzzy information in practical problems, we can obtain the internal relations among fuzzy sets, for example, five levels "water quality is better", "water quality is good", "water quality is medium", "water quality is bad", "water quality is worse" are different fuzzy sets in this paper;

(2) Establish factor set and evaluation set:

$$\text{factor set } U = \{u_1, u_2, ..., u_t\}, \text{evaluation set } T = \{h_1, h_2, ..., h_p\} \ ;$$

(3) Find relations among different evaluation levels. In this paper "water quality is better" is the opposite negation of fuzzy set "water quality is worse"; "water quality is good", "water quality is medium" and "water quality is bad" are different λ-medium negation sets; "water quality is good" is the opposite negation of fuzzy set "water quality is bad" and "water quality is medium" is their λ-medium negation as well;

(4) Determine membership degree function and parameter λ and membership degree matrix:

$$R = \begin{bmatrix} r_{11} & r_{12} & \cdots & r_{1p} \\ r_{21} & r_{22} & \cdots & r_{2p} \\ \vdots & \vdots & \vdots & \vdots \\ r_{t1} & r_{t2} & \cdots & r_{tp} \end{bmatrix},$$

r_{ij} represents membership degree of factor u_i to evaluation set h_j ($i=1,2,...,t; j=1,2,...,p$).

At last evaluation result is determined by compositional operation and the principle of maximum degree of membership.

4   Conclusion

After analyzing the disadvantages of previous fuzzy comprehensive evaluation methods of water quality level, fuzzy set FSCOM theory with three kinds of negations is applied to the fuzzy comprehensive evaluation of water quality grade. Through analysis relations among different water quality levels, membership function and evaluation system are established. And Beijing Beiyunhe Jiuxian bridge section water belongs to V level, the evaluation and calculation process are better than traditional water quality evaluation method mentioned in [3], which shows it is effective and practical using fuzzy set FSCOM theory to solve practical fuzzy comprehensive evaluation problems. Based on recent application of fuzzy set FSCOM theory, the general steps and method of fuzzy set FSCOM theory in practical problem evaluation application are given at last.

Acknowledgments. This work was supported by the National Natural Science Foundation of China (61375004) and the Fundamental Research Funds for the Central Universities (JUSRP51317B).

References

1. Zhou, X., Fan, Z., Zhan, J.: Application of Fuzzy Mathematical in Chemistry, pp. 61–72. National University of Defense Technology Press, Changsha (2002). (in Chinese)
2. Qiu, M., Liu, L., Zou, X., Pan, Z.: Comparison of Source Water Quality Standards and Evaluation Methods between China and Some Developed Countries. Journal of China Institute of Water Resources and Hydro-power Research **11**(3), 176–182 (2013). (in Chinese)
3. Li, L.: Research on Assessment of Surface Water Pollution and Controlling Strategies in Beijing, pp. 25–43. School of Resources and Environment, China Agricultural University, Beijing (2004). (in Chinese)
4. Pan, Z.: Three Kinds of Negation of Fuzzy Knowledge and Their Base of Set. Journal of Computers **35**(7), 1421–1428 (2012). (in Chinese)
5. Pan, Z., Yang, L., Xu, J.: Fuzzy set with three kinds of negations and its applications in fuzzy decision making. In: Deng, H., Miao, D., Lei, J., Wang, F.L. (eds.) AICI 2011, Part I. LNCS (LNAI), vol. 7002, pp. 533–542. Springer, Heidelberg (2011)
6. GB 3838-2002, The Surface Water Environmental Quality Standards
7. Yang, L., Pan, Z.: Fuzzy Comprehensive Evaluation Based on the Fuzzy Set FScom with Three Kinds of Negations. Computer Engineering and Science **33**(9), 136–140 (2011). (in Chinese)
8. Liu, Y., Guo, Z., Pan, Z.: Applying Fuzzy System FScom with Three Kinds of Negations to Air Quality Evaluation. Computer Application and Software **30**(8), 21–24 (2013). (in Chinese)
9. Zhao, J., Pan, Z.: Application of Fuzzy Set with three kinds of negation FScom in Stock Investment. Computer Science **40**(12), 59–63 (2013)
10. Zhang, G., Zhang, H.: Fuzzy Mathematical Basis and Application, pp. 89–91. Chemical Industry Press, Beijing (2011). (in Chinese)
11. Yang, L., Pan, Z.: Fuzzy degree and similarity measure of fuzzy set with three kinds of negations. In: Deng, H., Miao, D., Lei, J., Wang, F.L. (eds.) AICI 2011, Part I. LNCS, vol. 7002, pp. 543–550. Springer, Heidelberg (2011)
12. Liang, T., Zhu, B., Pan, Z.: Typhoon disaster grade evaluation based on fuzzy Set FScom. Computer Engineering and Application **50**(15), 211–214 (2014). (in Chinese)

A New Evolutionary Fuzzy Instance-Based Learning Approach: Application for Detection of Parkinson's Disease

Huiling Chen[✉], Jiayi Lu, Qiang Li, Congcong Lou, Daopin Pan, and Zhe Yu

College of Physics and Electronic Information, Wenzhou University,
Wenzhou 325035, People's Republic of China
chenhuiling_jsj@wzu.edu.cn

Abstract. In this paper, we present a new evolutionary instance-based learning method by integrating the bacterial foraging optimization (BFO) technique with fuzzy k-nearest neighbor classifier (FKNN), termed as BFO-FKNN. In the proposed approach, the issue of parameter tuning problem in FKNN is tackled using the BFO technique. The effectiveness of the proposed method has been rigorously evaluated against the Parkinson's disease (PD) diagnosis problem. The simulation results have shown that the proposed approach outperforms the other two counterparts via 10-fold cross validation analysis. In addition, compared to the existing methods in previous studies, the proposed method can also be regarded as a promising success with the excellent classification accuracy of 96.39%.

Keywords: Fuzzy k-nearest neighbor · Parameter optimization · Bacterial foraging optimization · Parkinson's disease diagnosis · Medical diagnosis

1 Introduction

As an improved version of the classical k-nearest neighbor (KNN) classifier, fuzzy k-nearest neighbor classifier (FKNN) [1, 2] has gained great attentions since its first proposal. Now, FKNN together with other improved versions of KNN such as fuzzy rough sets, intuitionistic fuzzy sets, type-2 fuzzy sets and possibilistic theory based methods become a distinctive area in the field of nearest neighbor classification and instance based learning [3]. One important characteristic of FKNN is its capability of allowing imprecise knowledge to be presented and fuzzy measures to be introduced, which provide an enhanced way of describing the similarities between the instances. FKNN method introduces the fuzzy set theory into KNN, which assigns degree of membership to different classes while considering the distance of its k-nearest neighbors. In other words, all the instances are assigned a membership value in each class rather than the binary decision to be given. When FKNN is ready to make the final decision, the class with the highest membership function value is taken as the winner. Thanks to its good property, FKNN has found its applications in many practical

© Springer International Publishing Switzerland 2015
Y. Tan et al. (Eds.): ICSI-CCI 2015, Part II, LNCS 9141, pp. 42–50, 2015.
DOI: 10.1007/978-3-319-20472-7_5

problems such as medical diagnosis [4], protein identification and prediction [5] and bankruptcy prediction [6].

The classification performance of FKNN greatly depends on the parameters including the neighborhood size k and the fuzzy strength parameter m. Therefore, these two values of parameters should be carefully chosen in advance when FKNN is applied to the practical problems. Several attempts have been made to tackle the problem of parameter tuning of FKNN. In 2011, Chen et al [6] have first proposed to use the particle swarm optimization (PSO) technique to automatically tune the two parameters of FKNN, and the simulation results on the bankruptcy prediction problems have demonstrated the effectiveness and efficiency of the proposed method. Recently, Cheng et al. [7] have employed the differential evolution optimization approach to select the most appropriate tuning parameters of FKNN, and the resultant model was successfully applied to classify the grouting activities in construction industry. More recently, they have used the firefly algorithm to determine the FKNN model's hyper-parameters and applied it to predict slope collapse events [8], the simulation results have shown that the established method can outperform other benchmarking algorithms. As a relatively new swarm intelligence based algorithm, bacterial foraging optimization (BFO) [9] has shown its great ability of optimization in various fields. Therefore, this paper attempted to employ the BFO to optimize the parameters of FKNN via the simulation of the foraging behavior of E. coli bacteria and its interaction with the surrounding environment. Then we applied the resultant model BFO-FKNN for effective detection of Parkinson's disease (PD). In order to evaluate the effectiveness of the proposed BFO-FKNN approach, support vector machine (SVM) and kernel based extreme learning machine (KELM) were taken for comparison in terms of classification accuracy (ACC), area under the receiver operating characteristic curve (AUC) criterion, sensitivity and specificity.

The remainder of this paper is organized as follows. The related works on detection of PD is presented in Section 2. In section 3 the detailed implementation of the BFO-FKNN method is described. Section 4 describes the experimental design in detail. The experimental results and discussion of the proposed approach are shown in Section 5. Finally, Conclusions are summarized in Section 6.

2   Related Works on Parkinson's Disease Diagnosis

Parkinson's disease (PD) is one kind of degenerative diseases of the nervous system, it has become the second most common degenerative disorders of the central nervous system after Alzheimer's disease [10]. Till now, the cause of PD hasn't been uncovered. However, it is possible to alleviate symptoms significantly at the onset of the illness in the early stage [11]. It has also been proven that a vocal disorder may be one of the first symptoms to appear nearly 5 year before clinical diagnose [12]. The vocal impairment symptoms related with PD are known as dysphonia (inability to produce normal vocal sounds) and dysarthria (difficulty in pronouncing words) [13]. Therefore, dysphonic indicators may play essential role in the early stage of PD diagnosis. Little et al [14] have made the first attempt to utilize the dysphonic indicators in their study to help discriminate PD patients from healthy ones. In their study, SVM in combination with the feature selection approach was taken to diagnose PD, the simulation results has shown that the proposed method can discriminate PD patients from healthy ones with approximately 90% classification accuracy using only four dysphonic features. After then, various techniques have been developed to study the PD diagnosis problem from the perspective of dysphonic indicators, including Artificial Neural Networks (ANNs) [15], Dirichlet process mixtures [16], multi-kernel relevance vector machines [17], fuzzy k-nearest neighbor (FKNN) [4].

3   Proposed BFO-FKNN Model

This study proposes a novel BFO-FKNN model for automatically tuning the two parameters of FKNN. The proposed model was comprised of two procedures as shown in Figure 1, the one is the inner parameter optimization, and the other is the outer performance evaluation. During the inner parameter optimization procedure, the parameter neighborhood size k and fuzzy strength parameter m of FKNN are determined dynamically by the BFO technique via the 5-fold cross validation (CV) process. And then the obtained optimal parameter pair (k, m) is fed to FKNN prediction model to perform the classification task for PD diagnosis in the outer loop using the 10-fold CV strategy. The classification error rate is considered in designing the fitness function:

$$fitness = (\Sigma_{i=1}^{K} testError_i)/k \tag{1}$$

where $testError_i$ represents the average test error rates achieved by the FKNN classifier via 5-fold CV within the inner parameter optimization procedure.

The main steps for the BFO to select the parameters of FKNN are described as follows:

step 1.   Initialize parameters $p, S, Nc, Ns, Nre, Ned, Ped, C(i)$, where p is the number of dimension of the search space, S is swarm size of the population, Nc is the number of chemotactic steps, Ns is the swimming length, Ned is the number of elimination-dispersal events, Nre is the number of reproduction steps, Ped is elimination-dispersal probability, and $C(i)$ is the size of step taken in the random direction specified by the tumble.

step 2.   Elimination-dispersal loop: $l=l+1$.

step 3.   Reproduction loop: $k=k+1$.

step 4.   Chemotaxis loop: $j=j+1$.

    (a) **For** $i=1,2,…,S$, take a chemotactic step for bacterium i as follows.

    (b) Train FKNN and compute the fitness $J(i, j, k, l)$

        Let, $J(i, j, k, l)=J(i, j, k, l)+J_{ar}(\theta)$ where J_{ar} is defined as $J_{ar}(\theta) = exp(M-J_{ar}(\theta))J_{cc}(\theta)$, $J_{cc}(\theta)$ is the fitness function value to be added to the actual fitness function to present a time-varying fitness function.

    (c) Let $Jlast=J(i, j, k, l)$ to save this value since we may find a better cost via a run.

    (d) Tumble: generate a random vector $\Delta(i) \in R^p$ with each element

$\Delta_m(i), m=1,2,\cdots,p$, a uniformly distributed random number on [-1, 1].

(f) Move: let

$$\theta^i\left(j+1,k,l,di\right) = \theta^i\left(j,k,l,di\right) + C(i)\frac{\Delta(i)}{\sqrt{\Delta^T(i)\Delta(i)}} \tag{2}$$

(g) Train FKNN and compute the fitness $J(i,j+1,k,l)$, and let
$$J(i,j+1,k,l) = J(i, j, k, l) + Jar(\theta).$$

(h) Swim.

    i) Let $n=0$;

    ii) **While** $n<Ns$

    iii) Let $n=n+1$;

    iv) **If** $J(i,j+1,k,l)<Jlast$, let $Jlast=J(i,j+1,k,l)$ and let

$$\theta^i\left(j+1,k,l,di\right) = \theta^i\left(j,k,l,di\right) + C(i)\frac{\Delta(i)}{\sqrt{\Delta^T(i)\Delta(i)}} \tag{3}$$

and use this $\theta^i(j+1,k,l)$ to train FKNN, and then compute the new fitness
$$J(i, j+1,k, l) \text{ as did in (g)};$$

    v) Else, let $n = Ns$.

(i) Go to next bacterium $(i+1)$ if $i \neq S$.

step 5. If $j<Nc$, go to step 4.

step 6. Reproduction:

Rank all of the individuals according to the sum of the evaluation results in this period, and then removes out the last half individuals and duplicates one copy for each of the rest half.

step 7. If $k<Nre$, go to step 3.

step 8. Elimination-dispersal:

For $i=1,2,\ldots,S$ with probability Ped, eliminate and disperse each bacterium.

If $l<Ned$, then go to step 2; otherwise end.

4   Experimental Studies

4.1   Data Description

The Parkinson's data was taken from UCI machine learning repository (http://archive. ics.uci.edu/ml/datasets/Parkinsons, last accessed: December 2014). The objective of this dataset is to discriminate healthy people from those with Parkinson's disease (PD). In the medical experiment, various biomedical voice measurements were recorded for 23 patients with PD and 8 healthy controls. The time since diagnoses ranged from 0 to 28 years, and the ages of the subjects ranged from 46 to 85 years, with a mean age of 65.8. Each subject provides an average of six phonations of the vowel (yielding 195 samples in total), each 36 seconds in length. It should be noted that there is no missing values in the dataset, and the whole features are real valued.

4.2   Experimental Setup

The BFO-FKNN, SVM and KELM classification models were implemented using MATLAB platform. The computational analysis was conducted on Windows 7 operating system with AMD Athlon 64 X2 Dual Core Processor 5000+ (2.6 GHz) and 4GB of RAM. LIBSVM [18] developed by Chang and Lin was taken for SVM classification. The implementation by Huang available from http://www3.ntu.edu.sg/home/egbhuang was used for KELM classifier. We implemented the BFO-FKNN method from scratch. The data were scaled into [-1, 1] before classification was performed. In order to guarantee the valid results, the k-fold CV analysis was employed to evaluate the classification performance. The detailed parameter setting for BFO is shown in Table 1. In the table, d_{att}, w_{att}, h_{repe} and w_{repe} are different coefficients in swarming strategy, and other parameters are described in section 3. For SVM and KELM, the determination of penalty parameter C and kernel bandwidth γ is done by the grid search method, and the searching ranges are set to $C \in [2^{\wedge}(-5), 2^{\wedge}(15)]$ and $\gamma \in [2^{\wedge}(-5), 2^{\wedge}(15)]$.

Table 1. Parameter setup for BFO

S	Nc	Ns	Ned	Nre	Ped	datt	watt	wrepe	hrepe	C(i)
8	12	2	2	2	0.25	0.1	0.2	10	0.1	0.25

4.3   Evaluation Metric

Four common performance metrics including ACC, AUC, sensitivity and specificity were used to test the performance of the proposed BFO-FKNN model. ACC, sensitivity and specificity are defined as follows:

$$ACC = TP + TN / (TP + FP + FN + TN) \times 100\% \qquad (4)$$

$$Sensitivity = TP / (TP + FN) \times 100\% \qquad (5)$$

$$Specificity = TN / (FP + TN) \times 100\% \qquad (6)$$

where TP is the number of true positives, FN is the number of false negatives, TN is the number of true negatives and FP is the number of false positives. AUC is the area under the ROC curve, which is one of the best methods for comparing classifiers in two-class problems. In this study the method proposed in [19] was implemented to compute the AUC.

5   Experimental Results and Discussions

The detailed results of classification accuracy, sensitivity, specificity, and optimal pairs of (C, γ) for each fold obtained by BFO-FKNN are listed in Table 2. As shown in the table, it can be observed that the parameter pairs (k, m) are automatically determined by BFO for each fold of the data. With the optimal combination of k and m, FKNN obtained different best classification accuracy in each fold. The explanation lies in the fact that the two parameters are coevolved adaptively by the BFO algorithm according to the specific distribution of the training data at hand.

Table 2. The detailed results of BFO-FKNN on the PD data set

Fold No.	BFO-FKNN					
	Accuracy	AUC	Sensitivity	Specificity	k	m
1	1.0000	1.0000	1.0000	1.0000	1	9.75
2	1.0000	1.0000	1.0000	1.0000	1	8.33
3	1.0000	1.0000	1.0000	1.0000	1	7.65
4	0.9500	1.0000	0.8333	0.9167	1	7.08
5	0.9500	0.9412	1.0000	0.9706	1	6.97
6	0.9500	1.0000	0.8000	0.9000	1	8.27
7	0.9474	0.9167	1.0000	0.9583	1	4.13
8	0.9474	0.9333	1.0000	0.9667	1	9.75
9	0.9474	0.9333	1.0000	0.9667	1	7.70
10	0.9474	0.9286	1.0000	0.9643	1	6.74
Avg.	0.9639	0.9643	0.9653	0.9633	1	7.64

Avg. represents the average result of 10-fold CV process.

In order to observe the evolutionary behavior of BFO, we have plot the classification error rate versus the iterations. Figure 1 shows the evolutionary process of the BFO-FKNN for fold 1 in 10-fold CV. It can be observed that the fitness curves gradually improved from iteration 1 to 10 and exhibited no significant improvements after iteration 10, eventually stopped at the iteration 50 where the bacteria reached the stopping criterion (maximum iteration number). The fitness of error rate decreased rapidly in the beginning of the evolution, after certain number of generations, it started decreasing slowly. During the latter part of the evolution, the fitness kept stable until the stopping criterion is satisfied. This phenomenon demonstrates that BFO-FKNN can converge quickly toward the global optima, and fine tune the solutions very efficiently.

To evaluate the effectiveness of the proposed BFO-FKNN method for PD diagnosis, the comparisons against with SVM and KELM were conducted. As shown in Figure 2, we can see that BFO-FKNN has dominated SVM and KELM in most folds within the 10-fold CV procedure, namely, BFO-FKNN has achieved the high classification accuracy equal to or better than that of the other two models obtained for 9 folds in the whole 10 folds. The average classification accuracy of BFO-FKNN is 96.39%, while the average classification accuracy of SVM and KELM are 93.24% and 93.82%, respectively. Figure 3 shows the training classification accuracy surface for one fold achieved by SVM and KELM via the grid search strategy, where the x-axis and y-axis are $\log_2 C$ and $\log_2 \gamma$, respectively. Each mesh node in the (x, y)

plane of the training accuracy stands for a parameter combination and the z-axis denotes the obtained training accuracy value with each parameter combination. The
better performance of the proposed method is owing to the fact that the BFO has
aided the FKNN classifier to achieve the maximum classification performance by
automatically detecting the optimal neighborhood size and the fuzzy strength parameter. It is also interesting to be noted that, the standard deviation obtained by BFO-
FKNN is 0.025, which is the smallest among the three methods, followed by KELM
with 0.049 and SVM with 0.055. It verifies the robustness and reliability of the proposed BFO-FKNN method.

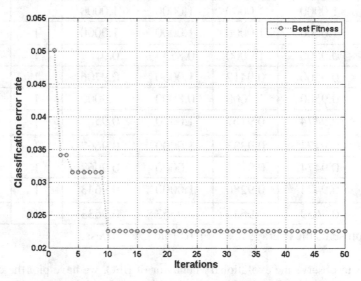

Fig. 1. Best fitness during the training stage for fold 1 in 10-fold CV

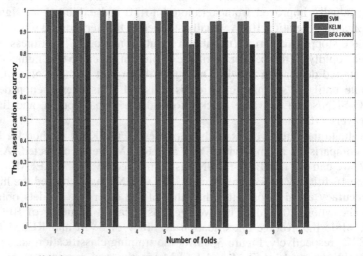

Fig. 2. The cross-validation accuracy obtained for each fold by BFO-FKNN, SVM and KELM

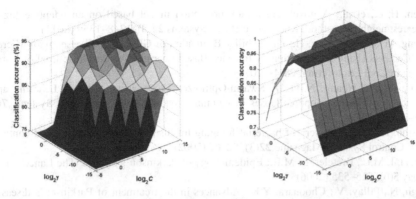

(a) Training accuracy surface with parameters obtained by the SVM in 1 fold   (b) Training accuracy surface with parameters obtained by KELM in 1 fold

Fig. 3. Training accuracy surface of SVM (left) and KELM (right) with parameters obtained by the grid search method on the PD dataset for 1 fold

6   Conclusions

This work has explored a new evolutionary FKNN approach, and successfully applied to the early detection of PD. The main novelty of the proposed method lies in the proposal of using BFO technique for exploring the full potential of FKNN by automatically determining k and m to exploit the maximum classification performance for early detection of PD. The empirical experiments have demonstrated the superiority of the proposed BFO-FKNN over SVM and KELM in terms of various performance metrics. It indicates that the proposed BFO-FKNN method can be a promising computer aided diagnosis tool for clinical decision making.

Acknowledgments. This research is supported by the National Natural Science Foundation of China (NSFC) under Grant Nos. of 61303113 and 61402337. This work is also funded by Zhejiang Provincial Natural Science Foundation of China under Grant Nos. of LQ13G010007, LQ13F020011 and LY14F020035.

References

1. Jóźwik, A.: A learning scheme for a fuzzy k-NN rule. Pattern Recognition Letters **1**(5), 287–289 (1983)
2. Keller, J.M., Gray, M.R., Givens, J.A.: A fuzzy k-nearest neighbours algorithm. IEEE Trans. Syst. Man Cybern **15**(4), 580–585 (1985)
3. Derrac, J., Garcia, S., Herrera, F.: Fuzzy nearest neighbor algorithms: Taxonomy, experimental analysis and prospects. Information Sciences **260**, 98–119 (2014)
4. Chen, H.-L., et al.: An efficient diagnosis system for detection of Parkinson's disease using fuzzy k-nearest neighbor approach. Expert Systems with Applications **40**(1), 263–271 (2013)
5. Sim, J., Kim, S.Y., Lee, J.: Prediction of protein solvent accessibility using fuzzy k-nearest neighbor method. Bioinformatics **21**(12), 2844–2849 (2005)

6. Chen, H.-L., et al.: A novel bankruptcy prediction model based on an adaptive fuzzy k-nearest neighbor method. Knowledge-Based Systems 24(8), 1348–1359 (2011)
7. Cheng, M.-Y., Nhat-Duc, H.: Groutability Estimation of Grouting Processes with Microfine Cements Using an Evolutionary Instance-Based Learning Approach. Journal of Computing in Civil Engineering, 28(4) (2014)
8. Cheng, M.-Y., Hoang, N.-D.: A Swarm-Optimized Fuzzy Instance-based Learning approach for predicting slope collapses in mountain roads. Knowledge-Based Systems 76, 256–263 (2015)
9. Passino, K.M.: Biomimicry of bacterial foraging for distributed optimization and control. IEEE Control Systems Magazine 22(3), 52–67 (2002)
10. de Lau, L.M.L., Breteler, M.M.B.: Epidemiology of Parkinson's disease. The Lancet Neurology 5(6), 525–535 (2006)
11. Singh, N., Pillay, V., Choonara, Y.E.: Advances in the treatment of Parkinson's disease. Progress in neurobiology 81(1), 29–44 (2007)
12. Harel, B., Cannizzaro, M., Snyder, P.J.: Variability in fundamental frequency during speech in prodromal and incipient Parkinson's disease: A longitudinal case study. Brain and Cognition 56(1), 24–29 (2004)
13. Baken, R.J., Orlikoff, R.F.: Clinical measurement of speech and voice, 2nd edn. Singular Publishing Group, San Diego, CA (2000)
14. Little, M.A., et al.: Suitability of Dysphonia Measurements for Telemonitoring of Parkinson's Disease. IEEE Transactions on Biomedical Engineering 56(4), 1015–1022 (2009)
15. Das, R.: A comparison of multiple classification methods for diagnosis of Parkinson disease. Expert Systems with Applications 37(2), 1568–1572 (2010)
16. Shahbaba, B., Neal, R.: Nonlinear models using Dirichlet process mixtures. The Journal of Machine Learning Research 10, 1829–1850 (2009)
17. Psorakis, I., Damoulas, T., Girolami, M.A.: Multiclass Relevance Vector Machines: Sparsity and Accuracy. IEEE Transactions on Neural Networks 21(10), 1588–1598 (2010)
18. Chang, C.-C., Lin, C.-J.: LIBSVM: a library for support vector machines. ACM Transactions on Intelligent Systems and Technology (TIST), 2(3) (2011)
19. Fawcett, T.: ROC graphs: Notes and practical considerations for researchers. Machine Learning 31, 1–38 (2004)

Data Mining Approaches

Temporal Pattern of Human Post and Interaction Behavior in Qzone

Bing Li[1,3], Nan Hu[4(✉)], Wenjun Wang[1,3], and Ning Yuan[1,2,3]

[1] School of Computer Science and Technology, Tianjin University, Tianjin 300072, China
[2] Department of Basic Courses, Academy of Military Transportation,
PLA, Tianjin 300161, China
[3] Tianjin Key Laboratory of Cognitive Computing and Application, Tianjin 300072, China
[4] Tianjin Emergency Medical Center, Tianjin 300011, China
824712118@qq.com

Abstract. The quantitative analysis of human pattern is an effective way to understand the complex social system. Relevant empirical studies reveal that human behavior in time follow a power-law distribution rather than a Poisson distribution. This paper aims at conducting statistical analyses based on the records of Qzone posts and interactive messages. The results show that the inter-event time distribution of posts and interaction follows a power-law distribution. Additionally, the time intervals of post comments differ from the time intervals of interact and in that there exists a clear cut-off point in the distribution of posting time, which indicates the subjection to a two-stage power-law. At the individual level, the posting time distribution exhibits fat tails. The analysis of post behavior indicates that there is a monotonous and negative relationship between the activity level and power-law exponent. The characteristics of local peaks also illustrate the burstiness and memory of post behavior.

Keywords: Temporal Pattern · Qzone · Inter-event time distribution · Activity

1 Introduction

Hawking believes that complexity science is the science in the twenty-first century. Researches on complex systems are actively expanding the depth and breadth of people's understanding about the real world. Many large complex systems are related to humans or constituted by humans directly. Barabási's paper published on Nature in 2005 brings about a breakthrough turning point of the analysis of human pattern [1]. This paper overthrows the views that the inter-event time distribution of human patter follows a Poisson distribution. Instead, empirical data in this research shows that inter-event time distribution of human pattern follows a power-law distribution. This paper has motivated and stimulated numerous researches into dynamics of human pattern. Researchers have studied and analyzed a variety of human behaviors. For example, studies have covered the structure and dynamics of large-scale human communication networks [2] and the laws of mobility [3], as well as the motifs of individual behavior [4]. In conclusion, as indicated in these studies, Poisson distribution shows the occurrence of an incident is random and unpredictable, but the power-law distribution implies burstiness and predictability in their results.

© Springer International Publishing Switzerland 2015
Y. Tan et al. (Eds.): ICSI-CCI 2015, Part II, LNCS 9141, pp. 53–62, 2015.
DOI: 10.1007/978-3-319-20472-7_6

The rapid development of the Internet has a major impact on people's access to and sharing of information and it also promotes the development of research about people's behavior on Internet. Stefan et al., who used 149,441 email records of 1,052 managers in a large consulting firm in the time between July 2006 and January 2007, analyzed the time interval between an email received and replied. Results of this research indicate that people tend to reply to his/her friends who have a social relationship with him/her in a shorter time [5]. In terms of social network, Chun et al. studied the message records of 17,788,870 users on South Korea's largest online social network from June 2003 to October 2005, the result of which shows that the interval time distribution at the group level has three sections of power law: when the inter-event time is less than 36 min, power-law exponent is 1.696; When the inter-event time between 36 min to 1 day, power-law exponent is 0.910; The inter-event time greater than one day, power-law exponent is 2.276 [6]. Ren et al. also studied 1,627,697 post records of 20,349 users on BBS in Nanjing University and found out that the inter-event time distribution of posts follow as heavy-tailed and the power-law exponent is 1.98 at the group level [7]. The same phenomenon also occurs when Song et al. analyzed the posting behavior of users in Sina blog and Weibo. The power-law exponents were respectively 1.3 and 2.0 in this research [8]. As to the human behavior of QQ chatting, Han et al. using data of five volunteers' QQ chatting found that power-law exponent is about 2.0 to 2.5 at the individual level [9].

However, fewer researches have analyzed the human communication pattern of Qzone. According to statistics, by the end of July 2014, QQ have 1 billion registered users and 500 million daily active users. Qzone, created by Tencent in 2005, is a social network website based on QQ users. It allows users to write blogs, keep diaries, send photos, listen to music, watch videos and so on. As of July 2014, it already had 645 million users. There are 150 million Qzone users who update their accounts at least once a month. This makes Qzone become the third largest social network in the world. Different from phone and instant messaging software, the interaction of Qzone has a time delay. Based on the data of main events such as the MsgFeeds, Blogs, MsgBoards and Comments Interaction, this paper aims to exploring the different distribution characteristics of different types of Qzone events.

2   Data Description

The Qzone dataset was from 571 volunteers' Qzone (468 effective users), with time-stamped records over a period of 3,315 days starting from July 2005 to August 2014. With QQ as the most successful live chat software, Qzone as a related application of QQ also has a large number of active users, and their behavior will be recorded (active promoters, passive sponsor, time and content). This paper mainly focused on inter-event time distribution and burstiness in human activities by user's behavior in Qzone such as MsgFeeds, Blog, MsgBoard and comments interaction. The exact time of each events happening can be accurate to second, see Table 1 for details of Qzone dataset.

Table 1. Data of Qzone dataset

Type	MsgFeeds	Blog	MsgBoard	Whole
Post	140,996	40,809	91,599	273,404
interactive	630,111	202,790	87,718	920,619

During the process of data cleaning, we removed some abnormal data, such as when the time is 1970-01-01 00:00:00 or time is null. During the formal experiment, we analyzed based on the unit of days, hours, minutes and seconds as inter-event time. Ex perimental results show that the interval in seconds fits more with people's normal work and rest time, so we chose second as our unit of inter-event time.

3   Results and Discussion

3.1   Inter-event Time Distribution for Post Behavior

In analyzing the inter-event time distribution of Posts, we considered four types of posting behavior: whole, MsgFeeds, Blog, MsgBoard. Inter-event time τ is defined as the time between two posts. $P(\tau)$ is the probability distribution of τ. In order to elim-inate the volatility and scattered phenomenon occurring the tail of $P(\tau)$ in log-log scales, we applied the method of Logarithmic binning [10] to conduct the statistical analyses. In addition, we added a corresponding fitting line in the figure. This method could help us observe the attenuation of power-law exponent α in log-log scales, the same below. As can be seen from Figure 1, there is a significant turning point as indicated in the experimental results. The curve could be segmentally fitted. We focused on the main part of the curve, because human behavior exhibits in days periodicity. One day has 86,400 seconds, and we chose the fitting range is [0, 86,400]. Figure 1 also shows that all kinds of inter-event time distribution of Posts show varying degrees of heavy-tailed characteristics, with corresponding α values as 0.823, 0.679, 1.036 and 0.957, respectively. These four posting behavior all exhibited the phenomenon that short periods of activity time followed by long periods of inactivity. Compared with the existing research results, including posting behavior of blog and micro-blog [8], cell phone text messaging behavior [11], QQ chat behavior [9], our results are basically similar to the previous findings. In comparison, although the amount of MsgFeeds is the largest, but the value of α is lower than that of MsgBoard and Blog, indicating a lower level of activity. By analyzing the raw data, we found that MsgFeeds is users' voluntary posting, but MsgBoard is mainly from QQ friends. At the beginning of MsgBoard, it was highly welcomed by many people. However, due to its single function, fewer and fewer people used it. To take a user (ID 9338) as an example, the number of MsgBoard messages in 2007 was 2,862, accounting for 85.13% of the total messages. So the MsgBoard posts showed strong intensity. The α of blog is higher because people often reprint some blogs when they browse Qzone of QQ friends, while MsgFeeds is mainly used to express ones' feelings about recent events. Thus, the rate of MsgFeeds is lower than that of the blog.

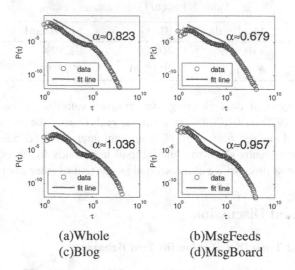

(a)Whole   (b)MsgFeeds
(c)Blog   (d)MsgBoard

Fig. 1. Inter-event time distribution for post behavior

3.2   Inter-event Time Distribution of Interact Behavior

Comments record and inter-event time represent the influence of different types of Post. In order to show the influence of various types of Post, this paper conducted fitting analysis based on the unit of inter-event time in second. The inter-event time of interaction distribution is shown in Figure 2, mapping the distribution of each of the four Post comments interval, including MsgFeeds, Blog, MsgBoard and whole. The corres-

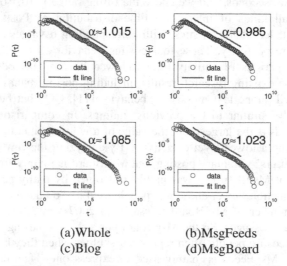

(a)Whole   (b)MsgFeeds
(c)Blog   (d)MsgBoard

Fig. 2. Inter-event time distribution of interact behavior

ponding α values were 1.015, 0.985, 1.085 and 1.023. Among these four types, the Blogs have a long quiet period. Reasons may be that fewer people in Qzone write the blogs themselves, and many of the blogs are reproduced or reprinted. In addition, the power-law exponent of Blogs is the highest, indicating that people are more concentrated on Blogs commenting than the others at the time, except for the reprinted blogs. If the blogs are written personally by the Qzone lord concerning a summary of recent experience and what he have done, such as the annual summary of 2014 and so on, MagFeeds comments is the highest, but time is dispersed. This is mainly because MsgFeeds is a reflection of someone's recent situation, attracting his QQ friends to comment his MsgFeeds. Meanwhile, people will interact with his QQ friends through MagFeeds, which results in a higher number of comments in MsgFeeds.

3.3   Inter-event Time Distribution for Interaction Between Two Friends

In this section, the interaction comment between QQ1 and QQ2 is marked as QQ1↔QQ2. After deleting the records of self-commenting, we obtained a total of 72,162 pairs of users in the Qzone data. The statistical results analyzing and comparing the QQ1 ↔ QQ2 interactive comments are shown in Table 2.

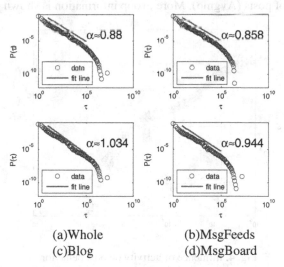

(a)Whole   (b)MsgFeeds
(c)Blog   (d)MsgBoard

Fig. 3. Inter-event time distribution for interaction between two friends

Table 2. Interaction between two User

Type	MsgFeeds	Blog	MsgBoard	whole
interactive	573,379	57,813	138,573	769,765

As can be seen from Figure 3, although the number of comments of MsgFeeds is more than the others, but the time of comments are more dispersed. It also has a lower value of power-law exponent, which is similar to the results of one-way comments. The results of Blog and MsgBoard are also accord with one-way comments and the α value is lower than that of one-way comments, which is consistent with the actual situation. The decline of power-law component of MsgFeeds is the highest, mainly because that although MsgFeeds receives more comments, the interaction with friends is less than the other two types.

3.4   Influence of Activity on Post Behavior

In previous empirical studies, the difference of individual activity will lead to statistical characteristics changes directly. Tao Zhou's [12] analysis of people's Online movie watching records found that there exists a monotonous relation between the activity level and power-law exponent changes in the group level, with the power-law exponent ranging from 1.5-2.7. Its changes reached 80%. Wei Hong, et al. [13] studied the inter-event time of cellphone messages, which also found a similar monotonous phenomenon. But Wang et al. [14] found the power-law exponent is not monotonic dependence on the activity by inter-event time of online order. The exponent actually has a slight decline after it achieves the peak. The above studies show that individual activity will lead to users' or a group of users' various characteristics. In the analysis of Qzone, we divided 468 users into five groups based on the average monthly quantity of posts (Avgmp). More group information is shown in Table 3.

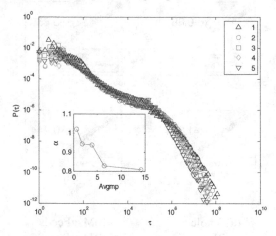

Fig. 4. Influence of activity on post behavior

From Table 3, group1 has the lowest level of activity and the number of average monthly posts is 0.695. In addition, the group 5 has the highest activity level, with 14.234 average posts in a month. This number is 20.5 times of that in group 1. In inter-event time of posts distribution, there is a negative correlation between the activity and power-law exponent at the group level. When the average monthly number of user posts increases from 0.695 to 14.234, the power-law exponent has dropped from

Table 3. Group information of users

Type	group1	group2	group3	group4	group5
Number	93	93	93	93	96
Avgmp	0. 695	1. 928	3. 889	6. 498	14. 234
α	1. 018	0. 941	0. 934	0. 847	0. 808

1.108 to 0.808. The trends of power-law exponent changes and each inter-event time of posts distribution are shown in Figure 4.

3.5   Burstiness and Memory in Post Behavior

In many systems including web browsing [15], earthquakes [16] and email communication [17], we could observe the common point that the pattern often indicates a short time period of activity followed by a long time period of non-actions. This kind of situation is called burstiness. Most of the time intervals are less than the average time interval, but it may be otherwise, resulting in a greater standard deviation of the distribution of time interval. In order to study the burstiness of human behavior of Qzone posts. We grouped all users into five groups base on the number of average posts in a month. The burstiness parameter D is defined as follows [18].

$$B = \frac{(\sigma_\tau/m_\tau - 1)}{(\sigma_\tau/m_\tau + 1)} = \frac{(\sigma_\tau - m_\tau)}{(\sigma_\tau + m_\tau)} \tag{1}$$

Where σ_τ and m_τ are the mean and the standard deviation of $P(\tau)$. Seen from Equation 1, B has a value in the bounded range (-1, 1). $B = -1$ corresponds to a completely regular (periodic) signal. $B = 0$ represents a random behavior, $B = 1$ shows this kind of event have most bursty signal.

Human behavior is often changing with memories and interests. For example, in nature, the time and place of extreme weather show a strong memory. Cai, Shi-Min, et al. [19] also observed that Internet traffic time series also have relatively strong memory. In short, when we sorted the time interval between two events, if a long (short) time interval with high probability is following after a long (short) interval, this series of time shows a better memory. On the contrary, the memory is weak or we could call it anti-memory. This sequence contains a total of n_τ elements (meaning a total of $n_\tau + 1$ behavior), the first $n_\tau - 1$ constitute sequence 1, and the last following n_τ -1 are specified as sequence 2 (see Equation 2). The memory coefficient M can be defined using the Pearson correlation of these two sequences [18].

$$M = \frac{1}{n_\tau - 1} \sum_{i=1}^{n_\tau - 1} \frac{(\tau_i - m_1)(\tau_{i+1} - m_2)}{\sigma_1 \sigma_2} \tag{2}$$

Where $m_1 (m_2)$ and $\sigma_1 (\sigma_2)$ are sample mean and sample standard deviation of $\tau_i's (\tau_{i+1}'s)$, M as a autocorrelation function biased estimate, the memory coefficient has a value in the range $(-1, 1)$, when $M > 0$ indicates a l is positive when a short (long) inter-event time tends to be followed by a short (long) one, and it is negative when a short (long) inter-event time is likely to be followed by a long (short) while M

< 0, when   M = 0 means neutral. We made more specific analysis on burstiness and memories (see Table 4).

Table 4. B and M in different level activity

Type	group1	group2	group3	group4	group5
Number	93	93	93	93	96
Avgmp	0.695	1.928	3.889	6.498	14.234
M	0.244	0.211	0.247	0.261	0.252
B	0.484	0.494	0.470	0.428	0.419

As can be seen from Table 3, with the increase of Avgmp, the memory coefficient, M , shows a maximum value of 0.261 and a minimum of 0.211, indicating a good memory. As to the value of B, the value of group 2 has a continuous decline after the peak, which indicates that with an increase of Avgmp, the B is becoming weaker. Here is an interesting phenomenon in our experimental results. In group 2, the value of M is the lowest but the value of B is the largest, which might provide additional evidence of the negative correlation between Burstiness and memories. Figure 5 shows the (M, B) values of all users, where each blue dot represents a vale for a user. The inset, an enlarged view for the center part of these dots, shows (M_{avg} , B_{avg}) for each group and the average value for all users (asterisk).

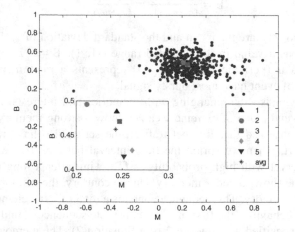

Fig. 5. The (M, B) phase diagram of all users

4   Conclusion

With the development and progress of Internet technology, an increasing number of people are communicating with each other through the social networks. The social network has replaced the previous communication tools such as letters, emails, text messages and other communications. Research on the Posts and interaction behavior of Qzone can help us better understand the temporal characteristics of human pattern. This paper analyzed inter-event time of 273,522 posts and 920,619 comments starting from July 2005 to August 2014, including analyzing inter-event time of posts, inter-event time of a single comment and bi-directional interaction, the trends of power-law exponents, Burstiness and memory on the different levels of activities. The study found out that each communication pattern of human individuals is known to be inhomogeneous or bursty, which is reflected by the heavy tail behavior in the inter-event time distribution. People's posting and interaction behavior in Qzone (inter-event time) follow the power-law distribution, which is similar to the first model raised by Vazquez et al [20]. However, there is an obvious turning point at a time in a certain day, showing two power-laws, which is different from the inter-event time of commenting behavior distribution. In contrast with previous studies [12], this paper found that there is a negative relationship between with the inter-event time of Posts distribution and user activity. The B and M of users with different levels of activity show local peaks, and they do not present a monotonous relationship. These findings not only help us have a better understanding of human behavior, and also provide guidance for future research into the reasons of a fat tail characteristic.

Acknowledgements. This work was supported by the Major Project of National Social Science Fund (14ZDB153), the National Science and Technology Pillar Program (2013BAK02B06 and 2015BAL05B02), Tianjin Science and Technology Pillar Program (13ZCZDGX01099,13ZCDZS F02700), National Science and Technology Program for Public Well-being (2012GS120302).

References

1. Barabasi, A.-L.: The origin of bursts and heavy tails in human dynamics. Nature **435**(7039), 207–211 (2005)
2. Onnela, J.-P., et al.: Structure and tie strengths in mobile communication networks. Proceedings of the National Academy of Sciences **104**(18), 7332–7336 (2007)
3. Song, C., et al.: Limits of predictability in human mobility. Science **327**(5968), 1018–1021 (2010)
4. Anteneodo, C., Dean Malmgren, R., Chialvo, D.R.: Poissonian bursts in e-mail correspondence. The European Physical Journal B-Condensed Matter and Complex Systems **75**(3), 389–394 (2010)
5. Wuchty, S., Uzzi, B.: Human communication dynamics in digital footsteps: a study of the agreement between self-reported ties and email networks. PloS one **6**(11), e26972 (2011)
6. Chun, H., et al.: Comparison of online social relations in volume vs interaction: a case study of cyworld. In: Proceedings of the 8th ACM SIGCOMM Conference on Internet Measurement. ACM (2008)
7. Xue-Zao, R., et al.: Mandelbrot Law of Evolving Networks. Chinese Physics Letters **29**(3), 038–904 (2012)
8. Song, Y., Chuang Z., Wu, M.: The study of human behavior dynamics based on blogosphere. In: 2010 International Conference on Web Information Systems and Mining (WISM), vol. 1. IEEE (2010)
9. Chen, G., Han, X., Wang, B.: Multi-level scaling properties of instant-message communications. Physics Procedia **3**(5), 1897–1905 (2010)

10. Milojević, S.: Power law distributions in information science: Making the case for logarithmic binning. Journal of the American Society for Information Science and Technology **61**(12), 2417–2425 (2010)
11. Zhi-Dan, Z., et al.: Empirical analysis on the human dynamics of a large-scale short message communication system. Chinese Physics Letters **28**(6), 068901 (2011)
12. Zhou, T., et al.: Role of activity in human dynamics. EPL (Europhysics Letters) **82**(2), 28002 (2008)
13. Wei, H., et al.: Heavy-tailed statistics in short-message communication. Chinese Physics Letters **26**(2), 028–902 (2009)
14. Wang, P., et al.: Heterogenous scaling in the inter-event time of on-line bookmarking. Physica A: Statistical Mechanics and its Applications **390**(12), 2395–2400 (2011)
15. Dezsö, Z., et al.: Dynamics of information access on the web. Physical Review E **73**(6), 066132 (2006)
16. Livina, V.N., Havlin, S., Bunde, A.: Memory in the occurrence of earthquakes. Physical review letters **95**(20), 208–501 (2005)
17. Harder, U., Paczuski, M.: Correlated dynamics in human printing behavior. Physica A: Statistical Mechanics and its Applications **361**(1), 329–336 (2006)
18. Goh, K.-I., Barabási, A.-L.: Burstiness and memory in complex systems. EPL (Europhysics Letters) **81**(4), 48002 (2008)
19. Cai, S.-M., et al.: Scaling and memory in recurrence intervals of Internet traffic. EPL (Europhysics Letters) **87**(6), 68001 (2009)
20. Vázquez, A., et al.: Modeling bursts and heavy tails in human dynamics. Physical Review E **73**(3), 036–127 (2006)

An Empirical Analysis on Temporal Pattern of Credit Card Trade

Bo Zhao[1,3], Wenjun Wang[2,3], Guixiang Xue[5], Ning Yuan[2,3,4], and Qiang Tian[6(✉)]

[1] School of Computer Software, Tianjin University, Tianjin 300072, China
[2] School of Computer Science and Technology, Tianjin University, Tianjin 300072, China
[3] Tianjin Key Laboratory of Cognitive Computing and Application, Tianjin 300072, China
[4] Deparment of Basic Courses, Academy of Military Transportation,
PLA, Tianjin 300161, China
[5] School of Computer Science and Software, Hebei University of Technology,
Tianjin 300130, China
[6] Office of the President, Tianjin Normal University, Tianjin 300387, China

Abstract. Credit card swiping is the behavior which people make a deal in POS with credit card. It is one of the simplest credit services. To research credit card swiping is helpful to understand temporal pattern of credit card swiping. In traditional views, human group behavior is random, irregular. However, the latest research presents that some human behaviors are predictable. Nowadays, researchers are analyzing different types of human behavior with human behavior dynamics methods. In this paper, we analyze credit card swiping temporal pattern. The empirical data is from a bank in Tianjin China. We analyze inter-credit time, and memory and burstiness phase diagram (M, B). Inter-credit time follows power-law distribution, and (M, B) mean value is at the range of human behavior (M, B) value.

Keywords: Credit card · Temporal pattern · Inter-credit time · Power-law

1 Introduction

Human society is a complex system. How to understand the evolution regularity and internal driving force of its development has been a hot topic explored by researchers. So far, the study of human behavior has become the focus of attention by a large number of disciplines, such as psychology, social psychology, sociology, physics, mathematics, information science, management, economics and finance etc. In traditional views, it was generally recognized that human behaviors obey Poisson distribution. However in 2005, Barabási's paper[1] published on Nature uncovered that there exist characteristics of human behavior that are inconsistent with the characteristics of Poisson process. He found that the characteristic of high-frequency outbreak in a short time and silence in a long time follows power-law distribution. In recent years, researches on the human behavior dynamic have been done based on a lot of empirical data, involving in correspondence letters[2][8], e-mail[1][3], phone calls[4][9], SMS[5][27], online communication[6][28], online on-demand movies and page views[7] etc. These studies analyzed the mode of communication, business practices, entertainment activities and other human behavior characteristics.

© Springer International Publishing Switzerland 2015
Y. Tan et al. (Eds.): ICSI-CCI 2015, Part II, LNCS 9141, pp. 63–70, 2015.
DOI: 10.1007/978-3-319-20472-7_7

Nowadays, it is convenient for financers[18][19][20] and statists[21][22] to analyze the data in quantification. A large number of scientists engaged in scientific research on complexity has joined in this field. Relying on massive empirical data, they try to make a new exposition of finance field. Currently, there is much econophysics analysis of human behavior, including stockbroker initiated transaction[10], ordering and stock trading in stock exchange[11], online futures trading[12], orders trading in JD Mall[13], China Baosteel's stock trading[15] and the S&P 500 Index trade changes[19].

In this paper, we adapt the theory of human behavior dynamics to analyze the credit card transaction data of bank in Tianjin China empirically. In the empirical data, there are 11460 credit cards and 525925 trade records. The duration is from March 2009 to November 2014. The analysis results showed that the credit card swiping inter-credit time probability follows power-law distribution. The power-law exponents are 1.09, 0.76 and 2.41 respectively. Then we analyzed (M, B) phase diagram of credit card swiping inter-credit time which (M, B) mean value is located in the range of human behavior. Comparing with existing communication behavior and social networking, credit card swiping behavior also follows distribution of human behavior. Our findings can provide theoretical support for credit card swiping temporal pattern.

2   Result

2.1   Inter-credit Time Probability Distribution

By analyzing credit card swiping inter-credit time, we can research its temporal pattern. Inter-credit time denoted as τ is defined as the time between consecutive credit card swiping by the same user, and $P(\tau)$ is probability distribution. The inter-credit time probability distribution in log-log plot is shown in Figure1. The power-law exponents of financial activities data set and credit card data set are shown in Table1.

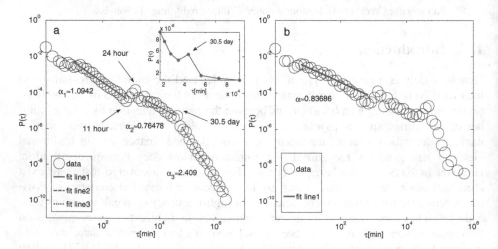

Fig. 1. Credit card swiping inter-credit time probability distribution. (a) All user Credit card swiping inter-credit time probability distribution. (b) Suspected fraud user Credit card swiping inter-credit time probability distribution.

Table 1. Comparison of the power-law distribution of financial activities data set and credit card data set

Data set	Power-law exponent
Central European Bank between June 1999 and May 2003, all 54374 transaction sponsored by a stockbroker [10].	Exponent is 1.3, and there is a truncated exponent.
In March, June and October 2002, GSK and VOD, nearly 800 000 order and 540 000 transaction in securities trading[11].	Fat-tail, but it do not follow power-law distribution.
In December 9, 2006, before Taipei mayoral elections, five candidates designed for online futures experiment lasting 30 days, more than 400 volunteers[12].	Exponent is 1.3, and there is a truncated exponent.
Between December 2, 2008 and January 12, 2010, 274148 users on the purchase of data 28620 items for JD Mall[13].	It follows power-law distribution, and different kinds of items have different power-law exponent.
Between January 2, 2001 and December 31, 2001, the Japanese yen against the US dollar data[14].	It follows power-law distribution, exponent is 2.07.
Between August 22, 2005 and August 23, 2006 , China Baosteel stock transactions in Shanghai Stock Exchange, including a total of 243 days over 25 million transactions[15].	It follows power-law distribution, exponent is 1.41.
A Fortune 500 company contains 44,000 data of purchase order at a quarter[16].	It follows power-law distribution, exponent is 1.91.
A bank in Tianjin China, 11460 credit card swiping record, between March 2009 and November 2014.	There are three power-law exponent 1.09, 0.76, and 2.41 respectively.

As is shown in Figure.1.a, we can divide all users' credit card swiping inter-credit time to three parts. We use least squares to fit them respectively. They all follow power-law. For the first part, the power-law exponent is 1.09 and the first lowest point in curve is at 11 hour which follows normal human circadian rhythm; After the lowest point, there is an upward trend, the next highest point is at 24 hour, which indicates the number of credit card swiping inter-credit time at 24 hour is more than that at 11 hour. It also follows normal human circadian rhythm. The second curve decays in power-law which power-law exponent is 0.76. After the lowest point, there is an upward trend. The next highest point is in 30.5 day approximately. To this highest point, we think it is an outlier. After querying data and consulting with bank expert, we affirm that a certain percentage of credit card users is fraud. This credit card swiping behavior is in periodicity highly. Thus, it may lead to which there are a lot of transactions occurring closing to 30 day inter-credit. The third part is the distribution which inter-credit time is more than a month. It also follows power-law.

Because of the outlier in 30.5 day, we re-analyze the data set. Through multi-step data processing, we choose the crowd who cause the outlier as suspected fraud crowd. Then we analyze their credit card swiping inter-credit time. The result is shown as Fig.1.b. The highest point in the decaying curve for suspected fraud crowd occur in close to 32 day. Comparing with credit card swiping behavior for all user, the outlier is higher obviously. Suspected credit card fraud crowd may choose 30 days as a cycle. In Fig.1.b, the first part follow power-law, then there is a highest point in close to 22.5 hour. Comparing with credit card swiping behavior for all user, it also follows human circadian rhythm. The fraud behavior did not occur in the beginning of credit card use.

2.2   The (M, B) Phase Diagram Characteristics

(M, B) phase diagram can be used to measure human behavior pattern of memory and burstiness. Fig.2.a shows the (M, B) phase diagram of all user, the blue dot is each credit card (M, B) value, the red dot is credit card (M, B) mean value. We compare it with other human behavior empirical result. Square is for personal print record data in a university[1], diamond is for email record data[26], erect triangle is for the bank call center data, dextrosinistral triangle is for the phone data in a mobile phone company[1], dextrorsal triangle is for library data[1]. Fig.2.b enlarges representation for Fig.2.a which is the distribution of different data sets (M, B) mean value and the red oval is the area of human behavior.

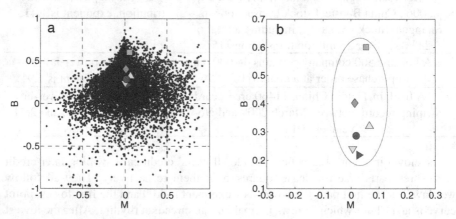

Fig. 2. (M, B) phase diagram. (a) The (M, B) phase diagram for credit card swiping inter-credit time. (b) The distribution of (M, B) mean value for some human behavior.

Fig.3.a is the burstiness distribution for all user. The highest proportion of burstiness is between 0.2 and 0.3 which value is 47.5%. Fig.3.b is the memory distribution for all user. The highest proportion of memory is in [-0.1, 0] and [0, 0.1] which values are 17.3% and 17.2% respectively. The burstiness mean value for all user is 0.287, and the memory mean value for all user is 0.018.

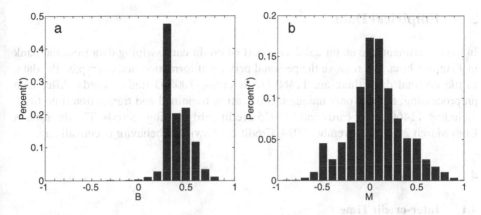

Fig. 3. Credit card swiping (M, B) value distribution proportion. (a) Burstiness value distribution proportion. (b) Memory value distribution proportion.

The empirical results show that credit card swiping is in burstiness, representing consecutive swiping in a short time and silence in a long time. Meanwhile, the credit card does not have a good prediction. Its memory is close to 0, indicating that credit card swiping for the overall population is irregular. As is shown in Fig.2.a, for only small part of crowd, memory is greater than 0.5, close to 1. The crowd whose bursti ness value is lower and memory value is higher may be fraud crowd, and another part of crowd may be frequent merchant swiping.

Table 2. The comparison of credit card swiping (M, B) mean value with human behavior existing

Data set	Mean value of memory	Mean value of bursti- ness
Library loans	0.028	0.219
Phone initiation record from a mobile phone company	0.006	0.241
Call center record at an anonymous bank	0.066	0.321
Activity pattern per- taining to email	0.014	0.402
Printing of individual in universities	0.054	0.601
Credit card record of a bank in Tianjin China	0.018	0.287

The comparison of credit card swiping (M, B) mean value with human behavior existing is shown in Table2.

3   Empirical Data

In the experiment, the empirical data is part of credit card swiping data from the bank in Tianjin China. We remove the personal privacy information and encrypted the data. In the original data, there are 11460 credit cards, 700803 trade records. After data preprocessing, there is only unique identification remained and transaction time field, including 11460 credit cards and 525925 credit card swiping records. The duration is from March 2009 to November 2014. Credit card swiping behavior occurs all day.

4   Methods

4.1   Inter-credit Time

Inter-credit time denoted as τ is defined as the time between consecutive credit card swiping by the same user. That is, $\tau = t_{[i+1]} - t_{[i]}$. $t_{[i+1]}$ is the $i + 1$ swiping moment, $t_{[i]}$ is the i swiping moment, and the probability distribution of τ is defined as $P(\tau)$.

4.2   The (M, B) Phase Diagram

Memory and burstiness (M, B) phase diagram[3], is used to measure memory and burstiness in human activity and phenomenon. The range of burstiness is between -1 and 1, which is as axis of ordinates in (M, B) phase diagram. It is defined as follows:

$$B = \frac{\sigma_\tau - m_\tau}{\sigma_\tau + m_\tau} \tag{1}$$

σ_τ and m_τ are represent as the standard deviation and the mean value of τ respectively. When the value of B is 0, it means that the standard deviation is equal to the mean value, which is Poisson distribution. When the value of B gets closer to 1, it means that the standard deviation is bigger than the mean value, which is fat-tailed distribution. And when the value of B gets closer to -1, it means that the standard deviation is smaller, which means the periodicity of event sequence is stronger.

The range of memory is between -1 and 1, which is as axis of abscissas in (M, B) phase diagram. It is defined as self-variable function of each individual inter-credit time τ_i sequence, which is defined as follows:

$$M = \frac{1}{n-1} \sum_{i=1}^{n-1} \frac{(\tau_i - m_1)(\tau_{i+1} - m_2)}{\sigma_1 \sigma_2} \tag{2}$$

n is the element number of two consecutive inter-credit time. m_1, m_2 and σ_1, σ_2 are the standard deviation and the mean value of sample inter-credit time τ_i, τ_{i+1} ($i = 1, 2, \cdots, n-1$).

5   Conclusion

By studying the empirical data in Tianjin China, we research credit card swiping inter-credit time distribution in group level, and find behavior exception. And we research the credit card inter-credit time of suspected fraud credit card and probability distribution. The power-law exponents are 1.09, 0.76 and 2.41 respectively. Then we explain it. Meanwhile, we analyze memory and burstiness of credit card swiping behavior and compare it with other human empirical result. The (M, B) mean value of credit card swiping is in human behavior area of the literature [3].

Acknowledgments. This work was supported by the Major Project of National Social Science Fund(14ZDB153), the National Science and Technology Pillar Program (2013BAK02B06 and 2015BAL05B02), Tianjin Science and Technology Pillar Program (13ZCZDGX01099,13ZCD ZSF02700), National Science and Technology Program for Public Well-being(2012GS120302).

References

1. Barabási, A.L.: The origin of bursts and heavy tails in human dynamics. J. Nature **435**, 207–211 (2005)
2. Zhou, T., Han, X,P., Wang, B H.: Towards the understanding of human dynamics. Science Matters: Humanities as Complex Systems, 207–233 (2008)
3. Goh, K.I., Barabási, A.L.: Burstiness and memory in complex systems. EPL (Europhysics Letters) **81**(4), 48002 (2008)
4. Jiang, Z.Q., Xie, W.J., Li, M.X., Podobnik, B., Zhou, W.X., Stanley, H.E.: Calling patterns in human communication dynamics. Proceedings of the National Academy of Sciences **110**, 1600–1605 (2013)
5. Karsai, M., Kaski, K., Kertesz, J.: Correlated dynamics in egocentric communication networks. Plos One **7**, e40612 (2012)
6. Zhao, Z.D., Zhou, T.: Empirical analysis of online human dynamics. Physica A. Statistical Mechanics and Its Applications **391**, 3308–3315 (2012)
7. Goncalves, B., Ramasco, J.J.: Human dynamics revealed through Web analytics. J. Physical Review E. **78**(2), 026123 (2008)
8. Oliveira, J.G., Barabasi, A.L.: Human dynamics: Darwin and Einstein correspondence patterns. Nature **437**, 1251–1251 (2005)
9. Jo, H.H., Karsai, M., Kertesz, J., Kaski, K.: Circadian pattern and burstiness in mobile phone communication. New Journal of Physics **14**, 013055 (2012)
10. Vázquez, A., Oliveira, J.G., Dezsö, Z.: Modeling bursts and heavy tails in human dynamics. J. Physical Review E. **73**(3), 036127 (2006)
11. Scalas, E., Kaizoji, T., Kirchler, M.: Waiting times between orders and trades in double-auction markets. J. Physica A. Statistical Mechanics and its Applications **366**, 463–471 (2006)
12. Wang, S.C., Tseng, J.J., Tai, C.C.: Network topology of an experimental futures exchange. J. The European Physical Journal B-Condensed Matter and Complex Systems **62**(1), 105–111 (2008)
13. Jinlong, W., Gao, K., Li, G.: Empirical analysis of customer behaviors in Chinese e-commerce. J. Networks **5**(10), 1177–1184 (2010)

14. Kim, K., Yoon, S., Kim, S.Y.: Dynamical Mechanisms of the Continuous-Time Random Walk, Multifractals, Herd Behaviors and Minority Games in Financial Markets. J. Journal Of The Korean Physical Society **50**(1), 182–190 (2006)

15. Ruan, Y., Zhou, W.: Long-term correlations and multifractal nature in the intertrade durations of a liquid Chinese stock and its warrant. J. Physica A. Statistical Mechanics and its Applications **390**(9), 1646–1654 (2011)

16. Gao, L., Guo, J., Fan, C.: Individual and group dynamics in purchasing activity. J. Physica A. Statistical Mechanics and its Applications **392**(2), 343–349 (2013)

17. Feng, L., Li, B., Podobnik, B.: Linking agent-based models and stochastic models of financial markets. J. Proceedings of the National Academy of Sciences **109**(22), 8388–8393 (2012)

18. Li, Y., Gao, J., Enkavi, A.Z.: Sound credit scores and financial decisions despite cognitive aging. J. Proceedings of the National Academy of Sciences **112**(1), 65–69 (2015)

19. Shefrin, H., Nicols, C.M.: Credit card behavior, financial styles, and heuristics. J. Journal of Business Research. **67**(8), 1679–1687 (2014)

20. Jambulapati, V., Stavins, J.: Credit CARD Act of 2009: What did banks do? J. Journal of Banking & Finance **46**, 21–30 (2014)

21. Sohn, S.Y., Lim, K.T., Ju, Y.: Optimization strategy of credit line management for credit card business. J. Computers & Operations Research **48**, 81–88 (2014)

22. Sobolevsky, S., Sitko, I., Combes, R.T.D.: Money on the move: big data of bank card transactions as the new proxy for human mobility patterns and regional delineation. In: 2014 IEEE International Congress on the Case of Residents and Foreign Visitors in Spain. C. Big data (bigdata congress), pp. 136–143. IEEE, Spain (2014)

23. Yan-li, Z., Jia, Z.: Research on Data Preprocessing In Credit Card Consuming Behavior Mining. J. Energy Procedia **17**, 638–643 (2012)

24. Ha, S.H., Krishnan, R.: Predicting repayment of the credit card debt. J. Computers & Operations Research **39**(4), 765–773 (2012)

25. Nie, G., Chen, Y., Zhang, L.: Credit card customer analysis based on panel data clustering. J. Procedia Computer Science **1**(1), 2489–2497 (2010)

26. Eckmann, J.P., Moses, E., Sergi, D.: Entropy of dialogues creates coherent structures in e-mail traffic. J. Proceedings of the National Academy of Sciences of the United States of America **101**(40), 14333–14337 (2004)

27. Wu, Y., Zhou, C., Xiao, J.: Evidence for a bimodal distribution in human communication. J. Proceedings of the national academy of sciences **107**(44), 18803–18808 (2010)

28. Rybski, D., Buldyrev, S.V., Havlin, S.: Communication activity in a social network: relation between long-term correlations and inter-event clustering. J. Scientific reports **2** (2012)

Discovering Traffic Outlier Causal Relationship
Based on Anomalous DAG

Lei Xing[1,2], Wenjun Wang[1,2], Guixiang Xue[3], Hao Yu[1,2], Xiaotong Chi[1,2],
and Weidi Dai[1,2 (✉)]

[1] School of Computer Science and Technology, Tianjin University, Tianjin 300072, China
davidy@tju.edu.cn
[2] Tianjin Key Laboratory of Cognitive Computing and Application, Tianjin 300072, China
[3] School of Computer Science and Software, Hebei University of Technology,
Tianjin 300130, China

Abstract. The increasing availability of large-scale trajectory data provides us more opportunities for traffic pattern analysis. Nowadays, outlier causal relationship among traffic anomalies has attracted a lot of attention in the research of traffic anomaly detection. In this paper, we propose a model of constructing anomalous directed acyclic graph (DAG) which is based on spatial-temporal density to detect outlier causal relationship in traffic. To the best of our knowledge, the graph theory of DAG is firstly used in this area and the algorithm with strong pruning is proved to have lower time complexity. Moreover, the multi-causes analysis helps reflect the causal relationship more precisely. The advantages and strengths are validated by experiments using large-scale taxi GPS data in the urban area.

Keywords: Traffic outlier causal relationship · Anomalous DAG algorithm · Multi-causes analysis

1 Introduction

With the increasing availability of trajectory data such as WIFI, GPS, many unusual traffic patterns and trends [1] have been studied [2], [3]. Recently, outlier causal relationship among traffic anomalies which can discover the causal relationship between unusual traffic patterns has been more and more critical in related domains including city planning and traffic management.

There are varieties of traffic anomaly detection approaches from several areas, including statistics and data mining. Linsey et al. [4] propose parametric anomaly detection techniques based on classical likelihood ratio test statistic. Another method is Tango et al. [5] where a space-time scan statistic based on negative binomial model is proposed to discover anomalies. These statistics approaches are very effective if there are sufficient knowledge of the data, however they are highly dependent on data.

Principle component analysis (PCA) has been used for network-wide anomaly detection [6]. The algorithm in [7] discovers traffic anomaly patterns using Grid-Based Traffic Model, which cannot reflect natural differences of regions in the traffic network. Wei Liu et al. have designed an OutlierTree Algorithm [8] to discover causal relationships among spatial-temporal anomalies based on minDistort. However, the

© Springer International Publishing Switzerland 2015
Y. Tan et al. (Eds.): ICSI-CCI 2015, Part II, LNCS 9141, pp. 71–80, 2015.
DOI: 10.1007/978-3-319-20472-7_8

most challenging problem, after identifying traffic anomalies, is how to infer outlier causal relationship among them efficiently, especially when the amount of traffic anomalies is extremely huge.

In this paper, we focus on the outlier causal relationship among traffic anomalies using massive taxi traces in traffic network. Traffic anomaly is defined as a traffic pattern that behaves differently from other traffic patterns. More specifically, the contributions of this paper are described as follows:

1. Region-Based Traffic Model: the urban area is partitioned into regions using road network. Compared with the CCL method used in [8], [9] which requires huge memory space to store traffic region information, Grid-Based PNPoly algorithm is proposed to map each GPS point to the corresponding region more efficiently.
2. Traffic anomaly detection based on spatial-temporal density: firstly we obtain the spatial-temporal density of each origin-destination pair of each time interval (which is defined "*TOD*"), then Density-Deviation method is proposed to identify traffic anomalies among different traffic flows in different time interval.
3. Anomalous DAG construction: this work proposes an algorithm with strong pruning to generate anomalous DAG which just traverses each traffic anomaly only once. Then multi-causes analysis is used to reflect the outlier causal relationship among traffic anomalies more precisely.

The rest of this article is organized as follows. In Section 2 we introduce the framework and definitions of our model. And Section 3 gives the detail materials and methods to discover outlier causal relationship. Experiments and analysis are reported in Section 4 and Section 5 concludes our work.

2   Model Description

In this section, the framework of model proposed in our work is introduced first, then we give several definitions used in this model.

2.1   Model Framework

The structure of our model mainly consists of three steps shown in Fig 1: Region-Based Traffic Network is built in the first step. Then we obtain the spatial-temporal density based on spatial-temporal feature. A depth-first search with pruning algorithm will be used to construct anomalous DAG during the last step.

2.2   Definitions

Definition 1 (*trajectory*): a *trajectory* is a sequence of time ordered GPS points generated by a taxi equipped with GPS in geographical space, e.g. $T: p_1 \rightarrow p_2 \rightarrow \cdots \rightarrow p_n$. Each GPS point is composed of a timestamp and geographic coordinates, e.g. $p = (timestamp, longitude, latitude)$.

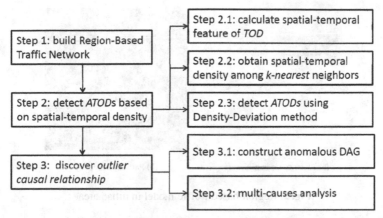

Fig. 1. Framework of anomalous DAG model

Definition 2 (*time interval*): *time interval* is the basic time unit in the model. If the *time interval* is 20 minutes, then there are 72 (24*60/20) *time intervals* per day. Each GPS point will be mapped into corresponding *time interval* according to its timestamp.

Definition 3 (*shift*): given a *trajectory* $T: p_1 \rightarrow p_2 \rightarrow \cdots p_n$, there exists a *shift* from region r_1 to another region r_2, $(r_1 \neq r_2)$ if two adjacent points p_i and p_{i+1} in *trajectory* T are mapped to region r_1 and r_2 respectively, denoted by $r_1 \Rightarrow r_2$.

Definition 4 (*TOD*): a *TOD* indicates a connection from an origin region (TOD_o) to a destination region (TOD_d) in a *time interval* (TOD_t) of a day. T represents the *time interval*. O and D represent the origin and destination region respectively. Each *TOD* has a spatial-temporal density which reflects the behavior similarity among its *k-nearest* neighbors. Here the *k-nearest* neighbors refer to the *TODs* with the same TOD_t, TOD_o, TOD_d as the *TOD* but are in different consecutive k days.

Definition 5 (*ATOD*): *ATOD* is an anomalous *TOD* the spatial-temporal density of which is extremely different from its k-nearest neighbors.

Definition 6 (*outlier causal relationship*): $ATOD_2$ is caused by $ATOD_1$ if and only if the following conditions are satisfied.

1. *time interval* $ATOD_1^t$ is adjacent to $ATOD_2^t$ and $ATOD_1^t$ is ahead of $ATOD_2^t$.
2. destination region of $ATOD_1$ is the same as the origin region of $ATOD_2$.

Anomalous DAG is generated from *ATOD* considering an *ATOD* may be caused by a series of other *ATODs*, which reflects abnormal traffic pattern more precisely.

3   Materials and Methods

In this section, materials and methods are provided about the model shown in Fig 1.

3.1   Building Region-Based Traffic Network

The map is partitioned into different regions bounded by major roads (high way, level-1, level-2, etc.) as shown in Fig 2 (We use the skill of generating polygon in Geographical Information System [10]). The key issue in this traffic network is how to map a GPS point to corresponding region efficiently.

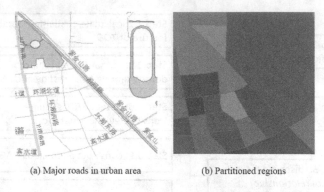

(a) Major roads in urban area   (b) Partitioned regions

Fig. 2. Region-based traffic model in urban area

As far as we know, PNPoly algorithm [11] can be used to judge whether a GPS point is in a specified region. One method is PNPoly with all regions traversal algorithm. However it works badly when dealing with billions of GPS points. Grid-Based PNPoly algorithm proposed in this paper works efficiently to solve this issue. Specifically, map is divided into grids based on longitude and latitude partition beforehand, then each region r is mapped to grid d if one of following conditions holds true:

1. $r \subseteq d$, that is, region r is within grid d.
2. $r \cap d \neq \emptyset$, that is, there are overlapping areas between region r and grid d.

As shown in Fig 3, region r_1 is mapped to grid d_1 because of condition 1, while region r_2 is mapped to both grid d_2 and grid d_4 in terms of condition 2. After preprocessing, each grid will associate with a collection of regions mapped to it. Then a GPS point can be mapped to the corresponding region by just traversing the regions associated with this grid.

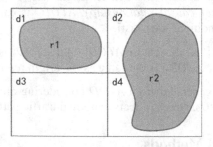

Fig. 3. Examples of mapping regions to grids

3.2   Detecting *ATOD*s Based on Spatial-Temporal Density

According to definition 3, each *trajectory* can be transformed to a sequence of *shift*s. For example, the *trajectory* (black curve) in Fig 4 can be transformed to 2 *shift*s: $a \Rightarrow b, b \Rightarrow c$, then *ATOD*s can be obtained by the following three steps:

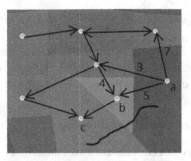

Fig. 4. Example of *shift* and spatial-temporal feature

Calculating the Spatial-Temporal Feature of Each *TOD*. The spatial-temporal feature of *TOD* is defined as a triple, denoted by $\overrightarrow{f_{TOD}}=<Tot, Pct^o, Pct^d>$, where *Tot* represents the total number of *shifts* from TOD_o to TOD_d in TOD_t. Pct^o represents the proportion of *Tot* in *shift* amount of leaving TOD_o, Pct^d represents the proportion of *Tot* in *shift* amount of arriving at TOD_d. The triple is created by feature selection which captures traffic flow more effectively.

The example of spatial-temporal feature is presented in Fig 4 in which the triple of $TOD(TOD_o = a, TOD_d - b)$ is $<5, \frac{5}{5+3+7}, \frac{5}{5+4}> = <5, \frac{1}{3}, \frac{5}{9}>$.

Obtaining the Spatial-Temporal Density. The spatial-temporal density is capable of capturing the behavior similarity among the k-nearest neighbors which is computed as follows:

$$den(TOD, k) = \left(\frac{\Sigma_{tod \in N(TOD,k)}\, distance(TOD, tod)}{|N(TOD, k)|}\right)^{-1} \tag{1}$$

Where $N(TOD, k)$ is the set containing the k-nearest neighbors of TOD, $|N(TOD, k)|$ is the size of this set. The distance between *TOD* and one of its *k-nearest* neighbors is computed by the following formula:

$$distance(TOD, tod) = \sqrt{\left\|\overrightarrow{f_{TOD}} - \overrightarrow{f_{tod}}\right\|^2} \tag{2}$$

Euclidean distance is used here and the spatial-temporal feature should be standardized.

Detecting *ATODs* Based on Spatial-Temporal Density of *TOD*. Formula 1 shows the density of each *TOD* is relative among *k-nearest* neighbors. If the density of *TOD* is small, this *TOD* will present big difference among *k-nearest* neighbors. The *TOD* which has extreme small density is defined as *ATOD*.

Here density-deviation method is proposed first step of which is to calculate the deviation between the density of *TOD* and average density of its k-nearest neighbors according the following formula:

$$deviation(den, k) = \frac{avgDen - den}{avgDen} \quad if \; avgDen > den \qquad (3)$$

Where $avgDen$ is the average density of the TOD's $k\text{-}nearest$ neighbors. den is smaller than $avgDen$ because $ATOD$s always have smaller density than others. Then the TOD whose deviation exceeds the pre-defined threshold s is deemed as $ATOD$.

3.3   Discovering Outlier Causal Relationship

Constructing Anomalous DAG. The core of Anomalous DAG algorithm is to do depth-first search with strong pruning for all the $ATOD$s. The algorithm is highly efficient especially dealing with large-scale data which is demonstrated in section 5. Specifically, the input of algorithm is a set of $ATOD$s, and output is a set of connected DAGs. Algorithm of $SearchMaxDepth$ showed below is a recursive function to find the max depth of $ATOD$ which is the basic to build anomalous DAGs. Pruning step is executed immediately if an $ATOD$ has been visited.

Compared with STOTree algorithm [8], Anomalous DAG is much more efficient. Suppose there are T time intervals in total and the amount of ATODs in i-th time interval is M_i. STOTree algorithm needs to search the amount of $\sum_{i=1}^{T-1} M_i \cdot M_{i+1}$ ATODs, but AnomalDAG algorithm just needs $\sum_{i=1}^{T} M_i$ ATODs.

```
Algorithm of SearchMaxDepth(atod)
maxDepth ← 1;
valueSet ← the value list of atod;
for each ATOD a in valueSet do
  if(a is visited before ) then
    dep ← a.depth;
  else
    dep ← SearchMaxDepth(a);
  end if
  if maxDepth < dep + 1 then
    maxDepth ← dep + 1;
  end if
end for
```

Multi-causes Analysis. *Outlier causal relationship* can be extracted based on the depth of each *ATOD* as shown in Fig 5: left part shows the example of connected anomalous DAGs, there are four consecutive *time intervals* from *t1* to *t4* in time order and *ATOD*s with depth value at top right corner. Two connected DAGs are extracted as shown in right part by multi-causes analysis which reflects the outlier causal relationship: *ATOD D* is caused by both *ATOD A* and *B*, continue influences *ATOD E* and *F*. *ATOD G* causes *H* and *I*, then lead to *ATOD J* finally.

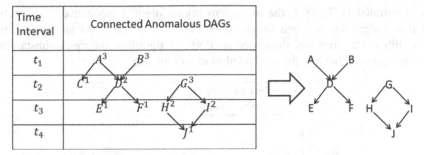

Time Interval	Connected Anomalous DAGs
t_1	
t_2	
t_3	
t_4	

Fig. 5. Example for process of discovering causal relationship

4   Results and Discussion

In this section, experiments are conducted using taxi trajectory data on the traffic network in the urban area of Tianjin.

4.1   Parameters and Data

The urban area is partitioned using Region-Based Traffic Model. Related information about parameter and data are presented in table 1.

Table 1. Basic information of parameter and data

Item	Value
Total Region Amount	1,351
Region Road Amount	12,886
Time Interval	20min
Data Type	taxi GPS data
GPS Record Time	2012.10.01-2012.10.31
Total GPS Amount	387,078,842

4.2   Analysis on Region-Based Traffic Network

In this subsection, experiments are conducted between PNPoly (all regions traversal) and Grid-Based PNPoly algorithms to map GPS points to corresponding regions and compare their time efficiency. In Grid-Based PNPoly algorithm, we divide the map into 400 grids equally and preprocess to obtain the regions mapped to each grid.

Table 2. Maximum region amount(MRA) associated with grid

Algorithm	Grid Amount	MRA
PNPoly	1	1,351
grid-based PNpoly	20*20	29

As illustrated in Table 2, the maximum region amount associated with grids in Grid-Based algorithm is much smaller than PNPloy algorithm. As shown in Fig 6, When GPS points increase, time used in PNPoly algorithm increases substantially, however time used in Grid-Based algorithm grows linearly.

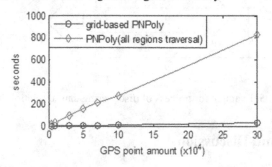

Fig. 6. Time used Comparison of mapping GPS points

4.3   Analysis on Anomalous DAG

Experiments between anomalous DAG and outlier tree are conducted in this subsection and density deviation threshold is set between 0.4 and 0.5. As shown in Fig 7, when the density deviation threshold decreases, time used of building anomalous DAG grows almost linearly, while outlier tree dramatically increases, showing exponential growth.

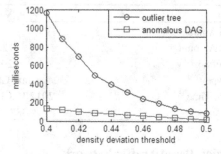

Fig. 7. Time used between building anomalous DAG and outlier tree under different settings of density deviation threshold

More importantly, the outlier causal relationship discovered by anomalous DAG better coincides with causal events in real life. A significant event is as follows:

According to People's Daily·Tianjin Windows report in October 4th, Tianjin Park is closed because of rebuilding during the National Day, townspeople prefer to play in Tianjin Water Park instead. On October 1st the traffic flow increases by forty percent than usual around Water Park. Traffic jams once occurred during rush hours on some roads.

Anomalous DAG algorithm successfully detects this event (Fig 8(a)), by multi-cases analysis we obtain an anomalous DAG from 7:00am to 9:30am, which shows the traffic jam in the Water Park East Road(A) caused the traffic jam in Water Park North Road(B), Tianta Street(C) and Cangqiong Street(D), and finally resulted in the traffic jam in Weijin South Road(E).

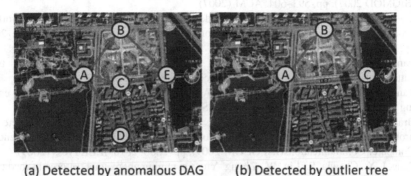

(a) Detected by anomalous DAG   (b) Detected by outlier tree

Fig. 8. Outlier causal relationship compared between anomalous DAG and outlier tree

While outlier tree algorithm only detects the event of Water Park East Road(A) to Water Park North Road(B), then to Weijin South Road(C), as shown in Fig 8(b). The main reason is that in anomalous DAG algorithm each traffic anomaly can be caused by more than one traffic anomaly by multi-causes analysis while outlier tree algorithm is single-cause. So the anomalous DAG is superior to outlier tree both in time complexity and effectiveness of model.

5   Conclusion

In this paper, Grid-Based PNPoly algorithm is firstly used to map each GPS point to the corresponding region in Region-Based Traffic Model which is proved to be efficient. Then we propose the anomalous DAG model to discover the causal relationship among traffic anomalies. Experiments based on large-scale taxi trajectory data are conducted to demonstrate the anomalous DAG algorithm is almost linearly and multi-causes analysis helps reflect the causal relationship more effectively. In the future, we will improve our model to detect outlier causal relationship in traffic network more precisely.

Acknowledgements. This work was supported by the Major Project of National Social Science Fund (14ZDB153), the National Science and Technology Pillar Program (2013BAK02B06 and 2015BAL05B02), Tianjin Science and Technology Pillar Program (13ZCZDGX01099, 13ZCD ZSF02700), National Science and Technology Program for Public Well-being (2012GS120302), Tianjin Research Program of Application Foundation and Advanced Technology (14JCTPJC 00517).

References

1. Zheng, Y., Zhou, X.F.: Computing with Spatial Trajectories. Springer (2011)
2. Lee, J., Han, J., hang, K.W.: Trajectory clustering: a partition-and-group framework. In: Proceedings of the 26th ACM SIGMOD International Conference on Management of Data (SIGMOD 2007), pp. 593–604. ACM (2007)
3. Lee, J., Han, J., Li, X.: Trajectory outlier detection: A partition-and-detect framework. In: Proceedings of the 24th International Conference on Data Engineering (ICDE 2008), pp. 140–149. ACM (2008)
4. Pang, L.X., Chawla, S., Liu, W., et al.: On detection of emerging anomalous traffic patterns us-ing GPS data. J. Data & Knowledge Engineering **87**, 357–373 (2013)
5. Tango, T., Takahashi, K., Kohriyama, K.: A space –time scan statistic for detecting emerging outbreaks. In: International Biometrics Society, pp. 106–115 (2010)
6. Brauckhoff, D., Salamatian, K., May, M.: Applying PCA for traffic anomaly detection: problems and solutions. In: Proceedings of the 28th IEEE International Conference on Computer Communications(INFOCOM 2009), pp. 2866–2870. IEEE Press (2009)
7. Pang, L.X., Chawla, S., Liu, W., Zheng, Yu.: On mining anomalous patterns in road traffic streams. In: Tang, J., King, I., Chen, L., Wang, J. (eds.) ADMA 2011, Part II. LNCS, vol. 7121, pp. 237–251. Springer, Heidelberg (2011)
8. Liu, W., Zheng, Y., Chawla, S., et al.: Discovering spatio-temporal causal interactions in traffic data streams. In: Proceedings of the 17th ACM SIGKDD International Conference on Knowledge Discovery and Data Mining, pp. 1010-1018. ACM (2011)
9. Yuan, N.J., Zheng, Y., Xie, X.: Segmentation of urban areas using road networks. R. MSR-TR-2012-65 (2012)
10. ArcGIS Information, http://help.arcgis.com/en/arcgisdesktop/10.0/help/index.html
11. PNPOLY Information,http://www.ecse.rpi.edu/~wrf/Research/Short_Notes/pnpoly.html

Text Classification Based on Paragraph Distributed Representation and Extreme Learning Machine

Li Zeng[1,2] (✉) and Zili Li[1,2]

[1] College of Information System and Management,
National University of Defense Technology Changsha, Changsha 410073, China
{crack521,zilili}@163.com
[2] Center for National Security and Strategic Studies (CNSSS),
National University of Defense Technology Changsha, Changsha 410073, China

Abstract. This paper implements a semi-supervised text classification method by integrating Paragraph Distributed Representation (PDR) with Extreme Learning Machine (ELM) training algorithm. The proposed Paragraph Distributed Representation-Extreme Learning Machine hybrid classification approach is named as PDR-ELM. Paragraph Distributed Representation is a recently proposed feature selection method based on neural network language model, while Extreme Learning Machine is well known as its high performance in classification. We propose PDR-ELM hybrid classification approach with the objective to minimize the training time and raise the classification accuracy meanwhile. We conduct experiments on a real research paper datasets crawled from Web of Science (WOS). Results show that the proposed PDR-ELM can achieve an accuracy of 81.01% and a training time of 5.1324 seconds on the datasets.

Keywords: Text classification · Paragraph distributed representation · Extreme learning machine

1   Introduction

Text Classification is a basic task in Natural Language Processing and plays an important role in many applications, e.g., web search, recommendation system and research paper categorization. One important pre-processing step before classification is text representation or feature selection. Perhaps the most common representation of text is the one-hot representation or bag of words [1] One-hot representation transforms the text into a feature vector which has the same length as the size of the vocabulary and only one dimension is on. Traditional methods on basis of one-hot representation such as Naive Bayes [2] or Latent Dirichlet Allocation [3] have been widely used for text classification for its efficiency and simplicity. However, such representation of text has many flaws: it overlooks the order and structure of text and cannot evaluate the similarity between documents. Moreover it always suffers from data sparsity when the vocabulary size increases.

Recent works in Nature Language Processing have shown that Neural Network language models [4] have demonstrated outstanding performance in a variety of tasks [5] [6].Instead of using traditional one-hot representation, these methods use the dis-

Y. Tan et al. (Eds.): ICSI-CCI 2015, Part II, LNCS 9141, pp. 81–88, 2015.
DOI: 10.1007/978-3-319-20472-7_9

tributed representations which map text or word into a dense real continuous vector using unsupervised approaches, and achieved a variety of state-of-art improvements in many tasks [7] [8]. Interesting thing is that these representations are less human interpretable than previous methods, they seem to work well in practice. Especially,Q.V. Le [9] show that their method, Paragraph Vectors, capture many document semantics in dense vectors and that they can be used in classifying movie reviews or retrieving web pages.

In this paper, we proposed a semi-supervised text classification method which is inspired by the recent work in learning paragraph representations of words using neural networks. In our model, the features of texts are generated by a forward neural network and then stacked into the Extreme Learning Machine [10].The ELM use parts of features to train the model and then to classify testing texts. Experiments results show that the proposed PDR-ELM can get a good performance on the datasets

The remainder of this paper is organized as follows. In Section 2, we discuss related work; Section 3 describes the pipeline of PDR-ELM, especially focuses on discussing the Paragraph Distributed Representation and Extreme Learning Machine algorithm; Section 4 describes the experiments, and we discuss in Section 5.

2   Related Works

Feature Selection is an important step to classify texts. Due to the success of deep learning method recently, distributed representation becomes a popular feature selection method in many task. A distributed representation of a symbol is a tuple (or vector) of features which represents raw symbol in a dense, low dimensional, and real-valued form. Word or Paragraph Distributed representations is also called word embeddings or paragraph embeddings. Each dimension of the embeddings represents a latent feature of word or paragraph, hopefully capturing useful syntactic and semantic properties. The idea of distributed representation was first proposed by Hinton [11].Bengio, Ducharme, Vincent and Janvin [4] proposed a Neural Probabilistic Language Model (NPLM) which can exploit distributed representations of symbolic data and character sequences. Bengio, Ducharme, Vincent and Janvin [4] demonstrated that distributed representations for symbols combined with neural network can surpass standard n-gram models in many tasks. Mnih and Hinton [12] speeded up model evaluation during training and testing by using a hierarchy to exponentially filter down the number of computations that are performed. The model, combined with this optimization, is called the hierarchical log-bilinear (HLBL) model. Mikolov, Karafiát, Burget, Cernocký and Khudanpur [13] proposed a Recurrent Neural Network Language Model(RNNLM) which can learn to compress whole history of text in low dimensional space, while Whereas feedforward networks model such as NPLM only exploit a fixed context length to predict the next word of a sequence. Sundermeyer, Schlüter and Ney [14] replaced the simple recurrent neural network with Long Short-Term Memory neural network for Language Modeling to overcome the vanishing gradient problem. Mikolov, Sutskever, Chen, Corrado and Dean [15] introduced continuous bag-of-words(CBOW) and skip-gram architectures for learning high quality distributed representations of words from large amounts of unstructured text data , Mikolov implemented these model into an open source toolkit called word2vec[16].Q.V. Le [9] proposed a Paragraph Distributed Representation named Paragraph vector model, an unsupervised algorithm that learns fixed-length feature representations from variable-length pieces of texts, such as sentences, paragraphs, and documents.

After doing the feature selection of text, Feedforward Neural Networks (FNN) [17] and Support Vector Machines (SVM) [18] are usually popular classifiers. It is clear that the feedforward neural networks has a far slower learning speed in general than required and it has been a major bottleneck in their applications for past decades. Due to their outstanding classification capability, support vector machine and its variants such as least square support vector machine (LS-SVM) have been widely used in binary classification applications [19]. The conventional SVM and LS-SVM cannot be used in regression and multi-class classification applications directly although different SVM/LS-SVM variants have been proposed to handle such cases [20]. While both face some challenging issues such as: intensive human intervene, slow learning speed, poor learning scalability. Recently, extreme learning machine (ELM) has been proposed for training single hidden layer feedforward neural networks (SLFNs) [10]. Compared to traditional FNN learning methods, ELM is remarkably efficient and tends to reach a global optimum. Theoretical studies have shown that even with randomly generated hidden nodes, ELM maintains the universal approximation capability of SLFNs [21].

3   Pipeline of PDR-ELM

Figure 1 describes the pipeline of our semi-supervised text classification method. After splitting the raw text into training set and testing set, we tokenize the raw texts, and sack them into the PV-DM model to get the paragraph distributed representation, thus every document d_i in the whole documents set is associated with a feature vector w_i, and then we use feature vectors of training set $W_{Training} = \{w_1, w_2, \ldots, w_n\}$ and its classid set $K = \{1, 2, \cdots k\}$ to train the ELM model. After the model converges, we use the feature vectors of testing set $W_{testing} = \{w_1, w_2, \ldots w_j, \cdots, w_m\}$ and the ELM model to find the class ID of every document in the testing set, we denotes the result as $k_j = \arg\max_k (p(k \mid w_j, \beta))$, where β is the parameters of ELM model learned in the training set.

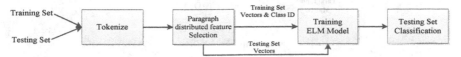

Fig. 1. Pipeline of proposed PDR-ELM model

3.1 Paragraph Distributed Representation

Paragraph Distributed Representation adds a temple memory vector to the stands neural network Language Model [15] which aims at capturing the semantic of the whole paragraph. And the authors name this model "Distributed Memory Model of Paragraph Vectors (PV-DM)".

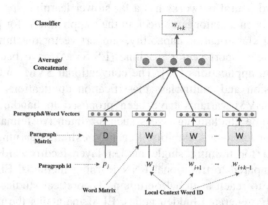

Fig. 2. The distributed memory model of Paragraph Distributed Representation

Figure 2 shows the framework of distributed memory model of Paragraph Distributed Representation. In the model, every word is mapped into a feature vector w which is a column of matrix W, and a paragraph id is also mapped into a vector which is a column of Matrix D. The concatenation or sum of the paragraph vector p_j and its local context word vectors $w_i, w_{i+1}, \cdots, w_{k-1}$ is then used as features by softmax function to predict the next word in a sentence.

Figure 3 shows another framework of paragraph distributed representation which is named "Distributed Bag of Words" (PV-DBOW) model:

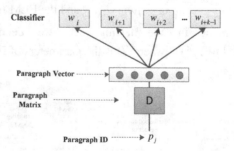

Fig. 3. The distributed bag of words model of Paragraph Distributed Representation

Unlike PV-DM, PV-DBOW not only trains the paragraph vectors but also trains word vectors to predict the adjacent words. The entire model is trained by stochastic gradient descent.

3.2 Extreme Learning Machine

Extreme Learning Machine (ELM) was proposed for generalized single-hidden layer feedforward networks where the hidden layer need not be neuron alike [10]. ELM was originally inspired by biological learning and proposed to overcome the challenging issues faced by BP learning algorithms.

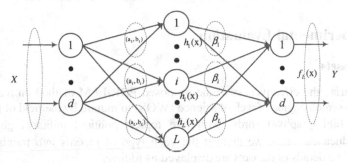

Fig. 4. The architecture of extreme learning machine [10]

The output function of basic ELM is:

$$f_L(\mathrm{x}) = \sum_{i=1}^{L} \beta_i h_i(\mathrm{x}) = h(\mathrm{x})\beta, \tag{1}$$

where $h(x) = [h_1(x), \cdots, h_L(x)]$ is output vector of hidden layer with regard to x, and $\beta = [\beta, \cdots, \beta_L]^T$ is weights of $h(x)$. Different output functions may be used in different hidden neurons. In particular, in real applications $h_i(x)$ can be:

$$h_i(\mathrm{x}) = G(\mathrm{a}_i, \mathrm{b}_i, \mathrm{x}) \qquad \mathrm{a}_i \in R^d, b_i \in R, \tag{2}$$

where $h_i(\mathrm{x}) = G(\mathrm{a}_i, \mathrm{b}_i, \mathrm{x})$ is a nonlinear piecewise continuous function satisfying ELM universal approximation capability theorems [21]. Usually, $h_i(\mathrm{x})$ can be a sigmoid、Sigmoid   function、Fourier   function、Hardlimit   function、Gaussian function、Multiquadrics function and etc.

ELM trains in two main stages: random feature mapping and linear parameters solving. In ELM, the hidden node parameters $(\mathrm{a}_i, \mathrm{b}_i)$ are randomly generated according to any continuous probability distribution instead of being explicitly trained, leading to remarkable efficiency compared to traditional BP neural networks. And the target of ELM is as follows:

$$\min_{\beta \in R^{Lxm}} \|H\beta - T\|^T, \tag{3}$$

where T is the training data target matrix, and $\|\bullet\|$ notes the Fresenius norm. The optimal solution to (3) is given by:

$$\beta^* = H^+T, \tag{4}$$

where H^+ denotes the Moore–Penrose generalized inverse of matrix H.

4   Experimental Evaluation

4.1   Datasets

To demonstrate the effectiveness of the proposed PDR-ELM method on real datasets, we used the search engine of web of science (WOS) to inquire by the field of topics, and got 25000 bibliography records[1] with five topics: politics、military、geography、biology、education，then we divided the each type of records into training set and testing set. The details of datasets are displayed as follows.

Table 1. Details of datasets crawled from the WOS

Category	#train	#test
politics	3750	1250
military	3750	1250
geography	3750	1250
biology	3750	1250
education	3750	1250

4.2   Results

In the experiments, we use three different output functions of Extreme Learning Machine: sigmoid function、radaus function and sin function. To evaluate the three output function's performances on the same datasets, we conducted series of experiments, and the results are listed as follows:

Table 2. Evaluation results of text classification by PDR-ELM

Model	Time(Seconds)		Number of Hidden Neurons	Accuracy	
	Train	Test		Train	Test
PDR-ELM (sigmoid)	**5.1324**	**0.3276**	**200**	**0.8238**	**0.8101**
	62.1508	1.0452	900	0.8574	0.8203
	81.0581	1.1544	1000	0.8633	0.8197
PDR-ELM (radaus)	5.2416	0.3900	200	0.6506	0.6195
	64.1320	1.2168	900	0.8055	0.7616
	79.6385	1.2168	1000	0.8145	0.7629
PDR-ELM (sin)	5.7720	0.3120	200	0.4812	0.4509
	64.4752	1.0764	900	0.6797	0.6139
	79.7789	1.1076	1000	0.6875	0.6213

---

[1] https://github.com/crack521/Datasets-of-PDR-ELM

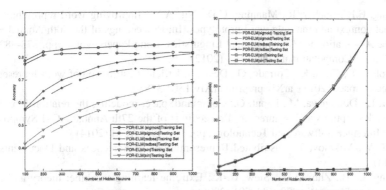

Fig. 5. Accuracy of PDR-ELM   **Fig. 6.** Training time of PDR-ELM

From table 2 we can conclude than PDR-ELM with a sigmoid output function has the best performance on the dataset, and PDR-ELM (sigmoid) with 200 hidden neurons can both have a little training time and better accuracy than other.

5   Discussion

In this paper, we propose a semi-supervised text classification method named PDR-ELM, aiming to incorporate Paragraph Distributed Representation and Extreme Learning Machine to improve the accuracy of text Classification and reduce the training time. We conduct experiments on a real dataset crawled from web of science database, and we evaluate three different types of PDR-ELM on the same datasets and find the best number of hidden neurons at last. Several promising directions remain to be explored. In the paper, we combine the Paragraph Distributed Representation and Extreme Learning Machine do to a text classification task, while [22] shows that Extreme Learning Machine also can do a better presentation learning. Using the ELM to learning the distributed representation of text is also worth investigating as it might result in a better text representations.

References

1. Harris, Z.S.: Distributional structure. Word (1954)
2. Eyheramendy, S., Lewis, D.D., Madigan, D.: On the naive bayes model for text categorization (2003)
3. Blei, D.M., Ng, A.Y., Jordan, M.I.: Latent dirichlet allocation. The Journal of Machine Learning Research **3**, 993–1022 (2003)
4. Bengio, Y., Ducharme, R., Vincent, P., Janvin, C.: A neural probabilistic language model. The Journal of Machine Learning Research **3**, 1137–1155 (2003)
5. Collobert, R., Weston, J.: A unified architecture for natural language processing: Deep neural networks with multitask learning. In: Proceedings of the 25th International Conference on Machine learning, pp. 160-167. ACM (2008)

6. Huang, E.H., Socher, R., Manning, C.D., Ng, A.Y.: Improving word representations via global context and multiple word prototypes. In: Proceedings of the 50th Annual Meeting of the Association for Computational Linguistics: Long Papers, vol. 1, pp. 873-882. Association for Computational Linguistics (2012)
7. Mikolov, T., Chen, K., Corrado, G., Dean, J.: Efficient estimation of word representations in vector space (2013). arXiv preprint arXiv:1301.3781
8. Adar, E., Dontcheva, M., Laput, G.: CommandSpace: modeling the relationships between tasks, descriptions and features. In: Proceedings of the 27th Annual ACM Symposium on User Interface Software and Technology, pp. 167-176. ACM (2014)
9. Le, Q.V., Mikolov, T.: Distributed Representations of Sentences and Documents ICML (2014)
10. Huang, G.-B., Zhu, Q.-Y., Siew, C.-K.: Extreme learning machine: theory and applications. Neurocomputing **70**, 489–501 (2006)
11. Hinton, G.E.: Learning distributed representations of concepts. In: Proceedings of the Eighth Annual Conference of the Cognitive Science Sociey (1986)
12. Mnih, A., Hinton, G.E.: A scalable hierarchical distributed language model. In: NIPS, pp. 1081-1088 (2009)
13. Mikolov, T., Karafiát, M., Burget, L., Cernocký, J., Khudanpur, S.: Recurrent neural network based language model. In: INTERSPEECH 2010, 11th Annual Conference of the International Speech Communication Association, Makuhari, Chiba, Japan, pp. 1045-1048 (2010)
14. Sundermeyer, M., Schlüter, R., Ney, H.: LSTM neural networks for language modeling. In: INTERSPEECH (2012)
15. Mikolov, T., Sutskever, I., Chen, K., Corrado, G.S., Dean, J.: Distributed representations of words and phrases and their compositionality. In: Advances in Neural Information Processing Systems, pp. 3111-3119 (2013)
16. Mikolov, T., Yih, W.-t., Zweig, G.: Linguistic regularities in continuous space word representations. In: HLT-NAACL, pp. 746-751 (2013)
17. Bebis, G., Georgiopoulos, M.: Feed-forward neural networks. IEEE, Potentials **13**, 27–31 (1994)
18. Joachims, T.: Text categorization with support vector machines: learning with many relevant features. In: Nédellec, C., Rouveirol, C. (eds.) ECML 1998. LNCS, vol. 1398, pp. 137–142. Springer, Heidelberg (1998)
19. Suykens, J.A., Vandewalle, J.: Least squares support vector machine classifiers. Neural Processing Letters **9**, 293–300 (1999)
20. Huang, G.-B.: An insight into extreme learning machines: random neurons, random features and kernels. Cognitive Computation **6**, 376–390 (2014)
21. Huang, G.-B., Chen, L., Siew, C.-K.: Universal approximation using incremental constructive feedforward networks with random hidden nodes. IEEE Transactions on Neural Networks **17**, 879–892 (2006)
22. Kasun, L.L.C., Zhou, H., Huang, G.-B., Vong, C.M.: Representational learning with ELMs for big data. IEEE Intelligent Systems **28**, 31–34 (2013)

Network Comments Data Mining-Based Analysis Method of Consumer's Perceived Value

Jiaming Lv[(✉)], Quanyuan Wu, Jiuming Huang, and Sheng Zhu

College of Computer, National University of Defense Technology, Changsha 410073, China
434757676@qq.com

Abstract. The user comments on e-commerce websites convey what users think of the goods sold. Thus when deciding the price of a goods, sellers should take many elements into consideration, including cost, price given by the competitors, estimated profit and, what's more, consumer's perceived value. Based on the analysis of the factors which influence consumer's perceived value, we introduces a term called consumer cognitive property of product for the first time. Meanwhile, a data mining method based on text clustering is proposed. We crawled millions of comments from e-commerce website. Then extracted the consumer cognitive properties by a variance cumulative method and cluster these properties based on K means. After these steps, most impressive influencing factors of the product on consumer's perceived value are extracted to assist the sellers to make final pricing strategies. An evaluation performed with the collections crawled from taobao shows that our algorithm works great.

Keywords: Consumer perceived value · Data mining · Variance cumulative method · Text clustering

1 Introduction

Perceived value is the worth that a product or service has in the mind of the consumer and the consumer's perceived value of a good or service affects the price that he or she is willing to pay for it. People understand perceived value mainly from two perspectives. One is raised by Zeithaml [1] that consumer's perceived value can be analyzed from the aspect of value comparison. According to Zeithaml, "Perceived value is the consumer's overall assessment of the utility of a product based on perceptions of what is received and what is given". The other one is raised by Sweeney and Soutar [2] who thought that multiple value dimensions explain consumer choice better, both statistically and qualitatively, than does a single 'value for money' item and should produce superior results when investigating consumption value.

If a company greatly overestimates its consumers' perceived value, leading to its overpricing, then the predicted sales quantity would be hard to achieve. Otherwise, if a company underestimates its consumers' perceived value, leading to its underpricing, then the income earned would be reduced. Thus before setting the initial price of a product, a company should do enough market research to get enough knowledge of its consumers' demand preference and then consumer's perceived value of the product

© Springer International Publishing Switzerland 2015
Y. Tan et al. (Eds.): ICSI-CCI 2015, Part II, LNCS 9141, pp. 89–97, 2015.
DOI: 10.1007/978-3-319-20472-7_10

should be determined by its various attributes such as general purpose, quality, brand and so on [3]. Then within the initial condition, feasible sales volume should be anticipated and target cost with sales income should be analyzed. After comparing the cost with income and the sales volume with the price, the company can ensure the feasibility of the pricing plan. Then the final price can be decided.

Online comments contain lots of valuable information, including consumers' attitude to the appearance and quality of a product as well as the feelings after use. They are all the factors which influence the consumer's perceived value. When the potential consumers see the comments, their perceived value of the product will be influenced. Using the method of data mining, this essay will extract the useful information from the comments to assist the companies and consumers to make decisions.

The main contribution made by this paper is to mix finance and computer science together and solve the economic issues by using data mining related techniques. First of all, we proposed the consumer's perceived value model analyzing factors which influence consumer's perceived value. Then extracts product information from the online trade comments using the variance cumulative method and analyzes the influential factors extracted from the comments. After that, we map the attribute words into vectors of a language model by word2vec model considering that the distance between similar semantic words in vector space is small. Finally, K-means clustering algorithm is used to cluster the vectors related to the words.

2   Research Status

In the past twenty years, consumer's perceived value has attracted people's long-term attention in the area of marketing. It plays a key role in predicting consumer purchase behavior. It can also explain consumer preference in a given situation very well.

Woodruff [4] has summarized three characteristics of consumer's perceived value (1) the relationship between consumer's perceived value and the use of the product; (2) consumer's perceived value is decided by consumer's perception of the product, not by the salesman; (3) the perception usually indicates what the consumer has to give up in return of what he or she can receive.

Based on the characteristics mentioned above, Woodruff proposes that consumer's perceived value is, in certain situations, the assessment and preference of the attributes of a product and, in addition, is the perception of the results achieved after using the product.

Woods [5] once indicated that consumers can experience the purchase in the atmosphere of imagination, emotion and appreciation and a product is just a representation of the consumption experience. What people really want is not the product itself but a satisfactory purchase experience and the feelings they love. The amazing experience and feelings are the value that the consumers perceive during the purchase.

Though the researches above propose the origin and definition of consumer's perceived value, the importance of perceived value and how the consumer purchase behavior is influenced by the perceived value, the components which form the consumer's perceived value are not introduced. Moreover the influences on perceived value caused by the components are not analyzed as well in today's popular B2C business model.

Taking the previous theories into account, this essay proposes the consumer's perceived value model. It holds the view that the basis of the consumer's perceived value theory is utility theory. Then the influence factors of consumer's perceived value are discussed from two aspects.

Meanwhile, the important technique used in this essay is text clustering. In the past, text clustering of a certain field is usually carried out by experts in that field. Though the accuracy is ensured, it is a waste of time and money. Thus, with the development of natural language processing and machine learning technology, a lot of researches on automatic text clustering of a certain field have been carried out in these years and many research results have been achieved. For example, Li Jie etc [6] put forward semantic automatic classification based on the "CNKI", refining its subclasses. Using the feature extraction method of text classification, Liu hua [7] clusters the words in the field to obtain a large scale of field acknowledge. However, attribution extraction is greatly disturbed by data sparseness and data noise. Sometimes obvious mistakes can appear. Kang Tiegang etc [8] proposed the word clustering method based on large-scale annotated corpus. According to Kang, they find out the words from the corpus of the qualifier and the words in the corpus with the same co-occurrence frequency. Then attribute vector is formed and clustering analysis is carried out. However this method needs enormous corpus of qualifier. Google has developed a toolkit named word2vec based on deep learning. The toolkit can map each word into a vector and cluster them by the similarity of different words. However this method doesn't work very well on small-scale corpus clustering. Therefore, this essay presents a new text clustering method. This method is a combination of word2vec and K-means. The combination makes the outcomes more accurate and reasonable.

3   Analysis of Relevant Knowledge

3.1   Consumer's Perceived Value Model

Profit is the main objective of a company. When pricing a product, companies always pursue the price to be as high as possible. To raise the price of the product, companies must use some effective methods, while the perceived value pricing is a good tool. From the definition of consumer's perceived value mentioned above, we can find that if the consumer's perceived value of a product is quantified, the consumers of the product can set an acceptable price range to guide the companies to price their new products. The level of consumer awareness is in a positive correlation with the price level [9], that consumer awareness of high-value products, product pricing can be set too high. Therefore, companies can influence consumer's perceived value by factors, such as improving product quality, improving product packaging, using the price of the psychological implications and other means to enhance the value of consumer awareness and to raise prices.

A prerequisite for measuring consumer's perceived value is the value of a clear mechanism of the formation of consumer awareness, to identify the drivers of consumer's perceived value, and then to determine the perceived value of the consumer based on the consumer judge of these factors.

So what are the factors affecting consumer's perceived value? First, we must define the consumer's perceived value on the theoretical basis of the utility theory [10]. This theory holds that the total value of a product is derived from its own utility and the utility is reflected in the ability to meet the desires and needs of people. Due to the ability to meet the physical and spiritual of human nature, a product becomes useful and worthy. According to this theory, utility is the basis of the formation of consumer's perceived value and the utility can be divided into two aspects. One is the utility on the material level to meet the needs of the human body areas and the other is the utility on the emotional level to meet the human spirit competing demands. Utility on material level of a product are mainly driven by the product's attribute itself. While utility on emotional level is driven by both the product itself and factors including brand, use results, level of satisfaction and other factors. Drivers of consumer's perceived value are shown in Figure 1.

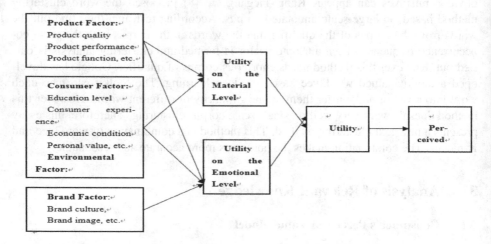

Fig. 1. Consumer perceived value model

3.2   Product Attributes Word Detection Method Based on Sales

On the commodity attributes extraction, we use the variance cumulative method. The main idea of the difference scores are: high frequency if a word or phrase appears in the article, and rarely appear in other articles, you think this word or phrase has a good ability to distinguish between categories, suitable for classification. The difference score approach is, if a keyword appears in a hot commodity, then add 1 point, if there is in unsalable merchandise, then subtract 1 point, finally get a score for each keyword. For example, as "the application of the algorithm", in Chinese, "Apply" is a very common word, and "algorithm" is a very professional word, the latter than the former in the relevance ranking in importance. So we need to give every Chinese word a weight. The ability to predict the theme of a word, the stronger the greater the weight, on the contrary, the smaller weights. We see the "algorithm" in a Web page, more or less understand the subject of the article. We see the "apply" once, but the subject is basically known nothing. Therefore, the "algorithm" weight should be larger than the applied.

Based on the sales of goods (the number of orders) in descending order, these product classification ranking list, which is a hot commodity before K per cent, the rest is unsalable merchandise. Reviews for the purchase of these goods and outlets of the publicity title word processing and other commodities, extract the noun part of speech the word set. Because selling goods relatively high degree of consumer focus, representative, while those with lower sales of merchandise are more likely to be some water or other non-military concerns normal buyers of goods, does not have a representative. Therefore, to prevent a large number of comments are not representative and advocacy word among a small number of key coverage and the right to have a representative misleading keyword weight, we used to distinguish between positive and negative attributes weighted word processing rules commodity extraction.Let Wi belong to Uk(w).

$$Q(Wi) = Q(Wi) + I(k) .$$ (1)

$$I(k) = \begin{cases} \alpha & k \in \text{Top } k \\ -\beta & \text{Others} \end{cases} .$$ (2)

Table 1. The Variance Cumulative method

Algorithm : The Variance Cumulative method
For Wi in Uk(w):
If k < K
I(k) = α
Else
I(k) = -β
End
Q(Wi) += I(k)
End

Wherein, Uk(w) stands for the k-th set of keywords and Q (Wi) is the weight of keyword W. α and β are the keyword weighting coefficient of positive and negative

3.3   Commodity-Based K-means Clustering Properties of Words

Online reviews by the user freedom to fill in, everyone used to vary, so the same kind of product attributes will be described by different words, for example, in the digital camera reviews in "shell", "look" of the "look" and "shape "and other words referring to the appearance of this camera about the same kind of property. Diversified property makes the word too trivial to extract attribute information is not easy to attribute the word frequency statistics and analysis. Therefore, this paper developed a method with the same, similar to the meaning of the word automatic clustering properties.

Word2vec is a form of the word is converted into a vector tool. You can simplify the handling of text content for the vector space vector calculation and calculate the similarity vector space to indicate the similarity of text semantics [11].

Prototype is thought Word2Vec NNLM language model based on the following structure, figure 2:

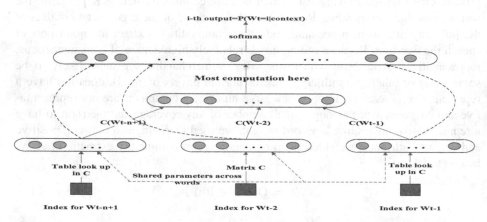

Fig. 2. NNLM language model

In the figure, by mapping between each input word is converted to a vector C, that C (wt-1) represents the word wt-1 vector. The output is a vector with the i-th element is the probability p ($wt = i \mid w1t$-1). Training objectives model are:

Max Likelihood : $1/T\Sigma tlogf(wt,wt-1,...,wt-n+2,wt-n+1;\theta)t+R(\theta)$

Where θ is the parameter, R (θ) as a regular item, the output layer using softmax function: ($wt\mid wt-1,..., wt -n+2,wt-n+1)=eywt/\Sigma ieyi$ .Where each output word yi is not normalized log probability i is calculated as follows: $y=b+Wx+U$tanh($d+Hx$).

Using stochastic gradient descent method to optimize this model, but after optimization, we get a certain word corresponding word from the output vector y in. After word to get to scale, we can directly calculate different word vector cosine distance, as the "distance" between different words. Because word2vec calculation is the cosine distance range is between 0-1, the larger the value the greater the representative correlation of these two words, so the closer the row above the word then the word entered. After this, the use of K-means clustering algorithm to come together almost like a word.

Purely from the perspective of word analysis of two product attributes similarity between words (on the appearance of the goods described in terms of colors, size, etc.), you need to analyze the semantic properties of words, the current data mining methods are difficult to effectively achieve. Therefore, we turn from the method of statistical analysis to measure the similarity of the properties of the word [12]. First, build a collection that contains the word of comment on each word to these comments aggregated into a document. Then, similar to the calculation of the similarity of web pages using TF-IDF we can comment document is mapped to a vector space, calculate the distance between the document vector cosine similarity resulting documents between the two comments. Comments document the use of the appropriate word similarity indicate similarity. Based on this, we have a corresponding word comments in the document vector space mapping algorithms use K-Means clustering process. Specifically:

1) Initialize K centers;2) Calculate distance between the corresponded document vector of each word and each centers;3) Assign each word to the nearest center, all words around each center point is marked as a cluster;4) Calculate the center of each cluster. i.e., the average value of the cluster document vectors;5) Calculate change values between new centers and centers of the last iteration;6) If the changing value is less than the threshold value set, then output the results of the clustering, otherwise repeat steps 2-5 until convergence.

4   Experimental Results and Analysis

4.1   Experimental Data

We crawled thousands of transactions from Taobao on 2014. Parsed pages and extracted data including online comment, transaction number, the seller account, purchase date and other information.

4.2   Experimental Process (Design)

Firstly, in the experimental preparation stage, this paper will be cleaned after the data format, retaining only comment content to remove punctuation. Then, put one hundred thousand comments were divided into three groups compared. Secondly, differences in scoring algorithm comment content word segmentation and obtain property, and do a word frequency statistics. (Table 2)

Table 2. Attribute word extraction results

Products	1903
Sellers	1666
Treasure	1287
Goods	1279
Effect	1276
Attitude	1224
Quality	1206
Brand name	1195

Secondly, the treated corpus use word2vec toolset depth application of neural network algorithm, converted to the corresponding word vector, the definition of the word vector cosine distance between the similarity between the words (Table 3). Finally, the word vector by K-means clustering algorithm to obtain the cluster can eventually return to the same input word corpus semantic closest word.

Table 3. Cosine distance between key words and attributes

Package	
Face job	0.53736144
Elegant-looking	0.5349589
High-Level	0.53384054
Top-ranking	0.50904995
Luxury	0.47808146
High-quality	0.47248584
First-rated	0.4723349

4.3   Experimental Results (Including Indicators)

Experiments using the classical evaluation: precision and recall rates to assess the validation results. Precision is to determine the number of words divided by cognitive attribute all judged as the ratio obtained by the cognitive attribute word; Recall rate is defined as the number of words and the words of cognitive attributes all feed ratio of the number of cognitive properties of words. They are calculated as follows:

$$P = A / (A + B). \qquad (3)$$

$$R = A / (A + C). \qquad (4)$$

Wherein, the meanings of A, B, C can be seen in Table 4.

Table 4. Binary classification adjacency table

	Words belong to the cognitive attributes	The words do not belong to the cognitive attributes
Determine the properties belonging to the cognitive words	A	B
Determine the properties do not belonging to the cognitive words	C	D

This paper makes the following comparison test: one hundred thousand comments were divided into three groups, the number was 20 000, 30 000, 50 000 comments were tested, the results in the following table5.

By comparing the experimental results found that when the corpus data less accurate detection of low cognitive attributes. But with the increasing amount of data, the precision and recall rates have increased. Overall, this method precision and recall rates are relatively high.

Table 5. Results from comments tested

Reviews	Precision	Recall rate
20000	87.42%	82.32%
30000	92.76%	85.37%
50000	95.49%	87.28%

5   Conclusion

This paper analysis the meaning and significance of consumer's perceived value. Moreover, we proposed to use a data mining method to measure the impact of a product on its consumer's perceived value. We also developed an automatic clustering method which utilities the language model based on Deep Learning. Since K-Means algorithm is sensitive to the initial cluster centers and the parameters of clusters should be set manually. Further research can be start from this aspect.

Acknowledgement. Sponsored by National Key fundamental Research and Development Program No.2013CB329601 and No.2013CB329600.

References

1. Zeithaml, V.A., Parasuraman, A., Malhotra, A.: Service Quality Delivery through Web Sites: A Critieal Review of Extant Knowledge. Academy of Marketing Science **30**, 362–375 (2002)
2. Sweeney, J.C., Soutar, G.N.: Consumer Perceived Value: The Development of a Multiple ItemScale. Journal of Consumer Research **77**, 203–220 (2001)
3. Wei, Z., Guo, C.: Based on analysis of consumer perceived value pricing strategy research. Management World. **89**, 162–163 (2007)
4. Woodruff, R.B.: The Next Source for Competitive Advantage. Journal of the Academy of Marketing Science. **25**, 139–1156 (1997)
5. Woods, W.A.: Consumer Behavior. North-Holland, NewYork (1981)
6. Jie, L., Xiedong, C., Fei, Y.: A Word Semantic Automatic Classification System Based Word on Similarity Computatio. Computer Simulation (2008)
7. Hua, L.: Clustering Field Words by Character Extraction in Text Classification. Applied Linguistics (2007)
8. Tiegang, K., Ruwei, D.: A Novel Approach for Word Clustering Based On Large Tagged Corpus. Computer Simulation (2003)
9. Li, G., Yuan, T.: Based on consumer perceived value of the enterprise product pricing. Bohai University **3**, 23–25 (2009)
10. Changhua, L.: Customer Perceived Value of New Product Pricing Strategies Used in the Application. Modern Busines (2012)
11. Wenchao, Z., Peng, X.: Research on Chinese word Clustering with Word2vec. Computer Engineering. **12**, 160–162 (2013)

Fruit Fly Optimization Algorithm Based SVM Classifier for Efficient Detection of Parkinson's Disease

Liming Shen, Huiling Chen[✉], Wenchang Kang,
Haoyue Gu, Bingyu Zhang, and Ting Ge

College of Physics and Electronic Information, Wenzhou University,
Wenzhou 325035, People's Republic of China
chenhuiling_jsj@wzu.edu.cn

Abstract. In this paper, we present a fruit fly optimization algorithm (FOA) based support vector machine (SVM) classification scheme, termed as FOA-SVM, and it is applied successfully to Parkinson's disease (PD) diagnosis. In the proposed FOA-SVM, the set of parameters in SVM is tackled efficiently by the FOA technique. The effectiveness and efficiency of FOA-SVM has been rigorously evaluated against the PD dataset by comparing with the particle swarm optimization algorithm (PSO) optimized SVM (PSO-SVM), and grid search technique based SVM (Grid-SVM). The experimental results demonstrate that the proposed approach outperforms the other two counterparts in terms of diagnosis accuracy as well as the fewer CPU time. Promisingly, the proposed method can be regarded as a useful clinical decision tool for the physicians.

Keywords: Support vector machine · Parameter optimization · Fruit fly optimization · Parkinson's disease diagnosis · Medical diagnosis

1 Introduction

Parkinson's disease (PD) is one kind of degenerative diseases of the nervous system, it has become the second most common degenerative disorders of the central nervous system after Alzheimer's disease [1]. Till now, the cause of PD hasn't been uncovered. However, it is possible to alleviate symptoms significantly at the onset of the illness in the early stage [2]. It has also been proven that a vocal disorder may be one of the first symptoms to appear nearly 5 year before clinical diagnose [3]. The vocal impairment symptoms related with PD are known as dysphonia (inability to produce normal vocal sounds) and dysarthria (difficulty in pronouncing words) [4]. Therefore, dysphonic indicators may play essential role in the early stage of PD diagnosis. Little et al [5] have made the first attempt to utilize the dysphonic indicators in their study to help discriminate PD patients from healthy ones. In their study, support vector machine (SVM) in combination with the feature selection approach was taken to diagnose PD, the simulation results has shown that the proposed method can discriminate PD patients from healthy ones with approximately 90% classification accuracy using only four dysphonic features. After then, various techniques have been devel-

© Springer International Publishing Switzerland 2015
Y. Tan et al. (Eds.): ICSI-CCI 2015, Part II, LNCS 9141, pp. 98–106, 2015.
DOI: 10.1007/978-3-319-20472-7_11

oped to study the PD diagnosis problem from the perspective of dysphonic indicators, including Artificial Neural Networks (ANNs) [6, 7], SVM [8, 9], Dirichlet process mixtures [10], multi-kernel relevance vector machines [11], similarity classifier [12], rotation forest [13], fuzzy k-nearest neighbor (FKNN) [14].

Among the proposed methods, SVM has shown to be a very promising tool for diagnosing PD. However, in our opinion despite its great potential, SVM has not received the attention it deserves in the PD diagnosis literature as compared to other research fields. SVM was first introduced by Vapnik [15], which has many good properties and has found its application in many fields. However, it has been pointed out that model parameter setting has great impact on the performance of SVM [16]. Values of parameters such as penalty parameter C and the kernel parameter g of the kernel function should be properly tuned before SVM applying to the practical problems. Traditionally, these parameters were handled by the grid-search method and the gradient descent method. However, one common drawback of these methods is that they are vulnerable to local optimum. Recently, biologically inspired global optimization methods such as genetic algorithm and particle swarm optimization (PSO) have been considered to have a better chance of finding the global optimum solution than the traditional aforementioned methods. As a new member of the swarm-intelligence algorithms, fruit fly optimization algorithm (FOA) [17] has been found to be a useful tool for real-world optimization problems such as the semiconductor final testing scheduling problem [18], continuous function optimization problems [19], parameter optimization of generalized regression neural network [20] and Least Squares SVM [21] for regression problems. This study attempts to employ FOA to tackle the parameter optimization of SVM and applied the resultant model for effective detection of PD. To the best of our knowledge, FOA has not been utilized to optimize the parameters of SVM classifier. Therefore, this study will be the first to report the FOA optimized SVM classifier and its application to PD diagnosis.

The remainder of this paper is organized as follows. The detailed implementation of the FOA-SVM method is presented in section 2. Section 3 describes the experimental design. The experimental results and discussion of the proposed approach are presented in Section 4. Finally, Conclusions and recommendations for future work are summarized in Section 5.

2   Proposed FOA-SVM Model

This study proposes a novel FOA-SVM model for parameter optimization problem of SVM. The proposed model was comprised of two procedures as shown in Figure 1, the one is the inner parameter optimization, and the other is the outer performance evaluation. During the inner parameter optimization procedure, the parameters of SVM are adjusted dynamically by the FOA technique via the 5-fold cross validation

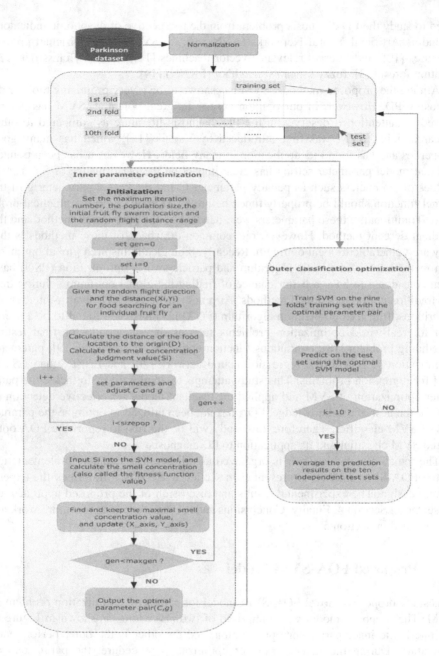

Fig. 1. The flowchart of the proposed FOA-SVM diagnostic system

(CV) strategy. And then the obtained optimal parameters are fed to SVM prediction model to perform the classification task for Parkinson's disease diagnosis in the outer loop using the 10-fold CV strategy.

The classification accuracy is taken into account in designing the fitness:

$$f = avgACC = (\Sigma_{i=1}^{K} testACC_i)/k \tag{1}$$

where $avgACC$ in the function f represents the average test accuracy achieved by the SVM classifier via 5-fold CV strategy.

3 Experimental Studies

3.1 Data Description

The Parkinson's data was taken from UCI machine learning repository (http://archive.ics.uci.edu/ml/datasets/Parkinsons, last accessed: December 2014). The objective of this dataset is to discriminate healthy people from those with Parkinson's disease (PD). In the medical experiment, various biomedical voice measurements were recorded for 23 patients with PD and 8 healthy controls. The time since diagnoses ranged from 0 to 28 years, and the ages of the subjects ranged from 46 to 85 years, with a mean age of 65.8. Each subject provides an average of six phonations of the vowel (yielding 195 samples in total), each 36 seconds in length. It should be noted that there is no missing values in the dataset, and the whole features are real valued.

3.2 Experimental Setup

The proposed FOA-SVM classification model was implemented using MATLAB platform. For SVM, LIBSVM implementation was utilized, which was originally developed by Chang and Lin [22]. We implemented the FOA algorithm from scratch. The computational analysis was conducted on Windows 7 operating system with AMD Athlon 64 X2 Dual Core Processor 5000+ (2.6 GHz) and 4GB of RAM. Before constructing the SVM models, the data was scaled to the range of [0, 1] to avoid the feature values in greater numerical ranges dominating those in smaller numerical ranges.

3.3 Measure for Performance Evaluation

Classification accuracy (ACC), area under the receiver operating characteristic curve (AUC) criterion, sensitivity and specificity were used to test the performance of the proposed FOA-SVM model. ACC, sensitivity and specificity are defined as follows:

$$Accuracy = TP + TN \; / \; (TP + FP + FN + TN) \times 100\% \tag{2}$$

$$Sensitivity = TP \; / \; (TP + FN) \times 100\% \tag{3}$$

$$Specificity = TN \; / \; (FP + TN) \times 100\% \tag{4}$$

where TP is the number of true positives, FN is the number of false negatives, TN is the number of true negatives, and FP is the number of false positives. AUC is the area under the ROC curve, which is one of the best methods for comparing classifiers in two-class problems. in this study the method proposed in [23] was implemented to compute the AUC.

4   Experimental Results and Discussions

The swarm size and number of generations play important role in controlling the search ability of FOA. Thus, we firstly investigated the impact of the three factors on the performance of FOA. Different sizes of swarm from 10 to 22 with step size of 2 were evaluated when the generation is fixed to 60, the detailed results is presented in Table 1. From the table, we can see that the best performance was achieved when the swarm size is 20, where the ACC, AUC, sensitivity and specificity are 95.95%, 92.81%, 98.52% and 87.10%, respectively. When the size of 20 is fixed, different generations from 20 to 120 with step size of 20 were tried. As shown in Table 2, we can see that the best performance of FOA-SVM is achieved when the generations are

Table 1. The detailed results of FOA-SVM with different swarm size

Swarm size	FOA-SVM			
sizepop (when gen = 60)	ACC (%)	AUC (%)	Sensitivity (%)	Specificity (%)
10	94.34(5.08)	91.06(7.99)	97.95(3.30)	84.17(15.02)
12	93.87(4.06)	88.58(7.64)	98.66(2.83)	78.50(15.58)
14	94.79(6.57)	91.95(7.62)	99.33(2.11)	84.57(14.62)
16	94.87(4.23)	91.54(8.57)	97.94(3.34)	85.14(17.94)
18	95.37(5.17)	92.18(11.59)	98.71(2.72)	85.65(23.17)
20	*95.95(4.61)*	*92.81(9.32)*	*98.52(3.13)*	*87.10(19.50)*
22	95.34(5.73)	91.54(122)	99.44(1.76)	83.86(19.32)

The best results have been shown in bold.

Table 2. The detailed results of FOA-SVM with different generations

Generation	FOA-SVM			
Gen (when sizepop = 20)	ACC (%)	AUC (%)	Sensitivity (%)	Specificity (%)
20	93.32(4.19)	85.92(8.86)	99.23(2.43)	72.62(17.5)
40	94.39(5.98)	91.75(8.47)	97.91(3.41)	85.58(16.52)
60	*95.84(5.92)*	*94.75(7.02)*	*99.09(2.87)*	*90.42(13.54)*
80	94.39(3.68)	92.05(9.81)	98.27(1.86)	85.83(19.18)
100	93.87(5.31)	92.01(6.18)	97.62(5.50)	86.39(13.95)
120	93.84(4.81)	89.04(7.16)	98.75(3.95)	79.33(12.43)

The best results have been shown in bold.

set to 60 with the ACC of 95.84%, AUC of 94.75%, sensitivity of 99.09% and specificity of 90.42%. In the above two tables, the average results of 10-fold CV are presented with the standard deviation described in the parenthesis. From the above analysis, we can see that FOA-SVM reaches the best performance when swarm size=20
and generations=60 in terms of ACC, AUC, sensitivity and specificity. Therefore,
these parameter values are adopted for the proposed FOA-SVM to implement the
subsequent experiments.

Apart from the FOA-SVM classifier, PSO-SVM and Grid-SVM classifiers were
implemented for the comparison purpose. For PSO-SVM, the number of the iterations
and particles are set the same as that of FOA-SVM, they are 60 and 20, respectively.
The maximum velocity v_{max} is set about 60% of the dynamic range of the variable on
each dimension for the continuous type of dimensions, acceleration coefficients $c_1 =$
2, $c_2 = 2$, maximum and minimum values of the inertia weight w_{max} and w_{min} are set to
0.9 and 0.4, respectively. The searching ranges of $C \in [2^{\wedge}(-5), 2^{\wedge}(15)]$ and g
$\in [2^{\wedge}(-15), 2^{\wedge}(3)]$ for PSO-SVM and Grid-SVM were set as the same. Table 3
shows the detailed results of ACC, AUC, sensitivity, specificity, and optimal pairs of
(C, g) for each fold obtained by FOA-SVM. The detailed comparison results among
PSO-SVM, Grid-SVM and FOA-SVM in terms of ACC, AUC, sensitivity and specificity are shown in Figure 2. It can be observed from Figure 2 that the performance of
FOA-SVM is superior over the other two competitors in most folds.

Table 3. The detailed results of FOA-SVM on the PD dataset

Fold No.	FOA-SVM					
	Accuracy	AUC	Sensitivity	Specificity	C	g
1	1.0000	1.0000	1.0000	1.0000	1.09637	1.09637
2	1.0000	1.0000	1.0000	1.0000	1.27755	1.27755
3	0.9474	0.9000	1.0000	0.8000	1.7315	1.7315
4	1.0000	1.0000	1.0000	1.0000	1.23075	1.23075
5	1.0000	1.0000	1.0000	1.0000	1.13659	0.988679
6	1.0000	1.0000	1.0000	1.0000	1.07801	1.07801
7	0.8421	0.8500	1.0000	0.7000	1.6111	1.6111
8	1.0000	1.0000	1.0000	1.0000	1.30109	1.30109
9	0.9000	0.8333	1.0000	0.6667	2.34289	2.34289
10	0.8947	0.8920	0.9091	0.8750	2.18101	2.18101
Avg.	0.9584	0.9475	0.9909	0.9042	1.49869	1.48390
Dev.	0.0592	0.0702	0.0287	0.1354	0.45641	0.47159

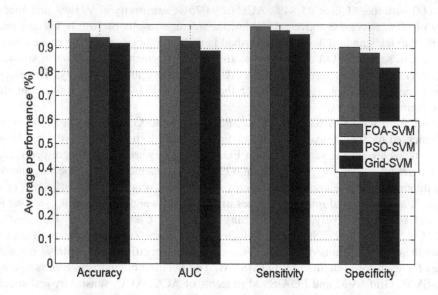

Fig. 2. ACC, AUC, sensitivity and specificity obtained for each fold by PSO-SVM, Grid-SVM and FOA-SVM

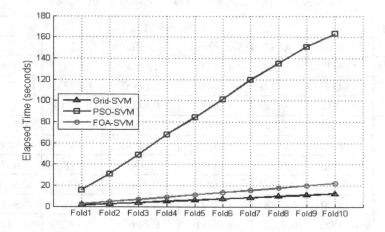

Fig. 3. The comparison results of CPU time for the three methods

In order to investigate the efficiency of the proposed method, we have compared FOA-SVM with other two methods in terms of CPU time. As shown in Figure 3, FOA-SVM needs almost the same CPU times as that of Grid-SVM, PSO-SVM appears to be the most time consuming among the three methods. Since the grid search technique needs no iterative procedure, it is comparatively fast when the search ranges are set properly. It is important to note that FOA-SVM needs only about 12.26 seconds on average during the whole 10 folds CV procedure, even when there are so many generations involved. However, PSO-SVM consumed 91.79 seconds when the same generations and swarm size were considered. From the above analysis, we can see that FOA-SVM has the evident superiority over PSO-SVM in terms of CPU time as well as the classification performance.

5   Conclusions and Future Work

This work has explored a new method, FOA-SVM, for effective and efficient detection of PD. The main novelty of this paper lies in the proposed FOA-based approach, which aims at maximizing the generalization capability of the SVM classifier by exploring the new swarm intelligence technique for optimal parameter tuning for PD diagnosis. The empirical experiments have demonstrated the evident superiority of the proposed FOA-SVM over PSO-SVM and Grid-SVM. It indicates that the proposed FOA-SVM method can be utilized as a valuable alternative clinical solution to PD diagnosis. In the future work, we plan to apply the proposed method to other medical diagnosis problems.

Acknowledgments. This research is supported by the National Natural Science Foundation of China (NSFC) under Grant Nos. of 61303113 and 61402337. This work is also funded by Zhejiang Provincial Natural Science Foundation of China under Grant Nos. of LQ13G010007, LQ13F020011 and LY14F020035.

References

1. de Lau, L.M.L., Breteler, M.M.B.: Epidemiology of Parkinson's disease. The Lancet Neurology 5(6), 525–535 (2006)
2. Singh, N., Pillay, V., Choonara, Y.E.: Advances in the treatment of Parkinson's disease. Progress in Neurobiology 81(1), 29–44 (2007)
3. Harel, B., Cannizzaro, M., Snyder, P.J.: Variability in fundamental frequency during speech in prodromal and incipient Parkinson's disease: A longitudinal case study. Brain and Cognition 56(1), 24–29 (2004)
4. Baken, R.J., Orlikoff, R.F.: Clinical measurement of speech and voice, 2nd edn. Singular Publishing Group, San Diego (2000)
5. Little, M.A., et al.: Suitability of Dysphonia Measurements for Telemonitoring of Parkinson's Disease. IEEE Transactions on Biomedical Engineering 56(4), 1015–1022 (2009)
6. Das, R.: A comparison of multiple classification methods for diagnosis of Parkinson disease. Expert Systems with Applications 37(2), 1568–1572 (2010)
7. Aström, F., Koker, R.: A parallel neural network approach to prediction of Parkinson's Disease. Expert Systems with Applications 38(10), 12470–12474 (2011)
8. Sakar, C.O., Kursun, O.: Telediagnosis of Parkinson's Disease Using Measurements of Dysphonia. Journal of Medical Systems 34(4), 1–9 (2010)
9. Li, D.C., Liu, C.W., Hu, S.C.: A fuzzy-based data transformation for feature extraction to increase classification performance with small medical data sets. Artificial Intelligence in Medicine 52(1), 45–52 (2011)

10. Shahbaba, B., Neal, R.: Nonlinear models using Dirichlet process mixtures. The Journal of Machine Learning Research **10**, 1829–1850 (2009)
11. Psorakis, I., Damoulas, T., Girolami, M.A.: Multiclass Relevance Vector Machines: Sparsity and Accuracy. IEEE Transactions on Neural Networks **21**(10), 1588–1598 (2010)
12. Luukka, P.: Feature selection using fuzzy entropy measures with similarity classifier. Expert Systems with Applications **38**(4), 4600–4607 (2011)
13. Ozcift, A., Gulten, A.: Classifier ensemble construction with rotation forest to improve medical diagnosis performance of machine learning algorithms. Comput. Methods Programs Biomed. **104**(3), 443–451 (2011)
14. Chen, H.-L., et al.: An efficient diagnosis system for detection of Parkinson's disease using fuzzy k-nearest neighbor approach. Expert Systems with Applications **40**(1), 263–271 (2013)
15. Vapnik, V.N.: The nature of statistical learning theory. Springer, New York (1995)
16. Keerthi, S.S., Lin, C.-J.: Asymptotic behaviors of support vector machines with Gaussian kernel. Neural Computation **15**(7), 1667–1689 (2003)
17. Pan, W.-T.: A new Fruit Fly Optimization Algorithm: Taking the financial distress model as an example. Knowledge-Based Systems **26**, 69–74 (2012)
18. Zheng, X.-L., Wang, L., Wang, S.-Y.: A novel fruit fly optimization algorithm for the semiconductor final testing scheduling problem. Knowledge-Based Systems **57**, 95–103 (2014)
19. Pan, Q.-K., et al.: An improved fruit fly optimization algorithm for continuous function optimization problems. Knowledge-Based Systems **62**, 69–83 (2014)
20. Li, H.-Z., et al.: A hybrid annual power load forecasting model based on generalized regression neural network with fruit fly optimization algorithm. Knowledge-Based Systems **37**, 378–387 (2013)
21. Li, H., et al.: Annual Electric Load Forecasting by a Least Squares Support Vector Machine with a Fruit Fly Optimization Algorithm. Energies **5**(11), 4430–4445 (2012)
22. Chang, C.-C., Lin, C.-J.: LIBSVM: A library for support vector machines. ACM Trans. Intell. Syst. Technol. **2**(3), 1–27 (2011)
23. Fawcett, T.: ROC graphs: Notes and practical considerations for researchers. Machine Learning **31**, 1–38 (2004)

A New Framework for Anomaly Detection Based on KNN-Distort in the Metro Traffic Flow

Yapeng Zhang[1], Nan Hu[2(✉)], Wenjun Wang[3], and Hao Yu[3]

[1] School of Computer Science and Technology, Tianjin University, Tianjin 300072, China
whiteshark@tju.edu.cn
[2] Tianjin Emergency Medical Center, Tianjin 300072, China
xiaoxiao@tju.edu.cn
[3] Tianjin Key Laboratory of Cognitive Computing and Application,
School of Computer Science and Technology, Tianjin University, Tianjin 300072, China
{wjwang,yuhao}@tju.edu.cn

Abstract. Anomaly detection is an important problem that has been well researched in diverse application domains. However, to the best of our knowledge, the anomaly detection for metro traffic flow has not been investigated before. In this paper, we proposed a new framework to solve two problems about anomaly detection in the metro traffic flow: obtaining the potential information by every passenger's trip and detecting anomalies among metro traffic flow. For the first problem, we proposed a novel encoding path model to infer the passing stations' information for each trip. For the second problem, we provide an improved K-Nearest Neighbor Distort (KNN-Distort) algorithm to quantify the anomalies in the metro traffic flow. We conduct intensive experiments on a large real-world metro dataset to demonstrate the performance of our algorithms.

Keywords: Anomaly detection · Encoding path model · KNN-Distort · Metro traffic flow

1 Introduction

Anomaly detection, the problem of finding patterns in data that do not conform to expected behavior [1, 2] is a hot research field both in GPS domain [3, 4, 5, 6, 11] and network domain [7, 8, 9, 10]. In the domain of GPS, with the increasing capability to track vehicle and a large volume of spatiotemporal GPS points leads to more and more studies on trajectory mining, especially anomaly detection in traffic data. Bu Y el at. [4] which partitions a trajectory into a set of line segments (sub-trajectories), and then, detects outlying line segments outliers based on an efficient algorithm TRAOD. Zheng el at. [5] have designed novel algorithms of minDistort and spatiotemporal outlier tree (STOtree) detecting the spatiotemporal abnormal regions and the major cause of abnormal region, where they use the map matching to locate the points to the graph of regions. A novel two-step mining and optimization framework, in which

© Springer International Publishing Switzerland 2015
Y. Tan et al. (Eds.): ICSI-CCI 2015, Part II, LNCS 9141, pp. 107–116, 2015.
DOI: 10.1007/978-3-319-20472-7_12

PCA is for the mining step and link-route incidence matrix is for optimization step, is proposed in [6] to infer the root cause of anomalies that appear in road traffic data.

In the domain of network, outlier detection is an increasingly fatal component of any network security infrastructure. The approach of PCA is used to diagnose outliers in [7, 8]. Brauckhoff D el at. [8] introduced an approach based on a predictive filter (kalman filter) and remove completely the curtain of black magic that draped the application of PCA to anomaly detection and detection results are significantly improved based this approaches. A new network signal modelling technique is proposed in [9] for detecting network anomalies, which combine the wavelet approximation and system identification theory.

However, to the best of our knowledge, none of those papers provide one useful and efficient method or model to process metro traffic data. In order to explain why those methods can not adapt to the metro data directly, we describe the challenges and our contributions.

Challenges and Contributions. Now, we describe our challenges which those algorithms could not meet, the following challenges need to be addressed: (i) Special metro OD records: unlike the traffic trajectory with a sequence of GPS points, the recording device of metro stations can only record the passengers trip with the original and destination metro stations' information, but can not record the passing stations' information, namely we can not obtain the exact passing stations ID and passing time directly. (ii)Metro lines segment anomaly: we not only need to detect anomalies of the real metro link, but also need to discover the exact point anomalies (the exact metro station), which is different from detecting anomalies of the region to region in [5, 6]. Thus, there is a challenge how to discover the outliers among metro traffic data by time.

In this paper, we design several steps to address the above challenges and present our solutions to the problem of inferring the waiting time and detecting anomalies among metro OD records. The followings are our contributions:

1. Urban metro traffic modeling: we rebuild the metro stops by their longitude and latitude collected from the Map of Google and draw lines between two geographically adjacent stops. We use a unique number as every station's ID.

2. Encoding metro records: we propose both passengers-waiting time model and path coding method to encode metro records, which can infer the passing stations' information of one passenger trip records.

3. Validating our framework: Our framework is validated by using nearly 15,300,000 metro OD records of all the four metro lines in Tianjin Metro. We are able to uncover real events which caused perturbations through our approach.

The rest of this paper is organized as follows: In Section 2 we define preliminary concepts and the problems. In Section 3 we propose methodology which contains three phases of building metro network, encoding path and anomaly detection with optimization techniques. The experimental setup and evaluation is described in Section 4. We conclude in Section 5 with a summary and direction for future works.

2   Preliminary Concepts

We define a directed graph $G = (G; E)$, representing the real metro traffic lines network; And V denotes a set of vertexes (metro stations) in the metro network and E denotes a set of directed edges (metro line segments) in the metro network.

Definition 1. OD: An OD is comprised of a pair of metro station, indicating a real spatial connection between the origin metro station and the destination metro station. This OD represents a real trip which one passenger is in and out the metro.

Definition 2. Time Bin: A time bin (TB) is a relatively fixed time intervals. Each time bin is 30 minutes.

Definition 3. Link: A link(Lnk) is comprised of a pair of metro stations ($<MS_i$, $MS_{i+1}>$) indicating the real spatial connection between two geographically adjacent metro stops. Fig. 2(b) gives examples of links.

Problem Definition. In order to define anomalies more accurately, in the paper, all timestamps are divided based on 30-minute time bins. A 17-hour metro running time of every day has 34 time bins. And the first time bin is from 06:00 to 06:30, the second time bin is from 06:30 to 07:00, and the last time bin is from 22:30 to 23:00. Then, based on days, all these time bins are partitioned into several groups. In this paper, we perform two partitioning strategies, namely the workday-weekend partitioning strategy and the day-of-the-week partitioning strategy. These two well-known strategies are also used in [5]. For instance, using the workday-weekend partitioning strategy, we create two groups where one group is for the workdays and the other is for the weekends, and each group contains a set of time bins. For this partitioning strategy, there are 34 x 2 = 68 possible time bins. In the following, for the sake of space, we just describe one way of partitioning strategy (e.g., the workday-weekend partitioning strategy).

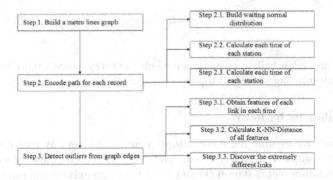

Fig. 1. The overall framework of our model for encoding path and detecting anomalies

To this end, we study the anomaly detection of metro OD with the following two goals. The first goal is metro path encoding. That is, given an OD record of one passenger, we infer all the passing stations and the time passing these stations by using a special method. The second goal is to find the metro lines segments with abnormal traffic flow based on the previous goal.

3   Methodology

In this section, we describe our model in detail as shown in Fig. 1. Firstly, we recon-struct the metro line network with a set of geographical coordinates of metro stations (Section 3.1). Secondly, we proposed our path-encoding algorithm based on shortest path method and passenger waiting sub-model for the first goal (Section 3.2). Thirdly, we introduce an improved K-NN method (KNN-Distort) for the second goal (Section 3.3).

3.1   Building Metro Network

In this paper, in order to rebuild the metro network usefully and simply, we obtain all the geographical coordinates of metro stations from Map of Google and formulate each metro station.

As show in the Fig. 2(a), we use a unique number as every station's ID, and for each metro line, the number ID is in descending order. And now we get a new metro network with four metro lines, eighty-six vertex (each vertex represent one metro station). In the following, for clarity, when we describe a metro network, we mean the sequence obtained after the station formulating operation.

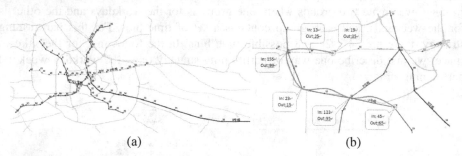

| (a) | (b) |

Fig. 2. Metro network is built by using unique number as every station's ID (subfigure (a)). An example of path encoding model on a trip OD (subfigure (b)).

3.2   Encoding Path Model

In this section, we assume the metro running time between any two geographically adjacent metro stops is relatively fixed. And one OD records the original metro sta-tion and destination metro station ID and timestamp, namely in and out of metro sta-tions' ID and timestamp.

In the recent studies, the method of map matching in the GPS domain can not be applied to metro recorded OD directly. So Instead of using map matching, we pro-posed a new path encoding model as illustrated in Fig. 2. In details, we build a metro segment graph according to the following two steps.

Building Waiting Sub-Model. According to the website of the traffic administrative department of the metro, the time interval between two adjacent metro trains is 6 to 9 minutes in one day, and passenger coming into metro station has the characteristics of random. In order to precisely estimate the waiting time of each passenger in entering station, we established our waiting time model. Firstly, we defined the time as bellow:

$$T_{od} = T_{wait} + T_{run} + T_{trans} + T_{out} + \varepsilon . \tag{1}$$

Where T_{od} represents the time interval between the entering station (original station) time and the out station (destination station) time, T_{wait} is time of waiting the metro train, T_{run} is the running time. T_{trans} is the time of passenger transfer from one metro line to another(i.e., transfer from line 1 to line 2). T_{out} is the time of out of the destination. ε is an infinitesimal. Although this method is simple, it is useful in practice in metro station (which will be shown in our experiments).

Secondly, Given a OD, the running time T_{run} and the time interval T_{od} is known, and we can assume the T_{out} is a relative small constant. So we can obtain:

$$T_{wait} + T_{trans} = T_{cons} - \varepsilon . \tag{2}$$

where $T_{cons} = T_{od} - T_{run} - \lambda$, is a constant, and ε is a small and negligible constant.

Thirdly, as we defined above, T_{wait} is the waiting time in the entering station on one metro line and the T_{trans} also can be considered as the waiting time in the transfer station on another station. So we can obtain:

$$T_{wait} + T_{trans\text{-}wait} = T_{cons} - \varepsilon . \tag{3}$$

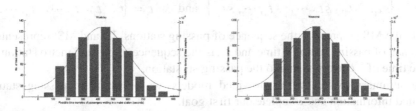

Fig. 3. Histogram on all possible time samples of passengers waiting in a metro station

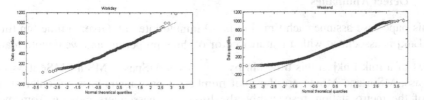

Fig. 4. Q-Q plot of all possible time samples of passengers versus a Gaussian Distribution

For these two waiting time (T_{wait} and $T_{trans-wait}$), we want to draw two figures to study the distribution of the passenger waiting time. Fig. 3 shows this histogram on our real training dataset. This figure is the histogram on all possible waiting time of this training dataset for one metro station. In this figure, the x-axis denotes all possible waiting time of this training dataset from 0 minute to 6 minutes, and the y-axis denotes the number of passengers which have the corresponding possible waiting time for this metro station. Fig. 4 shows the Q-Q plot on our real dataset. In the figure, the x-axis corresponds to the normal theoretical quantiles and the y-axis corresponds to the data quantiles. Each point (x, y) in the figure means that one of the quantiles from the Gaussian distribution is x and the same quantile from the data is y.

There are two following observations based on two figures above. Firstly, from the histogram, we observe that the passenger waiting time in our real dataset looks like a Gaussian distribution. Secondly, from the Q-Q plot, we observe that all points in the plot lie on a single line, which means that the distribution on the passenger waiting time is similar to a Gaussian distribution. Based on these two observations, we can model passenger waiting time as a Gaussian distribution.

In summary, based on all passengers waiting time for one station, we can obtain the Gaussian distribution. And based on this Gaussian distribution and EM, we can obtain the time (T_{trans}) of transfer from one line to another for each OD record.

Formulating OD. The passing stations' information can be calculated by using our passenger waiting sub-model and shortest path method. Given one OD with original station and destination station, we can use our Shortest path method to infer the most likely trip way and every passing stations along this trip way. Then use our passenger waiting sub-model to estimate the time of passing every station. Based on these methods, we can formulating every OD to two 86 - dimensional vectors (4 metro lines and 86 stations):

$$MSn = \{st_1, \ st_2, \ st_3, \cdots, \ st_n, \cdots, st_{86}\} \text{ and } MSt = \{t_1, \ t_2, \ t_3, \cdots, \ t_n, \cdots, t_{86}\}$$

where the MSn represents the sequence of passing stations' ID and MSt represents the sequence of passing stations' time and st_n is the sequence of passing metro station , t_n is the time of   corresponding to the passing st_n station.

In summary, based on our provided models, we obtain these passing stations' detailed information and complete our first goal.

3.3   Detect Anomalies

In this paper, we assume each time bin is 30-minute interval. Given a time bin tbi , a link Lnkj is associated with a feature vector of three properties and we denote feature values of a link Lnkj in this time bin by: $f\vec{v}_{i,j} = $ < MSpass , MSin , MSout >where MSpass, MSin and MSout are the total number of passengers passing, entering and out of the metro station, respectably. the links (i.e. passengers moving from metro station j to metro station j+1 in this time-bin tbi). And we use Fig. 2(b) as an example (where the number shown is the number of passengers relied on this link).

For each link (Lnkj) in each time bin tbi, we calculate the distort (denoted by KNN-Disti,j) by searching for the mean value of k-nearest neighbor distance (Euclidean distance) between tbi and the same time bins of the different days of workday or weekend . We use KNN-Disti,j obtained from Algorithm KNN-Distort as the score of distort for each link in each time bin. Extreme values among KNN-Disti,j all links are identified as anomalies. In the algorithm, we obtain distort by searching for the mean value of k-nearest neighbor distance:

$$KNN\text{-}Dist(tb_i, Lnk_j) = \frac{1}{k}\sqrt{\sum_{p=1}^{k}\left\|f\bar{v}_{i,j} - f\bar{v}_{p,j}\right\|^2} . \tag{4}$$

By subtracting the min and dividing by the max the feature values of the links are in the range of [0, 1]. The normalization removes the effect of different metro stations and number sizes. Another advantage of using KNN-Disti,j that it prevents the examination of many repeating patterns (where KNN-Disti,j≈ 0).

Based on the algorithm KNN-Distort, Given a predefined support threshold θ, we can find the metro links whose supports exceed from all links in each time bin. We see this metro links as anomalies in that time bin.

4   Experiments and Analysis

4.1   Data and Set-Up

We used passengers in and out of metro dataset which contains records of all metro lines in Tianjin City within 3 months (from March 1, 2014 to May 30, 2014). In this dataset, one OD record corresponding to one passenger trip only has original station ID, destination station ID and two times corresponding to these two stations.

We compared our algorithms that are Encoding Path Model and KNN-Distort, with the two state-of-the-art algorithms, namely PCA [5] and minDistort [6]. Encoding Path Model based on both passengers-waiting sub-model and Formulating OD algorithm is used to evaluate the passengers' waiting time in the original station. And Detect Anomalies based on our KNN-Distort with lower time complexity is used to diagnose extreme distortion in the metro traffic flow.

4.2   Performance Study

In this section, in order to prove that our methods are more effective and more efficient than others, we carried out experiments to compare our proposed algorithms (i.e., Encoding Path Model and KNN-Distort) with the two state-of-the-art algorithms (i.e., PCA and minDistort).

Firstly, we validate our Encoding Path model is effective and in high probability by comparing our two obtaining 86-dimensional vectors (Section 3.2) with the path of trip provided by Map of Google. We find our model performs perfectly due to the fixed running time between two geographically adjacent stops.

Secondly, our Encoding Path model can be realized and applied very well in of-fline setting, which has the application potential in practice. In order to evaluate and measure the good performance of these algorithms, we have to obtain the ground truth in advance. In these experiments, we regard two types of events as the ground truth of metro traffic anomalies, namely a metro traffic event, which is closely related to a metro operation equipment failure, and a holiday event, which is closely related to an extremely different metro traffic flow due to public holidays. We collected the events from four sources: (1) Tianjin News online edition (http://www.tianjinwe.com/), (2) The official micro-blog of Tianjin metro (http://www.weibo.com/tjdtyy) and (3) Wik-ipedia (http://zh.wikipedia.org/zh-cn/). There are altogether 21 traffic events and 10 holiday events collected from the above four sources.

For each event type, we partition the dataset of metro OD records into a number of groups according to the two partitioning strategies. And for each group, we use differ-ent algorithms to analysis and detect the extremely different metro traffic flows, name-ly anomalies. Then, we employ F1-Measure, an accuracy measure of the weighted harmonic mean of both precision rate and recall rate, to evaluate and measure the good performance of each algorithm. Specifically, we define F1-Measure as follows.

$$F1 = \frac{2 \cdot precision \cdot recall}{precision + recall}. \tag{5}$$

Fig. 5 illustrates our experimental results obtained from our F1-Measure scores. There are the following observations according to the Fig. 5. Our KNN-Distort algorithm based on Encoding Path Model consistently performs the best for all event types. This obviously indicates our novel framework based on Encoding Path Model and KNN-Distort method is effective and superior over the existing methods in capturing metro traffic anomalies.

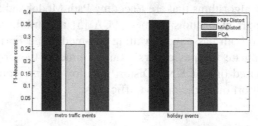

Fig. 5. The test F1-Measure scores of different algorithms for different event types

We can also observe that our method KNN-Distort usually performs better than the existing methods minDistort [6] and PCA [5] according to the results. For example, for the holiday event type, the average precisions (recalls) of KNN-Distort, minDistort and PCA are 0.280 (0.538), 0.267 (0.308) and 0.215 (0.240), respectively.

4.3   Case Study

In this section, we analysis two cases about real world events and show two promi-nent examples of known events as follows.

Event 1. The first event is the "Qingming Festival" (April 5, 2014). Qingming Festival is an important Chinese traditional festival for worshiping ancestors.

Event 2. The second event is an "a metro traffic event in the metro line 3" (April 10, 2014). This metro operation equipment failure happened in some metro line segment of metro line 3 in Tianjin during the morning rush-hour on April 10, 2014.

The results for Event 1 are illustrated in Fig. 6(a), where the color of a metro road segment represents the value of anomaly in this segment. And a darker color and thicker road segment indicates that the corresponding metro road segment has a higher anomaly value. As shown in this subfigure, the traffic of the metro lines segments in darker color near to some cemeteries on the Qingming Festival were identified as abnormal, which could be explained by the phenomenon that on the Qingming Festival, a lot of passengers went to cemeteries to worship their ancestors.

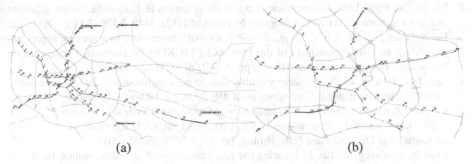

(a)   (b)

Fig. 6. Results between 10:00am to 10:30am, on April 5, 2014 for Case Study 1(subfigure (a)). Results between 7:30am to 8:00am, on April 10, 2014 in line 3 for Case Study 2(subfigure (b)).

The results for Event 2 are illustrated in Fig. 6(b). In this subfigure, we can observe that the traffic in some metro line segments in line 3 were abnormal during the morning rush-hour on April 10, 2014, which can provide us the information where the metro traffic event happened (Fig. 6(b)) and the information how the metro operation equipment failure affected the traffic of other metro lines segments.

5   Conclusion and Future Works

In this paper, we have proposed a novel framework to analyze the metro data, to build Encoding Path Model for calculating the waiting time, and to detect anomalies based on KNN-Distort algorithm. The proposed framework consists of two steps: 1) encoding path for every passenger's OD records, and 2) detecting anomalies on the metro network for each time bin. A novel passengers-waiting sub-model and an improved KNN-Distort algorithm were introduced to meet the two steps, respectively. With a large amount of the metro data collected from the metro lines in Tianjin, China about three months, we conducted comprehensive experiments to validate our methods. Based on these methods we were able to identify real and valid instances of anomalies in the metro traffic flow. This suggests that our approach has the potential of contributing to a new data driven approach towards metro OD data analysis. In the future, we plan to discover more abnormal patterns from the metro OD data.

Acknowledgements. We thank the support of the Major Project of National Social Science Fund (14ZDB153),the National Science and Technology Pillar Program (2013BAK02B06 and 2015 BAL05B02), Tianjin Science and Technology Pillar Program (13ZCZDGX01099, 13ZCD ZSF02700), National Science and Technology Program for Public Well-being (2012GS120302).

References

1. Chandola, V., Banerjee, A., Kumar, V.: Anomaly detection: A survey. ACM Computing Surveys (CSUR) **41**(3), 15 (2009)
2. Patcha, A., Park, J.M.: An overview of anomaly detection techniques: Existing solutions and latest technological trends. Computer Networks **51**(12), 3448–3470 (2007)
3. Bu, Y., Chen, L., Fu, A.W.C., et al.: Efficient anomaly monitoring over moving object trajectory streams. In: Proceedings of the 15th ACM SIGKDD International Conference on Knowledge Discovery and Data Mining, pp. 159–168. ACM (2009)
4. Lee, J.G., Han, J., Li, X.: Trajectory outlier detection: a partition-and-detect frame-work. In: IEEE 24th International Conference on ICDE 2008, pp. 140–149. IEEE (2008)
5. Liu, W., Zheng, Y., Chawla, S., et al.: Discovering spatio-temporal causal interactions in traffic data streams. In: Proceedings of the 17th ACM SIGKDD International Conference on Knowledge Discovery and Data Mining, pp. 1010–1018. ACM (2011)
6. Chawla, S., Zheng, Y., Hu, J.: Inferring the root cause in road traffic anomalies. In: ICDM (2012)
7. Lakhina, A., Crovella, M., Diot, C.: Diagnosing network-wide traffic anomalies. ACM SIGCOMM Computer Communication Review **34**(4), 219–230 (2004)
8. Brauckhoff, D., Salamatian, K., May, M.: Applying PCA for traffic anomaly detection: problems and solutions. In: IEEE INFOCOM 2009, pp. 2866–2870. IEEE (2009)
9. Lu, W., Ghorbani, A.A.: Network anomaly detection based on wavelet analysis. EURASIP Journal on Advances in Signal Processing **2009**, 4 (2009)
10. Muniyandi, A.P., Rajeswari, R., Rajaram, R.: Network anomaly detection by cascading k-Means clustering and C4. 5 decision tree algorithm. Procedia Engineering **30**, 174–182 (2012)
11. Lan, J., Long, C., Wong, R.C.W., et al.: A new framework for traffic anomaly detection. In: 2014 SIAM International Conference on Data Mining (SDM 2014), Philadelphia, Pennsylvania, USA, pp. 875–883 (2014)

Improving OCR-Degraded Arabic Text Retrieval Through an Enhanced Orthographic Query Expansion Model

Tarek Elghazaly

Institute of Statistical Studies and Research, Cairo University, Giza, Egypt
tarek.elghazaly@cu.edu.eg

Abstract. This paper introduces an Enhanced Orthographic Query Expansion Model for improving Text Retrieval of Arabic Text resulting from the Optical Character Recognition (OCR) process. The proposed model starts with checking the query word through two word based a word based error synthesizing sub-models then in a character N-Gram simulation sub-model. The model is flexible either to get the corrected word once it finds it from the early stages (in case of highest performance is needed) or to check all possibilities from all sub-models (in case of highest expansion is needed). The 1st word based sub-model that has manual word alignment (degraded & original pairs) alone has high precision and recall but with some limitations that may affect recall (in case of connected multi-words as OCR output). The second words based sub-model provides high precession (less than the 1st one) but also with higher recall. The last sub-model which is a character N-gram one, provides low precision but high recall. The output of the proposed orthographic query expansion model is the original query extended with the expected degraded words taken from the OCR errors simulation model. The proposed model gave a higher precision (97.5%) than all previous ones with keeping the highest previous recall numbers.

Keywords: Multi-Layer OCR errors degraded text · Arabic OCR degraded text retrieval · Arabic OCR-Degrade text · Orthographic query expansion · Synthesize OCR-Degraded text

1   Introduction

There are two ways for Improving OCR-Degraded Text Retrieval as per Darwish [1]. First, by correcting the OCR errors on the word or passage-level (post processing) through word n-grams, word collocations, grammar, conceptual closeness, passage level word clustering, linguistic context, and visual context but it may generate other mistakes. The second way is taken is to search the OCR-Degraded text without correction through Orthographic Query Expansion (by finding different misrecognized versions of a query word) which is the scope of this paper.

© Springer International Publishing Switzerland 2015
Y. Tan et al. (Eds.): ICSI-CCI 2015, Part II, LNCS 9141, pp. 117–124, 2015.
DOI: 10.1007/978-3-319-20472-7_13

2   Reviewing the Previous Work: Accuracy and Limitations

The main previous models introduced (up to the best knowledge of the author) are the character based Model (Darwish 2003) as described in [1, 3, 4, 5], the word based model (Elghazaly 2009) as described in [2,7,8], and the Word/Character based Model (Ezzat 2013-2014) as described in [9,10].

ElGhazaly model produced accuracy 84.74%, He trained model on dataset of 53,787 words, and then he tested his model of a test set contained 51,658 words [7]. Elghazaly introduced his own accuracy measure, which is the number of accurate replacements divided by the total number of OCR-Degraded words.

On the other hand Darwish model produced accuracy for 3- gram or 4 gram character indexing was 87% [1, 11]. Darwish Also illustrated that the best mean average precision of his model which was "0.56" [1, 12]. However, Darwish measurements was character based which could be decreased dramatically when measuring as word based (100 mistaken characters from 1000 character means 90% correct character based while means 50% correct word based if number of words is 200.

Although Ezzat's Model overcame many limitations of both Elghazaly and Darwish's Models, it lacks other advantages of them like the losing the guaranteed training set of Elghazaly and losing any combination of multi-words being one word in the degraded text.

3   The Proposed OCR-Degraded Arabic Text Retrieval Model

The proposed model works through two steps; first to synthesize the OCR-Degraded text, and the second is to expand the search query using the expected OCR. The synthesizing part works on three different layers (sub-models); a manually aligned word based synthesizing sub-model, an automated word based one, then a character based one.

3.1   The Manually Aligned Word Based OCR Degradation Synthesizing Sub-Model

In this Sub-model, words have been aligned manually to make sure of the pair (original work, degraded word). The model starts with checking every word and the corresponding degraded one. If they are the same, it ignores this pair as it has no value. Otherwise, it stores that pain in the degraded word training database. This Sub-Model is illustrated in Fig 1.

3.2   The Automatically Aligned Word Based OCR Degradation Synthesizing Sub-Model

This sub-model automatically aligns the degraded word/original word by calculating the edit distance between both words. In order to solve the challenge of Arabic words that have prefixes, it ignores the Arabic prefixes letters while computing the edit distance between any words.

This sub-model is built to be trained on different recognition accuracy returned from the OCR system; which means that user of the model can specify the deformation level he wants to be covered on his training data set by allowing specifying the min edit distance to consider the degraded word/its original word. The implemented alignment application is called the "Aligner" throughout this paper. Fig 2 illustrates the word based OCR-degradation synthesizing model.

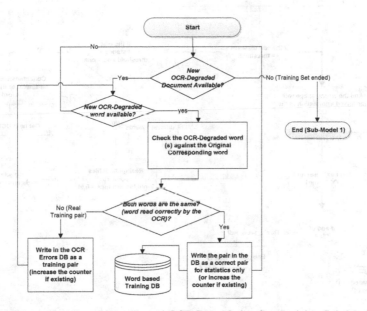

Fig. 1. The Manually Aligned Word based OCR Degradation Synthesizing Sub-Model

There are three cases should be handled. The first case is when the edit distance is within the accepted value specified by the model user in the application interface; so the model adds both words to the training database.

The second case means that either the degraded word is part of the original word but the OCR application split it during the recognition process, which happens sometimes for long words, or the recognition process was bad for this word and the system recognized only few characters of the word. It is decided to cover this limitation in the proposed model. This sub-model considers these parts as one word and stores them in the training database as one shape corresponding to the original shape.

Finally, the third case , if the edit distance between both words is larger than one, which means that both words are completely different, this means that the model has lost the correct position of anchors corresponding to each file, in this case this sub-model realigns the anchors until both files anchors are pointing on the same word. The realignment is done by fetching both words, the clean and degraded word, and if the edit distance is zero, this means that both words are identical, this means that the current anchors positions are correct, and otherwise words will be continuously fetched words until the edit distance is equal to zero.

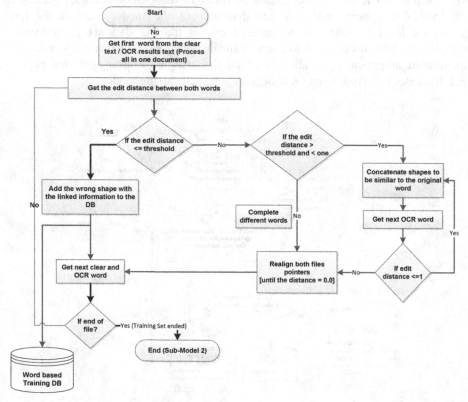

Fig. 2. The Automatically Aligned Word based OCR Degradation Synthesizing Sub-Model

3.3   The Character Based Sub-Model

Because getting the alternative degraded shapes of the word using the word based sub-model is limited by the training set size, this character based sub-model that covers the other cases where the user search query words don't exist in the training dataset is presented. The character based sub-model uses the same training dataset that is used in the word based degradation sub-models.

Using the modified alignment tool, the model compares every degraded word produced from the OCR operation with the corresponding clean word. And if the edit distance between both words is less than the specified value in the alignment tool, the model checks the word characters character by character. The model then stores the deformed character in the database. The tool stores the proceeding and succeeding character of the deformed character. Fig 3 illustrates this Sub-Model.

3.4   The Orthographic Query Expansion Model

In this model, the query builder gets each word in the user search query and checks if there are related degraded shapes in the training database generated from the word based model, and then adds these shapes to the search query, otherwise it uses the "Query Degrader" which generates the word degraded shapes based on the character based model. Fig 4 illustrates the model.

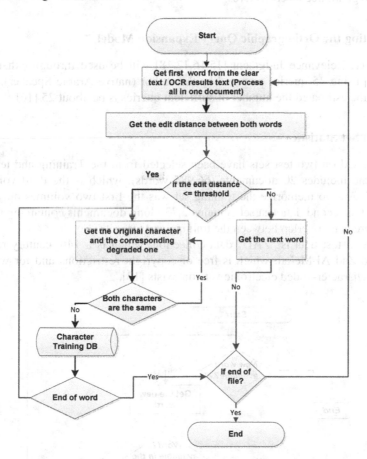

Fig. 3. The character based OCR Degradation Synthesizing Sub-Model

3.5   Training the OCR Degradation Synthesizing Model

The proposed model is trained on the electronic version of the Arabic three volume book "Abgad El Eloum" or "Alphabet of science" and called hereinafter "ABGAD". For the experiment the first two volumes of the book are stored as the original text, printed (containing 205,761 words), then scanned and OCR-ed the first two volumes using the Arabic OCR application, Sakhr Automatic reader version 10. This OCR application produces 99.8% accuracy for high-quality documents 96% accuracy for low-quality documents [14]. The generated database constructed from 1750 documents (305,586 words), consists of 397, 50 unique words, 193,756 words OCR-ed correctly and 230, 49 word OCR-ed wrong.

It worth mentioning that, the 1st word based sub-model has been trained on approximately only 20% of the training set. To do the manual alignment, every word has been put in one line and then the OCR done. Alignment has been reviewed manually to guarantee its correctives.

3.6   Testing the Orthographic Query Expansion Model

In this paper, Relevance Judgment [15,16,17,18] will be used through exhaustively search [16] using 35 queries prepared by the author (native Arabic Speaker). As per Voorhees, the estimated the number of sufficient queries is be about 25 [16].

3.7   Test Set Statistics

Tests are based on two test sets have been selected from the Training and test pool. The first one includes 20 documents (66,985 words), which is the third volume of "ABGAD". And to memorize the training set was the first two volumes only of the book. And the second data test set contains 2,730 long documents containing 621,763 words. There is no overlap between the training and test sets.

The second test data is "ZAD" data collection ,which is a 14th century religious book called Zad Al-Me'ad, which is free of copyright restrictions and for which an accurately character-coded electronic version exists [13].

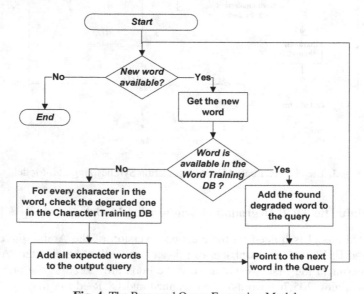

Fig. 4. The Proposed Query Expansion Model

3.8   Testing the Model Accuracy

For the experiments, the clean (original) and OCR-Degraded documents are indexed on the Arabic search engine, IDRISI 6.0[14].One of IDRISI features is creating separate search collection for different group of documents depending on the user requirements. So a separate collection is created for the clean text documents and another collection for the OCR-Degraded documents. The author of the paper, a native speaker of Arabic, developed 35 topics and exhaustively searched the collection for relevant documents.

The clean user search query is passed to the IDRISI clean text collection and the same query passed to the "Query Degrader" tool that takes the user clean text query and based on the training database generates the relevant OCR-Degraded text query.

Both the clean search query and the OCR degraded query are passed to IDRISI to search the clean text collection and the OCR-Degraded text collection separately to be able to compare the effectiveness of the retrieval of the degraded query on the degraded text against the clean query on the clean text.

Using this way, results of both collections could be compared and the precision and recall to check the proposed model accuracy could be measured.

After completing the experiment on the 35 queries, results are analyzed. For test data set 1, the number of relevant documents per topic ranged from one (for one topic) to eighteen, averaging 14. For test data set 2 the number of relevant documents per topic ranged from two (for one topic) to 224, averaging 121. The average query length used for the test data set 1 is 4 .1 words and for test data set 2 is 5.2words.

The proposed model produced mean average precision of 97.5% for test data set 1 and 94% test data set 2 is which means a significant improvement in the information retrieval effectiveness as it is higher than Ezzat's Model using the same Training and Test Sets (96%). It is also much more that Elghazaly (85% accuracy) and Darwish's model (56%) and. However, it was not possible to have same Training and Test Sets of the last to make sure of the exact difference.

4   Conclusion

The model is trained on 1500 documents containing 205,696 words, consists of 38,840 unique words, 182,647 words read correctly and 230,49 word read wrong. The model was tested on two different data sets. The first data set consisted only of 20 documents containing 66,985 words and the second test data set consisted of 2,730 separate documents containing 621,763 words. There were no intersection between the training data sets and test data sets.

As it was not possible get the same training & test sets of Darwish's Model and to unify it with Elghazaly's one, it is taken into consideration the same training & test set of Ezzat's Model in which he tried to re-implement Darwish's Model and enhance Elghazaly's one. And so, his numbers are taken as the baseline to check the enhanced results.

The proposed model produced mean average precision of 97.5% for test data set 1 and 94% test data set 2 is which means a significant improvement in the information retrieval effectiveness as it is higher than Ezzat's Model using the same Training and Test Sets (96%). It is also much more that Elghazaly (85% accuracy) and Darwish's model (56%).

References

1. Darwish, K.: Probabilistic Methods for Searching OCR-Degraded Arabic Text. A PhD Dissertation. University of Maryland, College Park (2003)
2. Elghazaly, T.: Cross Language Information Retrieval (CLIR) for digital libraries with Arabic OCR-Degraded Text. A PhD Dissertation, Cairo University, Faculty of Computers and Information (2009)
3. Darwish, K., Doermann, D., Jones, R., Oard, D., Rautiainen, M.: TREC-10 Experiments at University of Maryland- CLIR and Video. Technical Report. University of Maryland, College Park (2002)
4. Darwish, K.: Building a shallow morphological analyzer in one day. In: ACL Workshop on Computational Approaches to Semitic Languages (2002)
5. Darwish, K., Oard, D.: CLIR Experiments at Maryland for TREC 2002: Evidence Combination for Arabic-English Retrieval (2002)
6. Blando, L.: Evaluation of Page Quality Using Simple Features. A Master Thesis, University of Nevada, Las Vegas (1994)
7. Elghazaly, T., Fahmy, A.: Query translation and expansion for searching normal and OCR-degraded arabic text. In: Gelbukh, A. (ed.) Computational Linguistics and Intelligent Text Processing 2009. LNCS, vol. 5449, pp. 481–497. Springer, Heidelberg (2009)
8. Elghazaly, T., Fahmy, A.: English/Arabic Cross Language Information Retrieval (CLIR) for Arabic OCR-Degraded Text. Communications of the IBIMA **9**, 208–218 (2009)
9. Ezzat, M., Elghazaly, T., Gheith, M.: An enhanced arabic OCR degraded text retrieval model. In: Castro, F., Gelbukh, A., González, M. (eds.) Advances in Artificial Intelligence and Its Applications. LNCS, vol. 8265, pp. 380–393. Springer, Heidelberg (2013)
10. Ezzat, M., Elghazaly, T., Gaith, M.: A Word & Character N-Gram based Arabic OCR Error Simulation model. International Journal of Computers & Technology **12**, 3758–3767 (2014)
11. Darwish, K., Oard, D.: Term Selection for Searching Printed Arabic. SIGIR (2002)
12. Darwish, K., Oard, D.: Probabilistic Structured Query Methods. SIGIR (2003)
13. Ibn-al-Qayyim.: Zzad Al Ma'ad. AlResala, Damascus, Syria (1998)
14. Sakhr Software, http://www.sakhr.com
15. Soboroff, I., Nicholas, C., Cahan, P.: Ranking retrieval systems without relevance judgments. SIGIR (2001)
16. Voorhees, E.: Variations in Relevance Judgments and the Measurement of Retrieval Effectiveness. SIGIR (1998)
17. Wayne, C.: Detection & tracking: a case study in corpus creation & evaluation methodologies. In: Language Resources and Evaluation Conference (1998)
18. Tseng, Y., Oard, D.: Document image retrieval techniques for chinese. In: Symposium on Document Image Understanding Technology (2001)
19. Lesk, M., Salton, G.: Relevance Assessments and Retrieval System Evaluation. Information Storage and Retrieval **4**, 343–359 (1969)

Information Security

Framework to Secure Data Access in Cloud Environment

Rachna Jain[1(✉)], Sushila Madan[2], and Bindu Garg[3]

[1] Department of Computer Science, Banasthali Vidyapith University, Banasthali, India
rachnabtec@gmail.com
[2] Department of Computer Science, Delhi University, New Delhi, Delhi
sushila.madan@gmail.com
[3] Department of Computer Science, IP University, Dwarka, Delhi
bindugarg80@gmail.com

Abstract. Cloud computing is the key powerhouse in numerous organizations due to shifting of their data to the cloud environment. According to IDC survey, Security was ranked and observed first utmost issue of cloud computing. As a result, protection required to secure data is directly proportional to the value of the data. The major handicap of first level of security where cryptography can help cloud computing i.e. secure storage is that we cannot outsource the processing of the data without decryption. In this paper, a novel framework to secure data access in cloud environment is implemented. Here security is addressed for securing transaction in such a way that transaction should be encrypted and decrypted by data owners only. Server performs equality, addition and subtraction on encrypted data without decryption. Moreover, access should be provided to the users as per their access rights. Security is enhanced by utilizing the concept of multicloud.

Keywords: Cloud Computing · Security and Privacy · Homomorphic Encryption · Cellular Automata · AES · Hierarchical Encryption · Multi-Cloud

1   Introduction

Cloud computing is becoming pertinent technology due to its style of computing where user can use applications and software on the Internet that stores and protect the data while providing a service. Additionally, cloud computing is being attractive to business owners as it eliminates the imminent plan for provisioning of resources. At present, cloud computing is defined by numerous organizations in their own way such as National Institute for Standards and Technology (NIST) [1] describes the Cloud computing as "a model for enabling convenient, on demand network access to a shared pool of configurable computing resources (e.g., networks, servers, storage, applications and services) that can be rapidly provisioned and released with minimal management effort or service provider interaction". Berkeley [2] defined cloud computing as "to include application software delivered as services over the Internet and the hardware and systems software in the data centers that facilitate these services". There are different types of issues have been observed in cloud computing

© Springer International Publishing Switzerland 2015
Y. Tan et al. (Eds.): ICSI-CCI 2015, Part II, LNCS 9141, pp. 127–135, 2015.
DOI: 10.1007/978-3-319-20472-7_14

environment that need to be addressed. In the past, International Data Corporation (IDC) conducted a survey of 263 IT executives to gauge their opinion about the usage of IT cloud services in companies. Consequently, Security was ranked first and observed utmost issue of cloud computing. Ensuring the security of data is a major concern in cloud computing environment. Hence in this paper, a novel framework to secure data access in cloud environment is implemented and tested.

Based on type of service provided to users, cloud delivery models are exhibited as infrastructure as a service (IaaS), platform as a service (PaaS), and software as a service (SaaS) .Our implemented framework is offered as a SaaS. Depending on the purpose of setting up cloud and level of access to resources, there are four cloud deployment models: Public, Private, Community and Hybrid.We can install implemented framework on private and public cloud.

This paper addresses the concept of modularity to implement homo-hierarchical framework for client and cloud service provider security. The modules in implemented framework are client side application, authentication server and data server. The client side application is desktop based java application that is supported by java run time environment on all machines. Homomorphic encryption on client password is carried out and then, encrypted array of password will be generated. Client sends session key, username and encrypted array password information to authentication server. The communication between client and auhentication server is done with the help of simple object access protocol. Communication results encrypted array of random password and encrypted shared URL directory. Decryption of encrypted shared URL directory will be done and directory will be mounted on specified drive letter. Authentication server communicates to data server and retrieves user directory, decrypts it and share the encrypted shared URL directory with client. Section 1 is current part. Section 2 summarizes the related work for security of data. In Section 3, a novel framework is implemented which is designed to solve the security issue of cloud computing. Section 4 gives performance analysis. Section 5 concludes this paper.

2   Related Work

Every organization transfers its data on the cloud utilizes the storage service provided by the cloud provider. Therefore, there is arising need to protect the data against the unauthorized access, modification or denial of services etc. The Security of data includes Availability, Confidentiality and Integrity. Confidentiality of data in cloud is accomplished by cryptography. In today's time, cryptography is amalgamation of three types of algorithms i.e. (1) Symmetric-key algorithms such as DES, Triple-DES, AES and Blowfish algorithms. (2) Asymmetric-key algorithms such as RSA, Diffie-Helman Key Exchange, Elliptical curve cryptography and IBE. (3) Hashing i.e SHA1 and MD5. In this paper, we discuss Homomorphic encryption and its usage with AES and Cellular automata encryption for data owner and client security.

Often cloud users encrypt its data before sending to the Cloud provider and it will decrypt the data by using the private key of the user before performing any calculation which might influence the confidentiality of data stored in the Cloud. Homomorphic Encryption systems are needed to perform operations on encrypted data without decryption (without knowing the private key); only the user will have the secret key. When we decrypt the result of any operation, it is the same as if we had performed the calculation on the plaintext (or original data). For example, the encryption systems of Goldwasserand Micali [GM82], El Gamal [El-84] and Paillier [Pai99] support either adding or multiplying encrypted ciphertexts, but not both operations at the same time. Boneh, Goh and Nissim [BGN05] were the first to construct a scheme capable of performing both operations at the same time – their scheme handles an arbitrary number of additions and just one multiplication. More recently, in a breakthrough work, [Gen09, Gen10] constructed a fully homomorphic encryption scheme (FHE) capable of evaluating an arbitrary number of additions and multiplications on encrypted data.

The idea of homomorphic encryption scheme originally called privacy homomorphism was given by Rivest, R., Adleman, L., Dertouzos, M in [3]. Ideally, one should be able to transmit encrypted information to the server, process the encrypted data on the server and retrieve processed data from the server. This ideal situation, for long renown as the "Holly Grail of Cryptography", has finally got a brake through in 2009 by Craig Gentry in his Ph.D. thesis [4]. According to them any operation can be reduced to the basic addition and multiplication operations on bit level.Craig Gentry et al. [5] showed that "fully homomorphic encryption can, in principle, be constructed which was put forth by Rivest Adleman and Dertouzos [RAD78] in 1978. "According to this, an encrypted data can be processed without decrypting. Thus, we can get the cloud to perform a computation for us while revealing nothing of the input or output. Marten van Dijk et al. [6] presented a second fully homomorphic encryption scheme which uses many tools of Gentry's construction, but it does not require ideal lattices. "They illustrated that somewhat homomorphic component of Gentry's ideal lattice-based scheme can be replaced with a very simple homomorphic scheme by using just integers.The most efficient fully homomorphic encryption scheme has been implemented by the IBM research team conducted by S. Halevi and V. Shoup using ideas that can be found in [7], [8], and [9]. The implementation is called Homomorphic-Encryption Library (HELib) and can be found at:https://github.com/shaih/HElib.This software library implements the RLWE homomorphic encryption scheme, along with many optimizations to make homomorphic evaluation runs faster. HElib is written in C++ and uses the NTL mathematical library.

The main issue in this context is the question if fully homomorphic encryption schemes are efficient enough to be practical for cloud computing. Craig Gentry estimated in an article [10] that performing a Google search with encrypted keywords would multiply the necessary computing time by around 1 trillion. A more scientific analysis of Gentry's fully homomorphic encryption system was done in [11], but Gentry's estimation should make clear that the performance penalty of this scheme is a big way to use it in practice.

In [12], Lauter, Baehrig and Vaikuntanathan provided few concrete applications of homomorphic encryption and argued that there are many functions which could be useful for privacy preserving cloud services, which can be computed by many additions and a small number of multiplications on cipher-texts. For example, averages require no multiplications, standard deviation requires one multiplication, and predictive analysis such as logistical regression requires few multiplications. Smart et al. [13] in 2009 presented a specialization of Gentry's scheme that yielded a smaller cipher text size. They present a fully homomorphic encryption scheme which has both relatively small key and ciphertext size. Hongwei Li et al. [14] proposed a Hierarchical Architecture for Cloud Computing (HACC). "The presented method inherited attractive properties from IBC such as certificate-free and small key sizes.

3   Homo-Hierarchical Framework for Cloud Environment

The proposed framework is based on two fold approach, firstly secondary data has been reviewed from related work to evaluate cloud security by identifying unique security requirements and secondly an attempt to present a viable solution that eliminates the potential threats to cloud security which is based on functionality analysis. The outline of framework is based on homo-hierarchical encryption scheme which is shown in Fig1.The suggested framework comprises of three steps which are explained in 3.1, 3.2 and 3.3.

The framework aims to achieve the following objectives which enhance the security on data storage and it has been simulated on Microsoft IIS Express for Azure Cloud.

1. Develop a robust and progressive cloud computing security framework that provides trust to clients.
2. Subsequently, the implementation and testing of the developed framework has been carried out to validate its applicability for the public and private cloud.

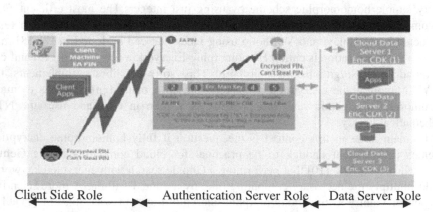

Client Side Role   Authentication Server Role   Data Server Role

Fig. 1. Homo-Hierarchial Framework for Cloud Environment

3.1   Description of Client Side Role

Client executes Homo-Hierarchy-Client.jar file, a network connection is created to get session key from authentication server. Client receives a session key which plays an important role for unique user/connection identification and expires on disconnection.

To access cloud data server, user needs to register an account with following input.

   a.) Username - unique identification of user

   b.) Password - required for authentication to access user data

On submit, Client application generates a Homomorphic encrypted array of password which is shown in Fig 2 and send encrypted array to authentication server.

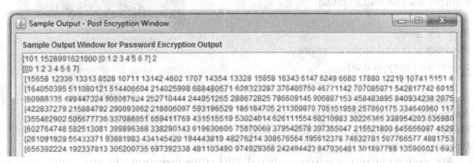

Fig. 2. Sample output of password encrypted array EA_PASS

At login, client provides session key, username and password encrypted array to authentication server and receives EA_RETURN_RANDOM_PASSWORD and ENCRYPTED_SHARED_DIRECTORY_URL shown in Fig 3, 4.

```
<authenticate xmlns="http://192.16
  <sessionKey>string</sessionKey>
  <username>string</username>
  <password>string</password>
</authenticate>
```

```
<authenticateResult>
  <string>string</string>
  <string>string</string>
</authenticateResult>
```

Fig. 3. Authenticate Request　　　　　　**Fig. 4.** Authenticate Response

Client subtracts EA_PASS from EA_RETURN_RANDOM_PASSWORD to obtain decryption key for ENCRYPTED_SHARED_DIRECTORY_URL and mount resultant URL directory on specified mount drive which is shown in Fig 5, 6.

Fig. 5. Client authentication window　　　**Fig. 6.** Client side Resultant output window

3.2   Description of Authentication Server

Authentication server acts as a session key, registration, authentication and URL provider. On authenticated client request, it connects to data server, retrieves user data directory, decrypts and responds URL directory with user privileges.

An authentication web service application named AuthService.asmx is published at Microsoft IIS Express web server on a LAN machine at 192.168.1.13:2534 which is shown in Fig 7.

Get session key
Returns a session id to client.

Register user
Register user and store username,
password encrypted array.

Authenticate
Authenticate client and return encrypted
Directory shared URL and return random
Password array.

Fig. 7. Authentication Server

Authentication server retrieves shared URL directory from database table active connection. Shared URL directory is a web location on data server which is encrypted using AES 128-bit. The private key for encryption and decryption is selected by administrator of authentication server. Authentication server generates pseudo random number key, encrypt it to generate PRN_EA_PASSWORD using homomorphic encryption. It adds EA_PASSWORD and PRN_EA_PASSWORD to generate EA_RETURN_RANDOM_PASSWORD. Authentication server encrypts shared URL directory using cellular automata encryption. The encryption-decryption key is pseudo random number key. Authentication Server generates a result list and adds EA_RETURN_RANDOM_PASSWORD and Encrypted-SHARED_DIRECTORY_ URL. Authentication server authorize client with correct session key. Authorized permissions provided to client are RWX.

3.3   Description of Data Server

Data server is a cloud web service which is published at mydataserver:2731/Data-Service.asmx. It is responsible for storing user data and includes following methods a.) Create, Access and Delete Directory   b.) Check Active Connection Data server receives a directory access request from authentication server. If directory exists, data server provides shared URL directory to authentication server. Shared directory is encrypted with AES 128 bit. On authentication server decryption request shared directory is decrypted in real time. Authorized access to data server is only allowed to authentication server. Framework interconnection and data model is shown in Fig 8.

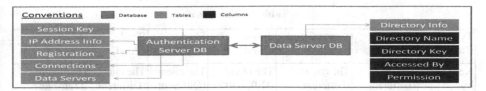

Fig. 8. Interconnection of framework with data model

4   Performance Analysis

To measure the security strength of implemented framework we compared it with current homomorphic encryption methods based on different parameters which are shown in Table 1.

Table 1. Comparison Of Implemented Framework With Current Homomorphic Encryption Methods

Character istics	RSA	El Gamal	Goldwass er-Micali	Boneh-Goh-Nissim	Gentry	Our Proposed Model(Homo-Hierarchiea l)
Homomorp hic Encryption Type	Multipl i-cative	either adding or multiplying (both not at same time)	either adding or multiplyin g (both not at same time)	arbitrary number of additions and just one multiplic ation.	arbitrary number of addition s and arbitrary number of multipli cation.	Arbitrary no of addition and subtraction + Equality testing on encrypted array returns 1 on true,128-bit AES, Cellular Automata Encryption
Security applied to	Cloud Service Provide r	Cloud Service Provider	Cloud Service Provider	Cloud Service Provider	Cloud Service Provider	Client side and Cloud Service Provider
Platform	Cloud Comput ing	Cloud Computing	Cloud Computin g	Cloud Computi ng	Cloud Computi ng	Cloud Computing

Table 1. (*Continued*)

Type Of Cloud	Single Cloud	Single Cloud	Single Cloud	Single Cloud	Single Cloud	Multi Cloud
Keys Used by	The client (Different keys are used for encryption and decryption)	The client (Different keys are used for encryption and decryption)	The client (Different keys are used for encryption and decryption)	The client (Different keys are used for encryption and decryption)	The client (Different keys are used for encryption and decryption)	The client (Same key Is used for encryption and decryption)
Privacy of data	communication and storage processes both	communication and storage processes both	communication and storage processes both	communication and storage processes both	communication and storage processes both	communication and storage processes both

5   Conclusion

We have designed a novel framework to store and secure data access from cloud storage by ensuring CIA – Confidentiality, Integrity and Authentication of data. This paper describes a brief concept on cloud computing and cryptography methods that have been used in cloud. A framework based on homomorphic encryption and its usage with AES and Cellular automata encryption for data owner and client security has been suggested and simulated on Microsoft IIS Express for Azure Cloud. Moreover, we articulated Interconnection of framework with data model. We also compared security strength level of our novel framework with other existing techniques in this domain. The suggested framework guarantees that no one except the authenticated user can access the data neither the cloud storage provider.

References

1. Mell, P., Grance, T.: Draft NIST Working Definition of Cloud Computing (2009)
2. Armbrust, M., et al.: Above the Clouds: A Berkeley View of Cloud Computing Technical report EECS-2009-28, UC Berkeley (February 2009). http://www.eecs.berkeley.edu/Pubs/TechRpts/2009/EECS-2009-28.html
3. Rivest, R., Adleman, L., Dertouzos, M.: On Data Banks and Privacy Homomorphisms. Foundations of Secure Computation, 169–180 (1978)
4. Gentry, C.: A Fully Homomorphic Encryption Scheme, PhD Thesis, Stanford University (2009). http://crypto.stanford.edu/craig
5. Gentry, C.: Fully homomorphic encryption using ideallattice. In: Proc. of STOC, pp. 169–178. ACM (2009)
6. Van Dijk, M., Gentry, C., Halevi, S., Vaikuntanathan, V.: Fully homomorphic encryption over the integers. In: Gilbert, H. (ed.) EUROCRYPT 2010. LNCS, vol. 6110, pp. 24–43. Springer, Heidelberg (2010)

7. Brakerski, Z., Gentry, C., Vaikuntanathan, V.: Fully homomorphic encryption without bootstrapping. In: Innovations in Theoretical Computer Science Conference, pp. 309–325 (2012)
8. Gentry, C., Halevi, S., Smart, N.P.: Homomorphic evaluation of the AES circuit. In: Safavi-Naini, R., Canetti, R. (eds.) CRYPTO 2012. LNCS, vol. 7417, pp. 850–867. Springer, Heidelberg (2012)
9. Smart, N., Vercauteren, F.: Fully Homomorphic SIMD Operations. Designs, Codes and Cryptography (2012)
10. Gentry, C.: Computing Arbitrary functions on encrypted Data. Communications of the ACM, 97–105 (2010)
11. Gentry, C., Halevi, S.: Implementing gentry's fully-homomorphic encryption scheme. In: Paterson, K.G. (ed.) EUROCRYPT 2011. LNCS, vol. 6632, pp. 129–148. Springer, Heidelberg (2011)
12. Naehrig, M., Lauter, K., Vaikuntanathan, V.: Can homomorphic encryption be practical? In: ACM Workshop on Cloud Computing Security Workshop, pp. 113–124 (2011)
13. Smart, N.P., Vercauteren, F.: Fully homomorphic encryption with relatively small key and ciphertext sizes. In: Nguyen, P.Q., Pointcheval, D. (eds.) PKC 2010. LNCS, vol. 6056, pp. 420–443. Springer, Heidelberg (2010)
14. Li, H., Dai, Y., Yang, B.: Identity-Based Cryptography for Cloud Security (2011). http://eprint.iacr.org/169.pdf

Mutual Authentication Protocol Based on Smart Card and Combined Secret Key Encryption

Guifen Zhao[⊠], Liping Du, Ying Li, and Guanning Xu

Beijing Key Laboratory of Network Cryptography Authentication,
Beijing Municipal Institute of Science and Technology Information, Beijing, China
gfzh@hotmail.com

Abstract. A mutual authentication scheme and secret key exchange based on combined secret key method is proposed. Use hardware including smart card, encryption cards or encryption machine to perform encryption and decryption. Hash function, symmetric algorithm and combined secret key method are applied at client and server. The authentication security is guaranteed due to the properties of hash function, combined secret key method and one-time mutual authentication token generation method. Mutual authentication based on smart card and one-time combined secret key can avoid guessing attack and replay attack. The mutual authentication method can be applied to cloud based application systems to realize mutual authentication and enhance security.

Keywords: Mutual authentication · Smart card · Combined secret key · Symmetric encryption

1 Introduction

Cloud users face with numerous security threats. Authentication technology is the foundation of network information security. Only when solve the problems of authentication, authorization, secrecy, integrity, non-repudiation and other security factors could be guaranteed. SafeNet Global Authentication Survey reveals over 20 percent suggest that 90-100 percent of users currently require strong authentication for mobile devices with access to corporate resources [1].

Static password is the most popular authentication method. However, it is usually easily broken. Hackers can use shoulder surfing, snooping, sniffing, guessing or other techniques to steal passwords [2]. Therefore, new authentication technologies are proposed.

Biometric authentication is more secure but needs special devices which is not convenient enough [3-4]. When using out of band authentication scheme, the user is authenticated in two tiers, including username and password authentication, and the code authentication which is received on mobile phone or email [5-6]. It is an accessorial method in general. One-time password can overcome a number of shortcomings, such as replay attack, social engineering, sniffing etc. However, one-time password token based on time synchronization or event synchronization technique must synchronize time or a counter value between client and server [7].

© Springer International Publishing Switzerland 2015
Y. Tan et al. (Eds.): ICSI-CCI 2015, Part II, LNCS 9141, pp. 136–143, 2015.
DOI: 10.1007/978-3-319-20472-7_15

Nowadays, most available authentication systems achieve one-way authentication. Therefore, attacks from artificial servers could not be avoided. More and more network applications need mutual authentication. Different mutual authentication solutions are proposed. Most adopt Exclusive OR operation and Hash Function [8-10]. [11] is an improved mutual authentication framework, but transfer the user identity and password via plaintext. [12] achieves mutual authentication and key agreement with smart cards. [13] uses Hash Function and Diffie-Hellman key exchange methods, and provides dynamic authentication.

The proposed mutual authentication method is based on smart card, combined secret key method and symmetric algorithm to achieve dynamic authentication and key agreement for cloud user.

2   Mutual Authentication Based on Combined Secret Key

Authentication servers perform identity authentication when users access cloud resources. Smart cards at client and encryption card or encryption machine at server are needed in the proposed mutual authentication scheme based on combined secret key encryption and decryption. If the proposed scheme applied to available applications, the client side mutual authentication module could be compiled as Dynamic Link Library under Windows or dynamic shared object under Linux, which is convenient for integration.

Users should enroll and complete initialization at server firstly, and then could login corresponding cloud resources servers after mutual authentication with smart cards. Users can select own username and password, and the plaintext of username and password are not be transferred over network in proposed scheme.

For describing the proposed mutual authentication protocol, define related symbols shown in Table 1.

Table 1. Symbol definition

Symbol	Definition
U_N	Username
U_{Pw}	Password
U_{ID}	User Identity
U_{Ks}	User Key Seeds
$k = C_{T,R}(\;)$	Combined Secret Key Algorithm, k is Combined Secret Key
$e = E_k(\;)$	Encryption algorithm, k is the encryption key, e is cipher text
$d = D_k(\;)$	Decryption algorithm, k is the decryption key, d is plaintext
$h = H(\;)$	Hash function, h is hashing value
T	Time factor
R	Random number factor

2.1   Enroll and Initialization

User should key in username and password, and enroll firstly before accessing cloud resources. After receiving the information, cloud server check the username. If it has not been registered, initialize the user's smart key, i.e. update user's smart key information, including user identity U_{ID}, key seeds U_{Ks}, enroll time factor T_R and hashing value h_0. Meanwhile store the parameters at server, including U_{ID}, cipher text of U_{Ks}, T_R and h_0.

$$h_0 = H\left(U_N \parallel U_{Pw} \parallel T_R\right). \tag{1}$$

2.2   Mutual Authentication Protocol

The proposed mutual authentication solution adopts time factor and random numbers as control parameters to generate one-time key and one-time authentication token when cloud user login. Each time generate different parameters.

The client and server use smart card and encryption cards or encryption machine to generate unique random number factor, and then use the cryptography technique to generate a one-time combined secret key and authentication token according to time factor and random number factor. The client and server complete mutual authentication according to the generated one-time password. Only the authenticated legal user can access cloud resources at server. Mutual authentication method can enhance system security.

The mutual authentication protocol is shown in Fig. 1.

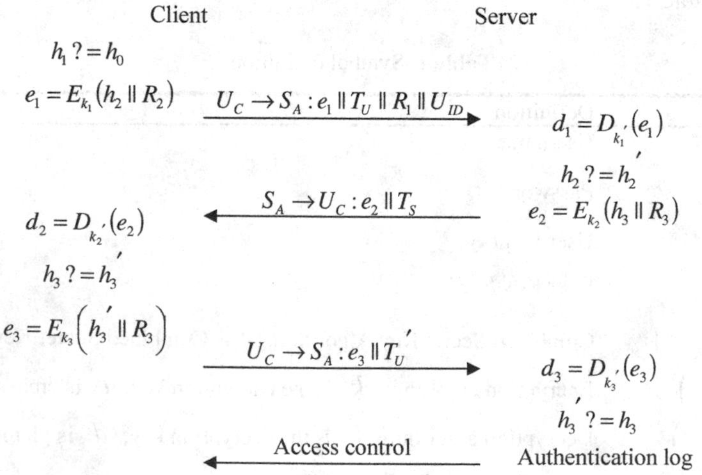

Fig. 1. The mutual authentication protocol based on smart key and combined secret key encryption and decryption

Check username by smart card and generate client token

While authenticating, users need input username and password firstly. Smart card will check username and password. Combine the input username $U_N{'}$, input password $U_{Pw}{'}$ and stored enroll time factor T_R, and compute hashing value h_1.

$$h_1 = H\left(U_N{'} \| U_{Pw}{'} \| T_R \right). \tag{2}$$

Check whether the hashing value h_1 is equal with the value h_0 stored in smart card or not. If they are the same, continue authentication, otherwise, terminate authentication.

Obtain local time factor T_U at client, and generate random numbers R_1 and R_2 by hardware random number generator. And generate a set of combined secret key k_1 using time factor T_R and random number R_1 according to combined secret key generation algorithm.

$$k_1 = C_{T_R, R_1} \left(U_{ID} \right). \tag{3}$$

Compute hashing value h_2 of user identity U_{ID}, local time factor T_U, stored enroll time factor T_R, random numbers R_1 and R_2. Use the generated combined secret key k_1, encrypt the hashing value h_2 and random numbers R_2 via symmetric encryption algorithm $E_k( )$, consequently cipher text e_1 is achieved at client, which is the client authentication token.

$$e_1 = E_{k_1} \left(h_2 \| R_2 \right) = E_{k_1} \left(H \left(U_{ID} \| T_U \| T_R \| R_1 \| R_2 \right) \| R_2 \right). \tag{4}$$

Send the client authentication token e_1, client time factor T_U, random number R_1 and user identity U_{ID} to authentication server.

Authenticate client token and generate server token

After receiving e_1, T_U, R_1 and U_{ID}, server check the validity of client time factor firstly. If it is valid, continue the authentication, otherwise, terminate authentication.

Regenerate a set of combined secret key $k_1{'}$ in encryption card using time factor T_R stored at server and received random number R_1 according to combined secret key generation algorithm.

$$k_1' = C_{T_R,R_1}(U_{ID}).$$ (5)

Authentication server decrypts the received cipher text e_1 using combined secret key k_1' according to symmetric decryption algorithm $D_k(\ )$.

$$d_1 = D_{k_1'}(e_1).$$ (6)

Therefore obtain hashing value h_2 and random number R_2. And then compute hashing value h_2' of U_{ID}, T_U, T_R, R_1 and R_2.

$$h_2' = H(U_{ID} \| T_U \| T_R \| R_1 \| R_2).$$ (7)

Verify whether h_2' is equal with h_2 or not. If they are the same, continue the authentication, otherwise, terminate authentication.

Obtain local time factor T_S at server, and generate a set of combined secret key k_2 using T_S and R_2 according to combined secret key generation algorithm.

$$k_2 = C_{T_S,R_2}(U_{ID}).$$ (8)

Generate random numbers R_3 by hardware random number generator in encryption card. Compute hashing value h_3 of U_{ID}, T_U, T_S, R_2 and R_3. Use the generated combined secret key k_2, encrypt h_3 and R_3 via symmetric encryption algorithm $E_k(\ )$, consequently cipher text e_2 is achieved at server, which is the server authentication token.

$$e_2 = E_{k_2}(h_3 \| R_3) = E_{k_2}(H(U_{ID} \| T_U \| T_S \| R_2 \| R_3) \| R_3).$$ (9)

Send the server authentication token e_2 and server time factor T_S to client.

Authenticate server token and return server parameter

After receiving e_2 and T_S, client regenerate a set of combined secret key k_2' using T_S and R_2 according to combined secret key generation algorithm.

$$k_2' = C_{T_S,R_2}(U_{ID}).$$ (10)

Decrypt the received cipher text e_2 using combined secret key k_2' according to symmetric decryption algorithm $D_k(\ )$.

$$d_2 = D_{k_2'}(e_2). \tag{11}$$

Therefore, obtain hashing value h_3 and random numbers R_3. And then compute hashing value h_3' of U_{ID}, T_U, T_S, R_2 and R_3.

$$h_3' = H(U_{ID} \parallel T_U \parallel T_S \parallel R_2 \parallel R_3). \tag{12}$$

Verify whether h_3' is equal with h_3 or not. If they are the same, continue the authentication, otherwise, terminate authentication.

Generate a set of combined secret key k_3 using T_S and R_3 according to combined secret key generation algorithm.

$$k_3 = C_{T_S, R_3}(U_{ID}). \tag{13}$$

Use the generated combined secret key k_3, encrypt h_3 and R_3 via symmetric encryption algorithm $E_k(\ )$, consequently cipher text e_3 is achieved at client.

$$e_3 = E_{k_3}\left(h_3' \parallel R_3\right). \tag{14}$$

Send the cipher text e_3 and a new local time factor T_U' at client to authentication server.

Complete mutual authentication

After receiving e_3 and T_U', server check the validity of client time factor firstly. If it is valid, continue the authentication, otherwise, terminate authentication.

Regenerate a set of combined secret key k_3' using T_S and R_3 according to combined secret key generation algorithm.

$$k_3' = C_{T_S, R_3}(U_{ID}). \tag{15}$$

Decrypt the received cipher text e_3 using combined secret key k_3' according to symmetric decryption algorithm $D_k(\ )$.

$$d_3 = D_{k_3'}\left(e_3\right). \tag{16}$$

Therefore, obtain hashing value h_3 and random numbers R_3. And then verify whether the received data are equal with original data or not. If they are the same, the mutual authentication succeeds, and then record log file and perform access control. Otherwise, the authentication is failed.

3   Analysis

The proposed scheme uses Hash function and combined secret key method, and performs 6 combined key methods, 5 Hash functions, 3 Encryption operations, and 3 Decryption operations. [13] has compared different authentication scheme, and uses Hash and Diffie-Hellman key exchange methods, and performs 12 XOR operations, 6 Hash functions, 2 Encryption operations, and 2 Decryption operations, totally 22 computations. The security of the proposed mutual authentication solution depends on how vulnerable the password and secret key is. Client and server use time factor and random numbers as parameters to generate one-time combined secret key, and then generate one-time token using symmetric encryption algorithm according to one-time secret key, which can't be guessed by attacker, consequently avoid guessing attack. Moreover, the replay attack is also avoided because of the time factor. The one-time token can be used only once. Therefore, repeated authentication token is refused.

4   Conclusions

Mutual authentication can avoid server spoofing attack compared with one-way authentication. The proposed authentication protocol based on smart card, combined secret key encryption and decryption provide higher security. The proposed scheme provides dynamic mutual authentication and secret key exchange. The security is guaranteed by the properties of hash function, combined secret key method and one-time authentication token generation.

Acknowledgments. The authors thank the helpful comments and suggestions from director and colleagues in Beijing Key Laboratory of Network Cryptography Authentication. This work is supported by Innovation Project II-2: Research and Development of Cryptographic Authentication System in Cloud Computing Security (No. PXM2014_178214_000011).

References

1. Annual Authenitication Survey. http://cn.safenet-inc.com/news/2014/authentication-survey-2014-reveals-more-enterprises-adopting-multi-factor-authentication/?LangType=2052#sthash.dSnZBDMn.dpuf

2. Chhabra, R.K., Verma, A.: Strong authentication system along with virtual private network: A secure cloud solution for cloud computing. International Journal of Electronics and Computer Science Engineering. **1**(3), 1566–1573 (2012)
3. Vallabhu, H., Satyanarayana, R.V.: Biometric Authentication as a Service on Cloud: Novel Solution. International Journal of Soft Computing and Engineering (IJSCE) **2**(4), 163–165 (2012). ISSN: 2231-2307
4. Rassan, I.A.L., AlShaher, H.: Securing Mobile Cloud Using Finger Print Authentication. International Journal of Network Security & Its Applications (IJNSA) **5**(6), 41–53 (2013)
5. Emam, A.H.M.: Additional Authentication and Authorization using Registered Email-ID for Cloud Computing. International Journal of Soft Computing and Engineering (IJSCE) **3**(2), 110–113 (2013). ISSN: 2231-2307
6. Singh, M., Singh, S.: Design and Implementation of Multi-tier Authentication Scheme in Cloud. International Journal of Computer Science Issues **9**(5), 181–187 (2012). No 2
7. Nafi, K.W., Kar, T.S., Hoque, S.A., Hashem, M.M.A.: A Newer User Authentication, File encryption and Distributed Server Based Cloud Computing security architecture. (IJACSA) International Journal of Advanced Computer Science and Applications **3**(10), 181–186 (2012)
8. Park, Y., Lee, Y.: Group-ID based RFID Mutual Authentication. Advances in Electrical and Computer Engineering **13**(4), 9–12 (2013)
9. Liu, A.-T., Chang, H.K.-C., Lo, Y.-S., Wang, S.-Y.: The Increase of RFID Privacy And Security With Mutual Authentication Mechanism In Supply Chain Management. International Journal of Electronic Business Management **10**(1), 1–7 (2012)
10. Le, X.H., Khalid, M., Sankar, R.: An Efficient Mutual Authentication and Access Control Scheme for Wireless Sensor Networks in Healthcare. Journal of Networks **6**(3), 355–364 (2011)
11. Nayak, S.K., Mohapatra, S., Majhi, B.: An Improved Mutual Authentication Framework for Cloud Computing. International Journal of Computer Applications (0975-8887) **52**(5), 36–41 (2012)
12. Li, C.-T.: An enhanced remote user authentication scheme providing mutual authentication and key agreement with smart cards. In: 2009 Fifth International Conference on Information Assurance and Security, pp. 517–520 (2009)
13. Baboo, S.S., Gokulraj, K.: An Enhanced Dynamic Mutual Authentication Scheme for Smart Card Based Networks. I. J. Computer Network and Information Security **4**, 30–38 (2012). MECS

Study of a Secret Information Transmission Scheme Based on Invisible ASCII Characters

Yuchan Li, Xing Hong, Guangming Cui, and Erzhou Zhu[✉]

Department of Software Engineering, Anhui University, Hefei 230601, China
{ycli,xhong,gmcui,ezhu}@ahu.edu.cn

Abstract. This paper proposes a secret information transmission scheme for English texts by using the characteristics of invisible ASCII codes. The scheme incorporates two stages, the information hiding stage and the information extracting stage. The first stage takes the prepared carrier documents and the secret information need to be embedded as the input. Then, the information hiding procedure of the scheme will synthesize the carrier document and the secret information into a stego document. In the second stage, an extracting procedure will pick up the secret information when it receives the stego document that generated by the first stage. The scheme achieves hiding effects by processing the bland spaces in the English texts. The experimental results have shown that our information hiding scheme is feasible, reliable, safe and efficient.

Keywords: Information hiding · Information extraction · ASCII codes

1   Introduction

Information hiding technology [1] is an important topic in the information security. In this field, a module that contains some specific information is invisible when the design and implementation of other modules that need not to access them takes place. This technology achieves the goal of communication security and copyright protection mainly by embedding the hidden information into known carriers.

The universal storage and transmission of information by text-based documents makes the text-based information hiding technology[2-3] as an important topic in computer security. On the basis of the different embedded methods for the hidden information, this technique can be divided into three categories:

- Typesetting-oriented information hiding [4]. It changes the subtle characteristics of texts by altering the character spacing, line spacing and font format in the texts to achieve the goal that the naked eyes cannot distinguish. But, this tech-nique is poor robustness, namely under the condition of rearranging the docu-ments' edition, the hidden information will disappear.
- Grammar-oriented information hiding [5]. It utilizes the grammatical structure of nature language to generate the implicit documents. This technique is suit for the grammar structure of the document, but their semanteme are unsystematic and can be easily distinguished by naked eyes.

© Springer International Publishing Switzerland 2015
Y. Tan et al. (Eds.): ICSI-CCI 2015, Part II, LNCS 9141, pp. 144–150, 2015.
DOI: 10.1007/978-3-319-20472-7_16

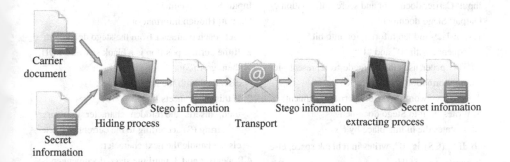

Fig. 1. Information security transmission

- Semantic-oriented information hiding [6]. It utilizes the synonym substitution to hide the target information. Common algorithms include steganography based on equivalent rules replacement, replacement based on the synonymous sentences, replacement based on synonym substitution, steganography based on machine translation. The defect of this method is low embedding rate. It can only hide information where meet the substitution rules.

To solve the above problems, this paper puts forward an information hiding scheme for English texts by using the characteristic of some invisible ASCII codes. By using the characteristic that blank spaces in text documents is imperceptible, coupling with the characteristic that the invisible characters of ASCII codes have no difference with blank spaces in a certain environment, our scheme can easily hide, detect and separate the information of the document.

2   Implementation

In the ASCII codes, the communicating character SOH (head, coded as 0000001) has different forms of display in different text. After performing many different tests, we concluded that this character is similar with space character. So, in this paper, we select SOH as a substitution character for hidden information. The scheme incorporates two stages, the information hiding stage and the information extracting stage, as shown in Fig.1.

2.1   Information Hiding

The information hiding stage firstly takes the prepared carrier documents and the secret information need to be embedded as the input. Then, the information hiding procedure of the scheme will synthesize the carrier document and the secret information into a stego document. Specifically, when the carrier document is opened and the secret information is received successfully, this procedure will analyze the character of the document one by one to look for the location that it is not only a blank space, but also satisfies the constraint function f(x) (this function will be described in Section

Input: Carrier document and secret information.
Output: Stego document.
1. Translates hidden information into bit
   sequence with "0" and "1";
   //This paper uses the Unicode to represent hidden information's sequence (S).
2. Gets each word (W$_i$) from carrier document;
3. Writes W$_i$ into Stego document
   a. Locates the hiding place by f(x);
   b. If S$_j$ (∈S) is "0", writes in a blank space, else
      writes in a SOH;
   c. If the place not meet the f(x), write in blank.
4. Repeat 2 and 3, until S is empty;
5. Copies remainders of the carrier document to
   secret document.

(a) Procedure of the information hiding.

Input: Stego document.
Output: Hidden information.
1. Gets each character from the stego document;
2. If the current position is a blank space or SOH,
   then, go to 3);
   else, deal with the next character;
3. If this place meets f(x),
   then, inserts the hidden character into target
         string (S) according to the current character;
   else, handle the next character;
4. Repeats 2 and 3, until the stego document ends;
5. Translates S into the Unicode sequence for the
   receivers.

(b) Procedure of the information extracting.

Fig. 2. Procedure of the information hiding and extracting

3.3). According to the bits of hidden information we can choose the hidden characters. It means, when the hidden information bit is "1", we write SOH, else write
space. Finally, the remainder of the carrier document is also copied to the stego document. This process can be described as Fig.2(a).

2.2   Information Extracting

In the second stage, the extracting procedure will pick up the secret information when
it receives the stego document that was generated by the first stage. When the stego
document is successfully received, the information extraction process will analyze the
character of the document one by one to look for the location that it is not only a
blank space or SOH, but also satisfies the constraint function f(x) (will be described in
Section 4.3). According to the ASCII code, we can decide whether the hidden character in this location is "0" or "1". By doing this, the hidden bits are inserted into the
target string. Finally, the characters of the target string are translated into Unicode for
the receivers. This process is shown in Fig.2(b).

3   Performance Analysis

The experiments in this section are carried out on the machine with Intel i3 4030U
CPU (1.9 GHz), 4GB DDR3 1333 RAM and 64-bits Windows 7 OS. The carrier
doc-uments are .doc and .tex files. Experiments will measure the performance of in-
forma-tion hiding scheme from the point views of robustness, security, stability and
efficien-cy respectively.

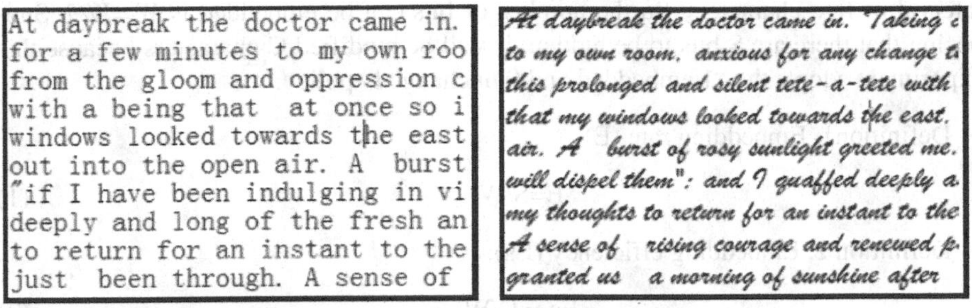

At daybreak the doctor came in.	*At daybreak the doctor came in. Taking c*
for a few minutes to my own roo	*to my own room, anxious for any change t*
from the gloom and oppression c	*this prolonged and silent tete-a-tete with*
with a being that at once so i	*that my windows looked towards the east.*
windows looked towards the east	*air. A burst of rosy sunlight greeted me.*
out into the open air. A burst	*will dispel them": and I quaffed deeply a*
"if I have been indulging in vi	*my thoughts to return for an instant to the*
deeply and long of the fresh an	*A sense of rising courage and renewed p.*
to return for an instant to the	*granted us a morning of sunshine after*
just been through. A sense of	

Fig. 3. Carrier document before and after the style changes

3.1   Robustness

This experiment adopts information hiding technology based on typesetting (compiled code or structure) information. Experimental results shown that our information hiding scheme is robust, which means, the changes for the font style cannot influence on the steganographic information's successful detection. This can be seen in Fig.3.

3.2   Safety

When the document has some subtle changes, the naked eyes cannot distinguish. So it has reached the transmission security. In the proposed algorithm, the secret document's format can be changed; the traditional attacks are nearly invalid. The traditional algorithms adopt the blank space replacement to hide hidden information, but, increasing the blank spaces may lead to a failure of the algorithm. However, in our improved algorithm, if the current location has at least two consecutive blank spaces without any SOH character, this place only takes one blank space. When the current locations have SOH character, in spite of how many the blank spaces have, it's regarded as one SOH character. As a consequence, attacks by inserting blank spaces into the target document cannot destroy the integrity of the hidden information. So, our scheme is safe. Table 1 lists the analyzing results by releasing our scheme and other popularly used schemes on different attacks.

Table 1. Security analysis by different attacks

attacks \ Schemes	change font	add blank space	add line	change font size
Our Scheme	no affect	no affect	no affect	no affect
Word Shift Coding	no affect	affect	affect	no affect
Line Shift Coding	no affect	affect	affect	affect
Front Hidden	affect	no affect	no affect	no affect

3.3   Embedded Ability

Supposing that a document has W characters, and this experiment adopts the constraint function $f(x)=5x$. It means embedding one bit of secret information at the interval of five characters. So the number of bits can be embedded is $W_f=W/5$. Supposing that there are S bits to be hidden. It will be modified C characters because this experiment adopts the "1 embedded and 0 unchanged" policy.

— Definition1. Embedding rate (E):

$$E=S/W_f \qquad (1)$$

— Definition 2. Embedding efficiency (Es):

$$Es = C/W_f \qquad (2)$$

— Definition 3: Embedding ability (E_x):
Supposing that "0" and "1" have average distribution in the hidden information, so $E_s=E/2$. When the proportion of "1" in S is L, $E_s=E/L$. L is the proportion of the characters need to be changed. Since a good way to embed should need high embedding rate and higher embedding efficiency. So embedding ability (E_x) is defined as:

$$E_x=E_s*E=SC/W_f^2 \qquad (3)$$

Table 2. E, Ex, Es comparisons

S	80			144		
L	E	Es	Ex	E	Es	Ex
0.1	0.500	0.050	0.025	0.900	0.090	0.081
0.2	0.500	0.100	0.050	0.900	180	0.162
0.3	0.500	0.150	0.075	0.900	0.270	0.243
0.4	0.500	0.200	0.100	0.900	360	0.324
0.5	0.500	0.250	0.125	0.900	450	0.405
0.6	0.500	0.300	0.150	0.900	0.540	0.486
0.7	0.500	0.350	0.175	0.900	0.630	0.567
0.8	0.500	0.400	0.200	0.900	0.730	0.648
0.9	0.500	0.450	0.225	**0.900**	**0.810**	**0.729**

From the formula we can see that when the SC product is bigger, the overall embedding capacity is higher. When "0" and "1" have average distribution, embedding ability $Ex= 2C^2/Wf^2 = 0.5S^2/Wf^2$; when the proportion of "1" in S is L, $Ex= C^2/(W_f^2 L)=LS^2/W_f^2$, Easy to know when L's ratio is bigger, embedding capacity is bigger. When W is 800, namely the $W_f = 160$, E, E_s, the E_x with L, S, as shown in Table 2.The table shows that the E_x is proportional to S and L.

3.4   Stability

For a specific document size, if the total tests times is N, the number of successful experiments is n, it is easy to know that the success rate is:

$$E_t = n/N \qquad (4)$$

Based on the formula (4), the stability analysis has been shown in Table 3. The table shows that the experiment's success rates are more than 90% in the experiment environment of many repeated tests and carrier documents of different sizes. It reaches a good ratio, namely a higher stability.

Table 3. Stability analysis

Group number	Document size (KB)	Test number	Success number	Success rate (%)
1	1	100	93	93
2	3	100	95	95
3	5	100	95	95
4	8	100	96	96
5	10	100	98	98
6	15	100	94	94

3.5 Efficiency Analysis

In the information hiding phase, the algorithm analyses carrier documents and secret information need to be hided one by one. Assuming that the size of carrier document is m and the size of the information need to be hided is n, the time complexity is $O(m+16*n)$. So the whole time cost has a tight relationship with m and n, as showing in Table 4 (the left values). In the information extraction phase, because the input is the stego document, the time complexity is also $O(m+16*n)$ and the whole time cost also associates with m and n. The experimental results shown in Table 4 (the right values).

Table 4. Overhead of information hiding/extracting (ms)

m \ n	5	10	15	20
10 K	5/3	6/3	6/4	6/5
15 K	6/8	7/8	7/8	9/8
20 K	8/15	9/15	8/15	8/15
25 K	10/24	10/24	10/24	12/24
30 K	12/24	13/35	12/35	12/35

4   Conclusion and Future Work

In order to solve the problems that the current popularly used text-based information hiding scheme has poor robustness, low embedding rate, semantic clutter and can be casily distinguished by unaided, this paper proposes an information hiding schema for English texts by using the characteristic of some invisible ASCII codes. The algorithm achieves hiding effects by processing the spaces in the English texts. Experiments have shown that the information hiding scheme which is proposed in this paper is feasible, reliable, safe and efficient. In the future, more efficient algorithms will be proposed to hide the secret information.

Acknowledgments. This paper is supported by the National Natural Science Foundation of China (61300169) and the Anhui University research project (Grant Nos. J10118520164).

References

1. Bennett, K.: Linguistic Steganography: Survey, Analysis, and Robustness Concerns for Hiding Information in Text. Technical Report 2004-13, Purdue University (2004)
2. Fu, D.L., Chen, G.X., Yang, Q.X.: A Covert Communication Method Based on XML Documents Slicing. Computer Applications and Software **28**(9), 106–108 (2011)
3. Gu, T.F., Yue, H.Y.: DCT Coefficient-based Encrypted Information Hiding Technology and its Implementation. Computer Applications and Software **28**(6), 173–175 (2011)
4. Udit, B., Deepa, K., Takis, Z.: Digital Video Steganalysis Exploiting Statistical Visibility in the Temporal Domain. IEEE Transactions on Information Forensics and Security **1**(4), 502–516 (2006)
5. Ehsan, N., Jane, Z.W., Rabab, K.W.: Robust Image Watermarking Based on Multiscale Gradient Direction Quantization. IEEE Transactions on Information Forensics and Security **6**(4), 1200–1239 (2011)
6. Murphy, B.: Syntactic information hiding in plain text. Master's thesis. Trinity College Dublin (2001)

Negative Survey-Based Privacy Protection of Cloud Data

Ran Liu and Shanyu Tang[(⊠)]

School of Computer Science, China University of Geosciences, Wuhan, China
ranliu_cug@foxmail.com, shanyu.tang@gmail.com

Abstract. Cloud platforms usually need to collect privacy data from a large number of users. Although the existing methods of privacy protection for cloud data can protect users' privacy data to a certain degree, there is plenty of room for improvement in efficiency and degree of privacy protection. Negative survey spired by Artificial Immune System (AIS) collects each user's unreal privacy information to protect users' privacy. This study focuses on the accuracy of the reconstructed positive survey from negative survey, which is one of the key problems in the Negative Survey-based privacy protection of cloud data.

Keywords: Artificial immune system · Privacy protection · Cloud data · Negative survey

1 Introduction

In recent years, users face cloud data privacy protection problems with the advent of cloud computing and intelligent computing techniques. Cloud computing is a development area and has various definitions. For example, cloud computing could be regarded as a large-scale distributed computing model [8], and massive ability on computation, storage, platform and service can be provided to users. Cloud computing has some advantages such as high speed of calculation, well availability and high compatibility. Traditional encryption method is not applicable sometimes because of complex calculation. Data confusing method can be used to reduce the computation. For example, SMART method [10] and PEQ [16] method were used to calculate the sum of the data. And the extremum values can be obtained by KIPDA method [9] or PriSense method [15].

The negative representation [4], which is inspired by Artificial Immune System (AIS), has some applications [12] in privacy protection. Negative selection principle [11] is one of the key mechanisms in Artificial Immune System. Inspired by negative selection principle, Forrest has proposed negative selection algorithm [7] which can be used in network security and virus detection [3,14].

Different from traditional representation, the negative representation always stores the information not consistent with the actual one. The negative survey [5], which is inspired by negative representation [4], is a novel and promising

© Springer International Publishing Switzerland 2015
Y. Tan et al. (Eds.): ICSI-CCI 2015, Part II, LNCS 9141, pp. 151–159, 2015.
DOI: 10.1007/978-3-319-20472-7_17

method for information security and enhancing privacy in collecting sensitive data and individual privacy.

In contrast to other traditional privacy protection methods, the negative survey method attains with lower power and higher degree of privacy protection, and boosts users' confidence. So the negative survey method is applicable to collecting large numbers of data in low-powered mobile devices (such as smart phones, tablets and so on [13]) and high-speed equipments.

Negative survey method divides the users' privacy information into $c(c \geq 3)$ categories for users to send. In order to protect the privacy information of each user, a category that does *not* agree with the fact [5,6] (i.e. a category from the other $c - 1$ unreal categories) is sent to the cloud platform. For convenience, the category that agrees with the fact is defined to be *positive category*, and the other $c - 1$ categories that does *not* agree with the fact [5] is defined to be *negative category*. Because only one negative category is sent to the cloud platform, the privacy degree can be protected effectively. Meanwhile, the distribution of privacy data (i.e. positive category) can be reconstructed relatively accurately. Negative survey method makes users like to participate the privacy information collection, because it doest not collect the privacy information directly. In Gaussian Negative Survey (GNS) [17], the probabilities of selecting negative categories follow a Gaussian distribution centered at the corresponding positive category. The GNS could attain higher accuracy but lower ability of privacy protection.

In [2], two reconstructing method solved that the traditional reconstructing method in [5] may lead reconstructed positive categories with negative values. In [1], an algorithm is used for calculating the confidence level with the analysis method being a generating function.

This study analyses the accuracy of the Negative Survey-based privacy protection of cloud data. The probability of being negative and the confidence level can then be obtained easily. The work indicates that the Negative Survey-based privacy protection method of cloud data is suitable for finding hot categories from a great numbers of cloud data users. Specially designed simulation experiments verify the proposed formulas.

In the remainder of this study, Section 2 introduces the related work of this study. Section 3 describes the calculation of the accuracy of the reconstructed positive survey, and two corollaries (the probability of being negative, and the confident level) are given. Section 5 discusses some problems and Section 6 concludes the whole study.

2   Related Work

In this section, the related work of negative survey [5,6] is introduced. Some definitions are described in Figure 1 for convenience.

Define n to be the number of users who use the negative survey method to send their privacy data, and c to be the number of categories. The results of the privacy data collected in the cloud platform are $R = (r_1, r_2, \cdots, r_c)$, where $r_i(1 \leq i \leq c, c \geq 3)$ represents the total number of users who send the i-th category to

n : the number of users sending privacy data
c : the number of categories in surveys
r_i : the number of users sending category i in negative survey method
t_i : the original number of users in positive category i
$\hat{t}_i$: the estimated number of t_i
R : the user vector, i.e. $R = (r_1, r_2, \cdots, r_c)$
T : the user vector, i.e. $T = (t_1, t_2, \cdots, t_c)$
p_i : the proportion of positive category $i (1 \leq i \leq c)$, i.e. $p_i = t_i/n$

Fig. 1. The definitions in this study

the cloud platform. Similarly, the real privacy data is $T = (t_1, t_2, \cdots, t_c)$, and $n = \sum_{i=1}^{c} r_i = \sum_{i=1}^{c} t_i$. In [5,6], the reconstructed positive survey of privacy data can be calculated by Formula (1).

$$\hat{t}_j = n - (c - 1)r_j \qquad (1)$$

Although $\hat{t}_j = E(t_j)$, it can be observed that $\hat{t}_i < 0$ when $r_i > n/(c-1)$. Therefore, this traditional method is not practical sometimes, and two methods [2] were proposed to solve the negative value problem.

This study analyses the accuracy of Negative Survey-based privacy protection of cloud data by central limit theorem, and two corollaries can then be easily obtained to calculate the probability of being negative and the confidence level. The theorem and the two corollaries indicate that the negative survey method is applicable to collecting users' privacy data.

3   Accuracy Analysis of the Reconstructed Privacy Data

This section analyzes the reconstructed accuracy of privacy data in Theorem 1. Based on Theorem 1, the probability that the negative values are producing in a reconstructed positive can be calculated in Corollary 1, and the confidence level of negative survey can be calculated in Corollary 2.

Theorem 1. *The accuracy can be quantified as*

$$P\left\{ \left| \frac{\hat{t}_i}{n} - p_i \right| < \varepsilon \right\} \approx 2\Phi\left(\frac{\sqrt{n}\varepsilon}{\sqrt{(c-2)(1-p_i)}} \right) - 1. \qquad (2)$$

where n is the number of participants for survey, ε is the estimated precision, $\hat{t}_i$ is the estimated number of category i, and p_i is the original proportion of category i.

Proof. Consider the negative category i and calculate the probability distribution of r_i. In the negative survey, $n - t_i$ interviewees are likely to select the i-th category. Define the random variable $X_j (j = 1, 2, \cdots, n - t_i)$. If the j-th interviewee selects the i-th category, $X_j = 1$, or else $X_j = 0$. Obviously, each X_j is independent and identically distributed, and follows the Binomial Distribution $B(n - t_i, 1/(c - 1))$. Let $X = \sum_{j=1}^{n-t_i} X_j$. So $r_i = X$, and

$$E(r_i) = \frac{n - t_i}{c - 1}, D(r_i) = \frac{\sqrt{(c - 2)(n - t_i)}}{c - 1}. \tag{3}$$

Owing to the De Moivre – Laplace central limit theorem, r_i follows the Normal Distribution as n goes to infinity, i.e.

$$r_i \sim N(\mu, \sigma^2) = N\left(\frac{n - t_i}{c - 1}, \frac{(c - 2)(n - t_i)}{(c - 1)^2}\right) \tag{4}$$

From Formula (1), it can be observed that

$$P\left\{\left|\frac{\hat{t}_i}{n} - p_i\right| < \varepsilon\right\} = P\left\{\frac{|r_i(c - 1) - n(1 - p_i)|}{n} < \varepsilon\right\} \tag{5}$$

According to Formula (3), Formula (5) can be normalized as Formula (6).

$$P\left\{\frac{|r_i - E(r_i)|}{D(r_i)} < \frac{\sqrt{n}\varepsilon}{\sqrt{(c - 2)(1 - p_i)}}\right\} \tag{6}$$

Combing Formula (4), Formula (5) and Formula (6), Formula (2) holds and Theorem 1 is valid.

From Theorem 1, the following three conclusions can be observed:

1. the greater p_i is, the higher accuracy the negative survey is.
2. the greater n is, the higher accuracy the negative survey is.
3. the less c (depending on privacy degree) is, the higher accuracy the negative survey is.

The projections of Formula (2) are shown in Figure 2. In this figure, the probability that $\varepsilon < 0.01$ increases with the increasing of n or p_i, and increases with the decreasing of c. Figure 2(a) illustrates the probability as $c = 5$ with different values of $n(n = 10^3, 10^4, 10^5, 10^6)$. Figure 2(b) shows the probability as $n = 1.2 \times 10^5$ with different values of $c(c = 3, 5, 7, 9)$.

The probability that a category of reconstructed positive survey is negative can be calculated by Corollary 1.

Corollary 1. *The probability that $\hat{t}_i < 0$ can be calculated as*

$$P(\hat{t}_i < 0) \approx 1 - \Phi\left(\frac{p_i\sqrt{n}}{\sqrt{(c - 2)(1 - p_i)}}\right). \tag{7}$$

where n is the number of interviewees for survey, t_i is the number of interviewees in the original positive category i, and $p_i = t_i/n$.

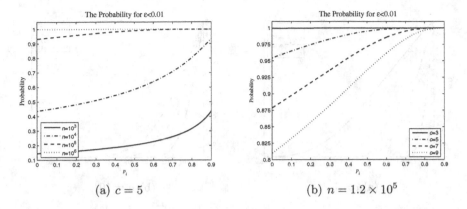

Fig. 2. The Probability for $\varepsilon < 0.01$ with different values of p_i and n

Proof. As $\hat{t}_i < 0$, $\frac{\hat{t}_i}{n} - p_i < -p_i$. According to Formula (2) in Theorem 1,

$$P(\hat{t}_i < 0) = 1 - \frac{P\{|\frac{\hat{t}_i}{n} - p_i| < p_i\}}{2} \approx 1 - \Phi\left(\frac{p_i\sqrt{n}}{\sqrt{(c-2)(1-p_i)}}\right) \qquad (8)$$

it can be observed that Formula (7) is valid, and Corollary 1 is valid.

Especially, $P(\hat{t}_i < 0) = 1/2$ if $p_i = 0$. This means the negative survey produces negative values with probability $1/2$ if the original positive category is empty.

From Corollary 1, the following three conclusions can be observed:

1. the greater p_i is, the less probability that the $\hat{t}_i$ is negative.
2. the greater n is, the less probability that the $\hat{t}_i$ is negative.
3. the less c (depending on privacy degree) is, the less probability that the $\hat{t}_i$ is negative.

Figure 3 shows the function graph of the projections of Formula (7). It is apparent from Figure 3 that the probability that $\hat{t}_i < 0$ decreases as the increasing of n or p, and decreases as the decreasing of c. Figure 3(a) illustrates the change trend of the probability of being negative as $c = 5$ and p_i ranging from 0 to 0.5. Figure 3(b) illustrates that when $n = 10^5$, p_i ranges from 0 to 0.09.

The confidence level can be calculated in Corollary 2.

Corollary 2. *If the confidence level is $1 - \alpha$, and the sampling error is ε, the number of participants n and that of categories c should satisfy*

$$n \geq \varepsilon^2 u_{\frac{\alpha}{2}}^2 (c - 2) \qquad (9)$$

where c is the number of categories, and $u_{\frac{\alpha}{2}}$ is the quantile $\Phi^{-1}\left(\frac{\alpha}{2}\right)$ of the Standard Normal Distribution.

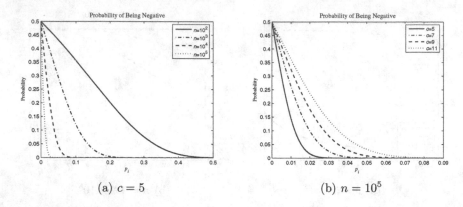

(a) $c = 5$ (b) $n = 10^5$

Fig. 3. Probability of being negative with different values of p_i and n

Proof. According to Formula (2) in Theorem 1,

$$\Phi\left(\frac{\varepsilon\sqrt{n}}{\sqrt{(c-2)(1-p_i)}}\right) \geq 1 - \frac{\alpha}{2} = \Phi\left(u_{\frac{\alpha}{2}}\right) \tag{10}$$

Backtrack the tables for statistical distributions – Normal Distribution function,

$$\frac{\varepsilon\sqrt{n}}{\sqrt{(c-2)(1-p_i)}} \geq u_{\frac{\alpha}{2}} \tag{11}$$

So n, c and p_i should satisfy the condition:

$$n \geq \varepsilon^2 u_{\frac{\alpha}{2}}^2 (c-2)(1-p_i) \tag{12}$$

In order that Formula (12) is identical valid for arbitrary p_i, n should be greater than the maximum. In consequence, n and c should satisfy Formula (9), and Corollary 2 is valid.

For example, if $\varepsilon = 0.01$ and $1 - \alpha = 0.95$ then $n \geq 38416(c-2)$.

4   Simulation Experiments

In this section, specially designed simulation experiments are used to verify the theoretic method in Theorem 1 and Corollary 1. There are 5 groups of negative survey for these experiments in Table 1, where the category number c is 5. Each group carries out negative survey for 10^5 times, and the frequencies are used as the experimental values of negative survey.

Figure 4 illustrates the probability of being negative. It can be observed that there is a small gap between the experimental values and the theoretic values.

Table 1. The Five Groups of Distribution for Negative Survey

	Category proportions
NS1	(0, 0.1, 0.2, 0.3, 0.4)
NS2	(0.05, 0.2, 0.5, 0.2, 0.05)
NS3	(0.2, 0.2, 0.2, 0.2, 0.2)
NS4	(0, 0.025, 0.075, 0.225, 0.675)

In Figure 4(a), the number of interviewees is 10^3, and the probability of being negative is 0 as $p_i > 0.2$. In Figure 4(b), the number of interviewees is 5×10^3, and the probability of that is 0 as $p_i > 0.1$.

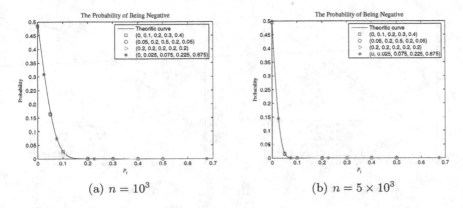

(a) $n = 10^3$ (b) $n = 5 \times 10^3$

Fig. 4. The Probability of Being negative for each category with different values of p_i and n

Figure 5 illustrates the probability that ε (the error between the reconstructed proportion and the original one) is less than 0.01. This figure demonstrates that the probability increases with the increasing of p_i or n. The discrepancy between the experimental values and the theoretic curves is negligible.

5   Discussion

In this study, we analyse and calculate the accuracy of the Negative Survey-based privacy protection of cloud data by central limit theorem, and there are some work for the future study.

Firstly, the analysis method is central limit theorem, which is valid when n is sufficiently large. Sufficiently large n should make np_i and $n(1 - p_i)$ to be sufficiently large, too. Otherwise, Poisson Distribution is the better approximation distribution rather than Normal Distribution. Secondly, the analysis method in this study analyzed the reconstructed category independently. The correlation of different categories should be taken into account.

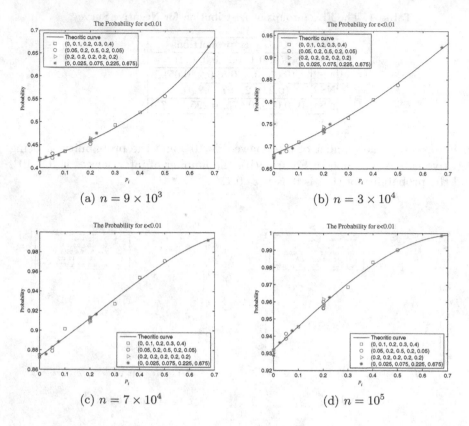

Fig. 5. The Probability for $\varepsilon < 0.01$ with different values of p_i and n

6   Conclusions

In this study, the accuracy of the Negative Survey-based privacy protection of cloud data is analysed and calculated. Depending on the analysis, the probability that the reconstructed values being negative and the confidence level of Negative Survey-based privacy protection can be calculated. According to this study, the accuracy of Negative Survey-based privacy protection increases with the increasing of n (the number of users) or p_i (the proportion of the category i), and increases with the decreasing of c (the number of categories). In other words, the Negative Survey-based privacy protection method is suitable for finding the hot category from large numbers of users, and the amount of calculation of each user is low.

Acknowledgments. The project was supported by the Fundamental Research Funds for the Central Universities, China University of Geosciences (Wuhan) under Grant CUGL 140840, and the National Natural Science Foundation of China under Grant 61272469.

References

1. Bao, Y., Luo, W., Lu, Y.: On the dependable level of the negative survey. Statistics and Probability Letters **89**, 31–40 (2014)
2. Bao, Y., Luo, W., Zhang, X.: Estimating positive surveys from negative surveys. Statistics and Probability Letters **83**(2), 551–558 (2013)
3. Du, G., Huang, T., Zhao, B., Song, L.: Dynamic self-defined immunity model base on data mining for network intrusion detection. In: the 4th International Coference on Machine Learning and Cybernetics, vol. 6, pp. 3866–3870 (2005)
4. Esponda, F.: Negative Representations of Information. Ph.D. thesis, University of New Mexico (2005)
5. Esponda, F.: Negative surveys. Arxiv: math/0608176 (2006)
6. Esponda, F., Guerrero, V.M.: Surveys with negative questions for sensitive items. Statistics and Probability Letters **79**, 2456–2461 (2009)
7. Forrest, S., Perelson, A.S., Allen, L., Cherukuri, R.: Self-nonself discrimination in a computer. In: the IEEE Symposium on Research in Security and Privacy, pp. 202–212 (1994)
8. Foster, I., Zhao, Y., Raicu, I.: Cloud computing and grid computing 360-degree compared. In: Grid Computing Environments Workshop, pp. 1–10 (2008)
9. Groat, M.M., Hey, W., Forrest, S.: Kipda: k-indistinguishable privacy-preserving data aggregation in wireless sensor networks. In: The Thirtieth Annual IEEE International Conference on Computer Communications, pp. 2024–2032 (2011)
10. He, W., Liu, X., Nguyen, H., Nahrstedt, K., Abdelzaher, T.: Pda: privacy- preserving data aggregation in wireless sensor networks. In: IEEE International Conference on Computer Communications, pp. 2045–2053 (2007)
11. Hofmeyr, S.A., Forrest, S.: Architecture for an artificial immune system. Evolutionary Computation **8**(4), 443–473 (2000)
12. Horey, J., Groat, M., Forrest, S., Esponda, F.: Anonymous data collection in sensor networks. In: The Fourth Annual International Conference on Mobile and Ubiquitous Systems: Computing, Networking and Services, pp. 1–8 (2007)
13. Horey, J.L., Forrest, S., Groat, M.: Reconstructing spatial distributions from anonymized locations. In: 28th International Conference on Data Engineering Workshops, pp. 243–250 (2012)
14. Kim, J., Bentley, P.J.: Towards an artificial immune system for network intrusion detection: an investigation of clonal selection with a negative selection operator. In: The 2011 Congress on Evolutionary Computation (CEC 2001), vol. 2, pp. 1244–1252 (2001)
15. Shi, J., Zhang, R., Liu, Y., Zhang, Y.: Prisense: Privacy-preserving data aggregation in people-centric urban sensing systems. In: IEEE International Conference on Computer Communications, pp. 1–9 (2010)
16. Vu, H., Nguyen, T., Mittal, N., Venkatesan, S.: Peq: A privacy-preserving scheme for exact query evaluation in distributed sensor data networks. In: 28th IEEE International Symposium on Reliable Distributed Systems, SRDS 2009. pp. 189–198, September 27–30 2009
17. Xie, H., Kulik, L., Tanin, E.: Privacy-aware collection of aggregate spatial data. Data & Knowledge Engineering **70**(6), 576–595 (2011)

Security Challenges and Mitigations of NFC-Enabled Attendance System

Cheah Boon Chew, Kam Chiang Wei, Tan Wei Sheng,
Manmeet Mahinderjit-Singh$^{(\boxtimes)}$,
Nurul Hashimah Ahamed Hassain Malim, and Mohd Heikal Husin

School of Computer Sciences, University Sains Malaysia, 11800, Penang, Malaysia
{cbchew.ucom12,kcwei.ucom12,twsheng.ucom12}@student.usm.my,
{manmeet,nurulhashimah,heikal}@usm.my

Abstract. Most of the universities or colleges, the lecturer has to take the attendance of the students manually by circulating a paper for them to register their names or calling the names. To date, there are various types of attendance systems that are applying different technologies such as biometrics, tokens and sensors such as RFID. The latest is by applying near-field communication (NFC), a sensor within the smartphone has been used as a mean for recording attendances. The aim of this paper is to list out the possible security attacks against NFC (Near Field Communication) enabled systems by focusing on a student-based attendance system. A brief overview over NFC technology and discussion on various security attacks against NFC in different media is presented. Overall, an attendance system is compromised mainly by tag swapping, tag cloning and manipulation of data occurring on the NFC device and operational server.

Keywords: Attendance system · Near Field Communication (NFC) · Security attack · Mitigations

1 Introduction

Attendance system is a system aim to record and track the attendance of a particular person and is applied in various locations such as the industries, schools, universities and business offices. There are various types of attendance systems that are applied in different technologies such as conventional, biometrics and tokens and sensors such as RFID. In most of the universities or colleges, the lecturer has to take the attendance of the students manually by circulating a paper for them to register their names or calling the names. This method waste time and lead to cheating issue done by the students as they can help their friends to sign for the attendance.

Latest technologies such as biometrics identifiers[1], tokens such as smartcards and sensors such as RFID [2] employed in attendance systems has reduced the weakness of conventional attendance method rapidly. To date, near-field communication (NFC) has been used as a mean for recording attendances. The fact of 1.75 billion users [14] worldwide carries a smartphone, sensor technologies such as NFC can be materialized. By using NFC based attendance system, data can be collected and process in a quicker way compared to manual system. In addition, all the data saved in the server

© Springer International Publishing Switzerland 2015
Y. Tan et al. (Eds.): ICSI-CCI 2015, Part II, LNCS 9141, pp. 160–167, 2015.
DOI: 10.1007/978-3-319-20472-7_18

eliminates redundancy of attendance record and increases tracking and tracing mechanism of attendances. In contrast to lecturers capable of checking the attendance, students can also check their attendance rate using their smartphones through login system from time to time to avoid any miss entering of attendance. This visibility effect increase system efficiency and effectiveness.

However due to contactless reading and identification of transponders without a line of sight, there is security concerns such as attacks on NFC devices can be performed without the acknowledgement of users. Attacks can be done to steal user's information and abuse the attendance record functionality that causes losses to the users.

Hence, the aim of this paper is to discuss the security attacks against NFC (Near Field Communication) that occur on NFC- based systems such as attendance system. The first objective is to present look into various security threats occurring on NFC based applications. Secondly, we will discuss the solutions to these issues. Finally we will look into attendance system as a case study in understanding the security and solutions. The outline of the paper is as follows. Section 2 introduces the background of NFC and NFC Research Framework. Section 3 shows the security attacks of NFC followed by the solutions in section 4. Next section demonstrates the NFC attendance system we developed and the security attack and its mitigation. Finally, we provide a conclusion in the last section.

2   Background of Near Field Communication

NFC can be used with a large diversity of devices like mobile phones, notebooks, printers, speakers, and consumer electronics. NFC enabled applications and services are developed in three different operating modes which are reader/writer mode, card emulation mode and peer-to- peer mode. Read and write mode provides NFC devices to read and modify data stored in NFC compliant passive transponders such as NFC tags while card emulation mode provides NFC devices to act as a standard smart card for payment and ticketing purpose. Lastly, peer-to-peer mode enables two NFC devices to establish a communication to exchange data or other kind of information.

Near Field Communication (NFC) is a short range, low bandwidth, high frequency and wireless communication technology based on Radio Frequency Identification (RFID) technology which can transfer data over a distance up to 10 cm only by simply touching with another NFC device or reader. In this NFC research framework, there are four group which are NFC theory and development, NFC infrastructure, NFC applications and services and NFC ecosystem.

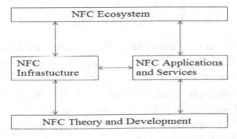

Fig. 1. Classified Framework for NFC Research [1]

Figure 1 shows NFC research framework. The usage of latest model of smartphone ensures the blooming of NFC research field. In this paper, focus is given on the security challenges which arise within the development of NFC technology. An overview of various attacks and how some of the attack contributes to the security vulnerabilities in application such as attendance system is discussed.

3   Security Concerns of NFC Technology

Due to contactless reading and identification of transponders without a line of sight and integration of wireless communication technology with applications such as payment and ticketing in one device may raise potential privacy issues and security risks [2]. There are 2 types of attack which are active attack and passive attack. Attacks against the NFC device can be performed anytime and anywhere and may not be noticed by the victim as the communication is contactless. Currently, the NFC Forum is standardizing the data format (NDEF; NFC Data Exchange Format) specifying the data types and data structure on the tag [3]. Such tags can store information for opening a URL in a web browser or pairing information for a Wi-Fi or Bluetooth connection. More generally speaking, by touching a tag containing an appropriate NDEF, the mobile device is about to launch an application to process the NDEF data [2]. Security measures of mobile are intentionally circumvented by the user via jail breaking or rooting to gain "improved" control over the device or to bypass Digital Rights Management (DRM) [4]. Jail breaking or rooting is not limited to experienced users.

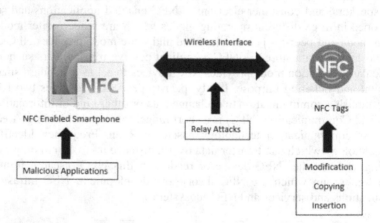

Fig. 2. Three Types of security concerned

After jail breaking or rooting event take place, malicious applications may be used since privileges elevated. Most of the users are negligence and carelessness, often install applications without review of requested permissions. Then, Apps installed even if they request dangerous combinations of privileges [4]. Figure 2 present three key elements of NFC security vulnerability points on which security attacks happen. The three elements NFC enabled devices, NFC tags and communication channel (over the air interface) will be discussed in next section.

3.1   Attack on NFC Device

The attack on device is the attacks on mobile phones such as mobile malware and privilege escalation and mobile phone theft. Jail breaking and rooting are the popular ways to avoid security measures set by mobile phone platform. These methods are used mainly for user or application to gain access it usually cannot access by breaking the restriction of the operating system. This will open further vulnerabilities that might be abused to gain access to restricted resources [4]. Stealing of data, damage on devices and alteration of confidential information can be done through the access gained by malware. Some examples of malware are Trojans, worms, botnets and viruses. Malwares can be affected by installing the applications from unofficial markets as these applications are not reviewed by the security team [5]. There are high risks of attacks such as the mobile phones are stolen as such devices are valuable to thief. According to the survey [6], there are approximately 10 % of the participants' mobile phones were stolen which means that the data stored in the devices could be facing high security risk of unauthorized access.

3.2   Attack on NFC Tag

The attack on tag shows an attacker target on the NFC reader and tag. The attacks included cloning/copying of tags, modification of data and tag swapping issue. A threat of NFC tags is the copying the tag's ID and any other associated data to the blank tag to create the tag with same content as the original [7]. This act of cloning or copying tags can easily to be done and leads to losses in money and trustworthiness towards the technology [8]. Since the security level of NFC tag is low and many NFC tags are reprogrammable, modification on the tag is based on the signal modulation is used [9]. The impact of data modification depends on the system scenario the information that was modified. Tag swapping is a quite simple tactics and is a threat in retail product tracking and automated sales processing [7]. Tag swapping is a process that involved in removing NFC tag from a tagged object and attaches it to another one NFC with modified data. For example, tag swapping is an attacker or thief that picks the higher price item but the tag has been switched with the tags of low price item. The integrity of back-end system would be violated due to its failure to track the correctly ID of the correlate items.

3.3   Attack over Air Interface

The attack over air interface means attack occurring without contact of NFC devices or NFC tags. Among attacks occurring at this stage is DOS, Man-in-the middle, relay attack, eavesdropping and data insertion. DOS attack is possible when an attacker generates collisions for every possible device address and simulates the existence of a high amount of devices in range of the reader. The attack of man-in the middle is when two parties are communicating with each other through a third participant with their knowledge. An authentication system would not help, because the attacker can also intercept and set up one secure channel to the first party and a second secure channel to the second one [10]. The attacker can launch a relay attack by using another communication channel as an intermediary to increase the range. The attacker needs no physical access to the device, but only an antenna and the relay in reading range [10]. In order to access the NFC tag, two proxy devices are set up between the tag and reader. Since NFC systems communicate over an open and accessible air medium with electromagnetic waves, hence, eavesdropping is a logical attack. Because the attacker does not need the power of the active part of the communication for answering, he is able to amplify weak signals received over a distance up to 30 - 40cm. The attacker could insert a message into the communication before the information is sent to the original receiver. This would be only successful if the transmission is finished, before the answering device starts with its answer. Otherwise the message would be corrupted.

4   Mitigation Solutions for NFC Attack

There are such many potential NFC attack discovered based on three type of attacks such as attack on NFC device, NFC tag and over air interface. The following are the suggested and recommendation possible solutions for those attacks.

4.1   Solution for Attack on NFC Device

In order to protect against secure element and mobile malware can be achieved in various ways such as secure element and antivirus solutions. *"Secure element is the combination of hardware, software, interfaces and protocols embedded in a mobile device that enable secure storage"*. Secure element can provide security for payment (e.g. payment data, key) and execution of applications. There are different forms of secure element classified as non-removable and removable such as Secure Memory Card, Universal Integrated Circuit Card (UICC), Baseband processor and embedded hardware [11]. Many antivirus products for mobile phones had been offered by many antivirus vendors like Kapersky, Symantec, F-Secure, ESET, McAfee and others. These products are still using traditional signature-based detection techniques that monitor execution traces and file accesses. The major challenges for these products is come from social engineering which may tempt the users skip the warning provided by antivirus software. Due to weak runtime privilege control on mobile systems, antivirus software could be neutralized once malware is launched [12].

4.2   Solution for Attack on NFC Tag

There are some possible countermeasures for attacks on NFC tag by the usage of digital signature [9], unique identifier and faraday cage approach [13].Every NFC tag has a unique identifier that is pre-programmed at the factory and constant throughout the lifetime of the hardware. The tag identifier is a low level serial number, it used for anti-collision and identification and this technique works perfectly in curbing cloning attack. Faraday cage is a metal mesh or foil container which is impenetrable by any kind of electromagnetic waves [13]. Faraday cages can use in preserving data privacy in NFC tag.

4.3   Solution for Attack over Air Interface

Man in the Middle attack is a challenging attack to be performed. For this to occur, three devices have to be in a single range in order to disturb each other. To get a stable working communication, the attacker in the middle has to shield the connection between the other two devices. This results in an attack if one of the parties is removed and replaced. Such an attack could be prevented by the use of authentication through a common, independent, trusted certification provider [10]. Passive eavesdropping attack can occur up to a distance of 30 - 40cm which limits the possibilities for an attacker to hide either himself or his equipment. However, in certain situations like a crowded underground train at rush hour, the attacking equipment can be placed in a bag to avoid suspicion and the owners of the NFC devices would, thus, be unaware that their device is being surreptitiously read from a passerby. To avoid this type of attack, the host device would need an application, which asks for permission, i.e. by entering a PIN code, before granting access to the data. As there are cases where the NFC function should also work even when the host device is short of energy or is switched off, there should also be the possibility to disable the NFC function. A simple mechanical switch would solve this requirement. Switching off the NFC functionality would then prevent an attacker from skimming the NFC data while walking by [10].

5   Case Study: NFC-Based Student Attendance System Security Attacks and Its Mitigations

We have implemented a NFC-based attendance system capable of collecting and processing data in a quicker way compared to manual system.NFC provides convenient means of collecting student/employees records. However, our scope of implementation is reduced to student based attendance system. There are three class users for this project which is the students, lecturers and system administrators. Each of the users has different privilege and the right given is based on their job scopes. For instances, a student is only able to view his records and nothing else in contrast to the lecturer who is capable of adding courses and students names into the system. Next we will present the system architecture of NFC-based attendance system and the security attacks and its mitigation.

5.1   System Architecture on Student-Based Attendance System

The overview on how the system function is described next. Firstly, the administrator of the school needs to create an account for the students and lecturers in order for them to login to the system. The admin are able to update the account and delete the account in case of wrong data is entered. Besides that, admin should generate a list of the students that enrolled in the particular subject for lecturers' reference. On the mobile app, students need to login to their account in order to register for the attendance for each class that they attend. They also can view the amount of attendance for respective subjects on their phones. For the lecturers, they need to login to the system first and select the subject every time they want to record the attendance. The lecturers will be able to calculate the total attendance of the class and generate a report about the attendance rate at the end of the semester. Based on Figure 3, the architecture of the attendance system is presented.

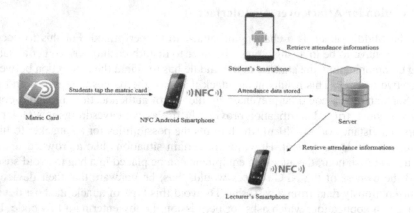

Fig. 3. NFC-based Attendance System Architecture

Next we will discuss the security attacks occurring on NFC attendance system and its mitigation.

5.2   Student-Based Attendance Systems Security Threats and Its Mitigation

There are two scenarios in which an attendance system can be attack in term of its security. First scenario is when a student is absence from a class but would trick the system in showing he has attended all are 80% of the course class. The second scenario is when a student that has not enrolls for the course but would like to join it without paying the fees for it. The first scenario in which an absent student could trick the system by the following technique such as; i) DOS attack – to compromised the server availability by performing jamming of server by multiple packet requests; ii) manipulation of data by either deleting and inserting additional data. This act can be done on the server by either an insider or an outsider (with the help of man in the middle attack) and finally iii) tag swapping in which genuine's tag data is replaced without the knowledge of its user. The second scenario on the other hand is an act of masquerading in which a genuine users information is manipulated by act such as eavesdropping and man-in-the middle. This valid information is then changed and added on blank NFC tag. The system will be unable to detect this kind of behavior unless detection and constant monitoring mechanism is planted in hand. Overall, the effect of the attacks taking place compromises the availability, confidentiality and integrity of the user, devices and servers. The mitigation for data manipulation on server and tag would be to ensure data is trusted to be alive by usage of random numbers mechanism. Attack such as man in the middle and eavesdropping could be mitigated as well with the use of either nonce or high level random number generators. In addition, key encryption between the server and NFC tag by using either symmetric or asymmetric key encryption will also solve the issue of data manipulation and insertion, tag swapping and tag cloning since the integrity of user digital identity can be maintained.

6   Conclusion

In this paper, NFC is analyzed through three main components which are attacks on air interface, NFC enabled devices and NFC tag. All possible and recommended solutions on tackling the attacks are also provided. An attendance system based on NFC is also presented and the security attacks and mitigation technique have also been discussed. In the future, we will venture into enhancing the security of NFC based attendance system by using multimodal authentication mechanisms.

References

1. Özdenizci, B., Aydin, M., Coskun, V., Kerem, O.K.: NFC Research Framework: A Literature Review and Future Research Directions. Information Technologies Department, ISIK University, Istanbul
2. Madlmayr, G., Langer, J., Scharinger, J., Kantner, C.: NFC Devices: Security and Privacy. The Third International Conference on Availability, Reliability and Security
3. Francis, L., Hancke, G., Mayes, K., Markantonakis, K.: Practical NFC peer-to-peer relay attack using mobile phones. In: Ors Yalcin, S.B. (ed.) RFIDSec 2010. LNCS, vol. 6370, pp. 35–49. Springer, Heidelberg (2010)
4. Roland, M.: Practical Attack Scenarios on Secure Element-enabled Mobile Devices. 4th International Workshop on Near Field Communication, Helsinki, Finland, March 13, 2012
5. Porter Felt, A., Finifter, M., Chin, E., Hanna, S., Wagner, D.: A Survey of Mobile Malware in the Wild. University of California, Berkeley
6. Breitinger, F., Nickel, C.: User Survey on Phone Security and Usage
7. Mitrokotsa, A., Beye, M.R.T., Lopez, P.: Classification of RFID Threats based on Security Principles. Security Lab, Faculty of Electrical Engineering, Mathematics and Computer Science, Delft University of Technology (TU Delft), Mekelweg 4, 2628 CD
8. Aigner, M., Dominikus, S., Feldhofer, M.: A System of Secure Virtual Coupons Using NFC Technology. In: Proceedings of the Fifth Annual IEEE International Conference on Pervasive Computing and Communications Workshops © (2007)
9. Kilas, M.: Digital Signatures on NFC Tags. Master of Science Thesis, March 18, 2009
10. Church, L., Moloney, M.: State of the Art for Near Field Communication: security and privacy within the field. Escher Group Ltd, Ireland, 3rd draft: May 10, 2012
11. Reveilhac, M., Pasquet, M.: Promising Secure Element Alternatives for NFCTechnology. 2009 First International Workshop on Near Field Communication
12. Yan, Q., Li, Y., Li, T., Deng, R.: Insights into Malware Detection and Preventionon Mobile Phones
13. Balitanas, O.M., Kim, T.: Review: Security Threats for RFID-Sensor Network Anti-Collision Protocol. Hannam University, Department of Multimedia Engineering, Postfach
14. Greeshma, M.: Global Smartphone Users to Touch 1.75 billion in 2014, January 20, 2014 (accessed on January 2, 2015)

Automation Control

Robotic Rehabilitation of Lower Limbs "Reproduction of Human Physiological Movements"

Mohamed Amine Mamou$^{(\boxtimes)}$ and Nadia Saadia

Bab Ezzouar, Electronic and Computing Faculty (FEI), Robotics Parallelism
and Embedded Systems (LRPE), University of Science and Technology
Houari Boumediene (USTHB), Algiers, Algeria
mamouma@gmail.com, saadia_nadia@hotmail.com

Abstract. Robotic devices are particularly useful applications in the field of functional rehabilitation following neurological injury such as the spinal cord and stroke which lead to motor impairments. Different rehabilitation devices members are under study to assist therapists in their work. In this paper we propose to reproduce human walking movements on a robotic rehabilitation chair of lower limbs, produced in the laboratory (LRPE). To do, we apply real walking signals on developed control law for this purpose. This law using kinematic model based on neural networks (Feed forward neural network (FFNN). Reproduction actual walking movement made from OpenSim database, is open-source software allows users to that develop, analyze, and visualize models of the musculoskeletal system, to generate dynamic simulations of movement, and compare with data base's movement. Proposed control law provides a high performance and fast convergence with extremely low error, and offers optimal reproduction of movement during human walking, and a secure rehabilitation.

Keywords: Robotic rehabilitation · Neuronal network control · Lowers limbs · Human physiological movements · Kinematic model · Path tracking

1 Introduction

The nervous system gradually building internal models by experience, they use them in combination with impedance and feedback control strategies, these models are developed and finalized in childhood [1]. The use of robotic rehabilitation for the practice of repetitive motion, using improved motor recovery, they eventually improve learning and motor rehabilitation beyond levels possible with conventional technology. This is possible because it can replace the physical training efforts of a therapist, allowing more intensive and repetitive motions delivering therapy at a reasonable cost for assessment and quantitatively the level of motor recovery by measuring strength and movement patterns [2]. Therefore, the interest to use it in therapy is increasingly growing, for example the need for functional rehabilitation after stroke or following neurological injuries such as lesions spinal cord [1,3]. Rehabilitation devices managed with its control algorithms, the last sound implemented so that the exercise to be executed by the participant cause motor plasticity, therefore, improves motor plasticity [4]. Many rehabilitation devices of human limbs and joints are the

© Springer International Publishing Switzerland 2015
Y. Tan et al. (Eds.): ICSI-CCI 2015, Part II, LNCS 9141, pp. 171–179, 2015.
DOI: 10.1007/978-3-319-20472-7_19

subject of study. In this context, we developed in LRPE laboratory a stationaries rehabilitation prototype of lower limbs. The device is a robotic chair for neuromeric rehabilitation of lower limb with two motorized orthotics (right and left leg), with three degree of freedom (3DOF) for each of them (hip, knee and ankle). Position sensors are mounted on each link to manage the flow of a predetermined motion in real time [5]

In order to have a good operation of the device, to obtain optimal reproduction of physiological movements such as walking, precise controller with rapid response to changes in the joints is required. Beforehand we need to define the control strategy to this end we have a state of the art non-exhaustive. Marchal group's people into four categories: Assisting, challenge-based, Simulating normal tasks, and non-contact [4].

To develop a controller we are interested in the laws of commands used on rehabilitation devices for lower limbs, and we group them into two categories "Gait pattern generator" and "finite state machine". In hierarchical control scheme the control laws are trying to imitate the behavior of human biological members during the march, from which they are inspired. To achieve this, they regulate the mechanical impedance, stiffness parameters, damping, speed and/or other criteria. These strategies require algorithms to detect movement phases, each of them and/or under phase controller [6 to 8]. In Gait pattern generator scheme control, the control, signal is preprogrammed, based primarily on the fact that walking is a periodic motion; the controller must manage the pace according to information, such as position, velocity, impedance, the electromyography (EMG) signals etc. [5,9 to 12].

Our work goes into the second category, it lies in the fact of reproducing the real motions of human walking and controlling the trajectory of a rehabilitation system of lower limbs (robotic rehabilitation chair) using the kinematic model of the robotic rehabilitation based on Feed forward neural network (FFNN). The use of neural networks for system control of lower limbs is not common; In addition we aim to improve the results obtained with a PID, and this because the error of the latter had strong variations on each joint, caused mainly by signal control. This type of signal is prejudicial for electronic component and actuators. The originality of this work is to use the error function and its derivative to control the system. The objective of this strategy is to cancel the error, especially when changing the direction of the trajectory and smoothing the control signal. As a result we have effective breeding movements of walking with a holy individual. To generate same walking signal, we used the open-source platform "OpenSim". This one permitted to modeling, Simulating, and Analyzing the neuro-musculoskeletal system [13]. These analysis tools calculate joint forces, muscle-induced accelerations, muscle powers, angle joints and other variables [14]. This software has a database containing analyzes of healthy individuals; we implement our controller to mimic the physiological gait movements.

This paper is arranged as follows. First, we present the rehabilitation system used and we establish the geometric and kinematic model before proposing its control law. Then, we present the control system for parameters identification based on a conventional PID and the scheme with FFNN controller. Then, we will implement the controller on the system. Finally, a discussion of results and conclusions are presented.

2   System Description

We performed a robotic rehabilitation device for lower members dedicated to patients with functional deficiency. The prototype shown below, consists in reality of functional orthoses, it allows the reproduction of physiological joint trajectories and take back loads of segmental body movement, especially walking. It comprises a seat mounted on a frame with two mechanical braces, placed on each side. Each orthotic work on sagittal plane, has three degrees of freedom, consists of three joints (hip, knee and ankle) and three segments (thigh, leg and foot). Joints ensure transmission of movement between different segments taking into account the factor of safety and patient's weight and waist using a mechanism consisting of rods, gears and DCM [15]

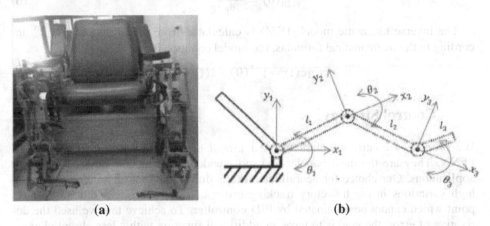

(a)   (b)

Fig. 1.a. Lower limbs rehabilitation chair robot. **b.** Link coordinate system of chair's frame.

l_1, l_2 and l_3 are the lengths corresponding to the thigh, leg and foot. They can be manually adjusted. They can be manually adjusted. We choose respectively the values *50, 60* and *25*cm, θ_i is the angle between Oxi and l_i

In our approach, we establish the basic coordinate system with chair's frame, represented by Ox_1y_1 (Fig 1.b). We usual introduce a fixed coordinate system (frame) in which all objects is referenced to create the geometric system model. The coordinates of the point are given according to the plan Ox_2y_2 which Ox_2 superimposed on l_1. In the same way, the coordinates of the last point are given on Ox_3y_3 plan; where Ox_3 is superposed on l_2.The tool's coordinates are expressed in this coordinate system [16]. The direct geometric model (DGM) is represented by the relation:

$$x = f(\theta) \tag{1}$$

Where θ is the vector of joint coordinates such as:

$$\theta = [\theta_1 \ \theta_2 \ \theta_3]^T \tag{2}$$

The vector x is defined by elements of the homogeneous transformation matrix $^\alpha A_\beta$ [17]. This matrix gives frame coordinate R_β from those of frame R_α. We denote:

$C_i = \cos(\theta_i)$; $S_i = \sin(\theta_i)$; $C_{ilk} = \cos(\theta_i + \theta_l + \theta_k)$; $S_{ilk} = (\theta_i + \theta_l + \theta_k)$;

Matrix which leads to frame coordinate R_3 from those of frame R_0 are given by $^0A_3$

$$^0A_3 = {}^0A_1 *{}^1A_2 *{}^2A_3 \tag{3}$$

$$^0A_3 = \begin{bmatrix} C_{123} & -S_{123} & 0 & l_1C_1 + l_2C_{12} + l_3C_{123} \\ S_{123} & C_{123} & 0 & l_1S_1 + l_2S_{12} + l_3S_{123} \\ 0 & 0 & 1 & 0 \\ 0 & 0 & 0 & 1 \end{bmatrix} \tag{4}$$

The Direct Kinematic Model (DKM) is given by [18]

$$\dot{x}(t) = J(\theta) * \dot{\theta}(t) \tag{5}$$

Where $J(\theta)$ denotes the (i x j) Jacobian matrix. The element «J_{ij} (θ) » is given by:

$$J_{ij}(\theta) = \frac{\delta f_i(\theta)}{\theta_j} \tag{6}$$

The inverse kinematic model (IKM) is calculated from the inverse matrix J^{-1} according to the mathematical formulas, the model equation is:

$$\dot{\theta}(t) = J^{-1}(\theta) * \dot{x}(t) \tag{7}$$

3   Control Strategy

We propose a control structure using a kinematic model of the device based on FFNN. They are the most popular and most widely used models in many practical applications. Our choice for neural networks is due to the fact of wanting decreased high variations in the trajectory tracking error; caused by changes direction of set point which cannot be eliminated by PID controller. To achieve this we used the derivative of error, the goal is to have an additional function with a less abrupt change, thanks to the specificity of the derivative. If $de/dt \leq 0$ then e is decreasing, if $de/dt \geq 0$ then e is growing.

To develop our controller law we done two steps: In the first we compute the NN parameters; in the second we test it to control the rehabilitation lower limb chair. The two orthoses have the same architecture, therefore the same models. Thereby, we present the command of a single one. To simplify the control scheme, we use a decoupled architecture, where each joint is controlled by an independent controller. This is possible in view of the mechanics of the device and the desired trajectories.

In the first step, to perform the learning process, the FFNN controller is trained through a classical PID controller. Identification of parameters was done by constructing a database from PID data controller, this step shown in Fig. 2.

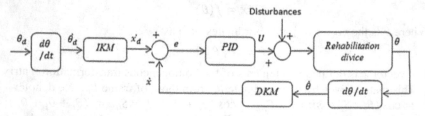

Fig. 2. Control low shame with PID

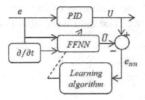

Fig. 3. Copy of PID controller. e is the error between the desired output trajectory and the trajectory of the device. U is the desired control; $\hat{U}$ is thee estimated control; e_{nn} is the NN error between the estimated control and the desired.

The PID is used to identify and provide the required training data. In training the network, its parameters are adjusted incrementally until the training data satisfy the desired mapping as well as possible; that is, until $(\hat{U})$ matches the desired output U as closely as possible up to a maximum number of iterations. To train the FFNN we use Levenberg-Marquardt optimization [19].

The second step an adaptation of the neural controller to regulate its parameters according to the task performed is used (Fig 4).

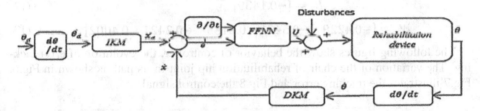

Fig. 4. FFNN control law Structure

The FFNN used is composed of three layers. The input layer consists of two neurons, the hidden layer of five neurons, and the output layer of one neuron. The matrix output « V » of hidden layer and the output « U » of network are given by:

$$V = f(W_{1ij}E + B_{1i}) \tag{8}$$
$$U = f(W_{2i}V + B_{21}) \tag{9}$$

Where (W_{Ji1}, W_{Ji2}) and (B_{Ji}, B_1) are respectively the weights and bias matrices to adjust. And $i = 0, ... ,5$ is the number of neurons in input layer and $j = 0, ... ,2$ is the number of neurons in hidden layer. f(.) and g(.) are the activation function. They are choosing as sinusoidal for hidden layer, and as linear for the neurons of output layer.

4   Implementation and Tests

The command structure that we have established is validated by generating gait of human walking, and implements it on lower limbs robotic rehabilitation. This trajectory is a challenge that is the subject of discussion within the scientific community. Marchal has made a detailed study that includes several works on this subject [3]. The figure below is this layout of the variation of hip angle, knee and ankle during two cycles of healthy human walking. The data were obtained from OpenSim library.

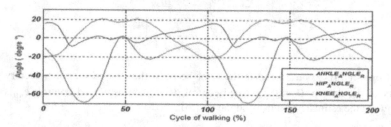

Fig. 5. Human walking cycle

Neural network's parameters (weights & biases) obtained after the learning are given:

$$B_1 = [-3.1364 \quad 0.5202 \quad -0.4258 \quad 1.7687 \quad 2.8117] \tag{10}$$

$$W_1 = \begin{bmatrix} 2.0986 & -1.3270 & 0.7758 & 1.8407 & 0.6019 \\ 2.3193 & -0.2283 & -2.6654 & 2.4756 & -3.3121 \end{bmatrix}^T \tag{11}$$

$$B_2 = [-0.1439] \tag{12}$$

$$W_2 = [-0.4228 \quad -0.5471 \quad 0.2135 \quad -0.2421 \quad 0.4002] \tag{13}$$

The following figures show the behavior of controller, the evolution of error tacking. The variation of the chair of rehabilitation hip joint in its path is shown in Fig.6. Fig.7 illustrates the tracking error and Fig.8 the control signal.

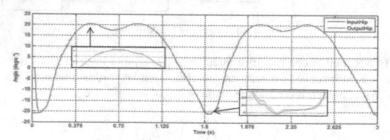

Fig. 6. Varying path of hip joint

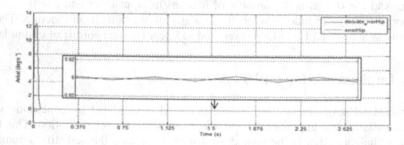

Fig. 7. Hip joint error tracking

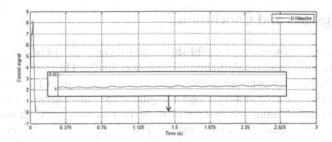

Fig. 8. Hip joint control signal (U)

We repeated the same thing, giving the motion tracking, error and the control signal successively, of the knee and ankle joint at the left and right respectively.

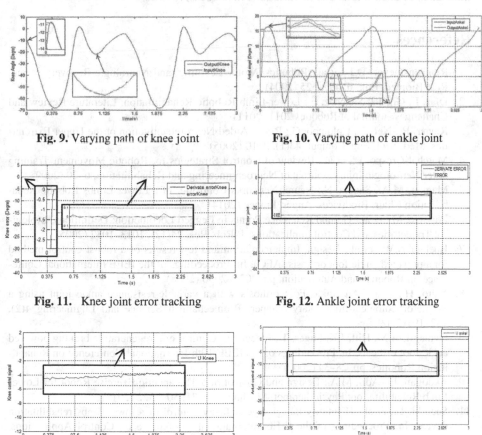

Fig. 9. Varying path of knee joint   **Fig. 10.** Varying path of ankle joint

Fig. 11. Knee joint error tracking   **Fig. 12.** Ankle joint error tracking

Fig. 13. Knee joint control signal   **Fig. 14.** Ankle joint control signal

5   Discussion and Conclusion

The objective of this work is to establish an intelligent control law using neural networks for a robotic chair rehabilitation of the lower limbs. The implementation of the control law is done by generating the signal physiological joint trajectories. For the smallest mistake possible and try to get a signal error with minimal oscillations, we conducted several training attempts, with more or less time. A high oscillation frequency of the error signal has a large variation of the control signal, which is harmful for electronic components, such as motors. Therefore, we introduced the derivative of the error in the second FFNN controller input. The tests of the proposed control show a good performance. Indeed all the errors joints are low and the control signal does not have a sudden change. However, the error signal shape of each municipality has thousands of oscillations due to variations in successive orders.

References

1. Reinkensmeyer, D.J., et al.: Robotics, Motor Learning, and Neurologic Recovery. Annu. Rev. Biomed. Eng. **6**, 497–525 (2004)
2. Díaz, I., Gil, J.J, Sánchez, E.: Lower-Limb Robotic Rehabilitation: Literature Review and Challenges. Journal of Robotics 2011 (2011)
3. Riener, R., Nef, T., Colombo, G.: Robot-Aided Neurorehabilitation of the Upper Extremities. Med. Biol. Eng. Comput. **43**(1), 2–10 (2005)
4. Marchal-Crespo, L., et al.: Review of Control Strategies for Robotic Movement Training After Neurologic Injury. Journal of Neuroengineering and Rehabilitation, 6–20 (2009)
5. Mamou, M.A., Saadia, N.: Neural Networks Controller of a Lower Limbs Robotic Rehabilitation Chair. Biodivices, Anger, France (2014)
6. Popovic, M., et al.: Ground Reference Points in Legged Locomotion: Definitions, Biological Trajectories and Control Implications. J. Rob. Res. **24**(12), 1013–1032 (2005)
7. Böehler A.W., et al.: Design, Implementation and test results of a robust control method for a powered ankle foot orthosis (AFO). In: Proceedings of the IEEE International Conference on Robotics and Automation, pp. 2025–2030 (2008)
8. Naito, H., et al.: An Ankle–Foot orthosis with a variable-resistance ankle joint using a magnetorheological-fluid rotary damper. Biomechanical Science and Engineering **4**(2), 182–191(2009)
9. Oymagil, A.M, Hitt, J.K, Sugar, T., Fleeger J.: Control of a regenerative braking powered ankle foot orthosis. In: Proceedings of the IEEE 10th International Conference on Rehabilitation Robotics, pp. 28–34 (2007)
10. Akdogan, E., Adli, M.A.: Design and Control of a Therapeutic Exercise Robot for Lower Limb Rehabilitation: Physiotherabot. Mechatronics **21**, 509–522 (2011)
11. Seddiki, L., et al.: H-Infinity takagi-sugeno fuzzy control of a lower limbs rehabilitation device. In: Proceedings of the IEEE International Conference on Control Applications, Munich, 06, pp. 927–932 (2006)
12. Yousef Tohidi, R., et al.: Design and Computer Simulation of a Software Controller for a Hardware Robot with 2 Degrees of Freedom for Lower Limb Physiotherapy. The Special issue in Management and Technology, 371–381 (2014)
13. OpenSim Workshop, University of Michigan, Ann Arbor, (2008)

14. Delp, S.L., et al.: OpenSim: Open-Source Software to Create and Analyze Dynamic Simulations of Movement. IEEE Transactions on Biomedical Engineering **54**(11), 1940–1950 (2007)
15. Merrouche, L.M.: Conception d'orthèses fonctionnelles pour les paraplégiques. Mémoire, USTHB, Alger (2011)
16. Spong, M.W., et al.: Robot Modeling and Control, vol. 3. Wiley, New York (2006)
17. Dombre, E., Wisama, K.: Modeling, Performance Analysis and Control of Robot Manipulators. ISTE, 31–33 (2010)
18. Jazar, R.N.: Theory of Applied Robotics: Kinematics, Dynamics, and Control. Springer Science & Business Media (2010)
19. Lourakis, M.I.A.: A Brief Description of the Levenberg-Marquardt Algorithm Implemented by Levmar. Foundation of Research and Technology **4**, 1–6 (2005)

Achievement of a Myoelectric Clamp Provided by an Optical Shifting Control for Upper Limb Amputations

Sofiane Ibrahim Benchabane[✉] and Nadia Saadia

Electronic and Computing Faculty (FEI), Robotics Parallelism and Embeded Systems (LRPE),
University of Science and Technology Houari Boumediene (USTHB), Algiers, Algeria
benchabane.ibrahim@gmail.com, saadia_nadia@hotmail.com

Abstract. The prehension is a complex biomechanical process; it involves nearly 200 muscles and a large number of joint and bones. Replicate this process is a complex and challenging technology. Nevertheless, it is possible to target apart of this process in order to develop systems to provide a degree of autonomy to people with handicap linked to an amputation of the hand, and the muscles of the forearm are still functional. This paper is about the design of a prototype myoelectric clamp, which can grasp and hold a wide range of usual objects. So the system is activated by EMG stimulations and relays on force and slip feedback to achieve the grasping task. In this work we show the feasibility of a low cost and efficient clamp that can maintain objects by avoiding accidental falls or damages caused by unregulated force applied over the grasped item.

Keywords: Myoelectric clamp · Humanoid prosthesis · EMG signal · Artificial hand · Force feedback

1 Introduction

Since the eighties, myoelectric systems are built, in order to give back some mobility, to persons suffering from handicap caused by an amputation of the upper or lower limbs. In 1986 a patent has been filed under the reference US4623354 [1] for the figment of a myoelectric clamp, operated by a myoelctric signal sensed over the muscles of the forearm. It was only able to grip or release items, without any control of the slid or the strength applied. In 1987 another patent was filed under the reference US4650492 [2], for achieving a myoelctric hand which can provide to gripe an items and hold it, using a slid control, the system is running after the myoelctric excitation, he uses a touch sensor to attest the begging of the contact between the hand and the object and then, controlling the position of the actuator according to the data provided by the slip sensor. Nowadays, myoelctric systems are made and marketed, for example the I-LIMB ULTRA [3], this hand has a humanoid design, with a carpe and five fingers independently animated by using encoding excitation, the hand grip is more human like. However this system can't support the slip control and there is no automatically grip or touch feedback. Other work is also carried out in this field, like myoelectric arm designed in APL John Hopkins university, this one is controlled by chest myoelctric signal, indeed the team has graft the nerves of the amputee arm on the chest [4,5,6,7,8]. This is more natural way for

© Springer International Publishing Switzerland 2015
Y. Tan et al. (Eds.): ICSI-CCI 2015, Part II, LNCS 9141, pp. 180–188, 2015.
DOI: 10.1007/978-3-319-20472-7_20

controlling prosthetic, but it needs lot of training in order to be able to activate some area of the chest rather than other, this system is also not provided with any slip control or sensitive feedback. So using such a system that is not equipped by a slipping control is very laborious for amputees, indeed in this case the user can only rely on a visual feedback, which often leads to accidents. Some myoelectric systems try to associate a vibrotactile feedback to the visual feedback [9], in fact the authors have equipped an anthropomorphic hand with a vibration device which is activated when a pressure is sensed over the fingers. Those vibrations are modulated according to the perceived strength. Such a system actually allows somehow the user to feel the object but it is not enough to prevent from grasping accidents, the user needs a lot of training to correlate the strength applied with the vibrations felt. Especially as a lot of studies shows that nerves ending stimulated in case of vibrotactile stimulations are different from those stimulated in case of a real grasp [10,11,12]. At the EPFL prosthesis with a haptic feedback has been developed. Indeed the patient may even evaluate the hardness of an object, this is achieved by electrical stimulation of the median and radial nerves. The method consist in surgically connect excitation electrodes to the mentioned nerves [13]. This way the patient would be able to regulate his grasp corresponding to what he feels. Even if this method allows the patient to somehow feel the grasp, its remains complicated to achieve because of the delicacy of surgery, also because of the haptic method used, in fact the electrical stimulation is a controversial approach [14,15,16]. Some robotic hands try to bring a mechanical response to those grasping control, in fact they are mechanically designed to keep holding items without any feedback or control system. The DEXTEROU HAND designed by SCHUNK Company is one example. This hand is provided by 03 fingers which can encompass the item, this prevent from slides but that do not prevent from damages because of no force control. The CANADARM used by NASA is also another example, he is fitted by a grasping ring which avoids any item skipping, it could be a good perspective when adapted to a hand proportion, but those solutions does not fit with an assistive anthropomorphic system. To summarise, to achieve an assistive myoelectric prosthesis it is necessary to consider the grasping control, in order to avoid damages that could be caused by the fall or an undue grasping force. That's why some systems try to incorporate a haptic feedback that allows the experimented user to regulate his grasp according to what he feels. However those approaches are still complicated and lead to a lot of work and spending. We can also consider mechanical approaches but that do not resolve the problematic. Our approach consists to set a unit that automatically controls the grasp. It is possible to proceed using a set of force sensors and merge data in order to find marker, but it involves a lot of sensors and data acquisition and fusion thus a relatively tedious process. It's also possible to proceed using a shifting sensor, and relay on the data provided to compute the strength of the grasp. This is in fact the goal of the work conducted and presented over this paper.

2   Proposed Structure

We designed and manufactured our own myoelctric platform in order to achieve and test the grasping control system, in this context we had to:

- Acquire the myoelectric signal;
- Design an efficient mechanical effector;
- Generate a command based on data collected.

Our work has focused on the development with a small cost, of a prototype of myoelectric system, which is able to allow the user to make easy grip for a lot of daily items, using EMG data and slip sensor. So this paper is about the realization of a clamp prototype, ordered to seizure and retention a range of items by electromyography. The system thus produced must ensure retention, this by analyzing the relative sliding of the requests object, by using an optical slide sensor. Indeed, for this purpose we will use a simple device for measuring relative displacements and very replied, with a good resolution: the pointing device or computer optical mouse. Finally developed system will be a myoelectric clamp, ensuring the maintenance object by analyzing their relative sliding with the assistance of a computer optical mouse. The system can be represented by the following diagram:

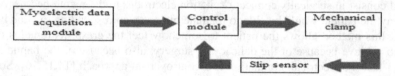

Fig. 1. Overall scheme of the system, shows how the system is conceived, the control module ensure the grasp according to the myoelectric data, and data provided by the slip sensor

3   Myoelectric Data Acquisition

The acquisition chain is operated like any other one, it is based on amplification, filtering and detection as shown in figure1 below. Indeed, the signal is detected by the electrode and amplified with a gain of 60db, and this one is then filtered and passed through a stage of detection up around an operational amplifier.

Emg Electrode. The electrode EMG are simple electrode Ag\AgCl, identical to the electrode used in ECG, our goal in this work is not to check what type of electrode is best to acquire myoelctric signal, however most studies bearing on myoelectric signals acquisition appeal to the electrode Ag\AgCl [17,18,19,20,21].

Amplification. The amplification as illustrate in figure 3. Is based on an instrumentation amplifier AD622 followed by a second amplifier OP177AP. The gain of this chain is about 60 db. This gain has been fixed experimentally, in fact the measurement performed confirm what found through literature. The myoelectrique signal delivered by the EMG electrode is of the order of microvolt (from 50 to 300) [22, 23].

Filtering. This process uses a band pass filter whose bandwidth is 70Hz to 200Hz, which included the majority of the spectral density of the myoelctric signal [24,25,26,27,28]. We used a 2nd order Butterworth filter with a slop of 40db/decade. This type of filter is known for its stable gain (see fig. 2).

The final step of the acquisition chain is the detection, at this level the amplified and filtered signal is detected if high enough, and the control circuit treats the data. At this point we have seen the acquisition chain of the myoelectric signal, in order to detect the will of the patient to close his hand and to grip an object.

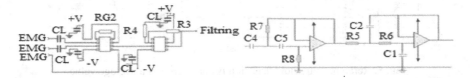

Fig. 2. Amplification circuit (left) and 2nd order Butterworth band pass filtering circuit (right)

4   Gripping Process

The gripping process is divided into two stages, the first step is to grasp the object and the second one is to ensure the maintenance.

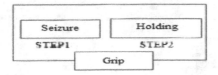

Fig. 3. Diagram of gripping process, the figure illustrates two steps, the first one concern the seizure (first contact), the second one the holding phase

Seizure Step. This step is to ensure the occurrence of an interaction between the clamp and the object. It comes to a person to feel that his hand is placed over an object, so the system needs to sense this step using a force sensor. The sensor used in this application is the FSR. This sensor is used with an amplifier AD620 doubling the amplitude of the signal outcome from the FSR, given the low amplification gain and the absence of major vibration phenomenon we don't use filter at this stage. Once the signal is amplified he enters into a detection unit as shown by the following figure:

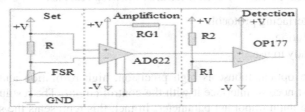

Fig. 4. Scheme of the conditioning chain designed to acquire data from the FSR sensor

The threshold of detection used in the detection unit is 1.2V; this value was fixed experimentally as detailed in the following paragraph. A FSR sensor was fixed over the thumb of six volunteers, and a gradually increasing stress was applied, the volunteers were asked to indicate when they feel characterized stress, the results are transcribed in the following figure:

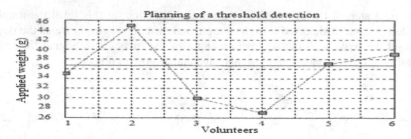

Fig. 5. Reference threshold detection is considered as a average value

The figure shows that the average stress felt for an effective touch corresponds to a weight of 36.6g. In our application we decided to match a weight of 40g as a contact detection threshold. To know the response of the FSR, we conducted a simple experience, which consists to put a calibrated weight of 40g over the sensor the obtained results are shown on the following figure.

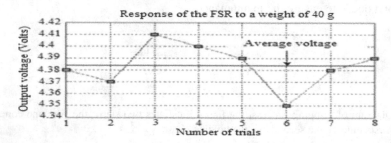

Fig. 6. Repeatability curve for the detection threshold of the FSR, which has led to an average voltage of 4.38 V, as shown by the blue dart on the figure

Bearing. During this step, the system keeps the object from sliding, and maintaining it. For this he uses data from the sensor, and if a shift whose amplitude is greater than the senor resolution is detected, the holding force increases.

Sliding Sensor. The device used to sense any shift during the bearing step, is a simple computer optical mouse, we used it for the following reasons:

- Simple interfacing protocol;
- High resolution;
- Device ready to use.

Indeed, the optical mouse is high precision, high resolution and already conditioned, we just need to interface it with the control circuit. The resolution of the optical mouse is a main important parameter; In fact, this represents the sensitivity of the clamp relative to the sliding as explained. The resolution of the optical mouse is given with the notice of the device and its value is 800 dpi (dot per inch), but for more precision we have to proceed to an experimental measurement. Experience conducted is summarized in the following. We lay the optical mouse over a linear graduated track, and then we read the data sent by the mouse corresponding to a shifting along the track. The displacement counters of the mouse are initialized to zero when we shift the mouse over a distance of 05mm, the X counter register is 132. That means the resolution measured is 05/132 = 0.0378mm. We can behold that we measure a difference of 0.00612 between the resolution given by the notice and what we found. From this experience we can deduce that the system may detect a sliding which amplitude is greater than 0.0378mm.

5   System Performance

The system developed shall be tried with a range of experiences; our tests will concern the following items:

- The span of the object may be held by the clamp;
- The maximum force that can be applied by the clamp;
- The precision of the system.

Span. The system where designed in order to ensure the kept of different usual objects. As a cup, a spoon, a cell phone or a pen. So the spacing between phalanxes is 7cm without incorporating the sensors, and it reduces to 5.5cm after.

Maximum Force. The maximum force applied by the clamp is depending from the potential landslides. Indeed more the slip is important, more the force required to maintain is high. To measure this force, we switched the optical mouse connected to the system by another one on the table, once the clamp is closed over an object. Than we slowly shift the second mouse to make the system increasing the force applied. And we visualize on a digital oscilloscope the output of the force sensor (figure7).

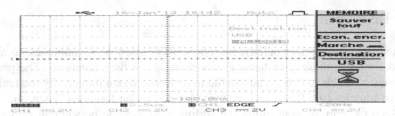

Fig. 7. FSR output voltage for a maximum stress is about 1.8Volts which corresponds to a stress induced by a weight of 1.6kg.

Ability to Grip. To test the gripping we will interest to the grasp of some usual objects as a raw egg, to exemplify gripping of brittle items. A screwdriver as usual tool, a deformable plastic cup, a cell phone and a plastic pen (photo 1, figure 8). The raw egg is in fact a very weak object it could be easily damaged, so the seizure of this kind of thinks proves that our system is precise and his touch is very soft. The experience focused on a raw egg, which weighs 45g (photo 2).

Photo. 1. Clamp firmly holding a range of usual items

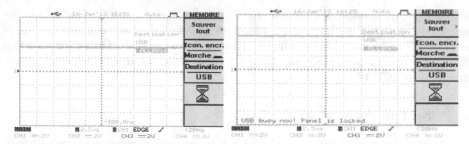

Fig. 8. (Left)FSR output voltage for a stress induced by holding a screwdriver. This figure shows the output of the FSR during the grip, we can see that the voltage delivered by the sensor is 3.8V that corresponds to a stress induced by a weight of 90g. (Right) FSR output voltage for a stress induced by holding the cup. As shown by the photo 3 the system has also succeeded to grip a plastic cup without major twist, the output voltage given by the FSR is 4.3V as shown by the figure. That corresponds to a weight about 50g.

Photos. 2. The clamp holding a raw egg. The egg was taken by horizontal sides successfully, without any damage, the test was repeated any the results were the same.

6   Discussion

Our goal is to perfect a myoelctric system so as to allow hand amputees, to achieve some daily tasks, those have become laborious, given their handicap. The proposed system must be able to detect and use the myoelectric signal induced by the tensing of involved muscles (the flexor and extensor located in the forearm). Then, the system must keep the grip, by controlling the sliding, this way the person can grip an item and hold it successfully without damages, as he shall do it before the amputation. The grip is realized using sliding information given by an optical sensor, this one has a resolution of about 800 dots per inch, which is sufficiently precise to make sure that the griped item is hold, so the system is running this way: When the will of closing hand is detected by the myoelectric module, the gripper is animated, the actuator make it close, until a touch is sensed trough the FSR. If a stress greater than the threshold is received the system enters the controlling mode. Indeed, at this point if a sliding which amplitude is over 0.0378mm occurs the force of the grip is increased. The system is constantly monitoring the sliding information or myoelectric excitation to decide the following actions to undertake. The tests realized show that the clamp built is able to gripe and hold a wide range of usual items, the grip is made softly to prevent frangible items from damaging. Most bionic systems available nowadays are devoid from any grip control, and the only way for the users to ensure the holding of a grasped object is to rely on a visual feedback. Unfortunately that requires a lot of training and attentiveness. Unless to incorporate a haptic feedback which can represent to the user the exact feeling of the grasp. The proposed approach provides those kinds of systems by an automatic holding control using an optical shift sensor to compute any sliding during the grasp and then increase proportionally the strength. This work also shows that a myoelectric prosthesis, easy to command and which can also provide a slip control can be developed with an approachable cost.

References

1. Childress, D., Strysik, J.: Myoelectrically Controlled Artificial Hand. Patent US 4623354 (1986)
2. Barkhordar, M., Nightingale, J.M., May, D.R.W.: Artificial Hands. Patent US 4650492 (1987)
3. TouchBionics Active Prosthesis. http://www.touchbionics.com
4. Kuiken, T.: Consideration of Nerve-Muscle Grafts to Improve the Control of Artificial Arms. Technology & Disability 15, 105–111 (2003)
5. Miller, L., Lipschutz, R., Stubblefield, K., Lock, B.A., Huang, H., Williams, T., Weir, R., Kuiken, T.: Control of a Six Degree of Freedom Prosthetic Arm After Targeted Muscle Reinnervation Surgery. Archives of Physical Medicine and Rehabilitation 89, 2057–2065 (2008)
6. Kuiken, T., Dumanian, G., Lipschutz, R., Miller, L., Stubblefield, K.: The Use of Targeted Muscle Reinnervation for Improved Myoelectric Prosthesis Control in a Bilateral Shoulder Disarticulation Amputee. Prosthetics and Orthotics International 28, 245–253 (2004)
7. Kuiken, T., Li, G., Lock, B.A., Lipschutz, R., Miller, L., Stubblefield, K., Englehart, K.: Targeted Muscle Reinnervation for Real-Time Myoelectric Control of Multifunction Artificial Arms. The Journal of the American Medical Association 301, 619–628 (2009)
8. Zhou, P., Kuiken, T.: Eliminating Cardiac Contamination from Myoelectric Control Signals Developed by Targeted Muscle Reinnervation. Physiological Measurement 27, 1311 (2006)
9. Cipriani, C., Zaccone, F., Micera, S., Carrozza, M.C.: On the Shared Control of an EMG-Controlled Prosthetic Hand: Analysis of User-Prosthesis Interaction. IEEE Transaction on Robotics 24, 170–184 (2008)
10. Johansson, R.S., Landstrom, U., Londstrom, R.: Responses of Mechanoreceptive Afferent Units in the Glabrous Skin of the Human Hand to Sinusoidal Skin Displacements. Experimental Brain Research 244, 17–25 (1982)
11. Johansson, R.S., Westling, G.: Signals in Tactile Afferents From the Fingers Eliciting Adaptive Motor Responses During Precision Grip. Experimental Brain Research 66, 141–154 (1987)
12. Johansson, R.S., Westling, G.: Responses in Glabrous Skin Mechanoreceptors During Precision Grip in Humans. Experimental Brain Research 66, 128–140 (1987)
13. Raspopovic, S., Capogrosso, M., Petrini, F.M., Bonizzato, M., Rigosa, J., Pino, J.D., Carpaneto, J., Controzzi, M., Boretius, T., Fernandez, E., Granata, G., Oddo, C.M., Citi, L., Ciancio, L.A., Cipriani, C., Carrozza, M.C., Jensen, E., Guglielmelli, T., Stieglitz, P.M., Rossini, S.: Restoring Natural Sensory Feedback in Real-Time Bidirectional Hand Prosthsis. SciTransl. Med. 6, 222 (2014)
14. Shannon, G.F.: A Comparison of Alternative Means of Providing Sensory Feedback on Upper Limb Prosthesis. Medical and Biological Engineering 14, 284–294 (1976)

15. Kaczmarek, K.A., Webster, J.G., Rita, P.B.Y., Tompkins, W.G.: Electrotactile and Vibro-tactile Displays for Sensory Substitution Systems. IEEE Transactions on Biomedical Engineering **38**, 1–16 (1991)
16. Pylatiuk, C., Kargov, A., Schulz, S.: Design and Evaluation of a Low-Cost Force Feedback System for Myoelectric Prosthetic Hands. American Academy of Orthotists and Prosthetists **18**, 5–61 (2006)
17. Zipp, P.: Effect of Electrode Geometry on the Selectivity of Myoelectric Recordings With Surface Electrodes. European Journal of Applied Physiology and Occupational Physiology **50**, 35–40 (1986)
18. HermensH, J., Freriks, B., Disselhorst-Klug, C., Rau, G.: Development of Recommendations for SEMG Sensors and Sensor Placement Procedures. Journal of Electromypgraphy and Kinesiology **10**, 361–374 (2000)
19. Roy, S., De Luca, G., Cheng, M., Johansson, A., Gilmore, L., De Luca, J.: Electro-Mechanical Stability of Surface EMG Sensors. Med. Bio. Eng. Comput. **45**, 447–457 (2007)
20. Ajiboye, A.B., Weir, R.: A Heuristic Fuzzy Logic Approach to EMG Pattern Recognition for Multifunctional Prosthesis Control. IEEE Neural Systems and Rehabilitation **13**, 280–291 (2005)
21. Hargrove, L., Zhou, P., Englehart, K., Kuiken, T.: The Effect of ECG Interference on Pattern Recognition Based Myoelectric Control for Targeted Muscle Reinnervated Patients. IEEE Trans. Biomed. Eng. **56**, 2197–2201 (2009)
22. Frank Netter, H.: Atlas d'Anatomie Humaine. MASSON, 4th edn., Translation of Pierre Kamina (2007)
23. Williams, M., Kirsch, R.: Evaluation of Head Orientation and Neck Muscle EMG Signals as Command Inputs to a Human-Computer Interface for Individuals With High Tetraplegia. IEEE Trans. Neural Syst. Rehabil. Eng. **16**, 48–496 (2008)
24. Zipp, P.: Effect of Electrode Geometry on the Selectivity of Myoelectric Recordings With Surface Electrodes. European Journal of Applied Physiology and Occupational Physiology **50**, 35–40 (1986)
25. Chu, J., Moon, I., Mun, M.: A Real-Time EMG Pattern Recognition System Based on Linear-Nonlinear Feature Projection for a Multifunction Myoelectric Hand. IEEE Biomedical Eng. **53**, 2232–2239 (2006)
26. Pan, T., Fan, L., Chiang, H., Chang, S., Jiang, J.: Mechatronic Experiments Course Design: A Myoelectric Controlled Partial Hand Prosthesis project. IEEE Transactions on Education **47**, 348–355 (2004)
27. Bolek, E.J.: Electrical Concepts in the Surface Electromyographic Signal. Applied Psychophysiology and Biofeedback **35**, 171–175 (2009)
28. Wirta, R., Taylor, D., Fonley, R.: Pattern Recognition Arm Prosthesis: A Historical Perspective-A Final Report. Bulletin of Prosthetics Research **10**, 8–35 (1978)

A Novel Control Approach: Combination of Self-tuning PID and PSO-Based ESO

Yanchun Chang, Feng Pan, Junyi Shu, Weixing Li, and Qi Gao$^{(\boxtimes)}$

School of Automation, Beijing Institute of Technology, Beijing 100081, China
changyanchunbit@gmail.com, andropanfeng@126.com,
piaoyi1435@163.com, {liweixing,gaoqi}@bit.edu.cn

Abstract. This study focuses on the steady speed control of brushless DC motor with load torque disturbance from the cam and spring mechanism. Due to the nonlinearity and complexity of the load torque, the control system proposed in this paper is divided into the inner-loop compensator, which is to feed-forward compensate the disturbance, and the outer-loop controller. The inner-loop compensator uses a nonlinear extended state observer (ESO) to compensate the actual system as a nominal model, and the outer-loop pole assignment self-tuning PID controller is used to stabilize the nonlinear nominal model. Since a set of suitable nonlinear ESO parameters are difficult to get normally, particle swarm optimization (PSO) is employed to optimize the observer. The simulation results with high precision verify the effectiveness of the proposed control system.

Keywords: Brushless DC motor · Nonlinear extended state observer · Particle swarm optimization · Self-tuning PID

1 Introduction

Brushless DC motor (BLDCM) has been used widely for its excellent speed regulating performance. As a kind of normal load mechanisms, cam mechanism transfers motor rotation into regular vibration of load connector. The nonlinear load torque disturbance from cam and spring mechanism is much larger than conventional disturbance such as friction torque, and this is a new challenge of motor steady speed control. If the disturbance action of load torque can be observed timely, it is possible to decrease the impact of periodic disturbance by feed-forward compensation. Han [1] proposed a novel observer, i.e. extended state observer (ESO), which treats the disturbance as an expansive state. Some scholars used compensation control to reject the uncertain external disturbance based on linear ESO [2,3]. However, if the disturbance changes in a great range just as the system studied this paper, the estimation performance of nonlinear ESO is much better than a lineal one [4]. Nevertheless, in the practical application, the lack of experiential knowledge makes it difficult to get a set of suitable nonlinear ESO parameters. Recently, some heuristic optimization algorithms are

© Springer International Publishing Switzerland 2015
Y. Tan et al. (Eds.): ICSI-CCI 2015, Part II, LNCS 9141, pp. 189–199, 2015.
DOI: 10.1007/978-3-319-20472-7_21

used to optimize nonlinear ESO, such as genetic algorithm, neural network and some other improved algorithms [5–7].

Particle swarm optimization (PSO) is widely used in pattern classification, optimization computation and controller parameter optimization, etc, for its good convergence, search capability and low computing complexity [8–10]. Moreover, many relative optimization algorithms are researched in different fields [11–14]. Liu [15] and Ye [16] obtained a positive effect in optimizing controller parameters by using PSO.

Motor speed control system requires explicit dynamic performance indicators, such as setting time, overshoot and risetime, etc. Pole assignment self-tuning PID controller is able to assign the system closed-loop poles to expected position, and make the system satisfy the dynamic response requirements. Moreover, the system is stable when control a non-minimum phase system [17].

The rest of the paper is outlined as follows. The BLDCM and load torque disturbance models are constructed in Section 2. In Section 3, in order to reduce the impact of cam-spring load mechanism on the speed control system, we first design a disturbance feed-forward compensator based on the nonlinear ESO and an optimization strategy by PSO. Then, a self-tuning PID outer-loop controller with parameter estimation is designed. Finally, we verify the effectiveness of the proposed control system by several simulation experiments.

2   System Modeling

In this section, we construct the models of BLDCM and load torque disturbance. It is worth noting that the load torque model constructed in this part is not total accurate. In fact, it is almost impossible to build a perfect load torque model due to its complexity and nonlinearity. The analyses of load torque is mainly to reflect its nonlinear characteristic.

2.1   BLDCM Model

Without lose of generality, considering the situation in which only two phases are working at arbitrary time, we analyze phases A/B here,

$$U_{\mathrm{d}} = u_{\mathrm{AB}} = R \cdot i + L\frac{di}{dt} + K_{\mathrm{e}}\omega \tag{1}$$

where $U_{\mathrm{d}}, u_{\mathrm{AB}}$ are the armature voltage and DC line voltage, respectively; i is the armature current; R is the armature resistance; $L = L_{\mathrm{s}} - M$ is the equivalent inductance, where L_{s}, M are the self-inductance and mutual-inductance, respectively; K_{e} is the voltage constant; ω is the motor rotational speed. Similarly, we can get line voltage u_{BC} and u_{AC}.

In addition the torque equation and motion equation are as follows,

$$\begin{cases} T_{\mathrm{e}} = K_{\mathrm{t}}i \\ T_{\mathrm{e}} - T_{\mathrm{l}} = J\frac{d\omega}{dt} \end{cases} \tag{2}$$

where T_e is the output torque of the motor; K_t is the torque constant; T_l is the load torque; J is the rotor inertia.

The transfer function of PWM amplifier can be approximated as proportional component [18], i.e. $U_d = ku$, where k is the amplification; u is control voltage of the trigger circuit. The BLDCM structure diagram is shown as Fig. 1.

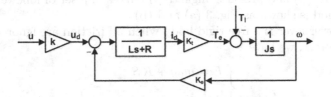

Fig. 1. The model structure diagram of BLDCM

From (1) and (2), the mathematical model of BLDCM can be obtained:

$$\ddot{\omega} = -\frac{KeK_t}{JL}\omega - \frac{R}{T_l}\dot{\omega} - \frac{R}{JL}T_1 - \frac{1}{J}\dot{T_1} + \frac{K_t k}{JL}u \tag{3}$$

2.2 Nonlinear Model of Load Torque

The load torque studied in the paper is a kind of specific mechanism which is made up of five groups of disc cams and springs. We randomly pick a set of cam-spring mechanism to analyse the load torque model, obviously, the total load torque is the linear superposition of five groups of mechanisms. The schematic diagram of one of the five mechanisms is shown as Fig. 2,

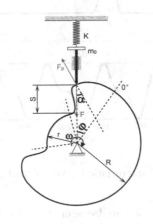

Fig. 2. The schematic diagram of load mechanism

where ω is the motor speed; α is the pressure angle; φ is the position angle of contact point under polar coordinates; r is the small circle radius of disc cam and the spring free length; R is the base circle radius of disc cam; $S \in [0, R-r]$ is the displacement of follower lever; m_0 is the mass of follower lever and platform; K is the elastic coefficient of spring.

According to [19], $\alpha = \arctan \frac{S'(\varphi)}{S'(\varphi)+r}$. The displacement $S(\varphi)$, velocity $S'(\varphi)$, acceleration $S''(\varphi)$ and pressure angle $\alpha(\varphi)$ curves of a set of follower lever in a rotation period is shown as Fig. 3 (a) and (b).

The follower lever is a free rigid body when all friction are ignored, analyses its vertical forces, we have,

$$\begin{cases} F = m_0 g + KS \\ \\ F_p \cos \alpha - F = m_0 a \end{cases} \tag{4}$$

where a is the linear acceleration and it is positively correlated to ω^2 and $S''(\varphi)$. Then the load torque can be described as,

$$T_1 = F_p \cdot (S + r) \sin \alpha \tag{5}$$

and the total load torque of the five load mechanisms in a rotation period is shown as Fig. 3 (c).

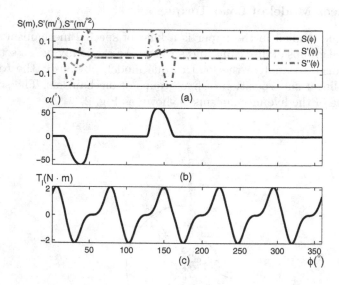

Fig. 3. The characteristic curves of load mechanism

From the above analyses, it can be seen that both the dynamic models of cam and spring are nonlinear, thus the load torque of motor be nonlinear, and the load is much greater than normal load torque disturbance, such as friction. Thus it's extremely significant to design a suitable control system.

3   Control System Design

For the control system designed in this paper, how to decrease the impact of the disturbance is the primary issue. Though the disturbance can be suppressed to some degree by adding negative feedback, the result is far from satisfactory. Another motivation is that if the disturbance action can be observed in advance, the disturbance can be compensated by feed-forward compensation.

3.1   Disturbance Feed-Forward Compensation Based on ESO

ESO is independent with disturbance mathematical model, and there is no need to measure the disturbances directly. For a n-order dynamic system, a $(n + 1)$-order ESO can be used to track output y and to estimate state variables and the real-time action of the total disturbance. Take the second order BLDCM system as an example, (3) can be rewritten as follows,

$$\begin{cases} \dot{\omega}_1 = \omega_2 \\ \dot{\omega}_2 = f_0(\omega_1, \omega_2) + w(t) + bu \\ y = \omega_1 \end{cases} \qquad (6)$$

where $f_0(\omega_1, \omega_2) = -\frac{K_eK_t}{JL}\omega_1 - \frac{R}{L}\omega_2$ is the acceleration of the known parts; $w(t) = -\frac{R}{JL}T_1 - \frac{1}{J}\dot{T}_1$ is the the disturbance action of load torque; $b = \frac{K_tk}{JL}$ is a gain of u. For such a second order system, Han proposed a particular nonlinear ESO in [20] as follows,

$$\begin{cases} e = z_1(k) - y(k) \\ z_1(k + 1) = z_1(k) + h(z_2(k) - \beta_1 e) \\ z_2(k + 1) = z_2(k) + h(z_3(k) - \beta_2 fal(e, \alpha_1, \delta) \\ \qquad\qquad + f_0(z_1(k), z_2(k)) + bu(k)) \\ z_3(k + 1) = z_3(k) - h\beta_3 fal(e, \alpha_2, \delta) \end{cases} \qquad (7)$$

where y is the output; z_1, z_2 are the estimated value of state variables ω_1, ω_2, respectively; z_3 is the real-time action of disturbance $w(t)$; $\beta_1, \beta_2, \beta_3$ are the adjustable parameters of observer, respectively; k is the sampling time; h is the sampling period. $fal(\cdot)$ is a nonlinear function of e described as (8), where $\alpha_1 = 0.5, \alpha_2 = 0.25, \delta = h$.

$$fal(e, \alpha, \delta) = \begin{cases} |e|^\alpha \, sign(e), & |e| > \delta \\ \frac{e}{\delta^{1-\alpha}}, & |e| \leq \delta \end{cases} \qquad (8)$$

Setting $u = u_c - \frac{z_3(k)}{b}$, the system can be approximated as follows,

$$\begin{cases} \dot{\omega}_1 = \omega_2 \\ \dot{\omega}_2 = f_0(\omega_1, \omega_2) + bu_c = -\frac{K_e K_t}{JL}\omega_1 - \frac{R}{L}\omega_2 + bu_c \\ y = \omega_1 \end{cases} \tag{9}$$

From the discussion above, we get that the ESO compensates dynamic disturbances by adding an estimated disturbance feed-forward to the front controller, and it is robust in the case that the system parameters vary in a large range. In this way, the outer-loop controller is simplified to design a nominal BLDCM model without conspicuous external load torque.

3.2 Nonlinear-ESO Parameter Tuning Based on PSO

The ESO can be divided into linear ESO and nonlinear ESO according to the function $fal(\cdot)$. If $fal(\cdot) = e$, (7) is a typical linear ESO. Gao proposed a simple and general parameter setting method of this kind of ESO in [21]. However, if the disturbance varies in a large range, just like the system discussed in this work, nonlinear ESO is much better than a linear one. To get a set of optimal nonlinear ESO parameters, PSO is employed.

In the continuous space coordinate system, the mathematical description of the PSO is as follows [22,23]: A group of n particles in d dimensional search space search at a certain speed, every particle update its speed and position (i.e. the solution) vectors according to its previous best position and the group's (or the neighborhoods') best position. The i-th particle of swarm is presented by three d-dimensional vectors: velocity vector $v_i(t)$, current position vector $x_i(t)$, and previous best position vector $p_i(t)$, x_i is a solution of each iteration. If the current position x_i is better than the previous best position p_i, then the p_i will be replaced by x_i. In addition, the best position of the whole particle swarm so far is denoted as p_g. For each particle, the j-th $(1 \leq j \leq d)$ dimensional speed and position vectors are updated as follows,

$$\begin{cases} v_{id}^{(k+1)} = \omega_p \cdot v_{id}^{(k)} + c_1 \cdot r_{id} \cdot (p_{id}^{(k)} - x_{id}^{(k)}) + c_2 \cdot r_{gd} \cdot (p_{gd}^{(k)} - x_{id}^{(k)}) \\ x_{id}^{(k+1)} = x_{id}^{(k)} + \eta \cdot v_{id}^{(k+1)} \end{cases} \tag{10}$$

where ω_p is the inertia factor; c_1 and c_2 are the accelerating factors; r_{id} and r_{gd} are random numbers that uniformity distribute over $[0,1]$, respectively; η is the restraint factor of speed ratio.

To tune the ESO parameters in this paper, PSO parameters are set as: number of particles $n = 15$; search space dimension (i.e. the number of adjustable parameters) $d = 3$; inertia factor $\omega_p = 0.79$; the accelerating factors $c_1 = c_2 = 1.49$ [24]; restraint factor of speed ratio $\eta = 1$. The fitness function of PSO should be able to reflect the quality of ESO parameters $(\beta_1, \beta_2, \beta_3)$, which is related to the error between the ESO estimation values (z_1, z_2, z_3) and the real values (ω_1, ω_2, w). We select the fitness function F_{fit} in this paper as follow,

$$F_{\text{fit}} = \sum_{i=1}^{L} (\omega_1(i) - z_1(i))^2 \sum_{i=1}^{L} (\omega_2(i) - z_2(i))^2 \sum_{i=1}^{L} (w(i) - z_3(i))^2 \qquad (11)$$

where L is the number of simulation steps. The variables of F_{fit} are z_1, z_2, z_3, ω_1, ω_2, w and L.

3.3   The Control Structure

After disturbance is compensated by feed-forward compensator based on ESO, the system considered now is a nominal one. Pole assignment self-tuning PID is employed here as the outer-loop controller. According to the nominal model which is identified by the parameter estimator online, the parameters of the ESO can be revised by off-line PSO. The control system structure diagram is shown as Fig. 4,

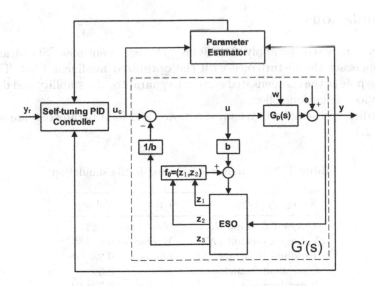

Fig. 4. The control system structure diagram

where w is the disturbance action from load torque; e is the sensor measurement error and environment noise after filtering; $G'(s)$ is the nominal system.

Discretizating the nominal system, we have,

$$y(k) + a_1 y(k-1) + a_2 y(k-2) = b_1 u(k-1) + b_2 u(k-2) + e(k) \qquad (12)$$

Parameter estimator is based on the recursive extended least squares (RELS) method with forgetting factor λ as follows,

$$\begin{cases} \hat{\theta}(k) = \hat{\theta}(k-1) + K(k)[y(k) - \hat{\phi}^T(k)\hat{\theta}(k-1)] \\ K(k) = \frac{P(k-1)\hat{\phi}(k)}{\lambda + \hat{\phi}^T(k)P(k-1)\hat{\phi}(k)} \\ P(k) = \frac{1}{\lambda}[I - K(k)\hat{\phi}^T(k)]P(k-1) \end{cases} \tag{13}$$

where $\phi(k) = [-y(k-1), -y(k-2), u(k-1), u(k-2)]$, $\theta = [a1, a2, b1, b2]^T$.

The stability analyses of the system includes inner-loop stability and outer-loop stability. Pole assignment self-tuning PID controller is employed in the outer-loop, thus we can ensure the stability of the system by assigning all poles into the left-half s plane. However, concerning the proofs of the convergence of inner-loop nonlinear ESO, it's a so attractive and challenging job that scholars worldwide are studying this issue. Zhao [25] has made some progress on this problem.

4   Simulations

In this section, firstly, we apply the PSO to get a set of optimal ESO parameters, and compensate the disturbance with the optimized nonlinear ESO. Then we assign the pole of the compensated system to guarantee the stability and dynamic performance of the controller.

According to the actual system, the motor and load parameters are selected as Table 1.

Table 1. The parameters selected in the simulation

Name	Units	Value
Torque Constant (K_t)	Nm/A	0.034
Voltage Constant (K_e)	V/(rad/s)	0.034
Terminal Resistance (R)	Ohms	0.28
Equivalent Inductance (L)	mH	0.56
Rotor Inertia (J)	kg $\cdot$ m^2	6.72e-04

In the following simulations, the simulation step size $h = 0.001\,\text{s}$, and the simulation time $t = 1\,\text{s}$. From Fig. 5, it shows that fitness value decrease quickly, and converge to a relatively tiny value after 50 iterations.

When the input signal is $u = \sin(t)$ and load torque is $T_1 = \sin(20\,t)$, the observed performance of linear ESO and nonlinear ESO are shown in Fig. 6. We can see that when the disturbance is large and change quickly, the observed performance of nonlinear ESO is significantly better.

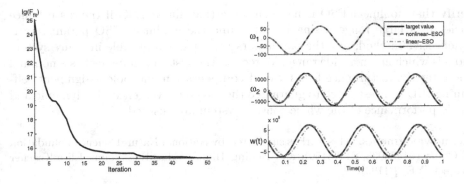

Fig. 5. The convergence of fitness function F_{fit}

Fig. 6. The comparison of observed performance with linear and nonlinear ESO

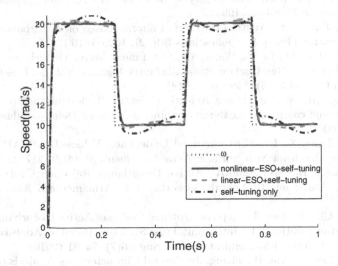

Fig. 7. The control performance comparison

Pole assignment self-tuning PID controller is insensitive to sensor measurement error and environment noise after filtering $e(t)$ and disturbance compensation residual, but is unable to restrain the impact of nonlinear strong disturbance $w(t)$. By using the control system designed in this paper can also eliminate the disturbance of $w(t)$ and $e(t)$, the control performance comparison is shown in Fig. 7.

5   Conclusion

This paper proposes an effective BLDCM steady speed control approach by feed-forward compensating disturbance based on ESO. The simulation results verify that nonlinear ESO is more suitable than linear ESO if the disturbance varies in a large range. Using PSO to tune the nonlinear ESO parameters is an effective method, and the parameters we get are applicable in many system models which are in a wide range. Moreover, the system can be seen as a nominal one after the disturbance feed-forward compensation and pole assignment self-tuning PID is investigated to guarantee the global convergence, the dynamic and static performance of the whole control system are ensured.

Acknowledgments. This work is supported by National Natural Science Foundation of China (61433003, 61273150), and Beijing Higher Education Young Elite Teacher Project (YETP1192).

References

1. Han, J.: The "Extended State Observer" of a Class of Uncertain Systems. Control and Decision **10**(1), 85–88 (1995)
2. Wang, L., Jian-bo, S.: Attitute tracking of aircraft based on disturbance rejection control. Control Theory & Applications **30**(12), 1609–1616 (2013)
3. Zhang, P., Shan, D., Li, C., Wang, Y.: New Linear Active Disturbance Rejection Control Design for Gun Control System Infantry Fighting Vehicle. Fire Control & Command Control **39**(6), 159–162 (2014)
4. Han, J.-Q.: Active Disturbance Rejection Control Technique-the technique for estimating and compensating the uncertainties. National Defense Industry Press, Beijing (2008)
5. Kun, H., Zhang, X., Liu, C.: Unmanned Underwater Vehicle Depth ADRC Based on Genetic Algorithm Near Surface. Acta Armamentarii **34**(2), 217–222 (2013)
6. Qi, X., Li, J., Han, S.: Adaptive Active Disturbance Rejection Control and Its Simulation Based on BP Neural Networks. Acta Armamentarii **34**(6), 776–782 (2013)
7. Chen, W., Chu, F., Yan, S.: Stepwise Optimal Design of Active Disturbances Rejection Vibration Controller for Intelligent Truss Structure Based on Adaptive Genetic Algorithm. Journal of Mechanical Engineering **46**(7), 74–81 (2010)
8. Pan, F., Chen, J., Xin, B., Zhang, J.: Several Characteristics Analysis of Particle Swarm Optimizer. Acta Automatica Sinica **35**(7), 1010–1016 (2009)
9. Li, Y., Zhan, Z., Lin, S., Zhang, J., Luo, X.: Competitive and Cooperative Particle Swarm Optimization with Information Sharing Mechanism for Global Optimization Problems. Information Sciences **293**(1), 370–382 (2015)
10. Shen, M., Zhan, Z., Chen, W., Gong, Y., Zhang, J., Li, Y.: Bi-Velocity Discrete Particle Swarm Optimization and Its Application to Multicast Routing Problem in Communication Networks. IEEE Transactions on Industrial Electronics **61**(12), 7141–7151 (2014)
11. Valdez, F., Melin, P., Castillo, O.: An Improved Evolutionary Method with Fuzzy Logic for Combining Particle Swarm Optimization and Genetic Algorithms. Applied Soft Computing **11**(2), 2625–2632 (2011)
12. Precup, R.-E., David, R.-C., Petriu, E.M., Preitl, S., Paul, A.S.: Gravitational Search Algorithm-Based Tuning of Fuzzy Control Systems with a Reduced Parametric Sensitivity. In: Gaspar-Cunha, A., Takahashi, R., Schaefer, G., Costa, L. (eds.) Soft Computing in Industrial Applications. AISC, vol. 96, pp. 141–150. Springer, Heidelberg (2011)

13. Zhou, W., Chow, T.W.S., Cheng, S., Shi, Y.-H.: Contour Gradient Optimization. International Journal of Swarm Intelligence Research **4**(2), 1–28 (2013)
14. El-Hefnawy, N.A.: Solving Bi-level Problems Using Modified Particle Swarm Optimization Algorithm. International Journal of Artificial Intelligence **12**(2), 88–101 (2014)
15. Liu, Z., Zhang, Y.: Coefficient diagram method based on PSO and its application in aerospace. Flight Dynamics **28**(6), 64–67 (2010)
16. Ye, Y., Lin, H.: Drying Room Temperature Predictive Functional Control Based on PSO Parameter Estimation. Machinery **50**, 21–24 (2012)
17. Dong, N.: Adaptive Control. Beijing Institute of Technology Press, Beijing (2009)
18. Li, H.: Electric drive control system, 100–109. Electronic Industry Press, Beijing (2006)
19. Zou, H., Dong, S.: Modern Design of Cam Mechanisms, 99–100. Shanghai Jiao Tong University Press, Shanghai (1991)
20. Han, J.: From PID Technique to Active Disturbances Rejection Control Technique. Control Engineering of China **9**(3), 13–18 (2002)
21. Gao, Z.-Q.: Scaling and Bandwidth-Parameterization Based Controller Tuning. In: Proceedings of the American Control Conference, pp. 4989–4996 (2003)
22. Kennedy, J.: The particle swarm: social adaptation of knowledge. In: Proceedings of the IEEE International Conference on Evolutionary Computation. IEEE Service Center, Piscataway (1997)
23. Pan, F., Zhang, Q., Liu, J., Li, W., Gao, Q.: Consensus analysis for a class of stochastic PSO algorithm. Applied Soft Computing **23**, 567–578 (2014)
24. Clerc, M., Kennedy, J.: The particle swarm: explosion, stability and convergence in a multi-dimensional complex space. IEEE Transactions on Evolutionary Computation **6**(1), 58–73 (2002)
25. Zhao, Z-L.: Convergence of Nonlinear Active Disturbance Rejection Control, 55–104. University of Science and Technology of China (2012)

Tools for Monitoring and Parameter Visualization in Computer Control Systems of Industrial Robots

Lilija I. Martinova, Sergey S. Sokolov$^{(\boxtimes)}$, and Petr A. Nikishechkin

Computer Control Systems Department, FSBEI HPE MSTU «STANKIN», Moscow, Russia
liliya-martinova@yandex.ru, sokolov@ncsystems.ru,
petrnikishechkin@gmail.com

Abstract. The article describes basic principles of creating software tools for monitoring and parameter visualization in computer control system of industrial robots. Concept of creating extendable data monitoring subsystem with open modular software architecture purposed. Architectural models, algorithms and software implementation of control system components on key levels of hierarchy such as collecting diagnosis data in real-time subsystem, data exchange with HMI devices and data visualization with help of predefined and user configured dialog screens are discovered. Implementation aspects of tools for monitoring and parameter visualization explained in application to different types of HMI devices like hand control panel, commissioning and programming software or remote terminal software.

Keywords: HMI · Monitoring · Visualization · Mechatronic equipment · Open architecture · XML

1 Introduction

Analysis of evolution of modern computer control systems reveals the trend to combine the different purpose software and create a common information environment for industrial automation projects. In such a project common software tools for implementing monitoring and parameter visualization subsystems should be used to support integration of equipment from different vendors [1].

2 Prerequisites for Creating Tools for Monitoring and Parameter Visualization

Most of modern computer control systems are based on dual-computer architecture concept: one computer serves the real-time task, executes NC part-program, controls actuators and processes PLC signals, while the other computer provides user interface for configuration, control and diagnosis of this system [2, 3]. Interface computer commonly provides the indication of limited amount of basic system parameters for monitoring purposes, such as the state of part program execution or current axes

© Springer International Publishing Switzerland 2015
Y. Tan et al. (Eds.): ICSI-CCI 2015, Part II, LNCS 9141, pp. 200–207, 2015.
DOI: 10.1007/978-3-319-20472-7_22

coordinates. All other system parameters are either not available for real-time monitoring at all, or displayed in table form which is not so much suitable for operator to interpret and evaluate parameter value. This issue raises the task of creating the concept for building extendable monitoring and parameter visualization subsystem with open modular software architecture [4, 5]. The requirements to the parameter monitoring and visualization subsystem are systematized in form of mind map displayed on Pic. 1.

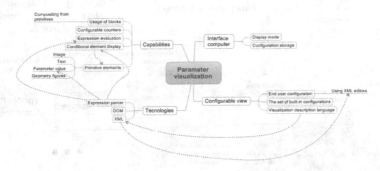

Fig. 1. Requirements to the parameter visualization subsystem

Parameter visualization subsystem should provide the following functionality:

- displaying of graphical elements (geometry figures, parameter values in text form, images, etc.);
- usage of block structure consists of primitive elements for creating visualization window content to make it easier to configure displaying of repeated elements and provide the possibility to reuse previously made visualization blocks;
- change the properties of visualization element (position, size, color, etc.) depending on certain conditions described in configuration file using the logical and numerical expressions;
- usage of counters with configurable intervals for creating of animation.

3   Architectural Model of Tools for Monitoring and Parameter Visualization

By the reason of control system separation into kernel and user interface parts monitoring and parameter visualization subsystem also consists of two parts: monitoring and parameter visualization agent acting in real-time part of system and user interface part (Fig. 2).

The basic function of monitoring agent is to interact with specific data sources inside the control system kernel [2, 6], such as configuration parameter sets, control system state parameters, internal operation data of motion or logic algorithms, or even directly communicate with executive devices. It also performs the initial data processing and afterwards transmits collected data to HMI applications using one of the available communication channels.

HMI part of monitoring and visualization subsystem is able to connect to one or several monitoring agents and provide the client application with services to read and write parameter values.

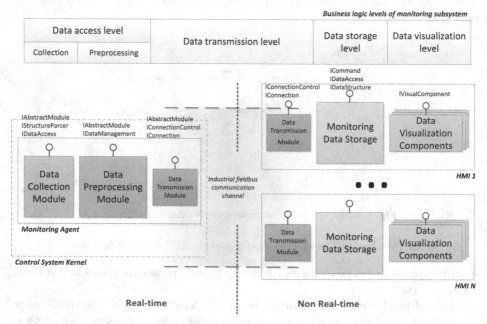

Fig. 2. Architectural model of monitoring and parameter visualization subsystem

4　Architectural Model of Monitoring Agent

The pipeline software architecture consists of three modules with predefined programing interfaces were proposed for monitoring agent (Fig. 3). This separation is caused by specific functions each module should serve.

Data collection module encapsulates the algorithms of interaction with data sources inside the control system kernel. This component should be made specific for each type of control system. It "knows" where inside the control system kernel the necessary data are stored and implements the desired methods of accessing that data. Instead of creating multifunctional data collection module, it can be made in task-oriented way being able to provide the access to the only data that are required by certain monitoring or visualization task.

Data preprocessing module is used to execute simple algorithms to buffer collected data and avoid excessive traffic through the communication channel, in example:

- monitor the parameter value changing and generate the notification about it for HMI clients;

- provide the functionality for synchronous and synchronous requests to parameter values;
- aggregate the incoming data and organize their packet transmission to visualization software together with different compression algorithms to optimize bandwidth utilization of communication channel.

Data transmission module performs communication between monitoring agent and different kinds of HMI devices using fieldbus specific to this device [7].

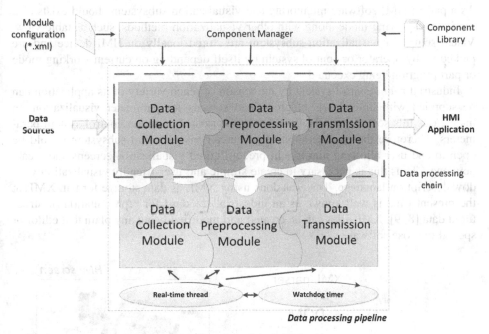

Fig. 3. Architectural model of monitoring agent

It is possible to create multiple data processing pipelines according to requirements of monitoring or visualization software being implemented. The list of modules in each data processing chain can be configured separately using XML configuration file which picks up the required modules from component library and organizes their interaction in desired order.

Such component structure gives the monitoring agent a list of advantages:

1. Encapsulating the specific of data access and transmission into separated modules allows specifying common interfaces of their interaction.
2. Using the common interfaces for cross-module interaction makes it possible to interchange modules of the same kind providing the possibility of flexible configuration of monitoring agent according to the task being implemented. Modules can be created not only by the control system manufacturer but also by the system end-user providing the possibility of system fine tuning to the current production needs.

3. Open programming interface of data transmission module allows the HMI application to communicate with different equipment control systems. Third-party visualization and monitoring software also can be made.

5   Tools for Creating User Configurable Visualization and Monitoring Screens

As a part of HMI software monitoring and visualization subsystem should exists as a separate displaying mode along with other visualization methods, such as table view. Visual setups for visualization subsystem are stored locally on HMI device and are picked up by operator or control system by itself depending on current working mode or part program being executed.

Industrial robot control system by the reason of reach variety of its application can be supplied with limited sets of predefined screens for parameter visualization. In most cases this screens are displaying only the sets of common control system parameters. Therefore the configuration of parameter visualization subsystem should be open to end user allowing him to edit preconfigured visualization screens and create new ones. Thus this is necessary to create simple and user-friendly visualization window description language. This was done using XML as data storage format. XML at the present time is well-known as an industrial standard for representation of structured data [8, 9]. XML-files can be created and modified using any plain text editor or special-purpose software.

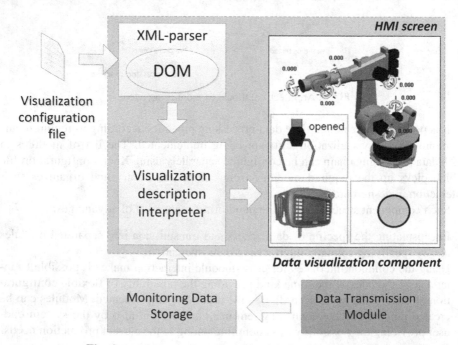

Fig. 4. Architecture of configurable visualization component

To reduce development time and costs visualization module was implemented as independent software module implementing standardized interface of data visualization component (see Fig. 2) and it can be used in various HMI applications. Object model of visualization component is shown on Fig. 4.

Working sequence of configurable parameter visualization component looks like following:

- HMI device stores a set of visualization description files;
- user choose one of them and it is being processed by XML parser into DOM document model;
- visualization description interpreter analyzes DOM document, determines visual elements in it and forms the set of control system parameters to be visualized;
- then, using monitoring data storage and transmission module the necessary requests are sent to control system kernel to set up monitoring of those parameter changes;
- when parameter update notifications are being received interpreter updates visualization screen contents according to given visualization description.

The structure of visual interface description file is shown on Fig. 5.

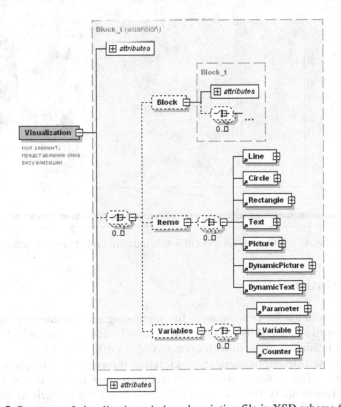

Fig. 5. Structure of visualization window description file in XSD scheme form

Main structure unit of visualization description file is Block (Block_t type). Main visualization window is a top-most Block item and is described by the same set of parameters as any other Block.

Blocks can contain a list of visual elements (Items node) which are the graphical primitives, such a Circle, Line, Picture or Text. Primitives inside Items section can be placed in any order but the later them defined in file the highest z-order they will get on screen. Block can also contain nested Blocks. This makes the system more flexible and allows reusing of previously configured blocks.

Inside Variables section of each block control system parameters used for visualization of that block should be described. Values of these parameters will be continuously updated from control system kernel while visualization is displayed on screen. Computable expressions and counters can also be defined in this section. All displaying properties of graphic primitives, such as size, location, colors, texts, etc., can be linked to the value of parameter or expression to make possible creation of animation.

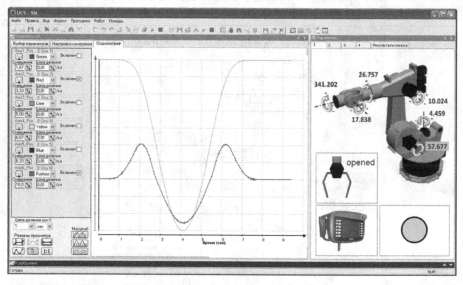

Fig. 6. Parameter visualization component inside industrial robot control system software

6   Conclusions

The proposed solution for creating tools for monitoring and parameter visualization in computer control system of industrial robots allows building extendable data monitoring and visualization subsystem with open modular software architecture. The task of increasing the self-descriptiveness and facilitating information perception by the operator is solved. The chosen software architecture is open to end-user and allows extending the functionality of different HMI devices software, such as hand control panel, commissioning and programming software or remote terminal software with minimal costs.

Acknowledgements. This research was supported by the Ministry of Education and Science of the Russian Federation as a public program in the sphere of scientific activity and Russian Federal Program of Supporting Leading Scientific Schools (grant NSh-3890.2014.9).

References

1. Grigoriev, S.N., Martinov, G.M.: Scalable open cross-platform kernel of PCNC system for multi-axis machine tool. Procedia CIRP **1**, 238–243 (2012)
2. Martinov, G.M., Kozak, N.V., Nezhmetdinov, R.A., Pushkov, R.L.: Design principle for a distributed numerical control system with open modular architecture. Vestnik MGTU Stankin **4**(12), 116–122 (2010). (in Russian)
3. Grigoriev, S.N., Martinov, G.M.: Decentralized CNC automation system for large machine tools. In: Proc. of COMA 2013, International Conference on Competitive Manufacturing, Stellenbosch, South Africa, pp. 295–300 (2013)
4. Nikishechkin, P.A., Grigoriev, A.S.: Practical aspects of the development of the module diagnosis and monitoring of cutting tools in the CNC. Vestnik MGTU Stankin 4, 65–70 (2013) (in Russian)
5. Martinova, L.I., Martinov, G.M.: Applied aspects of open CNC modules implementation. Avtotraktornoe Oborudovanie **3**, 31–37 (2002). (in Russian)
6. Grigoriev, S.N., Martinov, G.M.: Research and Development of a Cross-platform CNC Kernel for Multi-axis Machine Tool. Procedia CIRP **14**, 517–522 (2014)
7. Martinov, G.M., Lyubimov, A.B., Bondarenko, A.I., Sorokoumov, A.E., Kovalev, I.A.: An Approach to Building a Multiprotocol CNC System. Automation and Remote Control **76**(1), 172–178 (2015)
8. Simpson, J.E.: XPath and XPointer. Locating Content in XML Documents. O'Reilly Media (2002)
9. Vlist, E.: XML Schema. The W3C's Object-Oriented Descriptions for XML. O'Reilly Media (2002)
10. Zoriktuev, V.Ts: Mechatronic machine tool systems. Russian Engineering Research **28**(1), 69–73 (2008)
11. Nezhmetdinov, R.A., Sokolov, S.V., Obukhov, A.I., Grigoriev, A.S.: Extending the functional capabilities of NC systems for control over mechano-laser processing. Automation and Remote Control **75**(5), 945–952 (2014)
12. Pritschow, G., Altintas, Y., Jovane, F., Koren, Y., Mitsuishi, M., Takata, S., et al.: Open controller architecture - Past, present and future. Cirp Annals - Manufacturing Technology **50**(2), 463–470 (2001)

DMS-PSO Based Tuning of PI Controllers for Frequency Control of a Hybrid Energy System

Li-Li Wu[✉], J.J. Liang, Tie-Jun Chen, Yao-Qiang Wang,
and Oscar D. Crisalle

School of Electrical Engineering, Zheng Zhou University, Zheng Zhou, China
wuli1988@126.com, {liangjing,tchen}@zzu.edu.cn, crisalle@google.com

Abstract. A dynamic multi-swarm particle swarm optimizer (DMS-PSO) (This research is partially supported by National Natural Science Foundation of China (Grant No.61473266), and China Postdoctoral Science Foundation (Grant No. 2013M541990)) is implemented to tune the gain settings of PI controllers for an autonomous hybrid energy system consisting of wind-turbine, solar photovoltaic, diesel engine, and fuel cell generator, as well as an aqua electrolyzer and energy storage system. The system performance has been verified under three different conditions using experimental data. Simulation results show that the DMS-PSO achieves better suppression of frequency fluctuation than basic particle swarm optimizer and differential evolution algorithm.

Keywords: DMS-PSO · PI control · Power deviation · Frequency fluctuation · Hybrid energy system

1 Introduction

Sustaining system reliability and continuous power service to satisfy the load demand are essential specifications of electrical power generation systems. These specifications are even more significant in *hybrid energy system* (HES) environments, since the intermittent nature of renewable sources strongly contributes to the complexity of operations. Therefore, multiple strategies that seek suppressing power deviation and frequency fluctuation and hence focusing on the feasibility, stability and reliability of HES designs are found in the literature. Various optimization approaches, such as genetic algorithm, particle swarm optimization, neural network, fuzzy logic are studied in the literature as venues to optimize the frequency fluctuation and power deviation. Majid *et al.* [1] propose a novel control strategy for frequency control, where a first-order transfer function is used to represent the dynamics of the wind turbine, photovoltaic panels, fuel cell, as well as power and system frequency variations. Dulal Ch Das *et al.* [2] advance a PI frequency controller used in conjunction with a particle swarm optimizer to reduce the frequency deviation. Furthermore, PI controller based on a genetic algorithm was also proposed by those authors [3]. Taghizadeh *et al.*

© Springer International Publishing Switzerland 2015
Y. Tan et al. (Eds.): ICSI-CCI 2015, Part II, LNCS 9141, pp. 208–215, 2015.
DOI: 10.1007/978-3-319-20472-7_23

[4] investigate a novel fuzzy-logic frequency controller for an HES, based on the optimization of rule bases by a teaching-learning optimization algorithm.

Recent developments in numerical optimization technologies include the *particle swarm optimization* method (PSO) [5,6] is first introduced in 1995 by Kennedy and Eberhart inspired by observations of the flocking behavior of birds in flight. More precisely, the technique is a stochastic algorithm based on the unpredictable group dynamics of birds social behavior. A plethora of documented results show that the PSO algorithm has many advantages. It is simple in concept, easy to implement, efficient to compute, and less dependent on experience parameters. Multiple successful industrial applications attest to the usefulness and effectiveness of the PSO method. Building upon the success of the PSO method, the *multi-swarm particle swarm optimizer* (DMS-PSO) technique [7] is now being investigated in the literature, featuring based on a local PSO version as well as a new neighborhood topology. The neighborhood topology has two important characteristics: it involves multiple but small-sized swarms and includes a randomly regrouping schedule. In contrast to basic PSO approach, the DMS-PSO technique has a neighborhood topology that is dynamic and randomly assigned.

In this paper, three numerical optimization techniques are investigated for the design of proportional-integral controllers that can effectively suppress power-frequency fluctuation in an HES that consisting of a wind turbine, a photovoltaic panel, a diesel engine, a fuel cell, an aqua electrolyzer, an ultra-capacitor, and a battery. More specifically, we consider a PSO method [5,6], a DMS-PSO [7] technique known to posses the advantages of fast searching speed and good global searching stability, as well as a *differential-evolution* (DE) optimization method [8]. A numerical simulation study is carried out to assess the efficacy and effectiveness of each optimization-based control design technique considered. The paper is organized as follows: Section 2 describes the HES configuration, Section 3 introduces the optimization strategies used finding the optimal tuning parameters for PI controllers, Section 4 discusses and analyzes the results, and Section 5 presents concluding remarks.

2   System Configuration

A general block diagram of the hybrid energy system considered in this work is shown in Fig. 1. The system consists of the following seven subsystems: a *wind turbine* (WT), a *photovoltaic* (PV) panel, a *fuel cell* (FC), a *diesel engine* (DEG), an *aqua electrolyzer* (AE), a *battery* (BESS), and an *ultra-capacitor* (UC). In the proposed system, the WT and PV components are the main sources used for satisfying load demands, and they operate in an intermittent fashion as dictated by changes is wind speed and solar irradiation. The FC is a long-term backup power supply for load response, and all the hydrogen it consumes is produced by the AE. In addition, the UC and BESS components are used for short-time backup operation, supplying fast transient power to supplement the intermittent output production of other components. The DEG is a standby power supply that is also used to supplement the output of the intermittent sources to ensure meeting all load demands. Table 1 shows that the transfer functions for the subsystems WT, PV, FC, AE, DEG, UC, and BESS are considered to be first-order rational functions.

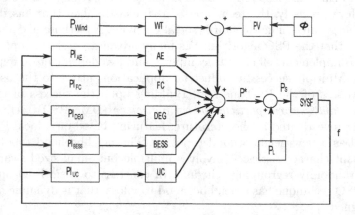

Fig. 1. Block diagram of the hybrid system considered, including WT, PV, FC, DEG, AE, BESS and UC. The wind power and solar irradiation are respectively denoted as P_{Wind} and ϕ. The HES frequency f is used as the input to five PI controllers used to adjust the subsystems seeking to achieve optimal performance.

Table 1. Transfer function nomenclature for the subsystems of the HES. where the symbol K is used to represent a gain and the symbol T a time constant of the subsystem identified by the subscripts of these symbols

Generation Subsystems	Fuel Cell Subsystems	Energy Storage Subsystems
$G_{PV}(s) = \dfrac{K_{PV}}{1 + sT_{PV}}$ (1)	$G_{FC}(s) = \dfrac{K_{FC}}{1 + sT_{FC}}$ (4)	$G_{UC}(s) = \dfrac{K_{UC}}{1 + sT_{UC}}$ (6)
$G_{WT}(s) = \dfrac{K_{WT}}{1 + sT_{WT}}$ (2)	$G_{AE}(s) = \dfrac{K_{AE}}{1 + sT_{AE}}$ (5)	$G_{BESS}(s) = \dfrac{K_{BESS}}{1 + sT_{BESS}}$ (7)
$G_{DEG}(s) = \dfrac{K_{DEG}}{1 + sT_{DEG}}$ (3)		

The PV power is a clean energy extracted from the solar irradiation ϕ (measured in power units, and defined as the product of radiance times the total active area of the panel). The first-order transfer function G_{PV} given in Equation (1) of the table describes the input-output relationship $G_{PV} = \frac{\Delta P_{PV}}{\Delta \phi}$. Analogously, the WT power depends on the wind power P_{Wind}, and the first-order transfer function G_{WT} shown in Equation (2) of the table describes the relationship $G_{WT} = \frac{\Delta P_{WT}}{\Delta P_{Wind}}$. The DEG, an important subsystem capable of fast dynamic response, is modeled as Equation (3). The dynamics of the fuel cell, whose electrochemical reaction consumes hydrogen generated by the AE subsystem, is modeled using Equation (4). The AE, using part of PV or WT generated power to produce available hydrogen for the FC, can be expressed by Equation (5). The UC subsystem and the batterym, which are valuable HES resources to deliver effective response to load changes under adverse sunlight irradiation variations or wind speed drops, are respectively given by Equations (6) and (7). [2,3] Fig. 1 shows that the system frequency f of the HES is related to the power balance P_S through the operator SYSF. In this study the dynamics of the frequency is described by the transfer function

$$G_{SYSF}(s) = \frac{\Delta f(s)}{\Delta P_S(s)} = \frac{1}{D + sM} \qquad (8)$$

where M is the equivalent inertia constant of the HES and D is damping constant of the HES [2,3]. The study is conducted in a numerical simulation environment using the MATLAB software, where the dynamics of all the transfer functions have the gain and time constants reported in Table 2 [2,3].

Table 2. Values for the parameters of the transfer functions used to describe the dynamics of the HES

Gain Constant	Time Constant	Gain Constant	Time Constant
$K_{WT} = 1.0$	$T_{WT} = 1.5$	$K_{DEG} = 1/300$	$T_{DEG} = 2.0$
$K_{PV} = 1.0$	$T_{PV} = 1.8$	$K_{UC} = -7/10$	$T_{UC} = 0.9$
$K_{AE} = 1/500$	$T_{AE} = 0.5$	$K_{BESS} = -1/300$	$T_{BESS} = 1.0$
$K_{FC} = 1/100$	$T_{FC} = 4.0$	$D = 0.03$	$M = 0.4$

3   Optimization Strategy

A previous literature contribution cited in Section 1 reports the HES performance obtained using PSO methods to adjust all the tuning parameters. In this study we investigate potential improvements that may be realized using the alternative DMS-PSO method as well as the DE approach for finding optimal controller tuning parameters. For comparison purposes, we also carry out optimization studies utilizing the closely-related PSO method as a reference.

The optimal tuning parameters for the controllers are found by taking into consideration both power-balance deficiencies (as captured as variations of power balance represented as P_S) and frequency changes (characterized thorough variations indicated by the symbol f). A set of 10 controller tuning parameters is considered optimal if they minimize the integral-of-the-squared-error (ISE) fitness function

$$ISE = K_{ISE} \int_0^{t_f} (f)^2 \, dt + (1 - K_{ISE}) \int_0^{t_f} (P_S)^2 dt \qquad (9)$$

where t_f is a final signal-duration time. The parameter K_{ISE} in Equation (9) is a scalar in the range $[0, 1]$ that weights the contribution of frequency and power-balance fluctuations. We minimize the ISE functional (9) using DMS-PSO, PSO and DE technique. We adopt the values $K_{ISE} = 0.5$ and $t_f = 150\,\text{s}$ for Equation (9).

4   Results and Analysis

The simulation results for Cases 1, 2, and 3 are respectively plotted in Fig. 2(a) to (d). Fig. 2(a) shows the time evolution of wind power P_{Wind}, solar irradiation ϕ and load signal P_L indicated in Case 1, 2, and 3. The graph groups in Fig. 2(b) shows the time evolution of the HES power generation $P^*$and frequency f for Case 1, 2, and 3. Inspection of Fig. 2(b) shows that in all the three Cases, the curves for the PSO, DE, DMS-PSO are not significantly different. However in Fig. 2(c), which has signal plots over a reduced time interval for the purpose of magnifying the wave features, there are some obvious differences among the plots. While Fig. 2(d) shows the power curves for DEG, UC, BESS and FC subsystems for Case 1, 2, and 3. Furthermore, graphs (a), (b) and (d) adopt the following line-type conventions: the black lines denote Case 1, the blue lines show Case 2, and the red lines designate Case 3. For graph (c), while a red solid line denotes the performance obtained by a PSO-optimized PI controller design, a blue dashed line denotes the DE-optimized PI controller design, and the black dotted line designates the performance of the PI controller design using the DMS-PSO approach.

4.1   Time-Domain Analysis of Case 1

In this case, only a step change of wind power is considered. Black lines in Fig. 2(a) shows the wind power change and load demand of Case 1. During 0 to 50 s, the average power wind turbine generated is 0.2 p.u., and load demand is 1 p.u., but at 50 s, the power of wind turbine generator suddenly increases to 0.7 p.u. and at 100 s, the load suddenly decreases from 1 p.u. to 0.8 p.u.. The black lines in Fig. 2(b) reveal that the abrupt changes of load and wind power of Case 1 indicated in Fig. 2(a) entail the generation of oscillatory patterns in the HES power generation P^* and frequency signal f. The wavy pattern is characterized by at least three easily visible local extrema (namely, wave peak or trough). An analysis of the first panel in Fig. 2(c) shows that the DMS-PSO approach leads to time-traces of lower amplitude and shorter settling times, hence outperforming the alternative PSO and DE methods.

4.2   Time-Domain Analysis of Case 2

This case assumes that only solar panel is connected to the system. As it is shown by the blue lines, the load demand increases to 1 p.u. at 30 s, the solar photovoltaic power keeps 0.2 p.u. before 50 s and increases at a steady rate between

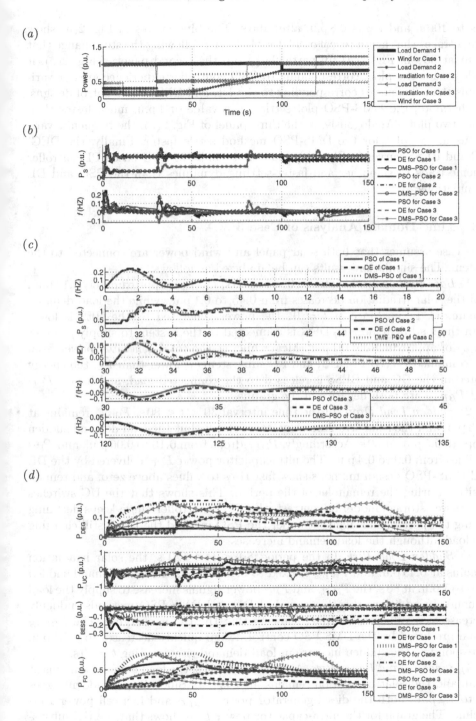

Fig. 2. Simulation results for three Cases

50 s to 100 s, and keeps 0.8 p.u. after 100 s. The blue curves in Fig. 2(b) show signal traces that again reveal oscillatory behavior following the step change that occur at 30 s in Fig. 2(a). In particular, from the second panel of Fig. 2(c), it is clear that the plots corresponding to PSO controller experience greater variations than the curves corresponding to the DE and DMS-PSO control designs. Furthermore, the DMS-PSO plot settles to a value of 1 p.u. much faster than other two plots. Analogously, in the third panel of Fig. 2(c), the frequency variations in the plot for the DMS-PSO method settle faster. Finally, the DEG, UC and BESS curves in Fig. 2(d) show that the DMS-PSO based PI controller scheme leads to smoother and faster-settling dynamics than the PSO and DE formulations.

4.3   Time-Domain Analysis of Case 3

This case assumes that both solar panel and wind power are connected to the system. The simulation results can be analyzed as follows:

1) *Base Regime.* In the interval $0 < t < 30$ s, P_{Wind} is approximately 0.13 p.u., and the solar irradiation decreases from 0.45 to 0.1 p.u., while the load demand remains at 1 p.u.. Since at $t = 0$ s the power generated by the HES is lower than the load demand, the DEG is connected to the system to supply a power P_{DEG} of approximately 0.1 p.u.. Meanwhile, as the total generated power from the combined PV, WT, DEG, and FC subsystems is lower than P_L, the energy stored in the UC and BESS is released to the connected load, and hence, P_{UC} and P_{BESS} are both negative in the graphs.

2) *Sudden Load Drop.* In the time interval $30 < t < 80$ s, P_{Wind} remains at 0.13 p.u. and the solar irradiation increases slowly, but the load demand sudden drops to 0.5 p.u. at 30 s. Accordingly, P_{DEG} drops from 0.16 to 0.06 p.u., and P_{FC} declines from 0.9 to 0.4 p.u.. The ultracapacitor power P_{UC} delivered by the DE and DMS-PSO design methods show fast rises to values above zero, and remain positive during the remainder of the period. This shows that the UC switches to a power storage mode. In contrast, the PSO-controller curve keeps declining during the period, showing that under the PSO method the UC is still charging the load although the load demand increases.

3) *Sudden Load Rise.* In the period $80 < t < 120$ s, the solar irradiation remains at 0.45 p.u. and P_{Wind} is fixed to 0.13 p.u.. As the load demand sudden rises to 1 p.u. at 80 s, the P_{DEG} and P_{FC} power signals increase to supply the load demand. In addition, the P_{UC} signal of DE and DMS-PSO methods suddenly decrease to a value of -0.6 p.u., showing that the UC is charging. The P_{BESS} signal produced by the DMS-PSO control design quickly decreases to -0.1 p.u. to respond to the sudden increase in load demand rise occuring $t = 80$ s.

4) *Sudden Wind Power Rise.* At $120 < t < 150$ s, the load demand remains at 1 p.u. and the solar irradiation remains at 0.45 p.u.. Since P_{Wind} suddenly goes up to 1.2 p.u., both the diesel generator power P_{DEG} and fuel cell power P_{FC} decline. The graph for the ultra-capacitor power P_{UC} shows that the UC subsystem is storing power under each of the three design methods. The P_{BESS} curve for the PSO design dips below zero, showing that the battery is still charging

when P_{Wind} is greater than the load demand, whereas, in contrast, the P_{BESS} curve obtained by the DMS-PSO design increases from -0.02 to 0.05 p.u. indicating that the battery begins to store extra power.

5 Conclusions

This paper presents time-domain simulations for a hybrid energy system featuring five PI controllers tuned using different optimization methods. Optimal control tuning prescriptions obtained using the PSO, DE and DMS-PSO techniques are analyzed under three cases representing operating conditions that include sudden drops or rises on load or generation. The simulation results show that the three optimization methods for tuning lead to control designs show adequate frequency-fluctuation attenuation properties; however, the DMS-PSO based PI controller is found to outperform the other two techniques under the simulation conditions described in the three cases considered. Considerd the power deviation and frequency fluctuation are both important objects for the HES, Multi-object PSO and DE will be used in the following research.

References

1. Nayeripour, M., Hoseintabar, M.: Frequency deviation control by coordination control of fc and double-layer capacitor in an autonomous hybrid renewable energy power generation system. Renewable Energy 36(6), 1741–1746 (2011)
2. Das, D.C.: Pso based frequency controller for wind-solar-diesel hybrid energy generation/energy storage system. In: Proc. of the Int. Conf. of Energy, Automation, and Signal, pp. 1–6 (2011)
3. Das, D.C., Roy, A.K.: Genetic algorithm based pi controller for frequency control of an autonomous hybrid generation system. In: Proc. of the Int. Multi-Conf. of Eng. and Comput. Sci, vol. 2, pp. 953–958 (2011)
4. Taghizadeh, M., Mardaneh, M.: Frequency control of a new topology in proton exchange membrane fuel cell/wind turbine/photovoltaic/ultra-capacitor/battery energy storage system based isolated networks by a novel intelligent controller. Renewable and Sustainable Energy 6(5), 053121 (2014)
5. Eberhart, R., Kennedy, J.: A new optimizer using particle swarm theory. In: Proc. of 6th Int. Symp. Micro Machine and Human Science, pp. 39–43 (1995)
6. Kennedy, J., Eberhart, R.C.: Particle swarm optimization. In: Proc. of IEEE Int. Conf. on Neural Networks, vol. 4, pp. 1942–1948 (1995)
7. Liang, J.J., Suganthan, P.N.: Dynamic multi-swarm particle swarm optimizer. In: Proc. of IEEE Int. Swarm Int. Symp., pp. 124–129 (2005)
8. Price, K.: Differential evolution a fast and simple numerical optimizer. In: Proc. of 1996 Biennial Conf. of the North American Fuzzy Inf. Proc. Society, pp. 524–527 (1996)

Combinatorial Optimization
Algorithms

Using Discrete PSO Algorithm to Evolve Multi-player Games on Spatial Structure Environment

Wang Xiaoyang[1(✉)], Zhang Lei[2], Du Xiaorong[2], and Sun Yunlin[2]

[1] Zhongshan Insitute, University of Electronic Science and Technology of China,
Guangdong, China
wxy_lele@163.com
[2] School of Physics and Engineering, Sun Yat-sen University, Guangdong, China

Abstract. Mechanisms promoting the evolution of cooperation in two-player, two-strategy evolutionary games have been discussed in great detail over the past decades. Understanding the effects of repeated interactions in multi-player social dilemma game is a formidable challenge. This paper presents and investigates the application of co-evolutionary training techniques based on discrete particle swarm optimization (PSO) to evolve cooperation for the n-player iterated prisoner's dilemma (IPD) game and n-player iterated snowdrift game (ISD) in spatial environment. Our simulation experiments reveal that, the length of history record, the cost-to-benefit ratio and group size are important factors in determining the cooperation ratio in repeated interactions.

Keywords: Evolution · Iterated Prisoner's Dilemma (IPD) · Iterated Snowdrift Game (ISD) · Discrete particle swarm optimization (PSO) · N-player

1 Introduction

The evolution of cooperation is a hot topic among natural and social sciences [1-5]. Cooperative behaviors have been observed in microbes, animals, plants, and humans. Five mechanisms have been proposed by Nowak [24] to evolve these cooperative behaviors: direct and indirect reciprocity, spatial selection, kin selection, and multi-level selection. Among them, direct and spatial reciprocity have perhaps been the most widely researched [6, 7].

In the presence of spatial reciprocity, each individual interact with his/her neighborhood which is organized in certain kind of spatial structure. Over the past decades, the promoting of cooperation in two players and two strategies (2 × 2) social dilemma games have been investigated. In social dilemma games, the Prisoner's Dilemma (PD) and the Snowdrift (SD) games are the most important model [8-10] for investigating the coevolution of cooperation in spatial multiplayer systems. As with many other heuristic researches, PSO has also been used in the context of prisoner's dilemma game. Franken and Engelbrecht [23, 24] investigated two different approaches using PSO to evolve strategies, one is Binary PSO and the other is neural networks. Chio [25] integrated strategies from the prisoner's dilemma into the PSO algorithm. They use these strategies to represent different methods to evaluate each particle's next

© Springer International Publishing Switzerland 2015
Y. Tan et al. (Eds.): ICSI-CCI 2015, Part II, LNCS 9141, pp. 219–228, 2015.
DOI: 10.1007/978-3-319-20472-7_24

position. Wang [11] investigated the evolution of cooperation in multiple choice IPD game. Those studies mainly focused on the IPD game played by two players and with two choices, cooperation and defection. Use discrete PSO algorithm to evolve cooperation in IPD game and ISD game with multiple players were not adequately considered.

Motivated by the findings above, this paper presents a detailed study of using PSO approach to evolve cooperation in the IPD game and ISD game with multiple players in a spatial environment. The rest of this paper is organized as follows. An overview of the core IPD and ISD problem and relevant historic related work is presented in Section 2, followed by a short overview of the model in section 3. Section 4 explains the experimental procedure followed for this study, and the results are analyzed by various techniques. Section 5 concludes this paper by summarizing some of the major experimental findings.

2   Multiplayer Social Dilemma Games

The N-IPD game was first proposed by Boyd and Richerson [5], the N-IPD consists of N players ($N > 2$) making decision independently based on the other players' past actions. The more players are cooperating, the more social benefit b they are receiving. And for cooperators, they always should pay cost c, and $b > c$ as in the two-player game. For example, if a player who cooperates is pitted against k other players and i of those are cooperators, then the payoff is $bi - ck$. Hence, the payoff is bi. The utility values $\prod n$, where $n \in [1, ..., N]$, for this scenario can be formally defined as:

$$\prod n = \begin{cases} b \times i - c \times (N-1), & for\ cooperators \\ b \times i & , & for\ defectors \end{cases} \quad (1)$$

In the N-player SD game, the payoff of a cooperator is dependent examined by Zheng[20]. If there is only one cooperator in the group, the payoff is $b-c$. If two cooperators exist, then the payoff is $(b - c/2)$. With three cooperators, the payoff becomes $(b - c/3)$, and so on. The utility can be summarized as equation (2). To normalize the range of cost and benefit, we define $r=c/b$ as cost-to-benefit ratio.

$$\prod n = \begin{cases} b - \frac{c}{i}, & for\ cooperators \\ b, & for\ defectors\ when\ i > 0 \\ 0, & for\ defectors\ when\ i = 0 \end{cases} \quad (2)$$

3   The Model

The spatial version of the PSO algorithm based n-player IPD games is considered in this section. In the following subsections, we describe the individual components of our model in detail.

3.1 Discrete Particle Swarm Optimization (PSO) Approach for Social Dilemma Games

Primarily PSO can successfully solve continuous problems. For discrete optimization problems, Kennedy and Eberhart [22,24-26] presented a binary version of PSO in 1997. Each particle is considered as a position in a D-dimensional space and each element of a particle position can take the binary value of 0 or 1 in which 1 means "included" and 0 means "not included". Each element can change from 0 to 1 and vice versa. Also each particle has a D-dimensional velocity vector the elements of which are in range $[-V_{max}, V_{max}]$.

In the PSO algorithm, The velocity of the particle is calculated using equation (3):

$$V_{id}^{k+1} = \omega V_{id}^k + c_1 rand(0,1)\left(P_{id}^k - X_{id}^k\right) + c_2 rand(0,1)(P_{gd}^k - X_{id}^k). \qquad (3)$$

For particle i, its position x_{id} is changed according to:

$$X_{id}^{k+1} = \begin{cases} 1, & if\ sig\left(V_{id}^{(k+1)}(j)\right) > r_{ij} \\ 0, & otherwise \end{cases}. \qquad (4)$$

Where,

$$sig\left(V_i^{(t+1)}(j)\right) = \frac{1}{1+exp\left(-V_i^{(t+1)}(j)\right)}. \qquad (5)$$

where for particle i, the position vector can be represented by $X_i = (X_{i1}, X_{i2}, ..., X_{id})$, V_{id} presents the velocity of the ith particle on the specific d-dimension. ω is the inertia weight [14-18], c_1 and c_2 are acceleration coefficients. r_{ij} in Equation (4) is a random number in range $[0, 1]$. P_{id}^k is the best position of particle i on dimension d at iteration k, and P_{gd}^k is the dth dimensional best position in the whole particle swarm at iteration k.

3.2 The Spatial Environment for Strategy Co-evolution

In spatial evolutionary environment, each agent participates in an interactive game with l-1 other agent drawn from its local neighborhood. Fig. 1 depicts the four examples of neighborhood structure for interaction, and single agent is located in the node (vertex) of the grid world. In Fig. 1(a) and (b), the two examples show that agent can only interact with the agents in dotted box. The regular-network (Fig. 1 (d)) is two-dimensional and agents are connected by the edges. As for the small-world network, we use a version similar to the one introduced by Watts and Strogatz [25]. From the two-dimensional regular-network substrate we rewire each link with probability ρ (Fig. 1 (c)). When $\rho=0$, it's essentially a regular-network. As $\rho=1$, it is defined as full-connected network.

(a) Von-Neumann   (b) Moore   (c) Small-World   (d) Regular-Network

Fig. 1. Examples of neighborhood structures

The calculation of the neighborhood best particle depends on the spatial neighborhood structure [13, 23, 26] agents used. Various spatial neighborhood structures have been defined [23], of which the four neighborhood structures of Fig. 1 are used in this study. In general, the neighborhood best position is calculated as

$$P_{gd}^{k+1} \in \left\{ NE_i | f\left(P_{gd}^k\right) = \max \left\{ payoff(x), \forall x \in NE_i \right\} \right\} . \tag{6}$$

with the neighbourhood defined as

$$NE_i = \left\{ P_{gd}^k \left(i - \frac{l}{2}\right), \dots, P_{gd}^k(i-1), P_{gd}^k(i), P_{gd}^k(i+1), \dots, P_{gd}^k \left(i + \frac{l}{2}\right) \right\} . \tag{7}$$

for neighbourhood size of l.

4   Simulation Experiments and Results

The experiments in the first part are conducted to analyse the influence of spatial structure to the cooperation learning in the population. The second part of the experiments is conducted to compare the frequency of cooperation by varying the cost-to-benefit ratio r and the number of players in all the spatial structures.

4.1   Evolution of Cooperation by Using Discrete PSO Approach in Spatial Structure Environment

The experiments of this part are testing the co-evolutionary process of behaviour performances in n-player IPD game within spatial structure environment. The following parameters are used in the PSO: the history record L_H; r=0.2 is fixed in this section.

From Table 1 and Table 2, we can find that the average cooperation ratios in two structures are similar; they all increase as the length of history record increase. Players in the random rewired network, such as the small-world and full-connected network, are easier to cooperate than in the regular networks, such as Von-Neumann and Moore networks. The players in ISD game are not willing to cooperate than in IPD game no matter in which spatial structure; most because of the rule of the game.

Table 1. Average cooperation ratio in Moore and Von-Neumann structure

History Record	Moore	Von-Neumann
$L_H=50$	(a)	(b)
$L_H=10000$	(c)	(d)

Table 2. Average cooperation ratio in Small-World and Full-Connected structure

History Record	Small-World	Full-Connected Network.
$L_H=50$	(a)	(b)
$L_H=10000$	(c)	(d)

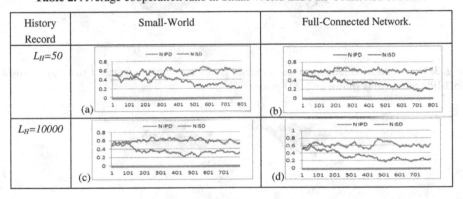

4.2 Evolution of Cooperation by Using PSO Approach in Spatial Structure Environment

In this part of experiment, the cost-to-benefit ratios, r is observed to find the influence to the evolution of cooperation in multi-player spatial game. Payoffs of each agent are calculated based on Equation (1) and Equation (2).

Fig. 2 to Fig. 5 show the influence of different cost-to-benefit ratios to average cooperation in 2-player games. Fig. 6 to Fig. 9 show the influence of different cost-to-benefit ratios to average cooperation in 3-player games. Fig. 10 to Fig. 13 show the influence of different cost-to-benefit ratios to average cooperation in 4-player games.

As r increases, the average cooperation ratio drops regardless of the spatial structures. When the value of r is sufficiently high, defectors would dominate the agent population; on the other hand, when the value of r is small it would be worth to cooperate. As the number of player n increases, the average cooperation ratio drops regardless of the spatial structures and the type of game. From the figures we can learn the more players join in one game, the more hard to create mutual trust and finally lead to the defect. For n-player IPD game, the average cooperation ratios are above 0.2; while for n-player ISD game, the average cooperation ratios are below 0.2. In the n-IPD game, players can keep higher cooperation ratio within smaller group sizes

persisted for the change of r. It seems that the influence of the number of player is less obvious in the spatial n-ISD population.

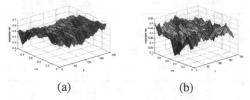

(a)   (b)

Fig. 2. The cooperation ratio in Von- Neumann structure gamws as r increase from 0.1 to 1. (a) is shown in 2-IPD game, (b) is shown in 2-ISD game.

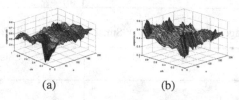

(a)   (b)

Fig. 3. The cooperation ratio in Moore structure games as r increase from 0.1 to 1. (a) is shown in 2-IPD game, (b) is shown in 2-ISD game.

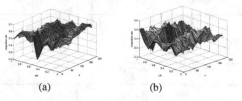

(a)   (b)

Fig. 4. The cooperation ratio in Small-World structure games as r increase from 0.1 to 1. (a) is shown in 2-IPD game, (b) is shown in 2-ISD game.

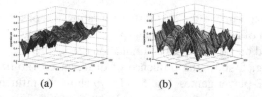

(a)   (b)

Fig. 5. The cooperation ratio in Full-Connected structure games as r increase from 0.1 to 1. (a) is shown in 2-IPD game, (b) is shown in 2-ISD game.

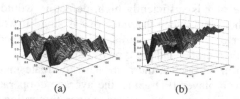

(a)   (b)

Fig. 6. The cooperation ratio in Von- Neumann structure games as r increase from 0.1 to 1. (a) is shown in 3-IPD game, (b) is shown in 3-ISD game.

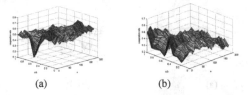

(a)   (b)

Fig. 7. The cooperation ratio in Moore structure games as r increase from 0.1 to 1. (a) is shown in 3-IPD game, (b) is shown in 3-ISD game.

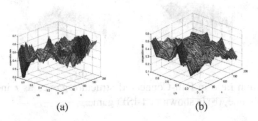

(a)   (b)

Fig. 8. The cooperation ratio in Small-World structure games as r increase from 0.1 to 1. (a) is shown in 3-IPD game, (b) is shown in 3-ISD game.

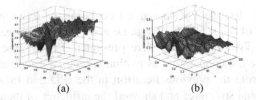

(a)   (b)

Fig. 9. The cooperation ratio in Full-Connected structure games as r increase from 0.1 to 1. (a) is shown in 3-IPD game, (b) is shown in 3-ISD game.

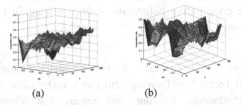

(a)   (b)

Fig. 10. The cooperation ratio in Moore structure games as r increase from 0.1 to 1. (a) is shown in 4-IPD game, (b) is shown in 4-ISD game.

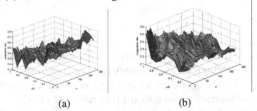

(a)   (b)

Fig. 11. The cooperation ratio in Von-Neumann structure games as r increase from 0.1 to 1. (a) is shown in 4-IPD game, (b) is shown in 4-ISD game.

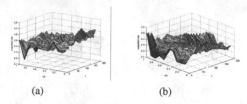

(a)   (b)

Fig. 12. The cooperation ratio in Small World structure games as r increase from 0.1 to 1. (a) is shown in 4-IPD game, (b) is shown in 4-ISD game.

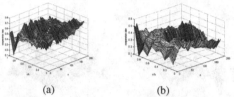

(a)   (b)

Fig. 13. The cooperation ratio in Full-Connected structure games as r increase from 0.1 to 1. (a) is shown in 4-IPD game, (b) is shown in 4-ISD game.

5   Conclusion and Discussion

This paper applied discrete PSO-based co-evolutionary training techniques to the nonzero-sum game of the n-player IPD game and n-player ISD game in spatial structure environments. Two main issues were presented. Firstly, discrete PSO algorithm is used to co-evolve the cooperation as evaluators, chose cooperate or defect according to the fitness from the previous iteration in the IPD and ISD game. The second introduced four spatial structures and showed the influence of them to the cooperation evolution. With the increasing cost-to-benefit ratio r and group sizes n, the cooperative strategies become difference.

Future work will include an investigation of the effects of using different payoff matrices to influence cooperative behaviour, competing against more neighbours of the spatial environment.

Acknowledgments. This work is supported by two research program of China. One is the Zhongshan Science and Technology Development Funds under Grant no. 2014A2FC385 and also supported by Production, learning and research of Zhuhai under Grant no. 2013D0501990003.

References

1. Chong, S.Y., Yao, X.: Behavioral Diversity, Choices and Noise in the Iterated Prisoner's Dilemma. IEEE Transactions on evolutionary computation **9**(6), 540–551 (2005)
2. Chong, S.Y., Yao, X.: Multiple Choices and Reputation in Multiagent Transactions. IEEE Transactions on evolutionary computation **11**(6), 689–711 (2007)
3. Chong, S.Y., Tiño, P., Yao, X.: Measuring Generalization Performance in Coevolutionary Learning. IEEE Transactions on evolutionary computation **12**(4), 479–505 (2008)

4. Chong, S.Y., Tiño, P., Yao, X.: Relationship Between Generalization and Diversityin Coevolutionary Learning. IEEE Transactions on computational intelligence and AI in games **1**(3), 214–232 (2009)
5. Chong, S.Y., Tiño, P., Ku, D.C., Yao, X.: Improving Generalization Performance in Co-Evolutionary Learning. IEEE Transactions on evolutionary computation **16**(1), 70–85 (2012)
6. Ishibuchi, H., Takahashi, K., Hoshino, K., Maeda, J., Nojima, Y.: Effects of configuration of agents with different strategy representations on the evolution of cooperative behaviour in a spatial IPD game. In: IEEE Conference on Computational Intelligence and Games (2011)
7. Axelrod, R.: The evolution of cooperation. Basic Books, New York (1984)
8. Nowak, M.A., May, R.M.: Evolutionary games and spatial chaos. Nature **359**(6398), 826–829 (1992)
9. David, B.F.: On the relationship between the duration of an encounter and the evolution of cooperation in the iterated prisoner's dilemma. Evolution of computation **3**(3), 349–363 (1996)
10. Hauert, C., Doebeli, M.: Spatial structure often inhibits the evolution of cooperation in the snowdrift game. Nature **428**(6983), 643–646 (2004)
11. Wang, X.Y., Chang, H.Y., Yi, Y., Lin, Y.B.: Co-evolutionary learning in the N-choice iterated prisoner's dilemma with PSO algorithm in a spatial environment. In: 2013 IEEE Symposium Series on Computational Intelligence, pp. 47–53. IEEE press, Singapore (2013)
12. Darwen, P.J., Yao, X.: Co-evolution in iterated prisoner's dilemma with intermediate levels of cooperative: Application to missile defense. International Journal of Computational Intelligence and Applications **2**(1), 83–107 (2002)
13. Ishibuchi, H., Namikawa, N.: Evolution of iterated prisoner's dilemma game strategies in structured demes under random pairing in game playing. IEEE Transactions on evolutionary computation **9**(6), 552–561 (2005)
14. Zheng, Y., Ma, L., Qian, I.: On the convergence analysis and parameter selection in particle swarm optimization. In: Processing of International Conference of Machine Learning Cybern., pp. 1802–1807 (2003)
15. Franken, N., Engelbrecht, A.P.: Comparing PSO structures to learn the game of checkers from zero knowledge. In: The 2003 Congress on Evolutionary Computation, pp. 234–241(2003)
16. Franken, N., Engelbrecht, A.P.: Particle swarm optimization approaches to coevolve strategies for the iterated prisoner's dilemma. IEEE Transactions on evolutionary computation **9**(6), 562–579 (2005)
17. Di Chio, C., Di Chio, P., Giacobini, M.: An evolutionary game-theoretical approach to particle swarm optimisation. In: Giacobini, M., Brabazon, A., Cagnoni, S., Di Caro, G.A., Drechsler, R., Ekárt, A., Esparcia-Alcázar, A.I., Farooq, M., Fink, A., McCormack, J., O'Neill, M., Romero, J., Rothlauf, F., Squillero, G., Uyar, A., Yang, S. (eds.) EvoWorkshops 2008. LNCS, vol. 4974, pp. 575–584. Springer, Heidelberg (2008)
18. Kennedy, J., Eberhart, R.C.: Particle swarm optimization. In: Proc. IEEE International Conference of Neural Network, vol. 4, pp. 1942–1948 (1995)
19. Ishibuchi, H., Takahashi, K., Hoshino, K., Maeda, J., Nojima, Y.: Effects of configuration of agents with different strategy representations on the evolution of cooperative behaviour in a spatial IPD game. In: IEEE Conference on Computational Intelligence and Games (2011)

20. Zheng, D.F., Yin, H.P., Chan, C.H., Hui, P.M.: Cooperative behavior in a model of evolutionary snowdrift games with N-person interactions. Europhys. Lett. **80**(1), 18002 (2007)
21. Moriyama, K.: Utility based Q-learning to facilitate cooperation in Prisoner's Dilemma games. Web Intelligence and Agent Systems: An International Journal, IOS Press **7**, 233–242 (2009)
22. Chen, B., Zhang, B., Zhu, W.D.: Combined trust model based on evidence theory in iterated prisoner's dilemma game. International Journal of Systems Science **42**(1), 63–80 (2011)
23. Chiong, R., Kirley, M.: Effects of Iterated Interactions in Multi-player Spatial Evolutionary Games. IEEE Transactions on evolutionary computation (2013). doi:10.1109/TEVC.2011.2167682
24. Nowark, M.A.: Five rules of the evolution of cooperation. Science **314**, 1560–1563 (2006)
25. Watts, D., Stogatz, S.H.: Collective dynamics of small-world networks. Natrue **393**, 440–442 (1998)
26. Chiong, R., Kirley, M.: Iterated N-Player Games on Small-World Networks. In: GECCO 2011 (2011)

A Hybrid Algorithm Based on Tabu Search and Chemical Reaction Optimization for 0-1 Knapsack Problem

Chaokun Yan[1], Siqi Gao[1], Huimin Luo[1,2(✉)], and Zhigang Hu[2]

[1] School of Computer and Information Engineering, Henan University, Kaifeng, China
hmluo@henu.edu.cn
[2] School of Information Science and Engineering, Central South University, Changsha, China

Abstract. The 0-1 knapsack problem(01KP) is a well-known NP-complete problem in combinatorial optimization problems. There exist different approaches employed to solve the problem such as brute force, dynamic programming, branch and bound, etc. In this paper, a hybrid algorithm CROTS (Chemical Reaction Optimization combined with Tabu Search) is proposed to address the issue. One of the four elementary reaction of CRO is performed first, and after that tabu search is employed to search for the neighbors of the optimum solution in the population. The experimental results show that CROTS owns better performance in comparison with GA and the original CRO.

Keywords: 0-1 knapsack problem · Chemical reaction optimization · Tabu search

1 Introduction

The 0-1 knapsack problem is known as a combinatorial optimization problem and it is studied in many fields such as resource allocation, complexity, cryptography, and so on. Generally, the 0-1 knapsack problem can be described as follows: Given n items, each of them owns weight w_i and profit p_i. A set of items are to be selected and put into the knapsack with weight constraint W. The problem is how to pack the items to obtain the profits as much as possible considering the carried total weight is no more than the fixed number W. As items are indivisible, the 0-1 variable x_i is employed to decide whether the item is taken or not. The mathematically model of the 0-1 knapsack problem can be described as:

$$\text{Maximize } f = sum(x_i p_i), i=1,2,\ldots,n. \tag{1}$$
$$\text{Subject to } sum(x_i w_i) \leq W, i=1,2,\ldots n, x_i \in \{0,1\} \tag{2}$$

If $x_i = 1$, then ith item is selected and put into the knapsack, otherwise not.

Up to now, there are varies of methods employed to solve 0-1 knapsack problem, such as simulated annealing algorithm, genetic algorithms, branch and bound, etc. Recently, Lam and Li[1] proposed CRO to optimize combinatorial problems, which has been demonstrated an effective approach solving quadratic assignment problem,

© Springer International Publishing Switzerland 2015
Y. Tan et al. (Eds.): ICSI-CCI 2015, Part II, LNCS 9141, pp. 229–237, 2015.
DOI: 10.1007/978-3-319-20472-7_25

grid scheduling problem, and so forth. Although, CRO is an efficient problem in solving optimization problems, it can still be improved regarding its local search process. Comparably, simple TS own good performance in finding local optimum solutions. Considering compensatory characteristics of CRO and TS, a new algorithm CROTS combining both of the meta-heuristic algorithms (CRO&TS) is proposed.

The rest of the paper is organized as follows: Sect.2 presents the related works. In Sect.3, CROTS is introduced in detail, followed by Sect.4 where complexity analysis and proof on convergence are presented. Next, Sect.5 gives the experimental results and analysis. Finally, the conclusion and the future works are described in Sect.6.

2   Related Works

Existed methods for 01KP can be classified as two categories, exact technologies and heuristic algorithms. Exact technologies encompass the branch-and-bound algorithms, core algorithms, dynamic programming, etc. Lalami[2] proposed an efficient implementation of the branch and bound method for knapsack problem on a CPU-GPU system via CUDA. Ben-Romdhane[3] proposed an online algorithm based on dynamic programming to address online knapsack problem. And core concept for the knapsack problems is developed in literature [4], which is based on the "divide and conquer" principle.

Nevertheless, an exact algorithm for a specialized problem is difficult to extend to adapt to other problems, and memory consumption is also the shortcoming for exact algorithms. So there are many heuristic approaches designed for the knapsack problems. For instance, Leung[5] develops a simulated annealing algorithm to jump out of the local optimal trap then find a further improved solution. Lai, GM[6] employs a new hybrid combinatorial g netic algorithm and use a combinational permutation to address MKP. Besides, hill climbing, harmony search and other heuristic algorithms have been proposed to address the 0-1 knapsack problem.

3   CROTS for 01KP

CRO imitates the chemical reaction process and is governed by the thermodynamics laws. The energy in the reacting systems can't be created or destroyed and the reacting system tends to be stable when the potential energy drops to the minimum, which can be achieved by converting potential energy to kinetic energy and gradually losing the molecules' energy to surroundings [1].

There're three stages in CRO, initialization, iterations and the final stage. At the initialization stage, parameters are assigned and algorithmic settings are initialized. During the evolutions, four elementary reactions perform alternately and parameters α, β play important roles in the process. There are two monomolecular collisions, decomposition and on-wall ineffective, also two intermolecular collisions, synthesis and intermolecular ineffective collision. If a molecule's hits number is larger than α, then decomposition happens, otherwise, on-wall ineffective collision occurs. β decides whether synthesis or intermolecular ineffective collision would happen. If both of selected molecules' kinetic energy is less than β then synthesis occurs; if not, intermolecular ineffective collision is triggered. And the final stage output the optimum solution.

To guarantee the normal reaction, several other parameters are needed in the algorithm, ω, *PE, KE, NumHit, MinStrut, MinPE., MinHit* ,where ω represents the molecule structure, *PE* denotes potential energy, *KE* stands for kinetic energy, *MinStrut* is the molecule structure with the minimum *PE* in its reaction history and *MinPE* is the molecule's potential energy when it obtains the *MinStrut*, and *MinHit* is the hits number when a molecule attains *MinPE*. There're also other necessary parameters: *PopSize* decides the population size , *KELossRate* denotes the kinetic loss rate during the reaction, *MoleColl* decides whether the intermolecular collision would occur, *buffer* stores the energy transformed from a portion of kinetic energy, *InitialKE* is the molecule's initial kinetic energy. All of them are necessary and regulate the reaction system.

CRO is an effective optimization method and has been employed to tackle problems in both the discrete and continuous domains. Nevertheless, the performance of local search in CRO can be strengthened. The main idea of TS is to avoid searching solutions in cycles by adopting tabu list to storage the forbidden items, which have been visited recently or limited by users-provided rules. Also, the algorithm searches the solution's neighbors to obtain a new potential solution. The solutions in tabu list cannot be chosen unless they satisfy the Aspiration Criteria(AC). For algorithm CROTS, once one of the four elementary reactions performs, the optimum solution *bestSol* in the iteration would be checked, and tabu search is employed to search *bestSol*'s neighbors, which is a local search process. The overall flowchart of CROTS is described as Fig.1. To adapt to PE, the fitness function here is set as follows:

$$. fitness=Prof-sum(x_i p_i), i=1,2,...n \tag{3}$$

In equation (3), *Prof* is the items' entire profits sum. The operators of CROTS are depicted as follows.

3.1   On-Wall Ineffective Collision

On-wall ineffective collision replaces a random position with a binary number to change or maintain the item's condition, where 0 represents the item is not put into the knapsack, otherwise the opposite condition. For instance, example below put the 3th item into the knapsack.

$$\omega:[1,0,0,1,0,0] \rightarrow \omega':[1,0,1,1,0,0]$$

3.2   Decomposition

Decomposition duplicates the original molecule to get two new ones and change the half of each new molecule randomly with number in $\{0, 1\}$. If the difference between $PE_\omega + KE_\omega$ and $PE_{\omega'1} + PE_{\omega'2}$ is no less than 0 or sum of energy buffer and the difference is no less than 0, then ω'_1 and ω'_2 would be conserved in population and ω is destroyed, or else destroy ω'_1 and ω'_2.

String before decomposition: ω [1, 0, 0, 0, 1, 0]
String after decomposition:ω'_1 [1,**1**,0,**1**,0,0] ω'_2:[**0**,0,**1**,**1**,1,0]

3.3   Intermolecular Ineffective Collision

Intermolecular ineffective collision chooses two molecules in population and selects a random position then displaces the item by random number generated in {0, 1} to obtain new molecules. This operator contributes to the algorithm's intensification. The manipulation of molecules is simulated as follows:

Strings before Intermolecular Ineffective Collision:ω_1[1,0,0,1,0,1] ω_2:[0,1,0,0,1,0]

Strings after Intermolecular Ineffective Collision: ω_1' [1,**1**,0,1,0,1] ω_2':[0,1,0,**1**,1,0]

3.4   Synthesis

Synthesis generates fresh molecule structure by combining two existing solutions ω_1, ω_2 and contributes to algorithm's diversification. Half of the new generated molecule ω' is duplicated from ω_1 with the items in the corresponding position and the left is derived from ω_2. The operator is simplified as follows:

Strings before Synthesis: ω_1 [**0,1**,1,0,**0**,1] ω_2:[1,0,**1**,0,1,**0**]

String after Synthesis: ω'[0, 1, 1, 0, 0, 0]

3.5   Tabu Search

CROTS combines tabu search to get better local search performance. After each reaction is performed, the optimum solution in the population *bestSol* is obtained and tabu search is employed to search its neighborhoods and update *bestSol* by the structure with more profits. *fitness(ω)* is used to calculate molecule ω's fitness value, *neighborMol(ω)* changes an item of ω in a random position and returns a new molecule structure; *tabuTableUpdate(ω)* add molecule structure ω to tabu table, which is an FIFO data structure; *judge(ω)* is used to judge whether solution ω is in the tabu list, if ω is not in the tabu list, then return 0; or return 1; Parameters *tabuLength* is the length of tabu table and *numNeighbor* is number of the *bestSol*'s neighbors. More details are represented as Algorithm 1.

```
Input: molecule bestSol
itr•0
while(itr < numNeighbor)do
           ω'•neighborMol(bestSol)
           if(!judge(ω'))
           if(fitness(ω')<fitness(bestSol))
            update bestSol with ω'
            tabuTableUpdate(ω')
           endif
           itr•(itr+1)
           endif
endwhile
```

Algorithm 1. Tabu Search

4   Complexity Analysis and Proof on Convergence

4.1   Complexity Analysis

The 0-1 knapsack problem's variable dimension is N, and hypothesis the flexible population size is M. Tabu length is denoted as L while the optimum solution's neighborhood number is set as T. As described in Fig.1, CROTS mainly encompasses four elementary reactions and tabu search. In consideration of the worst circumstances, the time complexity of CROTS is analyzed as follows:

    1) Initialize the population size molecules, the time complexity is O (MN).

    2) Judge the stopping criteria, the time complexity is O(1)

    3) Generate a random number *ran* in the interval [0, 1] uniformly. If *ran* is larger than *MoleColl* or the population size is equal to 1, then monomolecular collision happens. If a molecule's hits number is larger than α, then decomposition happens, or the on-wall ineffective collision performs. If *ran* is no larger than *MoleColl*, then intermolecular collision occurs. If both of the selected molecules' kinetic energy is no larger than β, then synthesis is triggered, or intermolecular collision reacts. The time complexity is O(N)

    4) Check the optimum solution and update *bestSol*, the time complexity is O(N)

    5) Generate T *bestSol*'s neighbors, for each neighbor, the algorithm judges whether it is in the tabu list, the time complexity is O(TN), then calculate the neighbor's fitness value, the time complexity is O(N). If the neighbor's fitness is less than *bestSol*'s *MinPE*, update *bestSol* with the neighbor and update the tabu table list, the time complexity is O(TN). So, this step's time complexity is O(TN).

    6) Iteration continues and go to Step 2)

As the population size is flexible, the CROTS's time complexity is MAX {O (MN), O (TN)}.

4.2   Convergence Analysis

Because the four elementary reactions and tabu search process of CROTS is irrelevant from the former state and merely associated with current condition, then the algorithm can be modeled by Markov chain. Detailed proof is described as follows:

Definition: The optimal solution for the problem is denoted as $S_{opt}=\{s_{opt} \in S, f(s_{opt})=\min(f(s)|s \in S\}$, where f is the fitness function and S is the molecule structures. N(t) represents the optimal solution numbers at t iteration.

Theorem: CROTS converges to the global optimum solution with probability 1.

Detailed Proof: Since optimum solution would be reserved in the population and delivered to next iteration, the equation N (t+1) =0 is out of the question unless N (t) =0. So,

    P{N(t+1)=0} = P{N(t+1)=0|N(t)=0}×P{N(t)=0}

    Let φ=min{P{N(t+1)≥1|N(t)=0},t=0,1,2...}

    P{N(t+1)=0|N(t)=0}=1-P{N(t+1)≥1|N(t)=0}≤1-φ; Then,

    0≤P{N(t+1)=0}≤(1-φ)×P{N(t)=0}≤(1-φ)2×P{N(t-1)=0}≤...≤(1-φ)t+1×P{N(0)=0}

As P{N(0)=0}∈[0,1], so when t→∞, lim(1-φ)t+1=0, then limP{N(t+1)=0}=0

Then, when t→∞, limP{N(t+1)≥1} = 1-limP{N(t+1)=0}=1, therefore, CROTS converges to global optimum solution utterly.

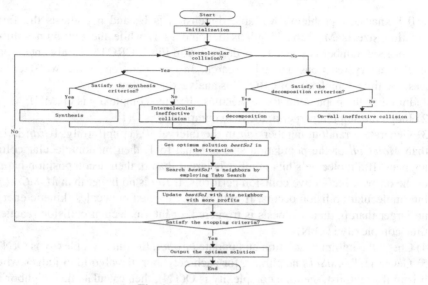

Fig. 1. Flowchart of CROTS

5   Experiments

5.1   Experimental Settings

All experiments are conducted on a PC with Intel i5-2450M at 2.5GHz CPU, 2G RAM, running on windows 7.0 and the algorithms are implemented with JAVA. Two experiments are involved. The first experiment (Exp1) is conducted on the fixed instance from [7]. The profits, weights and the weight constraint are as follows:

w_i= 80, 82, 85, 70, 72, 70, 66, 50, 55, 25,50, 55, 40, 48, 50, 32, 22, 60, 30, 32,40, 38, 35, 32, 25, 28, 30, 22, 50, 30,45, 30, 60, 50, 20, 65, 20, 25, 30, 10,20, 25, 15, 10, 10, 10, 4, 4, 2, 1

p_i= 220, 208, 198, 192, 180, 180, 165, 162, 160, 158,155, 130, 125, 122, 120, 118, 115, 110, 105, 101,100, 100, 98, 96, 95, 90, 88, 82, 80, 77,75, 73, 72, 70, 69, 66, 65, 63, 60, 58,56, 50, 30, 20, 15, 10, 8, 5, 3, 1

W=1000;

And the second experiment (Exp2) is based on random instances generated as literature[8].

$$_i = \text{uniformly random } [1, 10]. \qquad (4)$$
$$p_i = w_i+5 . \qquad (5)$$

The knapsack's weight constraint:

$$W=sum(w_i)/2, i=1,2,\ldots n \tag{6}$$

All the experiments are conducted over 30 runs and the maximum iteration is 1000 in each run. Parameters of GA and CRO are deduced from literature [7], [9]. As there's no fixed standard for the length of tabu list, based on different combination of tabu list length *tabuLength* and *numNeighbor* tried in experiments, *tabuLength* is set as 25 and *numNeighbor* is set as 100. The parameter settings of the algorithms are listed in Table 1.

Table 1. Parameter Settings

	GA	CRO	CROTS
popSize	100	20	20
crossover probability	0.9	-	-
mutation probability	0.1	-	-
KELossRate	-	0.8	0.8
initialKE	-	100	100
MoleColl	-	0.2	0.2
α	-	500	500
β	-	10	10
Buffer	-	0	0
tabuLength	-	-	25
numNeighbor	-	-	100

5.2   Results and Analysis

Two experiments were conducted based on fixed instances and random instances respectively. More details are listed in Table 2, where *b.*, *w.*and *m.* represents the average optimum profits, average worst profits and the average mean profits, and *bp.* is the optimum profit got in the 30 runs, *itr.* is the iteration the algorithm begins to converge. To test the performance of the algorithms further, convergence curves of the runs are given in Fig.2, and the average results are plotted Fig.3.

It's obvious that in any tested instances, CROTS obtains better results compared with GA and original CRO. Besides CROTS and CRO owns better convergence performance in comparison with GA. For instance, considering the 500 items in random instances, CRO begins to converge at the 34th generation and CROTS converges at 84th iteration while GA converges at 740th iteration. All the results show that CROTS owns better exploitation ability also pretty good convergence performance, which proves that CROTS is an effective approach to address 01KP.

Table 2. Experimental results of GA, CRO and CROTS

| | fixed instance | | | | | random instances | | | | |
| | | | | | | 100items | | | | |
	b.	w.	m.	bp.	itr.	b.	w.	m.	bp.	itr.
GA	2945.7	2545.73	2727.22	3006	934	566.83	519.67	540.38	573	568
CRO	2974.97	2685.63	2952.46	3031	42	574.97	537.6	571.89	580	61
CROTS	3013.07	2715.9	2983.1	3072	95	584.33	545.77	580.75	595	176

| | random instances | | | | | | | | | |
| | 250items | | | | | 500items | | | | |
	b.	w.	m.	bp.	itr.	b.	w.	m.	bp.	itr.
GA	1365.47	1288.07	1320.48	1384	527	2762.27	2655.17	2701.25	2780	740
CRO	1373.43	1318.37	1370.96	1385	127	2775.83	2703.03	2773	2808	34
CROTS	1396.17	1333.3	1391.91	1421	41	2801.83	2721.1	2797.82	2825	84

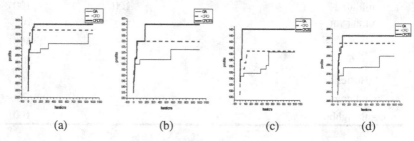

(a)   (b)   (c)   (d)

Fig. 2. Best profits of GA, CRO and CROTS,(a) fixed instance; (b) 100 items(random instances) (c) 250 items(random instances) (d) 500 items(random instances)

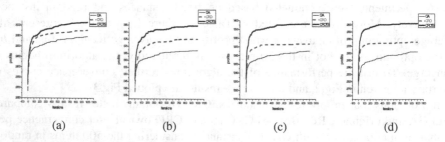

(a)   (b)   (c)   (d)

Fig. 3. Average results of the algorithms (a) fixed instance (b) 100 items (random instances) (c) 250 items (random instances) (d) 500 items (random instances)

6   Conclusion

A hybrid approach CROTS combining CRO framework and tabu search is proposed to address 01KP in this paper. The four elementary reactions in CRO framework is employed to carry out the local search and global search and tabu search is applied to search the solution's neighbors to find more excellent molecule structures. Tabu list used in tabu search makes the algorithm has an outstanding performance compared with GA and original CRO, also increase the algorithm's diversification and intensification, and the experimental results have demonstrated CROTS's effectiveness.

In the future, the study on the combination of the parameters to enhance the performance of the algorithm would proceed. Also, different collision schemes would be applied and tested to adapt to various problems. Besides, developing this algorithm for scheduling problems will be another key issue in the near future.

Acknowledgments. This work presented in this work is supported by the Natural Science Foundation of China (Grant No. 61272148), the Natural Science Foundation of Henan Province (Grant No. 14A520042) and the Science Foundation of Henan University, China (Grant No. 2012YBZR040).

References

1. Lam, A.Y.S., Li, V.O.K.: Chemical Reaction Optimization: a tutorial. Memetic Computing 4(1), 3–17 (2012)
2. Mohamed Esseghir, L., Didier, E.-B.: GPU implementation of the branch and bound method for knapsack problems. In: IEEE International Symposium on Parallel and Distributed Processing Workshops and Ph.D Forum-IPDPSW, pp. 1769–1777 (2012)
3. Ben-Romdhane, H., Ben Jouida, S., Krichen, S.: A dynamic approach for the online knapsack problem. In: Torra, V., Narukawa, Y., Endo, Y. (eds.) MDAI 2014. LNCS, vol. 8825, pp. 96–107. Springer, Heidelberg (2014)
4. George, M., Jose Rui, F., Kostas, F.: Solving the bi-objective multi-dimensional knapsack problem exploiting the concept of core. Applied Mathematics and Computation 215(7), 2502–2514 (2009)
5. Leung, S.C.H., Zhang, D.F.: A hybrid simulated annealing metaheuristic algorithm for the two-dimensional knapsack packing problem. Computers & Operations Research 39(1), 64–73 (2012)
6. Lai, G.M., Yuan, D.H.: A new hybrid combinatorial genetic algorithm for multidimensional knapsack problems. Journal of Supercomputing 70(2), 930–945 (2014)
7. Shen, W., Xu, B.: An Improved Genetic Algorithm for 0-1 Knapsack Problems. In: 2011 Second International Conference on Networking and Distributed Computing (ICNDC), pp. 32–35 (2011)
8. Han, K.-H., Kim, J.-H.: Quantum-Inspired Evolutionary Algorithm for a Class of Combinatorial Optimization. IEEE Transactions on Evolutionary Computation 6(6), 580–593 (2002)
9. Truong, T.K., Li, K.: Chemical reaction optimization with greedy strategy for the 0–1 knapsack problem. Applied Soft Computing 13(4), 1774–1780 (2013)

A New *Physarum*-Based Hybrid Optimization Algorithm for Solving 0/1 Knapsack Problem

Shi Chen[1], Chao Gao[1(✉)], and Zili Zhang[1,2]

[1] College of Computer and Information Science,
Southwest University, Chongqing 400715, China
cgao@swu.edu.cn
[2] School of Information Technology, Deakin University,
Geelong, VIC 3217, Australia

Abstract. As a typical NP-complete problem, 0/1 Knapsack Problem (KP), has been widely applied in many domains for solving practical problems. Although ant colony optimization (ACO) algorithms can obtain approximate solutions to 0/1 KP, there exist some shortcomings such as the low convergence rate, premature convergence and weak robustness. In order to get rid of the above-mentioned shortcomings, this paper proposes a new kind of *Physarum*-based hybrid optimization algorithm, denoted as PM-ACO, based on the critical paths reserved by *Physarum*-inspired mathematical (PM) model. By releasing additional pheromone to items that are on the important pipelines of PM model, PM-ACO algorithms can enhance item pheromone matrix and realize a positive feedback process of updating item pheromone. The experimental results in two different datasets show that PM-ACO algorithms have a stronger robustness and a higher convergence rate compared with traditional ACO algorithms.

Keywords: 0/1 Knapsack · *Physarum*-inspired model · ACO

1 Introduction

Many real world problems, such as investment decision-making and budget controlling, can be formulated as 0/1 Knapsack Problems (KP) [1], which are typical NP-complete problems [2]. Finding an effective approach to solve 0/1 KP has a certain practical and theoretical significance. Approximate algorithms (e.g., greedy algorithm) and deterministic algorithms (e.g., backtracking algorithm, dynamic programming) have been applied for solving 0/1 KP, but some shortcomings (e.g., the low convergence rate) affect their performances. Ant colony optimization (ACO), as a typical nature-inspired computing method proposed in [3], was originally applied in solving Traveling Salesman Problem (TSP) [4]. Ma et al. extend ACO algorithms for solving 0/1 KP [5]. Though ACO algorithms can obtain approximate solutions to 0/1 KP, there are limitations such as premature convergence and low stability. Moreover, some improved ACO algorithms for 0/1 KP, such as Ant Colony System (ACS) [6] and Elitist Ant System

© Springer International Publishing Switzerland 2015
Y. Tan et al. (Eds.): ICSI-CCI 2015, Part II, LNCS 9141, pp. 238–246, 2015.
DOI: 10.1007/978-3-319-20472-7_26

(EAS) [7], often fail into the local optimum because of inadequate and/or excessive exploration which lowers their computational efficiency.

Toshiyuki Nakagaki et al. find that *Physarum polycephalum*, an uncellular and multi-headed slime mold, can solve the maze problem by building pipelines in specific network [8]. Moreover, Tero et al. propose a mathematical model, named as *Physarum*-inspired mathematical (PM) model, to describe such intelligent behavior. Experiments show that the networks generated by PM model show strong stability and high transport efficiency [9].

Taking advantages of the critical paths reserved by PM model when constructing an efficient network, a new *Physarum*-based hybrid optimization algorithm, denoted as PM-ACO, is proposed in this paper. In order to overcome the limitations of ACO algorithms, PM-ACO algorithms improve the probability of selecting significant items through adding extra circulated pheromone that is optimized by PM model. Some experiments show that PM-ACO have a higher convergence rate and robustness compared with traditional ACO algorithms.

The structure of this paper is organized as follows. Section 2 formulates 0/1 KP problem. Section 3 presents the formulation of PM model and the PM-ACO algorithms for solving 0/1 KP. Section 4 provides experimental results between ACO algorithms and PM-ACO algorithms. Section 5 summarizes this paper.

2   Related Work

For a 0/1 KP, there are n items with various values and weights. We want to load some items into a given knapsack with determined capacity. Knapsack's capacity is less than the total weight of items. Therefore, items can be loaded into a knapsack on the condition that total weight does not exceed the capacity of the knapsack. And an item can be only selected once. Supposing c is the capacity of a knapsack, w_i ($w_i > 0$) and v_i ($v_i > 0$) are the weight and value of an item i ($i = 1, 2, \cdots, n$). Binary decision variable x_i ($x_i \in \{1, 0\}$) determines if the item i will be loaded into a knapsack. Therefore, the target of 0/1 KP is to maximize total values of items in a knapsack under capacity constraint, i.e.,

$$S_{\max} = \max \sum_{i=1}^{n} v_i x_i \tag{1}$$

$$s.t. \begin{cases} \sum_{i=1}^{n} w_i x_i \leq c, \\ x_i \in \{0, 1\} \ (i = 1, 2, \cdots, n) \end{cases} \tag{2}$$

In order to evaluate the performance of algorithms when solving 0/1 KP, some measurements are defined as follows:

1. *Iterative Steps* stands for the total steps of an iteration. If an algorithm converges to the theoretic optimal solution with a fewer *Iterative Steps* compared with other algorithms, this algorithm can obtain a higher convergence rate.

2. S_{max}, $S_{average}$, $S_{midvalue}$ and $S_{variance}$ represent the max value, average value, mid-value and variance of results, respectively. These measurements are obtained after repeating C times. For instance, $S_{average} = \sum_{i=1}^{c} S_{step(k)}^{i}/C$, where $S_{step(k)}^{i}$ represents the optimal solution of 0/1 KP in the kth *Iterative Steps* of the ith time. We can further estimate the efficiency and robustness of different algorithms through comparing these measurements.

Currently, ACO algorithms are the effective methods to obtain approximate solutions for 0/1 KP. Here, we take two typical ACO algorithms (e.g., ACS [6] and EAS [7]) as examples to present the process of solving 0/1 KP. Supposing there are n items and m ants. Ants select items based on the item pheromone matrix, which stores the pheromone of all items. An item cannot be selected again by the same ant in an iteration. For instance, ant^k selects an item j based on a certain probability $p_j(k)$ [3]. The definition of $p_j(k)$ is shown in (3), where $\tau_j(t)$ is the pheromone concentration of item j. The heuristic information $\eta_j(t) = v_j/w_j$, defined as (4), represents the expectation that ant^k chooses item j. Two preset positive parameters, α and β, control the relationship between heuristic information and pheromone concentration. $tabu_k$ is a list that stores the items ant^k chosen in the current iteration.

$$p_j(k) = \begin{cases} \dfrac{\tau_j^\alpha(t)\eta_j^\beta(t)}{\sum\limits_{s \notin tabu(k)} \tau_s^\alpha(t)\eta_s^\beta(t)} & j \in tabu_k \\ 0 & j \notin tabu_k \end{cases} \tag{3}$$

$$\eta_j(t) = v_j/w_j \tag{4}$$

Updating item pheromone matrix is the next step. ant^k will release definite amount of pheromone on selected items based on various pheromone updating strategies. For example, ACS algorithm updates item pheromone matrix with the global best ant in all iterations [6] . Meanwhile in the EAS algorithm, item pheromone matrix is updated by the elitist ants that have top m solutions in the current iteration [7]. Furthermore, there is also a local updating strategy for ACS algorithm to update item pheromone matrix. In ACS, the pheromone concentration of each item can be calculated by (5) and the local pheromone updating strategy is shown in (8). τ_j is the pheromone concentration of item j. The speed of pheromone evaporation is measured by parameter ρ ($0 < \rho < 1$). V^{best} and V^k stand for the solution of the global best ant and the elitist ants respectively. e is a parameter which represents the effect of the elitist ants.

$$\tau_j(t+n) = \begin{cases} (1-\rho)\tau_j(t) + \rho\Delta\tau_j^{best} & for\,ACS \\ (1-\rho)\tau_j(t) + \sum\limits_{k=1}^{m} \Delta\tau_j^k + e\Delta\tau_j^{best} & for\,EAS \end{cases} \tag{5}$$

$$\Delta\tau_j^{best} = \begin{cases} Q * v_j/V^{best} & j \in tabu_{best} \\ 0 & j \notin tabu_{best} \end{cases} \tag{6}$$

$$\Delta\tau_j^k = \begin{cases} Q * v_j/V^k & j \in tabu_k \\ 0 & j \notin tabu_k \end{cases} \tag{7}$$

$$\tau_j(t+n) = (1-\rho)\tau_j(t) + \rho\tau_0 \tag{8}$$

3 The Formulation of PM-ACO for Solving 0/1 KP

3.1 The Formulation of PM Model

Based on a maze-solving experiment [8], a computational characteristic of *Physarum polycephalum* is found. As shown in Fig.1(a), foods are placed at the start and the end node of a maze. The tubular pseudopodia of *Physarum* which cannot get foods will shrink and disappear, while others which obtain energy become thicker and thicker. Fig.1(b) shows the terminal network in which only pipelines which are on the shortest path can be reserved. Zhang et al. propose an optimized model with multi-pairs of inlet/outlet nodes to describe such network [10].

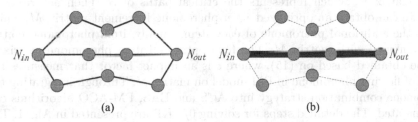

(a)   (b)

Fig. 1. The PM model with one pair of inlet/outlet nodes:(a) the initial network, (b) the final network

In Fig. 1, N_{in} and N_{out} represent the inlet and outlet node of a maze. D_{ij} measures the conductivity of tube which connects nodes i and j (denoted as $tube_{ij}$). Q_{ij} is the flux of $tube_{ij}$. According to the Kirchhoff Law, the flux of all tubes and pressure of all nodes can be calculated by (9)-(10), where d_{ij} represents the length of $tube_{ij}$. p_i and p_j are the pressure of node i and j. The conductivities change over time according to Q_{ij} based on (11).

$$Q_{ij} = \frac{D_{ij}}{d_{ij}}(p_i - p_j) \tag{9}$$

$$\begin{cases} \sum_i Q_{i,N_{in}} = -I_0 \\ \sum_i Q_{i,N_{out}} = I_0 \\ \sum_i Q_{i,j} = 0 \end{cases} \tag{10}$$

$$\frac{dD_{ij}}{dt} = \frac{|Q_{ij}|}{1+|Q_{ij}|} - D_{ij} \tag{11}$$

3.2   Solving 0/1 KP with PM-ACO

This section proposes a pheromone combination strategy to adapt PM model to the item pheromone matrix of ACO when solving 0/1 KP. The updated ACO algorithms are denoted as PM-ACO. PM-ACO algorithms can enhance the pheromone on significant items through filling circulated pheromone of PM model in a tubular network. Therefore, significant items can get more pheromone in the item pheromone matrix of PM-ACO algorithms. The details of this strategy can be described as follows.

The PM model with n nodes can be formulated as a $n * n$ matrix, which saves the conductivities of each path. The item pheromone matrix of ACO can be formulated as a $1 * n$ matrix when solving a 0/1 KP with n items. We first construct a matrix P. The matrix τ can be transformed into matrix P based on (12), where τ_i and τ_j are the pheromone concentrations on item i and j. After that, an intermediary $n * n$ constant matrix B is constructed. The constant term in B is larger than the maximum term in the matrix τ. After that, a new matrix S is generated based on (13). We set matrix S as the input matrix of PM model. Through optimizing matrix S by (9)-(11), we obtain an output $n * n$ matrix V, which represents the critical paths of S. Then according to (14), we can obtain an optimized $1 * n$ pheromone increment matrix M, which stores the additional pheromone of each item. Finally, item pheromone matrix τ is combined with matrix M. The final optimized item pheromone matrix τ' can be obtained based on (15), where ϵ is an impact factor that measures the effect of flowing pheromone in PM model on matrix τ'. Through integrating the pheromone combination strategy into ACS and EAS, PM-ACO algorithms can be generated. The detailed steps for solving 0/1 KP are presented in Alg. 1. The meaning and value of each parameter can be seen in Tab. 1.

$$\begin{cases} P_{ij} = P_{ji} = \frac{\tau_i + \tau_j}{n*(n-1)} & i, j \in [1, n] \\ P_{ii} = P_{jj} = 0 & i = j \end{cases} \tag{12}$$

$$S_{ij} = B_{ij} - P_{ij} \tag{13}$$

$$M_i = [(v_{i1} + ... + v_{in}) + (v_{1i} + ... + v_{ni})] + \tau_i \tag{14}$$

$$\tau_i' = [(1 - \rho)\tau_i + \frac{\rho}{V^{best}}] + \varepsilon M_i, i \in [1, n] \tag{15}$$

4   Experiments

Two datasets, D1 and D2, are used to verify the performance of PM-ACO algorithms. They have 50 and 100 items respectively. All experiments are undertaken in the same environment and are averaged over 100 times in order to eliminate the fluctuations. Some parameters used in these experiments are list in Tab. 1.

Algorithm 1. PM-ACO algorithms for 0/1 KP
1: Initializing parameters α, β, ρ, pheromone concentration τ_0
2: **for** $N := 0$ to T_{steps} **do**
3:     **for** $k := 1$ to m **do**
4:         Loading items into knapsack based on (3)
5:         Updating local item pheromone matrix with (8) (for PM-ACS)
6:     **end for**
7:     Finding elitist ants with top m solutions (for PM-EAS)
8:     $S_{optimal} :=$ The solution of ant^{best} //ant^{best} : the global best ant
9:     Using pheromone combination strategy based on (12)-(14)
10:     Updating item pheromone matrix τ with (15)
11: **end for**
12: Outputting the optimal solution $S_{optimal}$

Table 1. Main parameters and their values of PM-ACO algorithms

Parameter	Explanation	Value
α	The relative importance of pheromone concentration	1
β	The relative importance of heuristic information	1
ρ	The pheromone evaporation rate	0.4
n	The number of items	50
m	The number of ants	50
τ_0	The initial pheromone amount on each item	1/n
l_0	The fixed flux flowing in the *Physarum* network	20
T_{steps}	The total steps of iteration	50
D_{ij}	The initial value of the conductivity of each tube	1
ϵ	The impact factor which measures the effect of flowing pheromone on final item pheromone matrix	0.5

4.1 Experimental Results

Figure 2 illustrates the efficiency and robustness of PM-ACO algorithms in D1. Figs.2(a)(c) show that the $S_{average}$ of PM-ACO increases more obviously than that of ACO, which means PM-ACO has a higher efficiency and convergence rate. Figs.2(b)(d) plot the dynamic changes of $S_{variance}$. Results show that PM-ACO has a stronger robustness since the $S_{variance}$ of PM-ACO decreases sharply and gets a lower value. All above analyses prove that *Physarum*-based optimized strategy can enhance the efficiency and robustness of ACO algorithms.

Further experiments are conducted in D2 to verify the accuracy of PM-ACO algorithms. The S_{max}, $S_{average}$ and $S_{variance}$ calculated by PM-ACO and ACO are plotted in Fig. 3. We find that PM-ACO algorithms obtain a higher $S_{average}$ and a lower $S_{variance}$. The results indicate that PM-ACO algorithms show a higher computational efficiency when the problem scale increases.

4.2 Parameter Analysis

This section takes PM-ACS as an example for parameter analysis. Each experiment only investigates the effects of one parameter, and all results are averaged

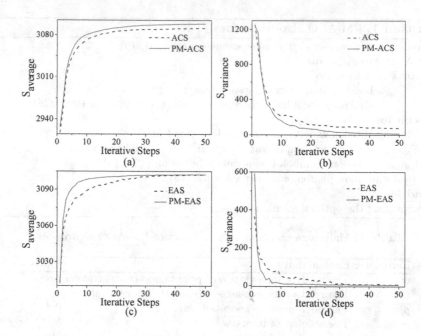

Fig. 2. The (a) $S_{average}$ and (b) $S_{variance}$ of PM-ACS and ACS for 0/1 KP in D1

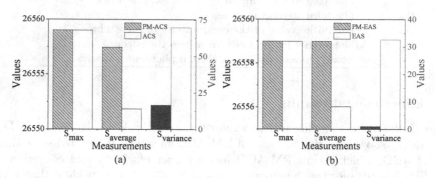

Fig. 3. The results calculated by PM-ACO and ACO for 0/1 KP in D2

over 100 times. Because α determines the impact of previous ants pheromone, PM-ACO algorithms will fall into local optimum if α is too large, and will obtain low efficiency if α is too small. Figs.4(a)(b) plot the dynamic change of $S_{average}$ with the change of *Iterative Steps* and α. Results show that $S_{average}$ gets the maximum value when $\alpha = 1$ and will decrease when $\alpha > 1$. On the other hand, the parameter β represents the relative importance of pheromone versus distance. Figs.4(c)(d) plot the value of $S_{average}$ with the change of *Iterative Steps* and β. $S_{average}$ is converges to a stable solution when $\beta \in [4, 10]$. The fluctuation which appears when $\beta \in [1, 3]$ is mainly caused by the characteristic of dataset.

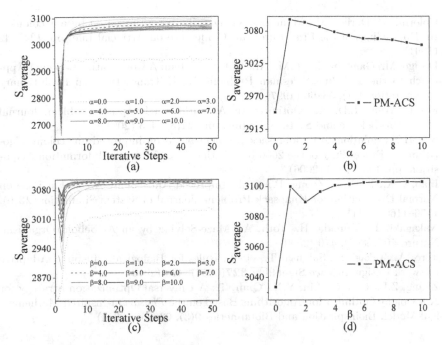

Fig. 4. The dynamic changes of $S_{average}$ with respect to the (a) *Iterative Steps* and (b) α in D1, respectively. The results show that $S_{average}$ increases when $\alpha \in [0,1]$ and decreases when $\alpha \in (1,10]$.

5   Conclusion

Making use of the critical paths reserved by PM model when constructing a high efficient network, a *Physarum*-based pheromone optimization strategy, denoted as PM-ACO, is proposed for solving 0/1 KP. PM-ACO algorithms strengthen the pheromone concentration of significant items through optimizing the item pheromone matrix. Experimental results show that PM-ACO algorithms can obtain a higher efficiency and stability compared with ACO algorithms.

Acknowledgments. This work was supported by National Natural Science Foundation of China (Nos. 61402379, 61403315), and Natural Science Foundation of Chongqing (Nos. cstc2012jjA40013, cstc2013jcyjA40022).

References

1. Li, K.L., Dai, G.M., Li, Q.: A genetic algorithm for the unbounded knapsack problem. In: Proceedings of the 2003 International Conference on Machine Learning and Cybernetics, pp. 1586–1590 (2003)
2. SysloM, M.: Discrete Optimization Algorithms, pp. 118–165. Prentice-Hall, Englewood Cliffs (1983)

3. Colorni, A., Dorigo, M., Maniezzo, V.: Distributed optimization by ant colonies. In: Proceedings of the First European Conference on Artificial Life, pp. 134–142 (1991)

4. Dorigo, M., Gambardella, L.M.: Ant Colony System: A Cooperative Learning Approach to the Traveling Salesman Problem. IEEE Transactions on Evolutionary Computation $1(1)$, 53–66 (1997)

5. Ma, L., Wang, L.D.: Ant Optimization Algorithm for Knapsack Problem. Journal of Systems Science and Systems Engineering $21(8)$, 4–5 (2001)

6. Shi, H.X.: Solution to 01 knapsack problem based on improved ant colony algorithm. In: Proceedings of the 2006 International Conference on Information Acquisition, pp. 1062–1066 (2006)

7. Liao, C.X., Li, X.S., Zhang, P., et al.: Improved Ant Colony Algorithm Base on Normal Distribution for Knapsack Problem. Journal of System Simulation $23(6)$, 1156–1160 (2011)

8. Nakagaki, T., Yamada, H., Toth, A.: Maze-Solving by an Amoeboid Organism. Nature $407(6803)$, 470 (2000)

9. Tero, A., Takagi, S., Saigusa, T., et al.: Rules for Biologically Inspired Adaptive Network Design. Science Signalling $327(5964)$, 439 (2010)

10. Zhang, Z.L., Gao, C., Liu, Y.X., Qian, T.: A Universal Optimization Strategy for Ant Colony Optimization Algorithms Based on the *Physarum*-Inspired Mathematical Model. Bioinspiration and Biomimetics $9(3)$, 036006 (2014)

A Discrete Ecogeography-Based Optimization Algorithm for University Course Timetabling

Bei Zhang[✉], Min-Xia Zhang, and Neng Qian

College of Computer Science and Technology, Zhejiang University of Technology,
Hangzhou 310023, China
zhangbei-zjut@outlook.com, {zmx,qn}@zjut.edu.cn

Abstract. Ecogeography-based optimization (EBO) is an enhanced version of biogeography-based optimization (BBO) algorithm borrowing ideas from island biogeographic evolution for global optimization. The paper proposes a discrete EBO algorithm for university course timetabling problem (UCTP). We first present the mathematical model of UCTP, and then design specified global and local migration operators for the problem. Computational experiment shows that the proposed algorithm exhibits a promising performance on a set of test problem instances.

Keywords: University course timetabling · Ecogeography-based optimization (EBO) · Biogeography-based optimization (BBO) · Migration

1 Introduction

As an important class of practical problems frequently encountered in education, manufacturing, transportation, etc., timetabling problems have received much attention in the areas of computer science for decades. However, most timetabling problems are NP-hard [15], with a large search space to be explored and a large number of constraints to be handled, and thus difficult to solve by classical mathematical programming methods. In consequence, it is important to design efficient heuristic methods for coping with such types of problems.

University course timetabling problem (UCTP) is one of the typical timetabling problems, in which a number of events of university courses need to be scheduled over a prefixed period of time (normally a week), satisfying a variety of soft and hard constraints on rooms, timeslots, students, teachers, etc. Such a timetabling process have to be done at each semester of a university, which is often very time consuming and often results in inefficient timetables that can significantly increase the cost and difficulty of implementations.

Biogeography-based optimization (BBO) [30] is a relatively new metaheuristic method that borrows ideas from island biogeographic evolution [25] for global optimization. As most other evolutionary algorithms (EAs), BBO searches for the global optimum in the solution space by continually evolving a population

© Springer International Publishing Switzerland 2015
Y. Tan et al. (Eds.): ICSI-CCI 2015, Part II, LNCS 9141, pp. 247–257, 2015.
DOI: 10.1007/978-3-319-20472-7_27

of solutions called habitats, where the solution components are analogous to a set of suitability index variables (SIVs) of the habitat, and the fitness of the solution is analogous to the species richness or habitat suitability index (HSI) of the habitat. High HSI habitats tend to share their features with low HSI habitats, and low HSI habitats are likely to accept many new features from high HSI habitats. In recent years, BBO has attracted much attention in the community and has been applied to many real-world problems [9,14,22,24,35].

In BBO, any two habitats in the population may communicate with each other. However, such a global topology easily causes premature convergence [34]. Therefore, Zheng et al. developed an improved version of BBO, called ecogeography-based optimization (EBO) [36], which enhances BBO with a local topology and two new migration operators called global migration and local migration: The former migrates features from both a neighboring habitat and a non-neighboring habitat, while the latter only migrates features from a neighboring habitat. The local topology can effectively avoid local optima, and the combination of two new operators can increase solution diversity and achieve a much better balance between exploration and exploitation.

In this paper, we propose a discrete EBO algorithm for efficiently solving UCTP. Under the EBO migration scheme, we design two specified global migration and local migration operators to effectively explore the solution space with low computational costs, and thus achieve high solution accuracy for the problem. To the best of our knowledge, it is the first work on BBO/EBO for UCTP, and experiments show that the proposed method outperforms a set of state-of-the-art methods on the test instances.

The rest of this paper is structured as follows: Section 2 reviews related work on metaheuristic methods for UCTP, Section 3 presents the mathematical model of UCTP, Section 4 describes our new discrete EBO algorithm for UCTP, Section 5 presents the computational experiments, and Section 6 concludes.

2   Related Work

The UCTP is a complex problem (as we can see in the next section) that is computationally difficult to solve by classical mathematical programming methods, and in recent years there are many efforts on the design of heuristic methods, in particular EAs, for efficiently solving the UCTP.

Tabu search (TS) is a metaheuristic search algorithm that continually moves from a current solution to the best non-tabu neighbor, using a tabu list to avoid cycling moves [12,13]. Alvarez et al. [5] presented a method based on TS for UCTP, which first generates a set of solutions for each student, then uses TS to improve the timetable quality, and finally assigns rooms and further improves the timetables. Aladag et al. [3] developed TS with two neighboring structures based on different types of moves, i.e., simple and swap moves, and tested the influence of various effects of TS in solving UCTP.

Abdullah et al. [1] solved UCTP by using variable neighborhood search (VNS) [26] to explore the solution space, and applying an exponential Monte

Carlo acceptance criterion for each solution in order to find a number of promised neighbors. In [2] the authors improved the algorithm by combining VNS with a local search method called RIICN, meanwhile using a tabu list for the penalty of inefficient neighboring structures after a given number of iterations.

In comparison with single-solution based metaheuristics, population based EAs evolve multiple solutions simultaneously and thus can exploit the potential of parallel computation and reduce the risk of premature convergence. Khonggamnerd et al. [20] proposed a genetic algorithm (GA) [16] that uses specified crossover, mutation, and selection operators for university timetabling arrangement. Simulation results show that the GA can achieve results satisfying all the hard constraints, but they do not evaluate the violation of soft constraints. Alsmadi et al. [4] improved the GA for UCTP by minimizing the number of violated soft constraints as well as teacher workload.

Tuga et al. [33] proposed a method combining simulated annealing (SA) [21] with Kempe chain neighborhood to solve UCTP, where the hard constraints are reformulated to relax soft constraints by first creating a feasible solution according to a heuristic based graph, and then applying SA with Kempe neighboring chain to loose soft constraints. In [6] Aycan and Ayav compared the performance of various neighborhood search methods including simple search, swapping, simple search-swapping, and their combinations. Results show that the most satisfactory timetable is achieved by SA with the combination of all the three methods.

Ant colony optimization (ACO) [11] is a metaheuristic inspired by ants' behavior to find a shortest route between formicary and food based on pheromone. Socha et al. [31] used a max-min ant system for UCTP, which generates for each path a constructive graph to allocate courses to timeslots affected by the amount of pheromone within a range to optimize the timetables. Ayob et al.[7] proposed two hybrid ACO, one combining with TS and the other with SA, where timeslots are probably allocated by ants to courses heuristically. The approach by Nothegger et al. [27] uses two distinct but simplified pheromone matrices to improve convergence, meanwhile providing enough flexibility for guiding the solution construction process. The approach exhibits promising performance in the International Timetabling Competition 2007 [18].

Unlike GA and ACO, particle swarm optimization (PSO) [19] is originally proposed for global optimization. Qarouni-Fard et al. [28] adapted PSO for UCTP by introducing the recombination operator of GA to enable learning timeslots of the local best timetable. The method by Shiau [29] designs an "absolute position value" for each particle, employs a process to repair infeasible timetables, and integrates a local search mechanism. The approach has been applied to timetabling in a university of Taiwan which allows teachers to lecture on flexible time. Tassopoulos and Beligiannis [32] developed a hybrid PSO with local search for high school timetabling, the main idea of which is the successive substitutions in the timeslots of the current particle with different subjects, at random, either from the personal best or the global best. The method has been applied to many high schools in Greek. Chen and Shih [10] presented two PSO methods named the inertia weight version and constriction version, which further utilize an interchange heuristic to explore neighborhood to improve solution quality.

Interested readers can refer to [8] for a detailed survey of UCTP solution methods, including mathematical programming methods and heuristic methods.

3   Description of the Problem

UCTP is to schedule a number of events of university courses in a set of timeslots and rooms. Formally, the problem inputs contain:

- $E = \{e_1, e_2, ..., e_n\}$ is a set of n events of courses (including lectures, speeches, laboratories, etc).
- $R = \{r_1, r_2, ..., r_m\}$ is a set of m rooms.
- $T = \{t_1, t_2, ..., t_o\}$ is a set of o timeslots.
- $S = \{s_1, s_2, ..., s_p\}$ is a set of p students.
- $F = \{f_1, f_2, ..., f_q\}$ is a set of q features (such as multimedia devices and experimental facilities) of the rooms.
- $c(e_i)$ is the capacity required by event e_i $(1 \leq i \leq n)$.
- $c(r_j)$ is the capacity of room r_j $(1 \leq j \leq m)$.
- $A_{(m \times q)}$ is a Room-Feature matrix where $A_{jl} = 1$ denotes that room r_j has feature f_l and $A_{jl} = 0$ otherwise $(1 \leq j \leq m; 1 \leq l \leq q)$.
- $B_{(n \times q)}$ is an Event-Feature matrix where B_{il} denotes whether event e_i requires feature f_l $(1 \leq i \leq n; 1 \leq l \leq q)$.
- $C_{(n \times n)}$ is an Event-Conflict matrix where $C_{ii'}$ denotes whether two events e_i and $e_{i'}$ can be held in the same timeslot $(1 \leq i \leq n; 1 \leq i' \leq n)$.
- $D_{(p \times n)}$ is a Student-Event matrix where D_{hi} denotes whether student s_h will attend event e_i $(1 \leq h \leq p; 1 \leq i \leq n)$.

Commonly, we have $o = 12 \times 5$, where 5 is the number of working days and 12 is the number of classes in each working day.

The decision variables of the problem is a three-dimensional Event-Room-Timeslots matrix $X_{(n \times m \times o)}$, where X_{ijk} indicates that event e_i will be held in room r_j and timeslot t_k $(1 \leq i \leq n; 1 \leq j \leq m; 1 \leq k \leq o)$.

For convenience, we use D_h to denote the set of events of student s_h, X_j the set of events assigned to room r_j, C_i the set of events conflicting to event e_i, and B_i the set of features required by event e_i.

The hard constraints of the problem include:

- Each event must be scheduled once and only once:

$$\sum_{j=1}^{m} \sum_{k=1}^{o} X_{ijk} = 1, \quad 1 \leq i \leq n \tag{1}$$

- Each student can attend one event in any timeslot:

$$\sum_{i \in D_h} \sum_{j=1}^{m} X_{ijk} \leq 1, \quad 1 \leq h \leq p; 1 \leq k \leq o \tag{2}$$

- Each room can hold one event in any timeslot:

$$\sum_{i \in X_j} X_{ijk} \leq 1, \quad 1 \leq j \leq m; 1 \leq k \leq o \tag{3}$$

- Any two conflicting events cannot be held in the same timeslot:

$$\sum_{i' \in C_i} \sum_{j=1}^{m} X_{i'jk} \leq 1, \quad 1 \leq i \leq n; 1 \leq k \leq o \tag{4}$$

- Each room allocated to an event should have adequate capacity:

$$c(e_i) \leq c(r_j), \quad \forall i, j : (\exists k : X_{ijk} = 1) \tag{5}$$

- Each room allocated to an event should have adequate features:

$$A_{jl} = 1, \quad \forall i, j, l : (\exists k : X_{ijk} = 1) \land l \in B_i \tag{6}$$

And the soft constraints of the problem include:

- Each student should not take more than four timeslots in a day:

$$\sum_{i \in D_h} \sum_{j=1}^{m} \sum_{k=a}^{a+11} X_{ijk} \leq 4, \quad 1 \leq h \leq p; a \in \{1, 13, 25, 37, 49\} \tag{7}$$

- Each student should not take more than two consecutive timeslots in a day:

$$\sum_{i \in D_h} \sum_{j=1}^{m} \sum_{k=a}^{a+2} X_{ijk} \leq 2, \quad 1 \leq h \leq p; 1 \leq a \leq 60 \land a\%12 \leq 10 \tag{8}$$

- Each event should not be assigned to the final timeslot of each day:

$$\sum_{i=1}^{n} \sum_{j=1}^{m} X_{ija} = 0, \quad a \in \{12, 24, 36, 48, 60\} \tag{9}$$

The problem is to find a schedule of events such that the hard constraints (1)–(6) are all satisfied, while the soft constraints (7)–(9) are violated as minimum as possible. Let $S_1 = \{1, 13, 25, 37, 49\}$, $S_2 = \{a | 1 \leq a \leq 60 \land a\%12 \leq 10\}$, $S_3 = \{12, 24, 36, 48, 60\}$, the objective function of the problem is defined as:

$$\min f = \sum_{a \in S_1} \sum_{h=1}^{p} \left(\sum_{i \in D_h} \sum_{j=1}^{m} \sum_{k=a}^{a+11} g(X_{ijk} - 4) \right)$$

$$+ \sum_{a \in S_2} \sum_{h=1}^{p} \left(\sum_{i \in D_h} \sum_{j=1}^{m} \sum_{k=a}^{a+2} g(X_{ijk} - 2) \right) \tag{10}$$

$$+ \sum_{a \in S_3} \left(\sum_{i=1}^{n} \sum_{j=1}^{m} X_{ija} \right)$$

where g is a function defined as:

$$g(x) = \begin{cases} x, & \text{if } x > 0 \\ 0, & \text{else} \end{cases} \qquad (11)$$

As we can see, for the considered UCTP, evaluating either the feasibility or the objective function value of a solution is often computationally expensive. Thus the design of an algorithm effectively exploring the solution space is of crucial importance for solving the problem.

4 A Discrete EBO Algorithm for the Problem

A directly encoding scheme uses a $n \times m \times o$ matrix to represent a solution to UCTP which is often very inefficient. In this paper we employ a two-stage problem solving tactic:

1. Determining a timeslot for each event;
2. Using a bipartite matching algorithm [17] to determine a room for each event with specified timeslot.

Thus in our algorithm we use a compact encoding such that each solution $\mathbf{x}$ is an n-dimensional integer valued vector, where $x(i)$ denotes the timeslot allocated for event e_i. Such a vector can be easily transformed to an assignment matrix according to [17]. Note that the encoding scheme ensures the satisfaction of constraint (1), and the result of the bipartite matching algorithm also satisfies the hard constraints (3), (5) and (6). Therefore we compute the following penalized objective function that takes the violations of constraints (2) and (4) into consideration:

$$f' = f + M\left(\sum_{h=1}^{p}\sum_{k=1}^{o}\left(\sum_{i \in D_h}\sum_{j=1}^{m} g(X_{ijk}) - 1 \right) + \sum_{i=1}^{n}\sum_{k=1}^{o}\left(\sum_{i' \in C_i}\sum_{j=1}^{m} g(X_{i'jk}) - 1 \right) \right) \qquad (12)$$

where M is a large positive constant (typically larger than the maximum possible number of soft constraint violations).

To solve UCTP, our EBO algorithm evolves a population of solutions (habitats) by continually migrating features probably from high fitness solutions to low fitness ones. The immigration rate λ_i and emigration rate μ_i of each solution $\mathbf{x}_i$ are respectively calculated as follows:

$$\lambda_i = \frac{f_i - f_{\min} + \epsilon}{f_{\max} - f_{\min} + \epsilon} \qquad (13)$$

$$\mu_i = \frac{f_{\max} - f_i + \epsilon}{f_{\max} - f_{\min} + \epsilon} \qquad (14)$$

where f_i is the (penalized) objective function value of $\mathbf{x}_i$, $f_{\max}$ and $f_{\min}$ are the maximum and minimum objective function values in the population, and ϵ is a small constant to avoid zero-division-error.

The original EBO is used for global optimization. Here we design novel migration operators for the discrete UCTP. When performing local migration on solution $\mathbf{x}_i$, we select a neighboring solution $\mathbf{x}_j$ with a probability in proportional to μ_j and select a random dimension $k \in [1, n]$, and set $\mathbf{x}_i(k)$ to $\mathbf{x}_j(k)$. Let $u = \mathbf{x}_j(k)$, we then perform the following procedure to repair $\mathbf{x}_i$:

1) Sort the set of timeslots other than u in increasing order of the number of events assigned to each timeslot, and let $T(\mathbf{x}_i)$ be the sorted list;
2) If $T(\mathbf{x}_i)$ is empty, stop the repair procedure.
3) Otherwise, check whether there exists another position k' of $\mathbf{x}$ such that $\mathbf{x}_i(k') = u$; If so, then:
   3.1) Check whether $C_{k,k'} = 1$; If so, goto Step 3.3);
   3.2) Otherwise, check whether there exists any student s_h such that $D_{h,k} = 1$ and $D_{h,k'} = 1$; If so, goto Step 3.3), otherwise stop the repair procedure.
   3.3) Let u be the first timeslot of $T(\mathbf{x}_i)$, remove u from $T(\mathbf{x}_i)$, set $\mathbf{x}_i(k') = u$, let $k = k'$ and goto Step 2).

In case that $\mathbf{x}_i$ is a feasible solution, if the above procedure stops at Step 3.2), the new solution is also feasible, and thus when evaluating its objective function we do not need to calculate the penalized part of Eq. (12); Otherwise the new solution is an infeasible one and thus Eq. (12) must be completely evaluated. In consequence, the procedure can reduce the computational burden on the evaluation of new solutions to a great extent.

When performing global migration on solution $\mathbf{x}_i$, we select a neighboring solution $\mathbf{x}_j$ and a non-neighboring solution $\mathbf{x}_{j'}$ with probabilities in proportional to their emigration rates, and select two different dimensions k and k' in the range of $[1, n]$. Then we:

1) Set $\mathbf{x}_i(k)$ to $\mathbf{x}_j(k)$, and then perform the above repair procedure at the kth dimension. If the result is infeasible then goto Step 3).
2) Set $\mathbf{x}_i(k')$ to $\mathbf{x}_{j'}(k')$, and then perform the above repair procedure at the k'th dimension. If the result is feasible then return the resulting solution.
3) Set $\mathbf{x}_i(k)$ to $\mathbf{x}_{j'}(k)$, and then perform the above repair procedure at the kth dimension. If the result is infeasible then return the resulting solution.
4) Set $\mathbf{x}_i(k')$ to $\mathbf{x}_j(k')$, and then perform the above repair procedure at the k'th dimension.

As the original EBO [36], we use a parameter η, called the immaturity index, to determine whether to perform global or local migration. The parameter is dynamically adjusted as follows:

$$\eta = \eta_{\max} - \frac{g}{g_{\max}}(\eta_{\max} - \eta_{\min}) \tag{15}$$

where g is the current generation number and $g_{\max}$ is the maximum generation number of the algorithm, and $\eta_{\max}$ and $\eta_{\min}$ are respectively the upper limit and lower limit of η.

Algorithm 1 presents the framework of our EBO for UCTP, where $rand()$ generates a random number uniformly distributed in [0,1]. Note in Step 11 of the algorithm, if $\mathbf{x}_i$ is feasible while $\mathbf{x}'_i$ is produced by a failed repair procedure, we do not need to evaluate $\mathbf{x}'_i$.

Algorithm 1.. The EBO algorithm for UCTP

1	Randomly initialize a population of solutions to the problem, let $g = 0$;
2	**while** $g < g_{max}$ **do**
3	Calculate λ_i, μ_i, and η according to Eq. (13)–(15);
4	**for each** $\mathbf{x}_i$ in the population **do**
5	**if** $rand() < \lambda_i$ **then**
6	**if** $rand() > \eta$ **then**
7	Use the local migration to produce a new solution $\mathbf{x}_i'$;
8	**else**
9	Use the global migration to produce a new solution $\mathbf{x}_i'$;
10	**if** $\mathbf{x}_i'$ is better than $\mathbf{x}_i$;
11	Replace $\mathbf{x}_i$ with $\mathbf{x}_i'$ in the population;
12	**return** the best solution found so far.

5   Computational Experiment

We test the performance of the proposed EBO algorithm on a set of 8 UCTP instances, which are constructed based on the data from two semesters of four universities in China. Table 1 presents a summary of the problem instances.

Table 1. Summary of the test UCTP instances

#	n	m	o	p	q
1	156	8	60	285	3
2	172	8	60	281	3
3	462	26	60	633	5
4	472	26	60	670	5
5	689	37	60	1475	6
6	706	37	60	1543	6
7	1711	72	60	5125	6
8	1860	72	60	5203	6

For EBO, we empirically set the population size to 50, neighborhood size to 5, $\eta_{max} = 0.95$ and $\eta_{min} = 0.05$. For comparison, we implement a BBO version that only uses the local migration for the UCTP described in Section 4, and three other competitive algorithms including the GA in [4], the ACO in [27], and the hybrid PSO in [32], the parameters of which are set as in the literature. The maximum running time is set to 3 min for instance #1 and #2, 10 min for #3 and #4, 15 min for #5 and #6, and 30 min for #7 and #8.

The experiment is conducted on a computer with Intel Core i7-4500M CPU, 8GB memory, and Windows 7. The algorithms are run for 30 times on each test instance, and the best, median and standard deviation of the resulting objective function values among the 30 runs are shown in Table 2. As we can see from the results, on the first two instances, EBO can obtain a solution that satisfies all the hard and soft constraints in all 30 runs (where both the mean and best objective values are zero), but among the other four algorithms only PSO achieves this on instance #1; on #3, all the algorithms can obtain a best objective value of 6, but only PSO, BBO and EBO can also obtain a median value of 6 which is much less than GA and ACO; on the remaining 5 instances, EBO not only always obtains the minimum median values among the five algorithms, but also always uniquely obtains the minimum best values; in the last four instances which are large-size, the results of EBO are much better than the other four algorithms. This demonstrates the effectiveness of our specified migration operators for UCTP.

Among the other four algorithms, BBO and PSO exhibit similar performance which is better than GA and ACO. This indicates that the local migration mechanism described in Section 4 is much more effective than the crossover & mutation operator of GA and the solution path construction method of ACO on the test instances. Nevertheless, EBO shows a significant performance improvement over BBO, which indicates that the combination of local and global migration is further more effective than the single local migration operator for UCTP.

Table 2. Comparative results on the test UCTP instances

#	Metric	GA	ACO	PSO	BBO	EBO
1	median	5	5	0	5	0
	best	0	0	0	0	0
	std	2.20	1.25	0	1.06	0
2	median	7	7	7	7	0
	best	0	0	0	0	0
	std	1.88	1.69	1.76	1.85	0
3	median	13	16	6	6	6
	best	6	6	6	6	6
	std	3.60	5.13	0	1.01	0
4	median	28	24	21	28	21
	best	21	21	21	21	12
	std	5.09	2.95	1.98	3.20	2.68
5	median	96	115	78	93	58
	best	69	70	63	69	53
	std	12.31	26.09	8.19	12.16	3.69
6	median	236	251	179	187	101
	best	169	154	89	89	75
	std	18.08	34.62	29.99	35.85	16.26
7	median	386	398	305	319	222
	best	251	272	228	248	156
	std	69.03	80.20	36.58	52.57	39.66
8	median	580	670	395	413	238
	best	413	416	325	306	193
	std	45.42	61.93	28.90	42.00	15.60

6   Conclusion

To efficiently solve UCTP, the paper proposes a new discrete EBO algorithm which designs two specified global and local migration operators, which can effectively explore the solution space with low computational costs and thus achieve high solution accuracy. Computational experiment shows that our algorithm outperforms a number of competitive algorithms on a set of test instances.

Acknowledgments. This work is supported by National Natural Science Foundation (Grant No. 61473263) and Zhejiang Provincial Natural Science Foundation (Grant No. LY14F030011) of China.

References

1. Abdullah, S., Burke, E.K., McColloum, B.: An investigation of variable neighborhood search for university course timetabling. In: Multidiscipl. Conf. Schedul., pp. 413–427 (2005)
2. Abdullah, S., Burke, E.K., McColloum, B.: A hybrid evolutionary approach to the university course timetabling problem. In: IEEE Congr. Evol. Comput., pp. 1764–1768 (2007)
3. Aladag, C.H., Hocaoglu, G.A., Basaran, M.: The effect of neighborhood structures on tabu search algorithm in solving course timetabling problem. Expert Sys. Appl. **36**, 12349–12356 (2009)
4. Alsmadi, O.M.K., Abo-Hammour, Z.S., Abu-Al-Nadi, D.I., Algsoon, A.: A novel genetic algorithm technique for solving university course timetabling problems. In: IEEE, Int. Conf. Syst., Signal Process. Appl., pp. 195–198 (2011)
5. Alvarez, R., Crespo, E., Tamarit, J.M.: Design and implementation of a course scheduling system using Tabu search. Eur. J. Oper. Res. **137**, 512–523 (2002)
6. Aycan, E., Ayav, T.: Solving the course scheduling problem using simulated annealing. In: IEEE Advance Computing Conference, pp. 462–466 (2008)
7. Ayob, M., Jaradat, G.: Hybrid ant colony systems for course timetabling problems. In: IEEE Conf. Data Mining Opt., pp. 120–126 (2009)
8. Babaei, H., Karimpour, J., Hadidi, A.: A survey of approaches for university course timetabling problem. Comput. Ind. Engin., in press (2014). doi:10.1016/j.cie.2014.11.010
9. Boussaïd, I., Chatterjee, A., Siarry, P., Ahmed-Nacer, M.: Biogeography-based optimization for constrained optimization problems. Comput. Oper. Res. **39**(12), 3293–3304 (2012)
10. Chen, R.M., Shih, H.F.: Solving university course timetabling problems using constriction particle swarm optimization with local search. Algorithms **6**(2), 227–244 (2013)
11. Dorigo, M., Caro, G.D.: Ant colony optimization: a new meta-heuristic. In: IEEE Congr. Evol. Comput., vol. 2, pp. 1470–1477 (1999)
12. Glover, F.: Tabu search, part I. ORSA J. Computing **1**, 190–206 (1989)
13. Glover, F.: Tabu search, part II. ORSA J. Computing **2**, 4–32 (1990)
14. Gong, W., Cai, Z., Ling, C.X.: DE/BBO: a hybrid differential evolution with biogeography-based optimization for global numerical optimization. Soft Comput. **15**(4), 645–665 (2010)
15. Garey, M.R., Johnson, D.S.: Computers and intractability: a guide to the theory of NP-completeness. W.H. Freeman (1979)
16. Holland, J.H.: Adaptation in natural and artificial systems: An introductory analysis with applications to biology, control and artificial intelligence. MIT Press (1975)
17. Hopcroft, J., Karp, R.M.: An $n^{5/2}$ algorithm for maximum matchings in bipartite graphs. SIAM J. Comput. **2**, 225–231 (1973)

18. ITC2007. Second international timetabling competition (2007). http://www.cs.qub.ac.uk/itc2007/
19. Kennedy, J., Eberhart, R.: Particle swarm optimization. In: IEEE Int. Conf. Neural Netw., vol. 4, pp. 1942–1948 (1995)
20. Khonggamnerd, P., Innet, S.: On improvement of effectiveness in automatic university timetabling arrangement with applied genetic algorithm. In: IEEE Int. Conf. Comput. Sci. Converg. Inform. Technol., pp. 1266–1270 (2009)
21. Kirkpatrick, S., Gerlatt Jr, C.D., Vecchi, M.P.: Optimization by simulated annealing. Science **220**, 671–680 (1983)
22. Lohokare, M., Pattnaik, S., Panigrahi, B., Das, S.: Accelerated biogeography-based optimization with neighborhood search for optimization. Appl. Soft Comput. **13**, 2318–2342 (2013)
23. Ma, H.: An analysis of the equilibrium of migration models for biogeography-based optimization. Inform. Sci. **180**(18), 3444–3464 (2010)
24. Ma, H., Simon, D.: Blended biogeography-based optimization for constrained optimization. Engin. Appl. Artif. Intell. **24**(3), 517–525 (2011)
25. MacArthur, R., Wilson, E.: The Theory of Biogeography. Princeton University Press (1967)
26. Mladenovic, N., Hansen, P.: Variable neighborhood search. Comput. Oper. Res. **24**(11), 1097–1100 (1997)
27. Nothegger, C., Mayer, A., Chwatal, A., Raidl, G.R.: Solving the post enrolment course timetabling problem by ant colony optimization. Ann. Oper. Res. **104**(1), 325–339 (2012)
28. Qarouni-Fard, D., Najafi-Ardabifi, A., Moelnzadeh, M.H., et al.: Finding feasible timetables with particle swarm optimization. In: IEEE Int. Conf. Innov. Inform. Technol., pp. 387–391 (2007)
29. Shiau, D.F.: A hybrid particle swarm optimization for a university course scheduling problem with flexible preferences. Expert Syst. Appl. **38**(1), 235–248 (2011)
30. Simon, D.: Biogeography-based optimization. IEEE Trans. Evol. Comput. **12**(6), 702–713 (2008)
31. Socha, K., Knowles, J., Sampels, M.: A max-min ant system for the university course timetabling problem. In: Dorigo, M., Caro, G.D., Sampels, M. (eds.) Ant Algorithms 2002. LNCS, vol. 2463, pp. 1–13. Springer, Heidelberg (2002)
32. Tassopoulos, I.X., Beligiannis, G.N.: A hybrid particle swarm optimization based algorithm for high school timetabling problems. Appl. Soft Comput. **12**(11), 3472–3489 (2012)
33. Tuga, M., Berretta, R., Mendes, A.: A hybrid simulated annealing with kempe chain neighborhood for the university timetabling problem. In: IEEE Conf. Comput. Inf. Sci. (2007)
34. Zheng, Y.J., Ling, H.F., Wu, X.B., Xue, J.Y.: Localized biogeography-based optimization. Soft Comput. **18**(11), 2323–2334 (2014)
35. Zheng, Y.J., Ling, H.F., Shi, H.H., Chen, H.S., Chen, S.Y.: Emergency railway wagon scheduling by hybrid biogeography-based optimization Comput. Oper. Res. **43**, 1–8 (2014)
36. Zheng, Y.J., Ling, H.F., Xue, J.Y.: Ecogeography-based optimization: Enhancing biogeography-based optimization with ecogeographic barriers and differentiations. Comput. Oper. Res. **50**, 115–127 (2014)

Constrained Optimization Algorithms

A Dynamic Penalty Function
for Constrained Optimization

Chengyong Si[1]([⊠]), Jianqiang Shen[1], Xuan Zou[1], Yashuai Duo[1],
Lei Wang[2], and Qidi Wu[2]

[1] Shanghai-Hamburg College, University of Shanghai
for Science and Technology, Shanghai 200093, China
sichengyong_sh@163.com
[2] College of Electronics and Information Engineering, Tongji University,
Shanghai 201804, China
wanglei@tongji.edu.cn

Abstract. Penalty function methods have been widely used for handling constraints, but it's still a challenge about how to set the penalty parameter effectively though many related methods have been proposed. In this paper, the penalty parameter is firstly analyzed systematically by introducing four rules. Based on this analysis, a new Dynamic Penalty Function (DyPF) is proposed by adjusting penalty parameter in three different situations during the evolution (i.e., the infeasible situation, the semi-feasible situation, and the feasible situation). The experiments are designed to verify the effectiveness of our newly proposed DyPF. The results show that DyPF presents a better overall performance than other five dynamic or adaptive state-of-the-art methods in the community of constrained evolutionary optimization.

Keywords: Constrained optimization · Constraint handling techniques · Differential evolution · Dynamic penalty function (DyPF) · Ranking methods

1 Introduction

Constrained Optimization Problems (COPs) are very important and common in real-world applications. The general COPs can be formulated as follows:

$$\text{Minimize} \quad f(\vec{x})$$

$$\text{Subject to:} \quad g_j(\vec{x}) \leq 0, \quad j = 1, \cdots, l$$

$$h_j(\vec{x}) = 0, \quad j = l+1, \cdots, m$$

where $\vec{x} = (x_1, \cdots, x_n)$ is the decision variable which is bounded by the decision space S. S is defined by the constraints:

$$L_i \leq x_i \leq U_i, \quad 1 \leq i \leq n \tag{1}$$

© Springer International Publishing Switzerland 2015
Y. Tan et al. (Eds.): ICSI-CCI 2015, Part II, LNCS 9141, pp. 261–272, 2015.
DOI: 10.1007/978-3-319-20472-7_28

Here, l is the number of inequality constraints and $m-l$ is the number of equality constraints.

The Evolutionary Algorithms (EAs), as the unconstrained search techniques and solution generating strategies, are not suitable enough to solve COPs without additional mechanisms to deal with the constraints. Consequently, many constrained optimization evolutionary algorithms (COEAs) are proposed [1]-[4].

Penalty function methods are the most widely used methods for handling constraints, in which some penalty parameters are adopted to balance the objective function values and constraint violation (i.e., to bias the search in the constrained search space [5]). It's clear that the performance of these methods is mainly determined by their parameters and the methods can be classified based on the form of these parameters.

If the penalty parameters keep constant throughout the evolution process, this method is called static penalty function method. Alternatively, if the penalty parameters are related with the current generation number, it is called dynamic penalty function method. As many parameters are required in dynamic penalty function method, some adaptive penalty functions which gather information from the search process, or self-adaptive approaches, which evolve both the penalty parameters and solutions, have also been proposed [6].

Some other approaches based on careful comparison among feasible and infeasible solutions are also developed.

For example, Deb [7] proposed a feasibility-based rule to pair-wise compare individuals:

1) Any feasible solution is preferred to any infeasible solution.

2) Among two feasible solutions, the one having better objective function value is preferred.

3) Among two infeasible solutions, the one having smaller constraint violation is preferred.

The stochastic ranking method (SR) proposed by Runarsson and Yao [8] is a very classical constraint handling technique, which tries to achieve a balance between objective function value and constraint violation stochastically. It compares pair-wise solutions using the following criteria: 1) if both individuals are feasible, the ranking of them is determined by the objective function value; else 2) the parameter P_f will determine the probability of ranking by objective function value or constraint violation. Deb's feasibility-based rule can be seen as a special case of SR with $P_f = 0$. The experimental results indicate that an overall better performance can be obtained when $P_f = 0.45$. However, this paper didn't provide the assurance that $P_f = 0.45$ is an optimal value.

And later, these two authors also pointed out that there should be some biases when solving single-objective optimization problems (SCOPs) [9].

Besides, some methods based on multi-objective optimization concepts are also presented. The main idea is to convert the single-objective constrained optimization problem into a bi-objective or multi-objective optimization problem taking the constraints as one or more objectives to be minimized.

After reviewing the three most frequently used constraint handling techniques, some approaches based on the "dynamic or adaptive" idea that are closely related with the work presented in this paper will be introduced in detail.

Wang *et al.* [10] proposed an adaptive tradeoff model (ATM) for constrained evolutionary optimization. In this model, to obtain an appropriate tradeoff between objective function and constraint violation, different tradeoff schemes during different situations in a search process (i.e., infeasible situation, semi-feasible situation and feasible situation) are designed. Based on this idea, an improved adaptive tradeoff model was proposed, in which each constraint violation is first normalized [11], and this model was combined with $(\mu+\lambda)$-DE with the name $(\mu+\lambda)$-CDE. To overcome the drawback of dynamic settings for tolerance value δ, Jia *et al.* [12] presented an improved version of $(\mu+\lambda)$-CDE, named ICDE. Unlike $(\mu+\lambda)$-CDE, in ICDE, the hierarchical non-dominated individual selection scheme is utilized in the infeasible situation and the feasibility proportion of the population is used to convert the objective function of each individual in the semi-feasible situation.

Besides, some other adaptive approaches or frameworks were also introduced [13]-[16]. As the solution's property (i.e., infeasible or feasible) plays an important role in solving COPs, some adaptive methods based on this were also presented.

Farmani and Wright [17] proposed a self-adaptive fitness formulation in which the infeasibility measure is used to form a two-stage penalty to the infeasible solutions. Venkatraman and Yen [18] presented a generic, two phase framework with the aim to find a feasible solution in the first phase. Based on Yao's stochastic ranking [8], Zhang *et al.* [19] proposed a dynamic stochastic selection (DSS) within the framework of multimember DE (DSS_MDE). Tessema *et al.* [20] introduced another adaptive penalty formulation which uses the number of feasible individuals to determine the amount of penalty added to infeasible individuals (i.e., to guide the search process toward finding more feasible individuals *et al.*).

Many of these methods mentioned above can get a relatively satisfying result, but the parameters used in these approaches are mainly determined by the experiments. The inner mechanism of Constraint Handling Techniques (e.g., the relationship of different CHTs, when and why some CHTs are more efficient) is few studied.

Herein, to overcome this drawback, in this paper we first studied the penalty parameter systematically, and then proposed a new dynamic penalty function (DyPF) based on the analysis.

The rest of this paper is organized as follows. Section 2 systematically analyzes the penalty parameter. Based on this, Section 3 presents a detailed description of the proposed DyPF. The experimental results and the comparison with some similar state-of-the-art methods are given in Section 4. Finally, Section 5 concludes with a brief summary of this paper and some future work.

2   Systematical Analysis of Penalty Parameters

The basic idea of the discussion is that by introducing four rules (i.e., A_1, A_2, B_1, B_2), if the penalty parameter is consistent with some rule's combination in certain range, then we can conclude that the penalty parameter value in this range has no effect on ranking the solutions [21].

As there are only two kinds of solutions in the evolutionary process (i.e., feasible solutions and infeasible solutions), the ranking or the selection process is mainly between these two kinds of solutions, including feasible solutions versus feasible solutions, feasible solutions versus infeasible solutions, and infeasible solutions versus infeasible solutions. As there is no constraint violation for feasible solutions, only the objective function values are used for the comparison between feasible solutions and feasible solutions. As to the comparison involving infeasible solutions, the following rules are introduced:

A_1: a feasible solution is superior to an infeasible solution.

A_2: a feasible solution is inferior to an infeasible solution.

B_1: two infeasible solutions will be ranked according to the constraint violation, and the one with less constraint violation will be preferred.

B_2: two infeasible solutions will be ranked according to the reciprocal of the constraint violation, and the one with more constraint violation will be preferred.

Additional rule: among two infeasible solutions with the same constraint violation, the one with less objective function value is preferred (for the minimization problems).

It should be noted that here the characteristics of the rules (i.e., good or bad) are not considered as the judgment of a rule is closely related with the problem characteristics (e.g., the location of the optimal solution, the topological properties of the constraints, etc.) and the solving goals (e.g., how to judge the solution's performance if no feasible solutions are found, etc.). And here, the main concern is that the ranking result following some rules' combination (A_1-B_1, A_1-B_2, A_2-B_1, A_2-B_2) is unique.

The introduction of penalty parameter enables us to transform a constrained optimization problem (A) into an unconstrained one (A') [8]. Here, we define the evaluation function L as follows.

For the given λ, $\delta > 0$, let

$$L(\vec{x}_i, \lambda, \delta) = f(\vec{x}_i) + \lambda G(\vec{x}_i, \delta) \quad i = 1, 2, ..., NP \tag{2}$$

where $\vec{x}_i$ stands for the NP n-dimensional real-valued vectors of the population. λ is the penalty parameter and δ is the tolerance value for the equality constraints. f is the objective function and G is the penalty function with the form as follows.

$$G(\vec{x}_i, \delta) = \sum_{j=1}^{m} G_j(\vec{x}_i, \delta) = \sum_{j=1}^{l} \max(0, g_j(\vec{x}_i)) + \sum_{j=l+1}^{m} \max(0, |h_j(\vec{x}_i)| - \delta) \tag{3}$$

As the effect of λ is mainly concerned, δ can be supposed as a constant. The formula (2) can be transformed as

$$L(\vec{x}_i, \lambda) = f(\vec{x}_i) + \lambda G(\vec{x}_i) \quad i = 1, 2, ..., NP \tag{4}$$

Given two population members, $\vec{x}_s$ and $\vec{x}_t$, where s and t are randomly selected from $[1, NP]$ and satisfying: $s \neq t$, the difference between their evaluation function values is:

$$
\begin{aligned}
\Delta(\vec{x}_s, \vec{x}_t, \lambda) &= L(\vec{x}_s, \lambda) - L(\vec{x}_t, \lambda) \\
&= [f(\vec{x}_s) + \lambda G(\vec{x}_s)] - [f(\vec{x}_t) + \lambda G(\vec{x}_t)] \\
&= [f(\vec{x}_s) - f(\vec{x}_t)] + \lambda[G(\vec{x}_s) - G(\vec{x}_t)]
\end{aligned} \tag{5}
$$

We define $\Delta f_{st} = f(\vec{x}_s) - f(\vec{x}_t)$, $\Delta G_{st} = G(\vec{x}_s) - G(\vec{x}_t)$, then formula (5) can be written as:

$$
\Delta(\vec{x}_s, \vec{x}_t, \lambda) = \Delta f_{st} + \lambda \cdot \Delta G_{st} \tag{6}
$$

Suppose the number of feasible and infeasible individuals in the population is p and q respectively, with $0 \leq p \leq NP, 0 \leq q \leq NP$ & $p + q = NP$.

1) Infeasible solution versus feasible solution

Similarly, given two population members, $\vec{x}_s$ and $\vec{x}_t$, where s and t are randomly selected from the p feasible solutions and q infeasible solutions, the formula (5) can be transformed as:

$$
\begin{aligned}
\Delta(\vec{x}_s, \vec{x}_t, \lambda) &= L(\vec{x}_s, \lambda) - L(\vec{x}_t, \lambda) \\
&= f(\vec{x}_s) - [f(\vec{x}_t) + \lambda G(\vec{x}_t)] \\
&= [f(\vec{x}_s) - f(\vec{x}_t)] - \lambda G(\vec{x}_t)
\end{aligned} \tag{7}
$$

According to rule A_1, member $\vec{x}_s$ is better than member $\vec{x}_t$. From formula (7), if $\lambda > \dfrac{\Delta f_{st}}{G(\vec{x}_t)}$, then $\Delta(\vec{x}_s, \vec{x}_t, \lambda) < 0$. In this case, rule A_1 and penalty function method have the same effect on ranking these two solutions.

2) *Infeasible solution versus infeasible solution*

According to rule B_1, two infeasible solutions will be ranked according to the constraint violation. So if $\Delta(\vec{x}_s, \vec{x}_t, \lambda)$ and ΔG_{st} have the same symbol (i.e., positive or negative), we can conclude that rule B_1 and penalty function method have the same effect on ranking these two individuals.

There are three different cases in this situation:

a) $\Delta G_{st} > 0$: In this case, if $\lambda > -\dfrac{\Delta f_{st}}{\Delta G_{st}}$, $\Delta(\vec{x}_s, \vec{x}_t, \lambda) = \Delta f_{st} + \lambda \cdot \Delta G_{st} > 0$.

b) $\Delta G_{st} < 0$: In this case, if $\lambda > -\dfrac{\Delta f_{st}}{\Delta G_{st}}$, $\Delta(\vec{x}_s, \vec{x}_t, \lambda) = \Delta f_{st} + \lambda \cdot \Delta G_{st} < 0$.

c) $\Delta G_{st} = 0$: This is the case when the two individuals have the same constraint violation, and $\Delta(\vec{x}_s, \vec{x}_t, \lambda) = \Delta f_{st}$, which is not related with λ. In this case, the ranking will follow additional rule.

In general, when ranking two individuals (e.g., $\vec{x}_s$ and $\vec{x}_t$), if $\lambda > -\dfrac{\Delta f_{st}}{\Delta G_{st}}, \Delta G_{st} \neq 0$

(here, s and t are randomly selected from the q infeasible solutions), then rule B_1 and penalty function method have the same effect.

Denote I_{fea} and I_{inf} as the set of the index of p feasible solutions and q infeasible solutions respectively.

Suppose

$$S_{inf} = \left\{ \begin{array}{l} \lambda_{ij} \mid \lambda_{ij} = -\dfrac{\Delta f_{ij}}{\Delta G_{ij}}, i = I_{inf}(1),...,I_{inf}(q);... \\ j = I_{inf}(1),...,I_{inf}(q) \,\&\, G(i) \neq G(j) \end{array} \right\} \quad S_{sem} = \left\{ \begin{array}{l} \lambda_{ij} \mid \lambda_{ij} = \dfrac{\Delta f_{ij}}{G(\vec{x}_j)}, i = I_{fea}(1),...,I_{fea}(p); \\ j = I_{inf}(1),...,I_{inf}(q) \end{array} \right\}$$

The set of λ can be described as $S = \{S_{inf}, S_{sem}\}$. As to S_{inf}, we define $\lambda_{inf}^{max} = \max(\lambda_{ij})$, $\lambda_{inf}^{min} = \min(\lambda_{ij})$, where $i = I_{inf}(1),...,I_{inf}(q)$; $j = I_{inf}(1),...,I_{inf}(q)$. Likewise, for S_{sem}, we define $\lambda_{sem}^{max} = \max(\lambda_{ij})$, $\lambda_{sem}^{min} = \min(\lambda_{ij})$, where $i = I_{fea}(1),...,I_{fea}(p)$; $j = I_{inf}(1),...,I_{inf}(q)$. Then the largest value of λ in S is $\lambda_{max} = \max(\lambda_{inf}^{max}, \lambda_{sem}^{max})$.

There are some different cases in this situation:

a) $\lambda > \lambda_{max}$: In this case, the ranking results is the same as rule A_1-B_1.

b) $\lambda_{sem}^{max} < \lambda \leq \lambda_{max}$: In this case, the results ranked by penalty function method can satisfy rule A_1, but the comparison among infeasible solutions may not satisfy B_1 or B_2.

c) $\lambda_{sem}^{min} \leq \lambda \leq \lambda_{sem}^{max}$: In this case, the ranking will not satisfy A_1 or A_2 fully.

d) $\lambda < \lambda_{sem}^{min}$: the ranking result can satisfy A_2 (i.e., infeasible solutions are better than feasible solutions).

Similarly, following rule B (B_1 and B_2),

e) $\lambda_{inf}^{max} < \lambda \leq \lambda_{max}$: In this case, the results ranked by penalty function method can satisfy rule B_1.

f) $\lambda_{inf}^{min} \leq \lambda \leq \lambda_{inf}^{max}$: In this case, the ranking will not satisfy B_1 or B_2 fully.

g) $\lambda < \lambda_{inf}^{min}$: the ranking result can satisfy B_2 (i.e., infeasible solutions are ranked according to the reciprocal of the constraint violation).

h) $\lambda < \lambda_{min}$: the ranking result can satisfy A_2-B_2 (i.e., all infeasible solutions are better than feasible solutions, and the infeasible solutions are ranked according to the reciprocal of the constraint violation).

The general results are illustrated in Fig.1, which can provide a limit but effective basis for analyzing and designing adaptive penalty function methods.

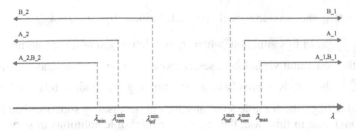

Fig. 1. The corresponding rule for penalty parameter λ

3   A Dynamic Penalty Function

3.1   Basic Idea

As mentioned in Section 2, a too large or too small penalty parameter value will have no influence on ranking the population, and the effect of penalty parameter is highly related with the solutions in the current generation. Based on this, a new DyPF considering the infeasible and semi-feasible situations is proposed.

If the solutions are experiencing the infeasible situation, as one of the main aims is to quickly find a feasible solution, the solutions will be ranked according to the degree of constraint violation, which is consistent with the rule B_1. This strategy is equivalent to the penalty function method when $\lambda > \lambda_{\max}$. And here, λ is set as follows:

$$\lambda = \frac{\max f - \min f}{\min(\Delta G_{ij})} \tag{8}$$

where $i = 1, 2, \ldots, NP; j = 1, 2, \ldots, NP$. Here, f is the objective function value, G is the constraint violation and $\Delta G_{ij} = G(i) - G(j) \, \& \, G(i) \neq G(j)$.

Similarly, if the solutions are experiencing the semi-feasible situation, as there are both feasible and infeasible solutions in the population, apart from considering the comparison between feasible and infeasible solutions, the proportion of feasible solutions should also be taken into account when setting the penalty parameters. And consequently a parameter r_f is introduced. λ is set as follows:

$$
\begin{aligned}
& \textit{if } r_f < \mu \\
& \quad \lambda = (1 + \alpha) \cdot \max\{\lambda_{\inf}^{\max}, \lambda_{sem}^{\max}\}; \\
& \textit{else} \\
& \quad \lambda = (1 - r_f) \cdot \lambda_{sem}^{\max} + r_f \cdot \lambda_{sem}^{\min}; \\
& \textit{end}
\end{aligned}
\tag{9}
$$

Here, μ is a threshold, determining using Deb's feasibility-based rule or penalty function method, which can be fixed or a rand number. α is a positive number, and here it's set as 0.01. The value of $\lambda_{\inf}^{\max}$, $\lambda_{sem}^{\min}$, $\lambda_{sem}^{\max}$ are set as Section 2.

From (9), it can be observed that the chance for an infeasible solution to survive in-to next generation is changing according to the proportion of feasible solutions in last generation. When $r_f < \mu$, the ranking will be based on Deb's feasibility-based rule;

when $r_f \geq \mu$, the ranking will be determined by r_f, λ_{sem}^{min}, λ_{sem}^{max}, and $\lambda_{sem}^{min} < \lambda < \lambda_{sem}^{max}$. In this situation, when r_f is relatively small, the ranking will mainly rely on the constraint violation to select more feasible solutions, as thc value of λ is near λ_{sem}^{max}; when r_f is relatively large, the ranking will mainly rely on the objective function value (e.g., the feasible solutions are not necessarily superior than the infeasible solutions), and in this case, the chance for infeasible solutions to survive into the next generation will increase.

3.2   Realization

Algorithm 1: Framework of DyPF

Input: NP: the size of population at each generation
　　　　Max_FES: maximum number of function evaluations
Output: $\vec{x}_{best}$: the best solution in the final population
Step 1 Initialization
　　　　Step 1.1 $t=0$;
　　　　Step 1.2 Generate an initial population $P_0 = \{\vec{x}_{1,0}, \cdots, \vec{x}_{NP,0}\}$ randomly.
　　　　Step 1.3 Evaluate the objective function values $f(\vec{x}_{i,0})$ and the degree of constraint violations $G(\vec{x}_{i,0})$.
　　　　Step 1.4 $FES=NP$.
Step 2 Dynamic penalty function model
　　　　Step 2.1 Update P_t using DE model to create offspring. These NP offspring form the offspring population Q_t .
　　　　Step 2.2 Evaluate $f(\vec{x}_{i,t})$ and $G(\vec{x}_{i,t})$ $(i=1,\cdots,NP)$.
　　　　Step 2.3 Compute the feasibility percent r_f of the combined population H_t (i.e., $H_t = P_t \cup Q_t$).
　　　　Step 2.4 Determinate the current situation of H_t according to r_f .
　　　　Step 2.5 Calculate the value of λ according to different situations.
　　　　Step 2.6 Compute the fitness function value with λ in **Step 2.5**.
　　　　Step 2.7 Rank the population through the value of fitness function and select the best NP individuals to constitute the next population P_{t+1} .
　　　　Step 2.8 $FES=FES+NP$.
Step 3 Set $t=t+1$.
Step 4 Stopping Criterion: If $FES \geq Max\_FES$, stop and output the best solution $\vec{x}_{best}$, otherwise go to **Step2**

The framework of DyPF is illustrated in Algorithm 1.

4   Experimental Study

4.1   Experimental Settings

This experiment is to verify the effectiveness of proposed DyPF. The details of these benchmark functions are reported in [23]. The evolutionary algorithm used in this paper is *DE/rand/1/bin* [22], and the boundary constraint is reset as [7].The parameters in DE are set as follows: the population size (*NP*) is set to 100; the scaling factor (*F*) is randomly chosen between 0.5 and 0.6, and the crossover control parameter (*Cr*) is randomly chosen between 0.9 and 0.95.

4.2   Experimental Results

25 independent runs were performed for each test function using 5×10^5 FES at maximum, as suggested by Liang *et al.* [23]. Additionally, the tolerance value δ for the equality constraints was set to 0.0001.

1) *The impact of parameter μ :* In this part, the impact of μ on the results generated by DyPF is evaluated. Six different values of μ are adopted, i.e., μ =0.1, 0.2, 0.3, 0.4, 0.5 and rand. For page limited, the detailed results are not listed here. The best overall performance can be obtained when $\mu = 0.5$.

2) *Comparison with some "dynamic" approaches in constrained evolutionary optimization:* In this part, we compare DyPF with five other state-of-the-art approaches using the concept of "dynamic" or "adaptive": SR [8]; SMES [5]; ATMES [10]; TPGA [18] and SaFF [17]. The experimental results of these five approaches are directly taken from the references and the comparative results are presented in Table 1.

To statistically compare the performance of different approaches, *t*-test results (h values) are presented in Table 1. Numerical values -1, 0, 1 represent that DyPF is inferior to, equal to and superior to other approaches respectively. It should be noted that the h values is determined considering the overall results, but for page limited, we just list the best value in Table 1.

For all the performance metrics, DyPF performs better than the other five approaches in g02, g05, g07, g09, and g10 as shown in Table 1.

All these six approaches have the same or similar performance in g08 and g12. As for g01 and g11, all approaches except TPGA can always reach the optimal value.

It should be pointed out that DyPF performs not so well on g03 and g13, though the other five approaches also can't obtain a satisfying result.

Besides, from the *t*-test results, it can be seen that DyPF is superior to, equal to and worse than other methods in 37, 21 and 7 cases, respectively out of the 65 cases. The worse cases are mainly from g03. Therefore, the overall performance of DyPF is highly competitive with the other five "dynamic" approaches, especially when considering that DyPF is simple and easy to realize.

Table 1. Comparison of DyPF with "Dynamic" approaches

Fun. & Optimal value		SR [8]	SMES [5]	ATMES [10]	TPGA [18]	SaFF [17]	DyPF
G01	best	-15.000	-15.000	-15.000	-14.9999	-15.0000	-15.0000
-15.0000	h	0	0	0	1	0	
G02	best	-0.803515	-0.803601	-0.803388	-0.803190	-0.802970	-0.803619
-0.803619	h	1	1	1	1	1	
G03	best	-1.000	-1.000	-1.000	-1.00009	-1.00000	-0.7412
-1.0005	h	-1	-1	-1	-1	-1	
G04	best	30665.539	30665.539	30665.539	30665.5312	-30665.50	30665.5387
30665.5387	h	0	0	0	1	1	
G05	best	5126.497	5126.599	5126.498	5126.5096	5126.9890	5126.4967
5126.4967	h	1	1	1	1	1	
G06	best	-6961.814	-6961.814	-6961.814	-6961.1785	-6961.800	-6961.8139
-6961.8139	h	1	1	0	1	1	
G07	best	24.307	24.327	24.306	24.410977	24.48	24.3062
24.3062	h	1	1	0	1	1	
G08	best	-0.095825	-0.095825	-0.095825	-0.095825	-0.095825	0.09582504
0.09582504	h	0	0	0	0	0	
G09	best	680.630	680.632	680.630	680.762228	680.64	680.630057
680.630057	h	1	1	1	1	1	
G10	best	7054.316	7051.903	7052.253	7060.55288	7061.34	7049.2480
7049.2480	h	1	1	1	1	1	
G11	best	0.750	0.75	0.75	0.7490	0.7500	0.7499
0.7499	h	0	0	0	1	0	
G12	best	-1.000000	-1.000	-1.000	NA	-1	-1.0000
-1.0000	h	0	0	1	1	0	
G13	best	0.053957	0.053986	0.053950	NA	NA	0.05562855
0.05394151	h	-1	0	-1	1	1	

5   Conclusion

In this paper, a new Dynamic Penalty Function (DyPF) has been proposed for con-strained evolutionary optimization, which is based on the systematical analysis of penalty parameter. Thus this enables DyPF to take advantage of different strategies by adjusting penalty parameters at different situations. To verify the effectiveness of the newly proposed DyPF, an experiment is carried out which is based on the 21 bench-mark functions collected in the IEEE CEC2006 special session on constraint real-parameter optimization.

The results show that DyPF has a high effectiveness and is very competitive com-paring with other five dynamic or adaptive state-of-the-art methods referred to in this paper in view of simplicity of DyPF. Nevertheless, DyPF can not find a successful solution in g13, which is due to the simplicity of the model.

DE and other evolutionary algorithms have shown a good performance on unconstrained problems, which means they are good at generating satisfying solutions. Thus key point in solving constraint problems is how to select or rank the solutions, especially how to keep the balance between objective function and constraint violations, which is also related with the problem's topological properties. Therefore, this will be our future work.mn.

Acknowledgments. This work was supported in part by the National Natural Science Foundation of China under Grants 71371142, 61174183. Chengyong Si would like to thank Prof. Dr. Robert Weigel for his great help in the life and research work, as this work is partially done when he was with the Institute for Electronics Engineering, University of Erlangen-Nuernberg in Germany as a joint doctor, and he is grateful to Dr. Y. Wang for the valuable suggestions. The authors also gratefully acknowledge Dr. T. Lan, Dr. J. Hu, and Dr. G. Gao for improving the presentation of this paper.

References

1. Michalewicz, Z., Schoenauer, M.: Evolutionary algorithm for constrained parameter optimization problems. Evol. Comput. **4**(1), 1–32 (1996)
2. Coello Coello, C.A.: Theoretical and numerical constraint-handling techniques used with evolutionary algorithms: A survey of the state of the art. Comput. Methods Appl. Mech. Eng. **191**(11/12), 1245–1287 (2002)
3. Cai, Z., Wang, Y.: A multiobjective optimization-based evolutionary algorithm for constrained optimization. IEEE Trans. Evol. Comput. **10**(6), 658–675 (2006)
4. Wang, Y., Cai, Z.: A dynamic hybrid framework for constrained evolutionary optimization. IEEE Trans. Syst. Man Cybern. B Cybern. **42**(1), 203–217 (2012)
5. Mezura-Montes, E., Coello Coello, C.A.: A simple multimembered evolution strategy to solve constrained optimization problems. IEEE Trans. Evol. Comput. **9**(1), 1–17 (2005)
6. Eiben, A.E., Hinterding, R., Michalewicz, Z.: Parameter control in evolutionary algorithms. IEEE Trans. Evol. Comput. **2**(2), 124–141 (1999)
7. Deb, K.: An efficient constraint handling method for genetic algorithms. Comput. Methods Appl. Mech. Eng. **186**(2–4), 311–338 (2000)
8. Runarsson, T.P., Yao, X.: Stochastic ranking for constrained evolutionary optimization. IEEE Trans. Evol. Comput. **4**(3), 284–294 (2000)
9. Runarsson, T.P., Yao, X.: Search bias in constrained evolutionary optimization. IEEE Trans. Syst. Man Cybern. C Appl. Rev. **35**(2), 233–243 (2005)
10. Wang, Y., Cai, Z., Zhou, Y., Zeng, W.: An adaptive tradeoff model for constrained evolutionary optimization. IEEE Trans. Evol. Comput. **12**(1), 80–92 (2008)
11. Wang, Y., Cai, Z.: Constrained evolutionary optimization by means of (μ+λ)-differential evolution and improved adaptive trade-off model. Evol. Comput. **19**(2), 249–285 (2011)
12. Jia, G., Wang, Y., Cai, Z., Jin, Y.: An improved (μ+λ)-constrained differential evolution for constrained optimization. Inform. Sci. **222**, 302–322 (2013)
13. Zhan, Z., Zhang, J., Li, Y., Chung, H.S.: Adaptive particle swarm optimization. IEEE Trans. Syst. Man Cybern. B Cybern. **39**(6), 1362–1381 (2009)

14. Gong, W., Cai, Z., Ling, C.X., Li, H.: Enhanced differential evolution with adaptive strategies for numerical optimization. IEEE Trans. Syst. Man Cybern. B Cybern. **41**(2), 397–413 (2011)
15. Minhazul, I.S., Das, S., Ghosh, S., Roy, S., Suganthan, P.N.: An adaptive differential evolution algorithm with novel mutation and crossover strategies for global numerical optimization. IEEE Trans. Syst. Man Cybern. B Cybern. **42**(2), 482–500 (2012)
16. Zhao, S.Z., Suganthan, P.N., Das, S.: Self-adaptive differential evolution with multi-trajectory search for large scale optimization. Soft Comput. **15**(11), 2175–2185 (2011)
17. Farmani, R., Wright, J.A.: Self-adaptive fitness formulation for constrained optimization. IEEE Trans. Evol. Comput. **7**(5), 445–455 (2003)
18. Venkatraman, S., Yen, G.G.: A generic framework for constrained optimization using genetic algorithms. IEEE Trans. Evol. Comput. **9**(4), 424–435 (2005)
19. Zhang, M., Luo, W., Wang, X.: Differential evolution with dynamic stochasitc selection for constrained optimization. Inform. Sci. **178**(15), 3043–3074 (2008)
20. Tessema, B., Yen, G.G.: An adaptive penalty formulation for constrained evolutionary optimization. IEEE Trans. Syst., Man, Cybern. A Syst. Hum. **39**(3), 565–578 (2009)
21. Si, C., Lan, T., Hu, J., Wang, L., Wu, Q.: Penalty parameter of the penalty function method. Control Decis. **29**(9), 1707–1710 (2014)
22. Das, S., Suganthan, P.N.: Differential evolution: A Survey of the state-of-the-art. IEEE Trans. Evol. Comput. **15**(1), 4–31 (2011)
23. Liang, J.J., Runarsson, T.P., Mezura-Montes, E., Clerc, M., Suganthan, P.N., Coello Coello, C.A., Deb, K.: Problem definitions and evaluation criteria for the CEC 2006. Technical report, Special Session on Constrained Real-Parameter Optimization (2006)

A New Physarum Network Based Genetic Algorithm for Bandwidth-Delay Constrained Least-Cost Multicast Routing

Mingxin Liang[1], Chao Gao[1], Yuxin Liu[1], Li Tao[1], and Zili Zhang[1,2(✉)]

[1] School of Computer and Information Science,
Southwest University, Chongqing 400715, China
zhangzl@swu.edu.cn
[2] School of Information Technology, Deakin University,
Geelong, VIC 3217, Australia

Abstract. Bandwidth-delay constrained least-cost multicast routing is a typical NP-complete problem. Although some swarm-based intelligent algorithms (e.g., genetic algorithm (GA)) are proposed to solve this problem, the shortcomings of local search affect the computational effectiveness. Taking the ability of building a robust network of *Physarum* network model (PN), a new hybrid algorithm, *Physarum* network-based genetic algorithm (named as PNGA), is proposed in this paper. In PNGA, an updating strategy based on PN is used for improving the crossover operator of traditional GA, in which the same parts of parent chromosomes are reserved and the new offspring by the *Physarum* network model is generated. In order to estimate the effectiveness of our proposed optimized strategy, some typical genetic algorithms and the proposed PNGA are compared for solving multicast routing. The experiments show that PNGA has more efficient than original GA. More importantly, the PNGA is more robustness that is very important for solving the multicast routing problem.

Keywords: Genetic algorithm · *Physarum* network model · Multicast routing

1 Introduction

Multicasting is one type of services in MANETs, which are very popular due to the no-restricted mobility and feasible deployment. With the growing of distributed multimedia application, the efficient and effective support of QoS (i.e., Quality of Service) has became more and more crucial for MANETs. The key issue in the design of network architectures of MANETs is how to manage the resources efficiently in order to meet the requirements of QoS during each connection. In general, to deliver the same data stream to different destinations efficiently, a tree structure is used for multicasting. More importantly, some constrains are often added to the entire tree for multicasting in order to meet

© Springer International Publishing Switzerland 2015
Y. Tan et al. (Eds.): ICSI-CCI 2015, Part II, LNCS 9141, pp. 273–280, 2015.
DOI: 10.1007/978-3-319-20472-7_29

the requirements of QoS. Thus, the object of multicast routing problem is to compute a tree structure (named as the multicast tree) under the conditions of the minimum communication resources and QoS requirements of a network [2], which is a typical NP-complete problem [3].

In the field of artificial intelligence, the genetic algorithm is a powerful tool for solving NP-complete problems. However, the shortcomings of local search affect the computational effectiveness. Recently, more and more scientists focus on the self-organization capability of a species of plasmodium, which is a 'vegetative' phase of *Physarum*. This plasmodium shows an amazing intelligence in the process of building a robust protoplasmic network for connecting food sources in order to deliver nutrients to all its body [4].

In this paper, we design an updated crossover operator, based on *Physarum* Network (PN) [5], for overcoming the shortcomings of traditional GA. Using this method, we incorporate PN into GAs for solving the multicast routing problem. Some experiments show that the hybrid algorithm has a stronger ability to exploit the optimal solution effectively.

The organization of this paper is as follows. Section 2 introduces the formulation and measurements of multicast routing problem. Section 3 formulates the hybrid algorithm. Section 4 provides some experiments to estimate the effectiveness of hybrid algorithm. Section 5 concludes this paper.

2   Problem Statement

In general, a QoS multicast routing problem involves in several constrains, such as delay jitter, packet loss, bandwidth, and cost. In this study, we simplify QoS constrains and present a feasible QoS multicast routing model. According to [1], we focus on three most important factors: cost, bandwidth and delay. As the cost is the most important metric for the effectiveness of a network, our research focuses on the bandwidth-delay constrained least-cost multicast routing problem.

A network is usually represented as a graph $G = (V, E)$, where $V = \{v_1, v_2, ..., v_n\}$ denotes a set of nodes representing routers or switches and $E = \{e_{ij} = (v_i, v_j) | v_i, v_j \in V, i \neq j\}$ denotes a set of edges representing physical or logical connectivity between nodes. Let a node $s \in V$ be the source and a set $DE \subseteq V - \{s\}$ be the set of multicast destinations. A multicast tree, denoted as $T(s, DE)$, is a sub-graph of G connecting a node s to each node in DE. The path from s to any destination node $d \in DE$ is denoted as $p_T(s, d)$. And the object of multicast routing problem is to find a $T(s, DE)$ with a minimum cost, which has a set of paths with acceptable bandwidth and delay from a node s to each node in DE.

The delay of a path from a node s to any destination node d in DE, denoted as $delay(p_T(s, d))$, is simply defined as the sum of delays in the $p_T(s, d)$, i.e., $delay(p_T(s, d)) = \sum_{e \in p_T(s,d)} delay(e)$. Meanwhile the bandwidth of a path from a node s to any destination node d in DE, denoted as $bandwidth(p_T(s, d))$, is defined as the minimum of bandwidth along $p_T(s, d)$, i.e., $bandwidth(p_T(s, d)) = \min\{bandwidth(e) | e \in p_T(s, d)\}$. And, according to existing studies in [6,7], we unify the cost of a multicast tree as the sum of costs in the tree, i.e., $cost(T) = \sum_{e \in T} cost(e)$, for comparing effects of different GAs.

Let Δ_d be the upper limit of delay constraint and Δ_b be the lower limit of bandwidth constraint for each path. And the bandwidth-delay constrained least-cost multicast routing problem is defined as (1). We want to minimize the cost under the condition of satisfying bandwidth-delay constraints.

$$
\begin{aligned}
&\min \cos t(T)\\
&st.\\
&\begin{cases} delay(p_T(s,d)) \leq \Delta_d & for \; \forall \, d \in DE \\ bandwidth(p_T(s,d)) \geq \Delta_b & for \; \forall \, d \in DE \end{cases}
\end{aligned}
\tag{1}
$$

3 Formulation of PNGA

This section introduces the basic idea of PNGA from two aspects: original PN model and PNGAs. In detail, Sect. 3.1 presents the original PN model. And, Sect. 3.2 shows how to improve GAs by PN.

3.1 *Physarum* Network Model

The *Physarum* network model is inspired by the maze-solving experiment [4]. Tero et al. capture the positive feedback mechanism of *Physarum* in foraging and build the PN model [5]. In addition, The model, designed for solving maze problem, can be used for building a multicast tree in our study. The details of PM are described as follows.

In PM, Q_{ij} represents the flux of pipeline, connecting nodes i and j, and D_{ij} stands for the conductivity of the pipeline. Moreover a node s and a set DE present the inlet and outlets of pipelines respectively. According to the Kirchhoff's law, the flux of input at node s is equal to the total flux of output at DE and, at any other nodes, the sum of flowing into that node is equal to the sum of flowing out of that node. This process can be denoted as (2) where N stands for the cardinality of DE.

$$
\sum_i Q_{ij} = \begin{cases} +1 & for \; j == s \\ \frac{-1}{N-1} & for \; j \in DE \\ 0 & for \; others \end{cases}
\tag{2}
$$

In each iteration step, Q_{ij} and p_i can be calculated according to Poiseuille's law based on (2) and (3), where L_{ij} represents the length of pipeline contacting nodes i and j, and p_i represents the pressure of node i. As the iteration going on, the conductivities of pipelines adapt to the flux based on (4). Then, the conductivities will feed back to the flux based on (3) at the next iteration step.

$$
Q_{ij} = (\frac{D_{ij}}{L_{ij}} + \frac{D_{ji}}{L_{ji}})(p_i - p_j)
\tag{3}
$$

$$\frac{dD_{ij}}{dt} = |Q_{ij}| - D_{ij} \tag{4}$$

After above processes, one iterative step is completed. This process will continue loop iteration until the terminal condition is satisfied. In this study, the terminal condition is $|D^t{}_{ij} - D^{t+1}{}_{ij}| < 10^{-6}$ for any i and j, where $D^{t+1}{}_{ij}$ stands for D_{ij} at iteration step $t+1$. After the loop iteration, critical pipelines will be reserved, and others will disappear. Finally, we can obtain a *Physarum* spanning tree. The description of PN model for a shortest path tree is shown in Alg. 1.

Algorithm 1. *Physarum* network model

Input: A graph NG, a source node s, a set of destination nodes DE.
Output: A shortest path tree connecting a source node s to each node in DE.
Step 1: Initializing with $D_{ij} = (0, 1]$, $Q_{ij} = 0$, $p_i = 0$.
Step 2: Computing Q_{ij} and p_i based on (2) and (4).
Step 3: If $Q_{ij} < 0$, **then** $Q_{ij} = 0$.
Step 4: Updating D_{ij} based on (4).
Step 5: **If** the terminal condition is not satisfied, **then** going to **step 2**.
Step 6: Outputting the shortest path tree.

3.2 PNGA

The GA is a powerful tool for solving PN-complete problems and a kind of searching algorithm that employs the ideas of natural selection and the genetic operators of crossover and mutation. Taking advantages of PN model and GAs, we propose a universal strategy for crossover operator in GAs. The new crossover operator is named as PNcrossover. And the novel hybrid algorithms with PNcrossover (denoted as PNGAs) are used to solve the multicast routing problem. The main ideal of PNcrossover is to reserve the same links between parent chromosomes and to integrate the offspring through PN model based on the reserved links. Other parts of PNGAs are same as the original ones. An example of crossover process is shown in Fig. 1. The details of PNcrossover are described as follows.

Firstly, for applying PN model, we need a new graph, denoted as NG, with the same topological structure as the network graph in the multicast routing problem. Because GAs may generate some unadaptable chromosomes, which do not satisfy with the constraints, there are two strategies to fit for different GAs. For the GAs, which do not generate unadaptable chromosomes, such as GAMRA [6], L_{ij} in NG is set equal to the delay of e_{ij}. For others, such as EEGA [7], L_{ij} in NG is set equal to the product of cost and delay of e_{ij}.

And then, PNcrossover selects the same links of parent chromosomes and reserves them in the offspring chromosomes. Since these same links may be in some separated sub-trees, PN model is used to transform these sub-trees into a multicast tree. In order to reserve the same links in PN model, the length of reserved links is set equal to zero in NG. Substituting the graph NG, source

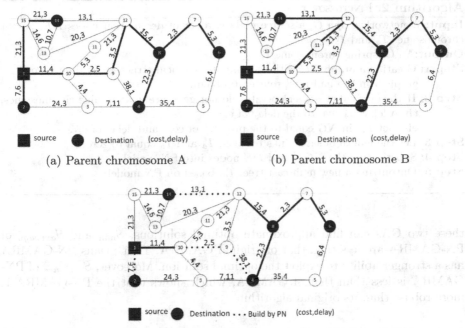

(a) Parent chromosome A   (b) Parent chromosome B

(c) A new offspring generated by parent chromosomes A and B through PNcrossover

Fig. 1. An example of crossover operation

and destinations into PN model, a complete multicast tree will be constructed. Alg. 2 describes detailed steps of crossover operator in PNGAs.

4   Simulation Experiments

4.1   Datasets

In order to estimate the effectiveness of PNcrossover scheme, we integrate PNcrossover into two different GAs [6,7] and implement them on two datasets. The first (denoted as $D1$) is a random graph with 20 nodes, which is constructed based on [6]. In $D1$, costs and delays of links are uniformly distributed between 0.3 and 1. The second (denoted as $D2$) is shown in Fig. 1, with Δ_d equaling to 24 [7]. All experiments are under the same environment, i.e., all parameters of PNGAs are same as these of original GAs. And all results in our experiments are averaged over 50 times.

4.2   Experiments Analysis

Figure 2 shows the minimum ($S_{\min}$), average ($S_{average}$) and variance ($S_{variance}$) of the results calculated by PN-GAMRA and original GAMRA [6]. Although

Algorithm 2. PNcrossover

Input: A network graph G, a source node s, a set of destination nodes DE, parent chromosomes T_a and T_b.

Output: An Offspring chromosome T_c.

Step 1: Creating a graph NG with the same topological structure as the network graph G in the multicast routing problem.

Step 2: **If** the original crossover operator do not generate unadaptable chromosomes, **then** let L_{ij} equal to the delay of e_{ij}.

else let L_{ij} in NG equal to the product of cost and delay of e_{ij}.

Step 3: Let the length of same links between T_a and T_b equal to zero.

Step 4: Substituting the NG, s and DE nodes into PN model.

Step 5: Outputting a new multicast tree, T_c, based on PN model.

---

these two GAs can find approximate optimal solutions, $S_{\min}$ and $S_{average}$ of PN-GAMRA are less than that of original GAMRA. That means PN-GAMRA has a stronger ability to exploit the optimal solution. Moreover, $S_{variance}$ of PN-GAMRA is less than that of GAMRA, which shows that the PN-GAMRA is more robust than its original algorithm.

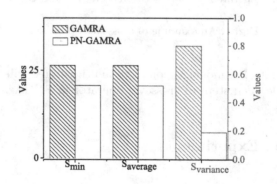

Fig. 2. Comparing results of PN-GAMRA and GAMRA on $D1$

As EEGA [7] may generate unadaptable solutions in the evolution, we compare the costs and delays of solutions. In order to further verify the accuracy and robustness of PNGA, Fig. 3 plots the convergent process of averages and variances with the increment of iterative steps. As shown in Fig. 3, average and variance of PN-EEGA decrease more obviously than that of EEGA. In detail, in the earlier iteration, there are slight difference between averages of PN-EEGA and EEGA. With the iterative steps going on, the average of PN-EEGA are less than that of EEGA obviously. PN-EEGA exhibits a better accuracy. Furthermore, the variance of PN-EEGA is also less than that of EEGA, which shows that PN-EEGA has more stronger robust. Moreover, Figure 4(a) and Fig. 4(b)

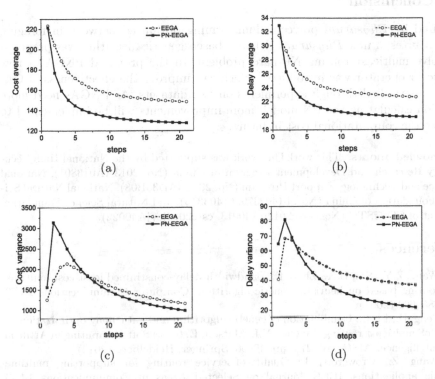

Fig. 3. Delays and costs calculated by EEGA and PN-EEGA on $D2$

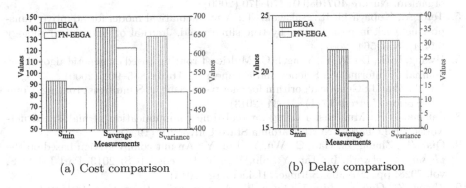

(a) Cost comparison   (b) Delay comparison

Fig. 4. Comparing results of EEGA and PN-EEGA on $D2$

plot $S_{\min}$, $S_{average}$ and $S_{variance}$ of PN-EEGA are less than that of EEGA in both costs and delays. These results show that the PNcrossover scheme can strengthen the searching ability for finding the optimal solution and improve the robustness of original GA.

5   Conclusion

Inspired by *Physarum* polycephalum forming optimized network in foraging food sources, a new *Physarum* Network based genetic algorithm was proposed to solve multicast routing NP-hard problem. In the proposed PNGA, a new crossover operator was introduced, which can improve the effectiveness of GA. This was verified through experiments on two datasets. As PNGAs need more time on calculating the PN model, more improvements will be implemented to reduce the computational cost in future.

Acknowledgments. This workThis work was supported by the National High Technology Research and Development Program of China (No. 2013AA013801), National Science and Technology Support Program (No. 2012BAD35B08), National Natural Science Foundation of China (Nos. 61402379, 61403315), and Natural Science Foundation Project of CQ CSTC (Nos. cstc2012jjA40013, cstc2013jcyjA40022).

References

1. Wang, Z.Y., Shi, B.X., Zhao, E.: Bandwidth-delay-constrained least-cost multicast routing based on heuristic genetic algorithm. Computer Communications **24**(7), 685–692 (2001)
2. Peng, B., Li, L.: Combination of genetic algorithm and ant colony optimization for QoS multicast routing. In: Cho, Y.I., Matson, E. (eds.) Soft Computing in Artificial Intelligence. AISC, vol. 270, pp. 49–56. Springer, Heidelberg (2014)
3. Wang, Z., Crowcroft, J.: Quality-of-service routing for supporting multimedia applications. IEEE Journal on Selected Areas in Communications **14**(7), 1228–1234 (1996)
4. Nakagaki, T., Yamada, H., Toth, A.: Intelligence: Maze-solving by an amoeboid organism. Nature **407**(6803), 470–470 (2000)
5. Tero, A., Kobayashi, R., Nakagaki, T.: A mathematical model for adaptive transport network in path finding by true slime mold. Journal of Theoretical Biology **244**(4), 553–564 (2007)
6. Hwang, R.H., Do, W.Y., Yang, S.C.: Multicast routing based on genetic algorithms. Journal of Information Science and Engineering **16**(6), 885–901 (2000)
7. Lu, T., Zhu, J.: Genetic algorithm for energy-efficient QoS multicast routing. Communications Letters **17**(1), 31–34 (2013)
8. Salama, H.F.: Multicast routing for real-time communication of high-speed networks. Ph D Thesis. North Carolina State University (1996)
9. Qian, T., Zhang, Z., Gao, C., Wu, Y., Liu, Y.: An ant colony system based on the *physarum* network. In: Tan, Y., Shi, Y., Mo, H. (eds.) ICSI 2013, Part I. LNCS, vol. 7928, pp. 297–305. Springer, Heidelberg (2013)
10. Zhang, Z., Gao, C., Liu, Y., Qian, T.: A universal optimization strategy for ant colony optimization algorithms based on the Physarum-inspired mathematical model. Bioinspiration & Biomimetics **9**(3), 036006 (2014)
11. Karthikeyan, P., Baskar, S.: Genetic algorithm with ensemble of immigrant strategies for multicast routing in Ad hoc networks. Soft Computing **19**(2), 489–498 (2015)
12. Liu, Y., Zhang, Z., Gao, C., Wu, Y., Qian, T.: A *physarum* network evolution model based on IBTM. In: Tan, Y., Shi, Y., Mo, H. (eds.) ICSI 2013, Part II. LNCS, vol. 7929, pp. 19–26. Springer, Heidelberg (2013)

Scheduling and Path Planning

Service Based Packets Scheduling for QoS of Mixed Mobile Traffic in Wireless Network

Zin Win Aye[✉] and Myat Thida Mon

University of Computer Studies, Yangon, Yangon, Myanmar
{zinwin2008,myattmon}@gmail.com

Abstract. Nowadays, a number of mobile users who use various mobile traffics are increasing rapidly in wireless environment. There are many researches concerning with traffic scheduling methods in order to have efficient Quality of Services (QoS) mechanisms for mobile users in Wireless Local Area Network (WLAN). The optimal traffic scheduling of mobile networks for various mixed traffic application is a challenging problem. Therefore, in this paper, a framework with a new scheduling method in WLAN, a two-step traffic scheduling method is proposed to satisfy efficient some QoS parameters: throughput, fairness and delay for mixed applications such as real time and non real time traffic in multicast and unicast applications. In the first step, service based scheduling adaptively balances between unicast and multicast applications and in the second step, compound scheduling combines Proportional Fair (PF) for non-real time traffic (NRT) and Delay Threshold (DT) Scheduling for real time traffic (RT).

Keywords: Quality of service · QoS · Wireless local area network · WLAN · Proportional fair scheduling · PF · Delay threshold scheduling · DT

1   Introduction

With the growing increases of internet users and their demands, the network shared the bandwidth among multiple traffic applications which consumes lots of network resources, namely, web browsing, email, voice data, and video in the wireless networks such as Wi-Fi and cellular networks is being paramount. Although many researches [1] focus on improving QoS with scheduling algorithms such as Proportional Fair, Round Robin, Opportunistic which emphasis only on each function such as throughput, fairness, delay, the resources sharing problem and the requirements of multimedia applications [2] and [3] for various mobile users are remained in wireless environment. Moreover, there are also drawbacks for employing isolated scheduling police to handle just RT or NRT traffic or employing mixed scheduling polices to handle both multicast and unicast traffic or both RT and NRT traffic.

Under isolated scheduling police, while simply considering guaranteeing delay for RT traffic, the policy will weak the target of throughput maximization and while simply considering maximizing the system throughput in a fair way for NRT traffic, the policy will ignore delay constraint on RT packets. In case of high probability of QoS violation, enjoying multimedia services will not be possible. Under mixed scheduling policies [4], although the system's throughput seems to be maximized for a

© Springer International Publishing Switzerland 2015
Y. Tan et al. (Eds.): ICSI-CCI 2015, Part II, LNCS 9141, pp. 283–289, 2015.
DOI: 10.1007/978-3-319-20472-7_30

NRT application and packets delay seems to be minimized for a RT application, the system cannot still solve to be fairness and traffic management problem for them and multicast and unicast traffic in the same Access Point (AP) or Base Station (BS).

Therefore, in this paper, QoS is considered as a main issue for mixed mobile traffic to be fairness and to arrive at their destination on time with the least delay and maximum throughput, limiting the throughput of users close to AP. It proposes two stages scheduling algorithms, namely, service based queuing and compound scheduling that can adapt four mixed traffic types such as RT multicast, NRT multicast, RT unicast and NRT unicast to get better traffic management and scheduling technique   solving their QoS requirements in WLAN.

2   System Design

In wireless systems of mixed traffic, how to design the packet scheduling algorithm to guarantee QoS requirements by providing methods of resource allocation and multiplexing at the packet level [8] is not only an important problem, but also a complex problem. Therefore, we analyze many queuing algorithms such as Round Robin (RR) [5] and [6], Strict Priority (SP) [6] and Class Based Weighted Fair Queue (CBWFQ) [7] and so on to solve above requirements.

Considering drawbacks of queuing algorithms, we consider Service Based Priority RR (SBPRR) and Service Based RR (SBRR) scheduling algorithm is proposed by contributing RR scheduling with priority or without priority based on multicast and unicast services solving the drawback of CBWFQ [7] and SP, which solves traffic management issue in order to improve the QoS of mixture traffic and meet fairness for multicast and unicast queues. The system presents a high level overview of three major parts: classification, the service based scheduling and the compound scheduling with priority value for a solution to QoS problem with an optimal resource allocation of mobile users.

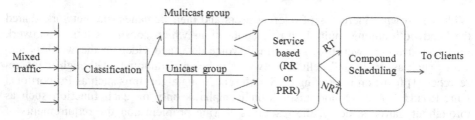

Fig. 1. Detail system architecture

In the detail system architecture in Fig. 1, it classifies incoming mixed applications according to their different parameters into two services such as multicast and unicast queue in which each has RT and NRT traffic due to mixed traffic of mobile users in WLAN simultaneously. Multicast types in the system are video streaming and News. Voice and web browsing belong to unicast type. Then, it fairly schedules each queue depending on priority scheduling methods such as SBRR or SBPRR for different scheduling time interval. Next, compound scheduling calculates priority of mixed RT and NRT traffic of mobile users for fair resource allocation.

3   First Stage Scheduling for Multicast and Unicast Stream

In first stage, it schedules queues of multicast and unicast services for mobile users who operate with heterogeneous mobile traffic to be fairness using SBRR or SBPRR scheduling algorithm. For queue fairness, the delay is unbalanced depending on the number of packets in multicast and unicast queue because resources are distributed unfairly. While multicast queue needs to send many packets to more than one node, unicast queue send only one packet to each node at a time. The unbalanced multicast and unicast queuing delay decrease the QoS of multimedia traffic for mobile users [9].

The dominant component of delay is on the queuing delay by considering the transmission delay and the transmission overhead are very small. Furthermore, the transmission delay will be the same for queues that use the same transmission rate. Therefore, we need to compute the queuing delay by multiplying the transmission time with the queue size according to little's law $D_{system} = Q_{system} / \mu_{system}$ [10] to balance multicast and unicast queue. Let it be the queuing delay of the multicast D_M and the unicast D_U, multicast transmission rate μ_M, unicast transmission rate μ_U as follows:

$$D_M = Q_M \cdot \frac{1}{\mu_M} \tag{1}$$

$$D_U = Q_U \cdot \frac{1}{\mu_U} \tag{2}$$

Where, $\mu_M = P \cdot \mu_U$ which means the multicast transmits P packets at a time interval while unicast transmits only one packet. Thus, D_M can be rewritten as:

$$D_M = Q_M \cdot \frac{1}{P \cdot \mu_U} \tag{3}$$

Then, we get the optimal P value for balancing the delay of both queues as follows:

$$D_M = D_U \tag{4}$$

$$Q_M \cdot \frac{1}{P \cdot \mu_U} = Q_U \cdot \frac{1}{\mu_U} \tag{5}$$

Then,

$$P = \frac{Q_M}{Q_U} \qquad \text{If } Q_M \neq Q_U \tag{6}$$

$$P = 1 \qquad \text{If } Q_M = Q_U \tag{7}$$

Priority P is the ratio of the queue size of multicast which is the number of multicast packets, and the queue size of unicast which is the number of unicast packets. When both queue size is equal, $Q_M = Q_U$, P results one which means each queue gets the same chance(same time interval) to transmit packet in each round because both multicast and unicast queue have the same queuing delay. In this case, the system uses SBRR to be fairness between queues. Otherwise, when the queue size of multicast and unicast are not equal, $Q_M \neq Q_U$, P results greater than or less than one which means each queue gets the more chance(different time interval) to transmit packet in each round. In meanwhile, in order to balance queuing delay and fair resource distribution between two services, the proposed scheduler utilizes SBPRR which is the property of RR with priority of the queue [11].

The priority of the multicast and unicast changes adaptively to balance between these services depending on both queue sizes. When queue size of multicast increases, the multicast priority will increase. When the queue size of unicast increases, the multicast priority will decrease. The service based scheduling algorithm with priority and without priority is as shown in Fig 2.

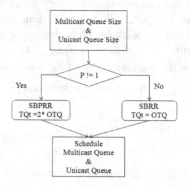

Fig. 2. Service based Scheduling algorithm

To achieve the fairness between the multicast and unicast stream in WLAN, in the same case of queue size, i.e, (P = 1), the system uses SBRR which gives the same time quantum (TQt) at time t as an original time quantum (OTQ) for both queues in round robin manner. Otherwise, the system will use SBPRR to schedule each queue within calculated TQ according to queue priority not to be much queuing delay, which will double OTQ according to the greater priority value of each queue and other will get OTQ. In our experiments, we use NS3 Simulation to test proposed algorithms for four traffic types of 10 mobile users in an AP in WLAN, and compare RT delay fairness and NRT throughput fairness using Jain's fairness index for users who use mixed multicast services as described in Fig 3. Fig 4 shows for users who use mixed unicast services according to their queue size. We observe that overall fairness of the system increases and stable for all users as the queue size increases which it means increasing mixed traffic in WiFi can cause high packet loss rate and long delay for users.

4   Second Stage Scheduling for Real Time and Non Real Time Traffic

The system schedules the multicast or unicast queue as first step scheduling as explained above and then implements the compound scheduling method for fairness of mixed traffic of mobile users by multiplying the corresponded priority value with PF for NRT traffic and DT for RT traffic respectively. It calculates priority value for RT and NRT traffic according to Bayesian Scheme with Two Priority Classes and

Proportional Priority Distribution [12]. The paper proposes a method for proportional priority calculation of mixed traffic types of each mobile user to get fair priority probability value in getting traffic scheduling time interval of second stage scheduling. The system introduces combined scheduling matrix such as PF for NRT traffic and DT for RT trpaffic described in (13) and (14) by using the priority calculation for each traffic type as in (8) to (12).

Therefore, it pays attention on PF for first part of equation which tries to increase the degree of fairness among connections by selecting those with the largest relative channel quality between the connection's current supportable data rate and its average throughput [13]. In second part of equation, It also uses the popular delay related scheduling metric [13] will be implemented with threadshold value not to be bound threadshold delay for RT. Then, it schedules both RT and NRT traffic according to P_{rt} and P_{nrt} in each associated time interval as in (13) and (14). Equation (13) is for unicast traffic and (14) is for multicast traffic to satisfy a user with least channel condition in multicast group. The compound scheduling is determined to be summation of individual policies by multiplying each police with associated real priority P_{rt} for first policy and non real priority P_{nrt} for the second policy [13]. As shown in Fig 5 and Fig 6, we compare throughput and delay between priority value of our proposed system and proposed priority value (0.3 and 0.7 for NRT and RT) described in [13] to get fairly maximum throughput and minimum delay for ten mixed traffic at an AP in WLAN.

$$\alpha_{nrt} = \frac{Q_{nrt}}{Q_M} \tag{8}$$

$$\alpha_{nrt} = \frac{Q_{nrt}}{Q_U} \tag{9}$$

$$P_{rt} + P_{nrt} = 1 \tag{10}$$

$$P_{rt} = \min\left(1, (1 + \alpha_{nrt}) * \rho\right) \tag{11}$$

$$P_{nrt} = 1 - P_{rt} \tag{12}$$

$$j = \max_{\substack{1 \le k \le K \\ 0 \le j \le 1}} \arg\left\{ P_{nrt}(D_{k,j}(t) / R_{k,j}(t)) + P_{rt}(d_{k,j} / d^{th}_{k,j} - d^{av}_{k,j}) \right\} \tag{13}$$

$$j = \min_{\substack{1 \le k \le K \\ 0 \le j \le 1}} \arg\left\{ P_{nrt}(D_{k,j}(t) / R_{k,j}(t)) + P_{rt}(d_{k,j} / d^{th}_{k,j} - d^{av}_{k,j}) \right\} \tag{14}$$

Where, $\arg\max$ denotes the argument of maximum, k is the index of user and K is the total number of user, $j = 0$ represents RT stream and $j = 1$ represents NRT stream. For NRT, $D_{k,j}(t)$ represents the instantaneous data rate that can be achieved by j stream of user k at time t and $R_{k,j}(t)$ represents the average data rate that j stream of user k perceived at time t. In RT, $d_{k,j}$ is the delay encountered by a packet at the head of the jth stream of the kth user. $d_{k,j}^{th}$ represents the delay threshold for packets at the jth stream of the kth user. $d^{av}_{k,j}$ represents the average delay for packets at j^{th} stream of the kth user.

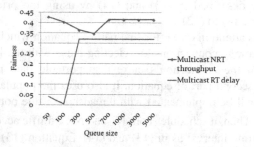

Fig. 3. Fairness between RT delay and NRT throughput for multicast users

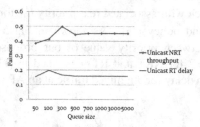

Fig. 4. Fairness between RT delay and NRT throughput for unicast users

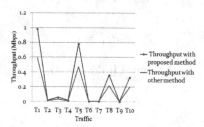

Fig. 5. Throughput between ten mixed RT and NRT traffic

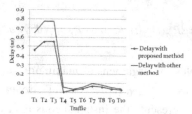

Fig. 6. Delay between ten mixed RT and NRT traffic

5   Conclusions

The paper proposes the framework of traffic management and packet scheduling to increase bandwidth usage and fairness without much delay for mobile users in WLAN and solve mixed traffic problem for four different traffic types and priority issue for them. The scheme has more performance in mixed applications in WLAN with a high degree of compatibility with existing scheduling methods. It can be applied not only WLANs but also other wireless and mobile networks. As a future work, we will investigate and implement the framework to be satisfied with QoS parameters with real wireless environment, many multicast groups and consider congestion avoidance for mobile traffic.

References

1. Nisar, K., Said, A.M., Hasbullah, H.: A Voice Priority Queue (VPQ) Fair Scheduler for the VoIP over WLANs. Department of Computer & Information Sciences, Universiti Teknologi PETRONAS, Bandar Seri Iskandar, 31750 Tronoh Perak, Malaysia, vol. 3, no. 2, Febuary 2011
2. Engin, M.A., Cavusoglu, B.: IEEE communication letter 15(11), 1234–1237, November 2011
3. Manvi, S.S., Venkataram, P.: Mobile agent based online bandwidth allocation scheme for multimedia communication. Electrical Communication Engineering Department, indian Institute of Science, Bangalore-560012, India
4. Qiao-yun, S., Ze-min, L., Wen-yan, C., Hui, T., Shu-guang, Z.: A novel scheduling scheme for multiple traffic classes. In: The Journal of China Universities of Posts and Telecommunications, 17(Suppl.), pp. 91–94, July 2010
5. Mushtaq, M.S.: QoS-Aware LTE downlink scheduler for VoIP with power saving. Image, signal and intelligent systems Laboratory-LISSI, Dept. of Networks and Telecoms, IUT C/V, University of Paris-Est Creteil (UPEC), France. In: Shahid, A., Fowler, S. (eds.) Mobile Telecommunications, Dept. of Science and Tech, Linköping University, Norrköping
6. Tsai, T.Y., Chung, Y.L., Tsai, Z.: Introduction to packet scheduling algorithms for communication network. Institute for Information Industry, Graduate Institute of Communication Engineering, National Taiwan University, and Taipei, Taiwan, R.O.C. http://www.intechopen.com
7. Vasiliadis, D.C., Rizos, G.E., Vassilakis, C.: Class-based weighted fair queuing scheduling on quad-priority delta networks. Department of Computer Science and Technology, University of Peloponnese, Tripolis, Greece, Technological Educational Institute of Epirus, Arta, Greece
8. Dai, C.Q., Liao, F.L., Zhang, Z.F.: Packet scheduling for downlink OFDMA wireless systems with heterogeneous traffic. In: Journal of Network, vol. 7, no. 6, Chongqing Key Lab of Mobile Communication Technology Chongqing University of Posts and Telecommunications Chongqing China, June 2012
9. Haas, H.J.: Queuing methods From FIFO to CB-WFQ/PQ, May 2005. http://www.perlhel
10. Schulzrinne, S.S.H.: Towards the quality of service for VoIP traffic in IEEE 802.11 wireless networks. Department of Computer Science Columbia University (2008)
11. Faisstnauer, C., Schmalstieg, D., Purgathofer, W.: Priority round-robin scheduling for very large virtual environments and networked games. Vienna University of Technology
12. Brand, A., Aghvami, H.: Multiple access protocols for mobile communication. http://www.wiley.co.uk or http://www.wiley.com, Swisscom Mobile, Switzerland, King's College London, UK (2002)
13. Haci, H., Zhu, H., Wang, J.: Novel scheduling for a mixture of real-time and non-real-time traffic. School of Engineering and Digital Arts, University of Kent (2012)

Network-Centric Approach to Real-Time Train Scheduling in Large-Scale Railway Systems

Alexander A. Belousov[1,2] (✉), Peter O. Skobelev[1], and Maksim E. Stepanov[1,2]

[1] Samara State Aerospace University (National Research University), Samara, Russia
{belousov,stepanov}@smartsolutions-123.ru, adark@narod.ru
[2] SEC "Smart Solutions", Samara, Russia

Abstract. Railway transport development leads to a rapid increase of passenger and cargo traffic. Train dispatcher often unable to cope with increasing traffic volume. So there is a great need in intelligent real-time systems of railway traffic control. We propose a network-centric approach to creating the system of real-time train scheduling on the basis of multi-agent technologies. The architecture of the network-centric multi-agent system consisting of base planning subsystems is described. Subsystem interaction protocols and protocols of agent interaction within each subsystem are presented. The example of schedule planning is presented. Productive characteristics of the developed system are presented. Good quality of train schedule planning and system performance is shown.

Keywords: Railway dispatching · Railway transport · Multi-agent systems · Network-centric approach · Method of coupled interactions · Real-time planning

1 Introduction

Modern development of railway transport is characterized by a constant increase of passenger and cargo traffic; new high-speed trains appear, train time intervals shorten, train-handling capacity of the railway system reaches limit value. Due to the high traffic intensity, trains are highly interconnected: changes in one train's schedule, or a conflict with this train, will affect the next train and can have impact on the whole train network in a very unpredictable way. In this case re-scheduling of all trains in the planning area might be required, which should be done quickly, in real-time. This is a time-consuming task given the whole variety of planning conditions, preferences and constraints as well as security requirements.

In spite of high level of railway network automation, today solving conflict situations in trains deviating from the standard schedule completely depends on the experience of the dispatcher in charge, which often leads to irrational placing of trains in the traffic, especially in stressful situations. The constantly growing scale of the task to be solved results in increasing complexity of disruptive situations and provokes a question as how to reduce dependency on the human factor by automating the decision-making process and introducing intelligent systems, enabling fast and effective adjustments in case of a disruptive event.

Y. Tan et al. (Eds.): ICSI-CCI 2015, Part II, LNCS 9141, pp. 290–299, 2015.
DOI: 10.1007/978-3-319-20472-7_31

An intelligent system should make decision "on the fly" under condition of constantly changing context of the situation, which requires correct changes to be made on time and the previously developed train schedule to be adjusted. Thus, to solve this task one has to reject the assumption of the equilibrium of the environment. Modern intelligent systems designed to help in real-time decision-making in complex and large-scale systems should use new methods and means of management automation, enabling to take into account the whole variety of factors, conditions, rules, and interactions, considering that limitations apply individually and can be regulated in the process of work, provide the high level of precision and quality of decisions as well as productivity.

This paper describes the network-centric approach, which solves the complex task of adaptive real-time train scheduling. The developed system was used in production by Russian Railways on a section of the high-speed rail Saint-Petersburg – Moscow and between Saint-Petersburg – Buslovskaya.

2   Network-Centric Approach to Real-Time Train Scheduling

2.1   Problem Statement

The task of adaptive real-time train scheduling is to create a detailed train schedule with minimum deviations from the master schedule, considering various constraints and requirements of continuously incoming events (trains movement in the infrastructure block sections, train failure, track occupation, maintenance, speed limitations, etc.).

The input data consists of: railway infrastructure with block sections (stations, railway switches, infrastructure block sections); requirements for train schedule (master schedule); maintenance requirements; updates on the current situation about trains and state of infrastructure block sections (busy signals or sections unavailability).

It is worth mentioning that the scale of the task is enormous which makes it a large-scale task.

Let us consider the Saint-Petersburg – Moscow railway direction. This planning area has 49 stations, 48 railway runs (there are ones with different number of tracks), planning area in infrastructure network consists of approximately 3700 infrastructure block sections. Every day there are more than 800 trains of different types passing through the planning area: cargo, maintenance, suburban, passenger, high-speed. The system can support up to 50 different train priorities. There can occur up to 50 different events (maintenance windows, track damage). Moreover, every train's position is updated continuously (up to 50 events at a time).

The main limitations of the system are traffic security requirements, normative route-building requirements, train priorities, dispatcher rules, etc.

However, besides the limitations listed above, there are ones which are hard to formalize (no thickening of lines on train schedules, no unjustified changing of tracks, no traffic jams between stations, no unjustified halts or trains stop on the main tracks, correct routing of trains by dispatcher schedule, etc.), which should be considered during scheduling. At the same time, the requirements implementation depends on the current situation, i.e. the decision-making is situational.

For instance, a train may not choose an opposite track, but if there is a busy block section on its route or a maintenance window, then it can do it, in order to bypass the obstacle and stay on schedule –only if its maneuvering does not affect other trains with higher priorities. On the other hand, it may stay on the same track and wait for a while, if this delay is not big and the train can catch up. However, waiting at a station means stop for a while, and a train can only stop at the infrastructure block sections of certain length. Therefore, a decision whether the route should be changed is connected to many conditions, which should be in balance.

Multi-agent approach allows to consider the whole set of diverse criteria, constraints and requirements for train management, including the ones that are hard to formalize, by agents negotiations and finding consensus satisfying all sides of conflict. Architecture of the multi-agent system is open and enables to introduce new conditions and criteria as well as change already existing ones and manage them dynamically.

The system for real-time train scheduling is implemented in the course of the "Vektor-M" software platform developed by ProgramPark Company [1,2], which allows to keep the dynamic infrastructure model of the planning area, get busy signals from block sections, appointed maintenance windows, satellite and other information.

2.2   Architecture of the Network-Centric System

The general architecture of the network-centric adaptive train scheduling system is depicted in Fig. 1[3].

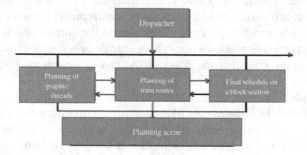

Fig. 1. The general architecture of the network-centric train scheduling system

The architecture of the developed system is built on the network-centric principles, where every subsystem has its own individual task and the final solution is reached through negotiating between individual decisions.

Dispatcher subsystem coordinates the work of all the other subsystems, reads and enters data into the planning scene. The planning scene is a set of different data storages where data required for scheduling is stored. The plan building takes places in 3 stages. Each subsystem builds a train schedule on its own level of data representation. A decision made at every stage, is conflict-free for its level of data representation (no converging train routes, the security requirements are intact). This layer-based train scheduling eliminates the combinatorial explosion of possibilities, makes the scheduling process more stable to disruptions due to reducing the scale of the task on higher levels and step-by-step considering all possible limitations according to the level of importance and impact on other layers. More detailed interaction of the components is depicted in Fig. 2.

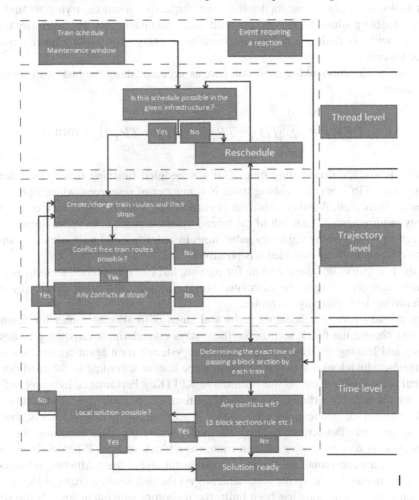

Fig. 2. Interaction of the main planning subsystems.

All events coming into the system can be divided into the two main types: new request and update on the current situation. Requests in their turn can be of the two basic types: request to let a train pass on schedule and request for infrastructure maintenance works. An update on the current situation can be either a train moving along infrastructure block sections or a state of an infrastructure block section (damage or busy condition). An event-request arrives in the thread subsystem, an event-update – in the time subsystem.

The thread planning in thread subsystem is similar to visual schedule analyzing. The main task is to build a new possible train schedule considering the normative schedule limitations and train priorities. The solution is based on the method of coupled interactions for managing resource allocation in real time [4]. On this level a train agent creates subtasks (operations) for each station in a train schedule. A subtask agent looks for a placement for itself in the respective resource, trying to find the most profitable position by negotiating with other subtask agents. High-priority trains are more active in finding a placement (have more energy for pushing other requests for resources).

The main decision-making condition here is accomplishing the task with minimal divergence:

$$DEV_T = \sum_{i=1}^{N} \left(|TD_S - TP_S| + |TD_F - TP_F| \right) \rightarrow \min, \tag{1}$$

where TDs – scheduled starting time, TDf – scheduled finishing time, TPs – actual starting time, TPf – actual finishing time, N – number of resources, where operations can be implemented. Another condition is compliance with the normative traffic and security requirements. The result of the thread system's work is an approximate schedule, which is sent to the trajectory subsystem in order to build train routes by infrastructure block sections, provided it is possible.

In the trajectory subsystem a route for passing infrastructure block sections is built for each train according to the calculated simplified schedule, conflicts are resolved by overtaking and changing the routes.

The primary task of scheduling paths and stops is to allocate routes for trains to take, and choose the block section for their stays considering overlapping routes of arriving and leaving for the parking. In this subsystem a train agent creates new subtask agents, which look for routes for passing the station according to the condition of minimal route costs. "Cost" is the cumulative KPI (Key Perfomance Indicator) of the route, which includes different normative requirements for train routes (correct or not, length, number of connections etc.). After scheduling with minimal KPIs, station route agents enter the active phase of life cycle, where the main condition for decision-making is no overlapping of train routes at a block section. When such an overlap is found, a station route agent will try to transmit one of the conflicting subtasks to other route agents. Route agents communicate via the task sharing protocol [5, 6].

When a successful route has been built, the trajectory subsystem spreads the schedule to the time subsystem among block sections, or sends a message to the thread subsystem with parts of the schedule which cannot be implemented in the current infrastructure.

In the time component, the final schedule is checked and corrected if required in order to comply with the normative requirements.

The time component as well as the thread one uses the method of coupled interactions for managing resource allocation in real time [4], however the resources are represented not by railway runs and station platforms, but by infrastructure block sections, whereas the train route creates many task agents for following the route's block sections. The conditions of cumulative minimal running time divergence in resources (1) and the normative requirements are extended by normative requirements, typical for depicting the planning area (traffic intervals, requirements for speeding up and slowing down and so on). The result consists of final concerned train schedules. If all limitations cannot be applied, the time subsystem sends a message stating the location in the schedule where the irregular event takes place and transfers it to the trajectory component.

Building and negotiating the final train schedule takes place in all subsystems (Fig. 2). In every planning subsystem there is a swarm of agents, representing the level, between the subsystems there are back links which come into play when a conflict cannot be resolved locally in the current subsystem. In this case a conflict is transmitted to a subsystem capable of solving it, negotiating takes places again after that.

For example, the system receives an event about a train behind the schedule time. This event enters the time subsystem. If the delay is not big, the conflict will be mitigated by changing the train speed on block sections. If the conflict cannot be resolved, because it would affect other trains, for example, this data is transferred to the trajectory system, which will try to resolve the conflict by changing the train routes. If this is not enough, the thread subsystem takes the conflict over.

It should be noted that back links are also activated when a schedule from the previous level cannot be implemented in the next one. For example, the thread subsystem does not consider possible failures on block sections at a station (in general, such failures may not affect the traffic capacity of a station), it only estimates the capacity of the station as a whole as long as there is no failure alert from the trajectory subsystem. Then scheduling trains in the thread scheduler will be done considering unavailable parts of the station tracks.

Swarm agents communicate concurrently and asynchronously. The primary allocation of tasks to resources is done based on the best decision possible independently and in concurrent threads, which allows for reduction of computing time by excluding the rest of possibilities. Such "greedy" allocation results in conflicts that are resolved by agents grouping together into structures within a swarm – domains. In each domain searching for a compromise takes place between agents in order to resolve the conflict. Communication within domains takes place independently, concurrently and asynchronously.

3   Software Implementation

3.1   Realization Features

The following time indicators for performance analysis of the system for real-time train scheduling were used: incoming data load time, rescheduling time for a new arrival, rescheduling time depending on duration of maintenance window, rescheduling time for added maintenance windows, rescheduling time depending on the number of tracks with maintenance windows, rescheduling time depending on speed limits on a block section, rescheduling time depending on the number of limitations, speed on a block section, rescheduling time depending on the number of tracks occupied due to a speed limit at a station.

A train has around 45 operations (lines of schedule), every train operation has its own agent. There are around 800 trains in total and around 36000 agents for train operations. Apart from that, there are train agents – around 800, station agents – 49, station route agents – 500, block section agents – 3700, maintenance request and availability agents – around 100-200. Dividing this many agents into levels and grouping them into isolated swarms of agents, which are active at certain points of time, allows for increasing the system performance.

In order to get implementation features for each index, an average value was defined, based on the results of system performance for two planning areas: Saint-Petersburg – Buslovskaya and Saint-Petersburg – Moscow. Cumulative decisions instead of ones made by separate schedulers have been taken into consideration.

Figures relevant for planning the features of the planning areas are represented in Table 1. The Moscow – Saint-Petersburg planning area has 2.8 times more infrastructure objects than the Saint-Petersburg – Buslovskaya planning area.

Table 1. Relevant features of planning areas

Planning area	Number of stations	Number of infrastructure objects	Number of turnouts
Saint-Petersburg – Buslovskaya	17	1293	133
Saint-Petersburg – Moscow	47	3640	304

Table 2 represents time planning features on the two planning areas. The data is represented according to the primary schedule and rescheduling after updating the train positions and other events.

Table 2. Time planning features

Planning area	Average scheduling time (ms)	Average rescheduling time (ms)
Saint-Petersburg – Buslovskaya	1552	1302
Saint-Petersburg – Moscow	7592	5049

The load time of the Saint-Petersburg – Moscow planning area is 4.9 times higher than on the Saint-Petersburg – Buslovskaya planning area. It is caused by the scale and complexity of the infrastructure.

According to Table 2, the rescheduling time is 15-40% less in comparison to the initial scheduling time. It is caused by adaptive rescheduling based on incoming events instead of rescheduling everything from scratch.

Table 3 represents the change of planning scheduling characteristics depending on the number of scheduled trains on the planning areas. According to the table, the difference between the scheduling times on the planning areas with around the same number of trains is 2.6 times, which is close to the difference between numbers of infrastructure elements. Thus, linear dependency can be observed between the number of infrastructure elements and the train load time on the planning area.

Increasing the number of trains increases the train scheduling time in direct proportion.

Table 3. Planning scheduling characteristics depending on the number of tasks

Planning area	Number of trains	Average scheduling time (ms)	Average re-scheduling time (ms)
Saint-Petersburg – Buslovskaya	12	1511	1185
	17	2180	1429
	30	2257	1799
	49	2300	2092
	67	2746	2356
Saint-Petersburg – Moscow	71	7202	5063
	86	8786	6164
	152	12691	11420
	241	22453	22899
	311	36143	37351

The following qualitative characteristics can be noted: no unjustified changing of tracks, no traffic jams between stations, keeping security intervals, almost no delays among intercity and high speed trains in conflict situations, average train delays less than 9% (20 trains engaged in one conflict).

This outcome has been achieved on such large-scale planning tasks for the first time.

3.2   Example of Resolved Conflict Situations

Let us consider a situation with a high number of disruptions as shown in Fig. 3., with 6 maintenance windows, two of which completely block the traffic between Roshino and Zelenogorsk for an hour.

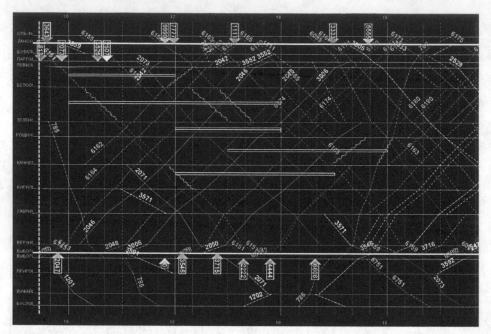

Fig. 3. Bypassing 6 maintenance windows, resuming after the disruptions have been eliminated

Resolving this situation required involving back links between the subsystems. Due to a high number of schedule lines after the window the time subsystem was unable to create the final schedule, since the traffic security requirements didn't allow the trains to stop and switch on to the alternative route. The time subsystem registers this mismatch as a conflict and sends a message to the trajectory subsystem, which, having failed to find a solution, sends the conflict to the thread subsystem. In order to resolve the situation, the thread subsystem must delay a few trains from previous stations (for example, 2048, 6155, 6163), taking the overload of the station limits into account. It sends the newly made decision to the trajectory subsystem which builds the route and transfers it further to the time subsystem. The time subsystem offers its own correct solution based on the received changed decision. As a result, the schedule has turned out to be more balanced and stable to possible further disruptions, the effect of the maintenance window has been localized, after which the schedules tend to be exemplary again.

It is important to note that if the time subsystem hadn't been able to offer its own solution due to the station overload, then this kind of conflict could have been resolved by additional bypasses in the trajectory subsystem and wouldn't have required any changes in the thread subsystem.

An overload between two stations cannot be resolved without the thread subsystem, because changing the route with switching on to the opposite tracks will cost much more than changing the train schedule.

4   Conclusion

The suggested network-centric system of adaptive train scheduling based on multi-agent technologies have been developed within the project of the unified intelligent train scheduling system for the Russian railways and now is in production usage [7,8].

Further development directions of the scheduler: explaining the made decisions, visualization of the decision space, demonstrating the decision logic to the dispatcher, interaction with the dispatcher for improving the quality of the schedule, modelling the future developments, teaching, evaluating the quality of the decision based on a flexible set of conditions and the common "satisfaction" level of resources, increasing performance by paralleling asynchronous planning processes.

The mentioned solutions will help increase the quality of decision-making and performance level of end-users.

Acknowledgments. The project has been funded by the Russian Foundation for Basic Research, the Ministry of Education and Science of Russia under the SSAU "Increase of competitiveness among the worldwide leading research and education centers" program for 2013-2020.

References

1. Shabunin, A.B., Markov, S.N., Dmitriev, D.V., Kuznetsov, N.A., Skobelev, P.O., Kozhevnikov, S.S., Simonova, E.B., Tsarev, A.V.: Integrative platform for the network-centric approach to creating allocated intelligent resource management systems at Russian Railways. Software engineering, Issue 9, pp. 23–28 (2012)
2. Shabunin, A.B., Kuznetsov, N.A., Skobelev, P.O., Babanin, I.O., Kozevnikov, S.S., Simonova, E.V., Stepanov, M.E., Tsarev, A.V.: Developing an ontology for the multi-agent resource management system for Russian railways. Information technologies, Issue 12, pp. 42–45 (2012)
3. Belousov, A.A., Skobelev, P.O., Efremov, G.A., Stepanov, M.E., Goryachev, A.A., Shabunin, A.B.: Multi-agent approach to a complex task of train scheduling in a large-scale railway management systems. Papers of the MLSD 2014, Moscow, pp. 252–263. ISBN 978-5-91450-161-4
4. Vittich, V.A., Skobelev, P.O.: Method of conjugate interactions in allocated resource management in real time. Autometria, Issue 2, pp. 78–87 (2009)
5. Abbink, E.J.W., Mobach, D.G.A., Fioole, P.J., Kroon, L.G., van der Heijden, E.H.T., Wijngaards, N.J.E.: Actor-Agent Application for Train Driver Rescheduling. In: Proceedings of AAMAS, Budapest, Hungary. pp. 513–520 (2009)
6. Skobelev, P.O., Belousov, A.A., Lisicin, S.O., Tcarev, A.V.: Development of the intelligent cargo logistics management system for Vostochny polygon. Papers of XV World Conference. Problems of management and simulation in complex systems, pp. 391–396. Samara Science Centre RAS, Samara, 25–28 June 2013. ISBN 978-5-93424-662-5
7. Shabunin, A.B., Kuznetsov, N.A., Skobelev, P.O., Babanin, I.O., Kozevnikov, S.S., Simonova, E.V., Stepanov, M.E., Tsarev, A.V.: Developing a multi-agent resource management system for Russian railways. Mechatronics, automation, management, Issue 1, pp. 23–29 (2013)
8. Matuchin, V.G., Shabunin, A.B.: IRMS. Concept and implementation. Papers from the first conference "Intelligent railway management systems IRMS-2012", Moscow, pp. 15–18, 15–16 November 2012

Knowledge Based Planning Framework for Intelligent Distributed Manufacturing Systems

Aleksandr Fedorov[✉], Viacheslav Shkodyrev, and Sergey Zobnin

Saint-Petersburg State Polytechnic University in Russia,
29, Polytechnicheskaya st., St. Petersburg 195251, Russia
aleksandr.v.fedorov@gmail.com

Abstract. This paper describes the different features and theoretical advantages of knowledge based framework in smart manufacturing. The main purpose of knowledge-based framework is smart reorganization of automation line based on expert's knowledge entered to the system. This knowledge leads to minimizing costs and maximizing of the efficiency of production line by preventing faults, dangerous situations and optimizing plans of production. In paper defined the main technics and technologies that used in prototype of the system. Also provided the modelling results on the test initial data for chemical automation line.

Keywords: Knowledge-based · Neuro-fuzzy · Stochastic planning · Smart manufacturing

1 Introduction

Today's Intelligent or smart manufacturing is one of the promising approaches of automation and robotization of big industrial distributed systems. Modern trends of industrial systems development show that such a development is tightly connected with adoption of artificial intelligence approaches and attempts to increase performance, especially in case of complex manufacturing. Usually information systems on plants have large-scaled very complex structures with high number of unknown or badly estimated parameters. Leading vendors of automation equipment have surpassed all imaginable expectations about development of complex robotic systems, means of information perception, increase in number of control channels, intellectualization of the low level of industrial automation such as PLC, fieldbus systems, sensors and actuators.

Knowledge usage in smart manufacturing can be divided into two main sections: 1) in domain for planning and execution domain-related actions 2) internally in manufacturing systems for optimization of manufacturing systems work i.e. independent actual application.

This paper presents such intelligent knowledge-based framework, showing how it is possible to adapt large interactive systems to user combining control theory and neuro-fuzzy-based learning as mentioned in [1].

In that paper reviewed first section – organizing and optimization of action related to the operational process. As an example of manufacturing plant viewed the laboratory in the SPbSTU – Festo Chemical station.

© Springer International Publishing Switzerland 2015
Y. Tan et al. (Eds.): ICSI-CCI 2015, Part II, LNCS 9141, pp. 300–307, 2015.
DOI: 10.1007/978-3-319-20472-7_32

Abstract dataflow diagram of the final system is depicted in Fig. 1.

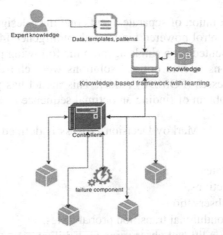

Fig. 1. Knowledge based system framework for smart manufacture

Some expert imports data in simple format such as Excel-file and system generates the representation of these rules, and then user can connect each term with some sensor or actuator from PLC or any input/output device that can be reached via OPC. Next step for user to define boundaries for sensor information. Another case of program is that user can generate rules and launch system in learning phase where it can generate training data and find some abnormal behavior if it will occur in production. The same approach described in [2].

Knowledge based system considered in that paper takes from some experts' data such as – plans for actions in emergency situations, and based on that rules and fuzzy logic algorithm system can produce output data for reorganizing parameters of the another components or stops them.

For generating and visualizing of fuzzy logic data will be used Java - framework for generating understandable and accurate models as explained by authors in [3]. Dataflow can be viewed on Fig. 2.

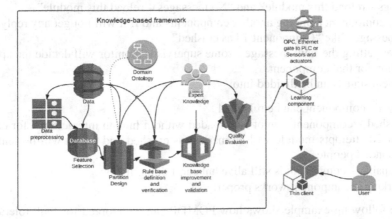

Fig. 2. Scheme of the proposed knowledge-based environment

2   Integrated Intelligence for Planning and Fault-Tolerance

For systems build as a union of separate agents one may design a layer of integrated self-supervision and control powered by internal intelligence [4]. A view of decision-making process is presented for a single agent in the following paragraphs.

For making decisions to find optimal solutions were chosen partially observable Markov decision processes (POMDP) model. This model has proven to be an effective solution to the problem of finding an optimal sequence of actions that lead to the highest winnings.

A Partially Observable Markov Decision Process is defined as a tuple M = (S, A, O, T, Ω, R), where

- S is a set of states,
- A is a set of actions,
- O is a set of observations,
- T is a set of conditional transition probabilities,
- Ω is a set of conditional observation probabilities,
- R:A×S→R is the reward function.

The example of system architecture with POMDP solver showed on Fig.3:

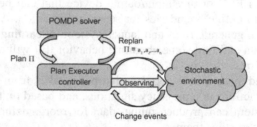

Fig. 3. Component architecture for planning and fault-tolerance

System actions in most basic view can be just leave component as it is, reload it, turn on or turn off. Systems states can be working or not. System read messages of type "No logs – reload this module" and "No messages – reload this module".

Any component can ping another components and if it will not get any reply, it will send messages like "Component 1 has crashed"

After getting the fault messages, some supervisor monitor will decide the appropriate action for that component.

System states can be divided into:

- Down – component can be reloaded
- Crushed – component cannot be reloaded without human interaction – for example after 20 attempts module will be marked as crushed and report will be sent to the responded people.
- Incapable – component is still alive but can't doing its job
- Working – component works properly

The following example shows how POMDP can be adopted for fault tolerant systems implementation.

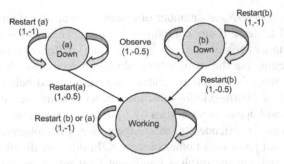

Fig. 4. Recovery model

Figure 4 shows a simple example of how an MDP might be used to model the recovery process of two components (a) and (b). In the figure, the different states represents is component works or it is down. Every component can observe its state with unavailability reward of 0.5 and if it down – restart it with reward of 0.5. But if components works good and they tried to restart – cost will be more – unavailability reward 1. The same situation for restarting component in a case when it will lead to the down of component.

For reacting on any changed states the POMDP need a special policy, which can maximize reward after achieving of goal. The optimal policy, denoted by p*, yields the highest expected reward value for each belief state, compactly represented by the optimal value function V*. This value function is solution to the Bellman optimality equation:

$$V^*(b) = \max_{a \in A}[r(b,a) + \gamma \sum_{o \in O} \Omega(o|b,a)V^*(\tau(b,a,o))], \tag{1}$$

where γ is the discount factor, τ is the belief state transition function and b is a belief state probability. This value function can be find with a variety of algorithms (for references see [4, 5]).

Another advantage of using POMDP in planning that it can optimize path of getting the final product on manufacture and can be scaled since it can be parallelized.

2.1   Algorithm Description

Algorithm for finding the best strategy for cases when system has many observers is described below:

```
Input: P[Failure] , Actions, ObserversNum, Observers, ε
do forever
   Observe(Observers)
   for i = 1 to ObserversNum
      If P[Failure] > ε then return ALERT
      If P[Failure] ≤ ε then break
   Action ← FindBestAction(Actions)
   ExecuteAction(Action)
   UpdatePF()
```

For system with *ObserverNum* – number of components at first algorithm goes throw all *Observers* and *Observe* if any component is fault (*P[Failure]* – denotes the joint probability of Failure of components, ε – is a constrain). After observing algorithm cat yield *ALERT*-function. For the set of actions algorithm decide which subset of actions optimal based on rewards for actions, states, observations and belief states(these parameters included in *FindBestAction*-function). After finding the best action consequence algorithm will update probabilities by *UpdatePF* – function.

This algorithm can be extended for multi-agent case, when observers can make decisions if they do not know about other observers. In this way distributed system will be more complicated, but the result – intelligent fault tolerance can outperform the shortcomings as mentioned by authors [6, 7].

2.2   Combining the Prescriptive Analytics and POMDP

Area of automated systems is one of the most needy in the systems in which all elements of the system are reliable. Reliability means that when an error occurs the system will react adequately.

System can adapt if no errors for some period not so much logs. System is distributed – every module looks throw every modules and modules can redeploy this module with some voting algorithm. System can stop affected modules, if they are using functionality of crushed component.

Prescriptive analytics in control systems can give an optimal strategy, which will give a list of features. In addition, program will not only make optimal decisions it also will generate reports for a special users and they can make their final decision. The simple example of report showed below:

Table 1. Number of actions for any component in case of a fall

Fault detected	Number of actions		
	Ping	Redeploy module	Restart OPC or module
Actuator1	12	2	0
Sensor 1	20	88	1

From the report system administrator can make decision that maybe that one component has problems with electricity or internet connection and call the appropriate service. If the system is decentralized, one component may generate such reports. In a best case, system will make the entire job and there is no need in a human interaction.

The system can make recommendations based on the statistics of components. For example, if one component continuously decreases, it may lack the memory, in this case, the system can analyze the log files, and if they are often referred memory supervision component can generate a message that it is desirable to increase the machine memory.

3   Low-Level Decision Making

While a high level planner, as presented in paragraph 2, may serve as a universal decision making tool for advanced fault-tolerance, it's characteristics are not suitable for all domains and environments. Local agents may require reduced decision latencies and ability to make partially optimal decision in absence of fully established connections between each other and the main decision making components. Required characteristics can be achieved by a lower level distributed control circuit, which is integrated into complete decision-making subsystem. Resulting hierarchical approach with the high level planning system on the top and a lower level distributed decision making system in the bottom is shown in Fig. 5.

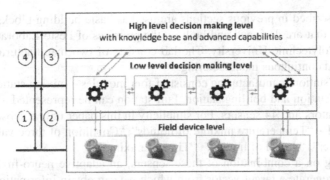

Fig. 5. A complete structure of a decision-making system

Logical flows of data between layers are visualized in the left part of the Fig. 5. Low-level decision-making layer connects to field device level to acquire the raw data (2) and sends control signals (1). Depending on usage of knowledge, statistics, learning capabilities and cross-observability, the low-level decision making system, being comprised of a set of agents, possibly interconnected, may serve different roles in the whole decision making contour. Exchange of data between low level decision making and high level decision making (3 and 4 in Figure 5) may be performed a) either via direct communication between low level agents and high level system or b) via a distributed approach with advanced fault tolerance characteristics. In this case agents organize themselves into distributed data structure with eventually consistent shared state, high level decision making then needs contacting only one low level agent to retrieve all information. Different low-level decision-making use cases are identified in the following paragraphs.

The low-level decision making circuit may have a trivial decision making component based on either hardcoded rules or an expert system. Such design best serves purposes of low-latency emergency situations detection an immediate reaction.

Low-level decision-making may contain a learning module that is capable of accumulating sensor statistics and aggregating it with basic assumptions on environment (basis functions in the resulting optimization criteria).

Information sharing capabilities can also be introduced. Mathematically an approach can be defined as a utility-based optimization problem.

$$R(a_n) + \gamma \sum R_j(a_{nj}) \to max, \tag{2}$$

$$R(a_n) = \sum_i P(i) * r(a_n) * L(a_n), \tag{3}$$

where $R(a_n)$ is a complete agent reward for performing action a_n, γ is a discount factor for decisions of neighbors, $P(i)$ is a predicate that holds true for current situation (depending on internals, that could be either simple if-else constructions or an expert system with advanced inference), $r(a_n)$ is a direct reward of an agent, $L(a_n)$ is a learner feedback (for example, a neural network).

4 Prototype Implementation

Concepts described in previous sections are used as basic building blocks of automation systems that are designed and built on the premises of Festo Laboratory in Saint Petersburg Polytechnic University. The lab consists of two manufacturing models – a discrete and continuous manufacturing.

Chemical station manufacturing consist of four modules – mixing station, filter station, reactor station and bottling station. Full station can be represented as small plant with 20 actuators and 44 sensors. For simplicity in this paper only a subset of sensors is considered – "Temperature mixture of liquids", "Condition of three valves", "Condition of mixer", " Level of liquid", "Time of mixing".

Depending on a combination of these sensors, an adaptive neuro-fuzzy inference system can generate a target vector from which we can obtain information – whether the temperature is exceeded or liquid level was too high or low-level system can report alarm if expert introduced the boundaries of parameters.

The main problem of ANFIS toolbox is that for 7 terms and all combinations of rules (6804 rules) system trains model for a long period. For example for 6804 rules the time consumption of training the model takes 6 minutes – which is unacceptable for that kind of small problem. For making training faster considered to use java neuro-fuzzy framework, which can be easily parallelized on the same PC.

5 Conclusion

Each day the number of unknowns in the production increases, as well as the number of controllable parameters. One way to effectively manage and optimize production is the proposed system knowledge bases with the possibility of self-study, which is part of the scheduler high and low level of production. Described techniques improve characteristics of resulting automated systems comparing to traditional approaches of design and implementation: fault-tolerance, scalability and latencies; that fact is proven by systems designed and implemented on the premises of Festo Laboratory in Saint Petersburg Polytechnic University and in customer project. Further development of the project involves the implementation and integration of the components of the optimization and refinement.

References

1. Arseniev, D.G., Shkodyrev, V.P., Potekhin, V.V., Kovalevsky, V.E.: Smart manufacturing with distributed knowledge-base control networks. In: Proceedings of Symposium on Automated Systems and Technologies, Hannover, PZH Verlag, pp. 5225–5230 (2014)
2. Legat, C., Neidig, J., Roshchin, M.: Model-based Knowledge Extraction for Automated Monitoring and Control, Preprints of the 18th IFAC World Congress Milano (Italy), pp. 85–89, August 28 – September 2, 2011
3. Alonso, J.M., Magdalena, L.: Guaje – a java environment for generating understandable and accurate models. In: XV Spanish Conference for Fuzzy Logic and Technology, Universidad de Huelva, Spain, pp. 399–404 (2010)
4. Fedorov, A.V., Zobnin, S.S., Potekhin, V.V.: Prescriptive analytics in distributed automation systems. In: Proceedings of Symposium on Automated Systems and Technologies, Hannover, PZH Verlag, pp. 43–49 (2014)
5. Milos, H.: Value-Function Approximations for Partially Observable Markov Decision Processes. Journal of Artifcial Intelligence Research **13** (2000)
6. Joshi, K.R.: "Stochastic-model-driven adaptation and recovery in distributed systems" dissertation, University of Illinois at Urbana-Champaign (2007)
7. Michał, N., Krzysztof R., Piotr B.: Asynchronous agent system for monitoring communication and system states based on the SOA paradigm. European Council for Modeling and Simulation, pp. 55–60 (2011)

An Event-Driven Based Multiple Scenario Approach for Dynamic and Uncertain UAV Mission Planning

Ke Shang[1,2], Liangjun Ke[1(✉)], Zuren Feng[1], and Stephen Karungaru[2]

[1] State Key Laboratory for Manufacturing Systems Engineering,
Xi'an Jiaotong University, Xi'an, China
[2] Department of Information Science and Intelligent System,
The University of Tokushima, Tokushima, Japan
kelj163@163.com

Abstract. In this paper, a Dynamic and Uncertain Unmanned Aerial Vehicle Mission Planning(DUUMP) is considered. New targets reveal stochastically during the mission execution and the surveillance benefit of each target is a random variable. To deal with this problem, an Event-driven based Multiple Scenario Approach(MSA) is developed. Experiment studies show that the Event-driven based MSA can solve the DUUMP effectively and efficiently with quick system responsiveness and high quality solution, which shows its practical value for real world applications.

Keywords: Dynamism · Uncertainty · UAV mission planning · Event-driven · Multiple scenario approach

1 Introduction

Unmanned Aerial Vehicle (UAV) has been playing an important role in military missions nowadays. Dispatch a UAV to execute surveillance tasks, leads an optimization problem called UAV Mission Planning(UMP) [1]. In UMP, there exists a set of targets that require surveillance. Each target has a surveillance benefit indicated by the degree of importance of that target. The UAV is dispatched from the base to execute the surveillance mission. Surveillance benefit of the target is obtained by the UAV if UAV reached that target. Meanwhile, the UAV has to return back to the base before its fuel drained. The objective of the UMP is thus to gather as much surveillance benefits as possible while safely return back to the base.

In real world situation, UMP is always in a dynamic and uncertain environment [1,2]. For example, when the UAV is executing the initial mission, new targets reveal as time lapses; therefore, the original mission plan needs to be updated. This factor causes the dynamism of UMP. There are two main framework to deal with the dynamism: time-driven reoptimization framework [3,4],

© Springer International Publishing Switzerland 2015
Y. Tan et al. (Eds.): ICSI-CCI 2015, Part II, LNCS 9141, pp. 308–316, 2015.
DOI: 10.1007/978-3-319-20472-7_33

and event-driven reoptimization framework [5]. [6] summarised the two frameworks in details.

In another situation, at the beginning of the mission, the planner doesn't know the accurate benefits of the targets but an interval data or a stochastic information, also because of the wind or other natural force, the UAV cannot keep a constant speed all the time. For these reasons cause the uncertainty of the UMP. The commonly used methods to deal with uncertainty are stochastic modelling and sampling [7]. In [6] more comparisons are given between the two methods.

The Event-driven based MSA uses the Event-driven framework and the Sampling procedure to deal with the dynamism and uncertainty. By using event-driven framework, the dynamic targets can be processed without delay; By Sampling, different scenarios are generated for decision making. One important consideration is the need of fast optimization algorithm to optimise scenarios, because the time lag between events can be small, time-consuming algorithms will influence the system responsiveness. A fast and efficient Variable Neighborhood Search(VNS) is used for optimization.

The remainder of the paper is organised as follows: Section 2 reviewed the Multiple scenario approach and the Event-driven framework which is the foundation of this paper, section 3 describes the dynamic and uncertain UAV mission planning. In section 4 the application of the event-driven based MSA to the dynamic and uncertain UAV mission planning is described. The experiment studies are given in section 5 and section 6 concludes the paper.

2   Event-Driven Based Multiple Scenario Approach

2.1   Multiple Scenario Approach

The Multiple Scenario Approach (MSA) is proposed by Van Hentenryck and Bent [7].

There are five components in MSA: Scenario Pool (SP), Scenario Generator (SG), Scenario Optimiser (SO), Decision Maker (DM), and Scenario Updater (SU). At the beginning, an empty SP is established. Then the main procedure of MSA starts from SG, SG generates scenarios based on sampling and then add the scenarios generated into the SP until the SP is full, each scenario generated represents an initial solution for the problem under the corresponding sampling. Then SO is used to optimise the scenarios in SP to get better overall scenarios. When the time of making a decision comes, the DM is used to make a decision according to the scenarios in SP, which tells the UAV the next target to visit. After the decision is made, the SU is used to update the scenarios in SP, the compatible scenarios are modified and the incompatible ones are discarded which leaves rooms for new scenarios. Until now, one iteration of the MSA procedure is finished, the next iteration starts from SG again for the preparation of the next decision making.

2.2 Event-Driven Framework

Event-driven architecture is a software architecture pattern promoting the production, detection, consumption of, and reaction to events. Here it is used to improve the responsiveness of the MSA with the MSA wrapped in the event-driven framework. The execution of each MSA component is triggered by corresponding event.

In UMP, the following events are defined: Decision, New target, Scenario generate, Scenario optimise, Pool update. Decision event is pushed into the event queue whenever the UAV arrives a target. New target event is added when a new target reveals. Pool update event is added after a decision is made and after a new target reveals. Scenario generate and Scenario optimise events are resident in the event queue, which makes the idle time can be completely used for generating and optimising the scenarios.

3 Dynamic and Uncertain UAV Mission Planning

3.1 UAV Mission Planning

The UMP consists of planning a path starting and ending at the starting and ending base 0 and n respectively and visiting other targets $i = 1, ..., n - 1$ to get corresponding surveillance benefits b_i, such that the total surveillance benefit of visited targets is maximized. l_{ij} represents distance between i and j. Each target can be visited at most once. For the UAV, a time limit T_{max} is defined because of the fuel limitation.

Let $x_i(i = 1, ..., n - 1)$ represents the binary variable of the UAV visiting target i. $x_i = 1$ if target i is visited by the UAV, otherwise $x_i = 0$. Let $y_{ij}(i, j = 0, ..., n)$ represents the binary variable of the UAV visiting arc (i, j), $y_{ij} = 1$ if arc (i, j) is visited by the UAV, otherwise $y_{ij} = 0$.

The UMP can be formulated as follows:

$$\max \sum_{i=1}^{n-1} x_i b_i \tag{1}$$

subject to the following constrains:

$$\sum_{j=1}^{n} y_{0j} = \sum_{i=0}^{n-1} y_{in} = 1 \tag{2}$$

$$\sum_{i<j} y_{ij} + \sum_{i>j} y_{ji} = 2x_j (j = 1, ..., n - 1) \tag{3}$$

$$\sum_{i=0}^{n-1} \sum_{j>i} l_{ij} y_{ij} \leq T_{max} \tag{4}$$

$$x_i \in \{0, 1\} (i = 1, ..., n - 1) \tag{5}$$

$$y_{ij} \in \{0,1\}(i,j = 0, ..., n) \tag{6}$$

where constraint(2) ensures that the UAV starts at vertex 0 and ends at vertex n, constraint(3) ensures the connectivity of the path, constraint(4) means the time of the UAV planning path is limited in T_{max}, constraint(5) and (6) ensure the binary variables value are integers.

3.2 Dynamism and Uncertainty

Dynamism. We usually use a parameter called Dynamism Degree to describe the dynamic level of the problem. The Dynamism Degree δ defined in [8] is the ratio between the number of dynamic targets n_d and the total number of targets n_{tot} as follows:

$$\delta = \frac{n_d}{n_{tot}} \tag{7}$$

Uncertainty. Suppose $\tilde{b}_i$ is the uncertain surveillance benefit of target i. The $\tilde{b}_i$ is a random variable in $[b_i{}^{LB}, b_i{}^{UB}]$ where $b_i{}^{LB}$ and $b_i{}^{UB}$ are the lower bound and upper bound of the benefit interval.

4 Application to Dynamic and Uncertain UAV Mission Planning

4.1 Scenario and Decision

To generate a scenario, the first step is sampling on the random variables. In UAV mission planning, the surveillance benefit of targets is uncertain, then the sampling procedure is conducted on the target surveillance benefit $\tilde{b}_i$.

After sampling, a static and deterministic UMP with sampled surveillance benefit is obtained, then a solution of this problem is called a scenario.

The Consensus algorithm [5] is used for decision making. The Consensus algorithm evaluates all the scenarios in the scenario pool and select the target with the highest appearance frequency as the next target [5].

4.2 Optimization Algorithm

In this paper, an Adaptive Variable Neighborhood Search(AVNS) is adopted to solve the UMP. The neighbourhoods are explored adaptively depending on their previous performance. The performance of a neighbourhood is measured by the ratio of the improvement to time. The neighbourhoods with better performance are more likely to be selected first.

Algorithm 1 gives the pseudocode of AVNS algorithm.

In the perturbation procedure, a random replace neighbourhood operator is used for perturbation. The random replace operator tries to exchange targets that belong to the current solution with non-included targets. In our experiment,

---

Algorithm 1. The Adaptive Variable Neighborhood Search algorithm

---

Input: An initial solution s_0, Neighborhood set $\mathbb{N} = \{N_1, ..., N_k\}$;
Output: The best solution found s_{best};
 1: $s \leftarrow s_0$
 2: $s_{best} \leftarrow s_0$
 3: **while** $\mathbb{N} \neq \emptyset$ **do**
 4:     $N^* \leftarrow NeighborhoodSelection(\mathbb{N})$
 5:     $s' \leftarrow Perturbation(N^*, s)$
 6:     $s' \leftarrow LocalSearch(s')$
 7:     $UpdateNeighborhoodPerformance(N^*, s, s')$
 8:     **if** $ObjectiveValue(s') > ObjectiveValue(s)$ **then**
 9:         $s \leftarrow s'$
10:     **else**
11:         $\mathbb{N} \leftarrow \mathbb{N} \setminus \{N^*\}$
12:     **end if**
13:     **if** $ObjectiveValue(s') > ObjectiveValue(s_{best})$ **then**
14:         $s_{best} \leftarrow s'$
15:     **end if**
16: **end while**

---

2 random replace and 3 random replace are used where 2 and 3 means the number of targets considered for replacement.

Randomly choose 2 for 2 random replace and 3 for 3 random replace non-included targets and insert them into the current solution, next the deletion will be performed to restore the feasibility of the solution. The insertion procedure is finished by inserting the targets into the best position in the solution without considering the feasibility. The deletion procedure is the opposite process of insertion. The deletion procedure will repeat until the solution is feasible again.

In local search procedure, three neighbourhood operators 2-opt, swap and replace are used to improve the solution. 2-opt and swap are used to reduce the route length. The replace operator tries to exchange one non-included target with included targets to improve the route benefit. All the non-included targets are evaluated for replacement and the best improving one is selected for replace operation.

4.3   Update

The update procedure need to be performed whenever a decision is made or a new target reveals. Both events change the state of the problem. After a decision is made, the assigned target is eliminated from the unvisited targets set. When a new target reveals, the new target is admitted into the unvisited targets set.

The update principles are:

1. For the decision event situation, if the next target in the scenario is just the target chosen in the decision procedure, then the next target is included in the visited targets list and excluded from the unvisited target, and this scenario will stay in the scenario pool. If the next target in the scenario is not the target

decided in the decision procedure, then this scenario is an incompatible scenario and will be eliminated from the scenario pool.

2. For the new target event situation, For each scenario in the scenario pool, the replace operation is considered to exchange the new target with targets in this scenarios.

5   Experiments

In this section, we present the computational experiments of the DUUMP. The proposed algorithm for solving the DUUMP is coded in Java, all the instances are tested on an Intel Core i7 with 2.2GHz, 16GB RAM.

5.1   Test Instances

The instances for the DUUMP tested in this paper are based on Novoa [9] which are initially used for Vehicle Routing Problem with Stochastic Demand (VRPSD). In the instances, all targets are uniformly distributed in a 1×1 square grid with discrete uniform benefits. Each target is associated with a reveal time which is generated randomly within the planning horizon. In order to simulate the real world UMP, the 1×1 square is amplified 20 times with the UAV speed set to 1. Table 1 shows the 3 test instance settings.

Table 1. Instances Information

Instance	Targets Number	Dynamism Degree	Time Limit
Instance 1	30	0.3	60
Instance 2	40	0.5	80
Instance 3	60	0.7	120

5.2   Performance Evaluation

In order to evaluate the performance of the Event-driven based MSA, we use the value of information [10] as a metric.

This metric can be written as

$$VI = \frac{OV_{sd} - OV_{du}}{OV_{sd}} \tag{8}$$

where OV_{du} is the objective value of the dynamic and uncertain instance I_{du}, OV_{sd} is the objective value of the static and deterministic problem I_{sd} when all information of I_{du} is known beforehand.

Each instance is run 50 times and the average objective value is calculated. The static and deterministic counterpart of the dynamic and uncertain problem is obtained by setting all dynamic targets as static targets, and all targets benefit

Table 2. The computational results

Instance	OV_{du}	OV_{sd}	VI
Instance 1	173	185	6.5%
Instance 2	220	243	9.4%
Instance 3	389	451	13.7%

is set to the actual benefit which is the real benefit the UAV get when the UAV reach the targets.

The results are shown in table 2. From table 2 we can see, the average objective value of the dynamic and uncertain problem is less than the static and deterministic counterpart, which is natural because when all the information is known beforehand, the planner can make a better overall plan by using all the known information. The value of information is increasing as the dynamism degree increasing. This phenomenon is caused because with the dynamism degree increasing, less information is known beforehand and it is more difficult to make the overall plan.

5.3 Simulation

Fig. 1 gives the simulation of Instance 1. The left panel is the real time information which gives the real time decision making and new targets arrival status. The right panel is the visualisation of the real time routing. In this example, the

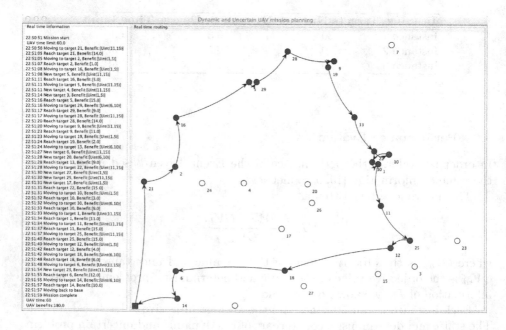

Fig. 1. Real time routing of instance 1

dynamic target 5 reveals when the UAV is moving to target 16, after the UAV arrive target 16, the dynamic target 5 is selected as the next target, we can see the Event-driven based MSA can make quick decision and modify the current plan to adapt the changing environment. In the simulation, the time between the UAV start and finish mission is 63 seconds which has a 3 seconds error, this is caused mostly by the delay between the UAV arrives a target and the next decision is made. But the error is small enough which can be ignored in the real world application. This shows the system's quick responsiveness.

6 Conclusion

The DUUMP is very challenging because of its dynamic and uncertain characters. Quick responses and fast optimisations are needed for solving this problem. The Event-driven based MSA solves it efficiently and effectively and it is applicable for real-world application because of its quick responsiveness and capabilities for acquiring high quality solutions. More future works need to be done to improve the approach including algorithm improvement and dynamic targets prediction.

Acknowledgments. This work was supported by the Fundamental Research Funds for the Central Universities, the Open Research Fund of the State Key Laboratory of Astronautic Dynamics under Grant 2014ADL-DW402, the Scientific Research Foundation for the Returned Overseas Chinese Scholars, State Education Ministry, and State Key Laboratory of Intelligent Control and Decision of Complex Systems. We are also thankful to the anonymous referees.

References

1. Evers, L., Barros, A.I., Monsuur, H., et al.: Online stochastic UAV mission planning with time windows and time-sensitive targets. European Journal of Operational Research **238**(1), 348–362 (2014)
2. Evers, L., Barros, A.I., Monsuur, H., et al.: UAV Mission Planning: From Robust to Agile. In: Military Logistics, pp. 1–17. Springer International Publishing (2015)
3. Montemanni, R., Gambardella, L.M., Rizzoli, A.E., et al.: Ant colony system for a dynamic vehicle routing problem. Journal of Combinatorial Optimization **10**(4), 327–343 (2005)
4. Chen, Z.L., Xu, H.: Dynamic column generation for dynamic vehicle routing with time windows. Transportation Science **40**(1), 74–88 (2006)
5. Pillac, V., Guret, C., Medaglia, A.L.: An event-driven optimization framework for dynamic vehicle routing. Decision Support Systems **54**(1), 414–423 (2012)
6. Pillac, V., Gendreau, M., Guret, C., et al.: A review of dynamic vehicle routing problems. European Journal of Operational Research (2012)
7. Hentenryck, P.V., Bent, R.: Online stochastic combinatorial optimization. The MIT Press (2009)
8. Lund, K., Madsen, O.B.G., Rygaard, J.M.: Vehicle routing problems with varying degrees of dynamism. IMM Institute of Mathematical Modelling (1996)

9. Novoa, C., Storer, R.: An approximate dynamic programming approach for the vehicle routing problem with stochastic demands. European Journal of Operational Research **196**(2), 509–515 (2009)

10. Mitrović-Minić, S., Laporte, G.: Waiting strategies for the dynamic pickup and delivery problem with time windows. Transportation Research Part B: Methodological **38**(7), 635–655 (2004)

An Improvement of Shipping Schedule Services: Focusing on Service Provider A in Korea

Jae Un Jung[1], Hyun Soo Kim[2(✉)], Hyung Rim Choi[2], and Chang Hyun Park[2]

[1] BK21Plus Groups, Dong-A University, Busan, Korea
imhere@dau.ac.kr
[2] Department of MIS, Dong-A University, Busan, Korea
{hskim,hrchoi}@dau.ac.kr, archehyun@naver.com

Abstract. In the shipping industry, there are commercial information providers who collect and advertise shipping schedules by global region between merchant shippers to attract cargo and forwarders to find tonnage. The service providers perform most jobs for data (shipping schedules) gathering and processing by hand. In addition, they are also challenged to extend the range of schedule collection and provision to both on- and off-line sites in terms of service quality and diversity. For these issues, we carried out a project to improve the existing business process of the service provider A (fictitious name) in Korea. Our research case introduces a significant commercial information service specialized in the shipping and logistics industries which is ordinary in the field but new in the realm of research. Also compared to the previous process, we found that our applications developed by the existing S/W technology improved the work efficiency by 25% and diversified revenue models with stability.

Keywords: Information service · Process reengineering · Shipping schedule

1 Introduction

It is a commonplace in the shipping market that ship operators fail to fill up the tonnage (shipping space) with cargo or forwarders experience difficulties to find the optimal shipping schedules. Such a mismatch in the shipping market is a typical issue caused by limitation of information sharing between demand and supply in common with other fields [1,2,3]. This issue is a good business chance for information service providers who provision shipping schedules like Indonesia Shipping Gazette [4], Korea Shipping Gazette [5], New Zealand Shipping Gazette [6], etc. Their business model is a weekly shipping news magazine or book which contains advertisements with vessel schedules of shipping companies, so these service providers are also called shipping news media. The service providers obtain revenue from both sides of shipping companies for advertising shipping schedules and forwarders for subscribing a weekly shipping schedule magazine. Like this, their business model looks attractive but there are endemic problems on heavy workload, low productivity and service instability because almost all of schedule data managed within each ship operator have different forms (types), so service providers should retype in everything for

© Springer International Publishing Switzerland 2015
Y. Tan et al. (Eds.): ICSI-CCI 2015, Part II, LNCS 9141, pp. 317–322, 2015.
DOI: 10.1007/978-3-319-20472-7_34

converting their raw schedule data received from shipping companies in a common format. Such an old process depending only on staffs' hands and eyes falls behind in the current business environment.

With regard to this issue, our research performed an industry-university project to improve the existing service process and model of the shipping schedule service provider A (fictitious name) representing in Korea. Through this research, we reengineered the overall service process in terms of process stability, productivity and revenue model with the conventional S/W engineering technologies. And we found that our improvements were sufficient to draw satisfaction from the A.

In section 2, business and system requirements for improving the shipping schedule service are analyzed and in section 3, our redesigned and implemented results are explained. Lastly, significance and limitation of our research are discussed in section 4.

2   Business and System Requirements

The following Fig. 1 represents the heart process of the service provider A from collecting ship operators' schedule data to provisioning a schedule service through the publication of a weekly magazine. Analyzed and required features of the existing process are explained in sections 2.1 to 2.5.

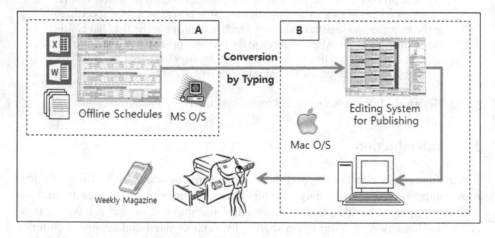

Fig. 1. Analyzed Systems for Provisioning a Shipping Schedule Service (Source: [7,8,9])

2.1   Productivity and Stability

Almost all schedules collected from shipping companies have different types or forms (code, length, table form, etc.) in different files (pdf, doc, excel, etc.) for different O/S (windows or Mac). This issue is represented in the part A of Fig. 1. In this work environment, every schedule would be retyped into a common form by editors' hands. The service provider A had a self-developed data editing system with 4D [10] but it just played a role of the database management system (DBMS) for QuarkXpress (desktop publication S/W) [11] to publish shipping schedules through the weekly magazine, not for supporting a file conversion in the part B of Fig. 1. For this reason, staffs in charge of editing schedule data would suffer from a heavy workload caused by routinely repeated retyping. When we analyzed the existing process and system, keystrokes of three editors for input and editing of schedule data would go on for 4 days a week. So our urgent priority was to improve the work productivity (load and time) by adopting data conversion technologies.

Also, the publishing system occasionally stopped with an unknown reason. An unexpected stoppage at the end of the publication process used to disconcert the service provider A as well as his or her advertisers and subscribers (ship operators and forwarders). What is worse, the service provider would suffered hardship to find a proper S/W engineer who could repair the system. Such instability required to replace the existing system with a new one. And it was a reasonable decision in terms of efficient and cost-effective maintenance.

2.2 Service Quality and Diversification

The service provider A had collected only offline data from advertisers but there were open schedule data on websites that ship operators who were not A's clients provided. As time goes by, they were expected to increase but the service provider had no idea how to gather free schedule data scattered on the web. For a schedule-rich service, an additional function to gather the online schedule data was required to develop by utilizing a web crawler [12].

On the other hand, even though a weekly magazine was a conventional revenue model of the service provider A and there was no counterpart in the same field, for subscribers hoping to save their time by searching data on the internet, this method was old fashion to decline. Looking at this angle, an online schedule query service was advanced further than downloading pdf files and it was expected to increase subscribers' satisfactions and to bring new revenue by diversifying business models at the same time.

2.3 Technical Constraints

In the field of publishing, a publication system for Mac O/S was preferred than that for Windows O/S because of a custom of long standing that a Mac PC was better in terms of print quality [13]. For the same reason, the service provider A also operated his or her publication system on a Mac O/S. But there was an issue on incompatibility between the data handling process on Windows O/S and the publication process on Mac O/S. On this account, a new schedule DBMS to develop was required to be independent or compatible from/with both O/Ss.

3   Process Reengineering and Implementation

Based on the requirements and considerations drawn in section 2, we carried out the redesign of the overall process for the shipping schedule service as shown in Fig. 2. Implemented results and the consequent benefits are explained in section 3.1 to 3.4.

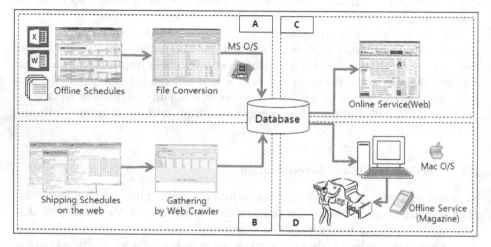

Fig. 2. Improved Systems for Provisioning a Shipping Schedule Service (Source: [7,8,9])

3.1   Part A: Manual to Semi-auto Data Conversion

First, we replaced the existing schedule DBMS with a new one that a file conversion function was added as shown in the part A of Fig. 2. In developing the new system, there were more than 100 formats for shipping schedules in different files but we could neither integrate all types into one standardized nor convert all files automatically because atypical or exceptional syntax included images and copy for advertising shipping schedules in each file. So we decided to concentrate on semi-automatically converting Excel (spreadsheet) files which accounted for about 70% of all files. The new system developed in Java (programming language) consists of two parts; file conversion support system and DBMS. They utilized the Apache POI (Java API for Microsoft documents) [14] for a file conversion function and MS SQL for a new DBMS.

The new application is compatible with Windows PCs as well as Apple Macs. And after it was implemented on the scene, keystrokes for input and editing of shipping schedules were reduced by 25% from 4 days to 3 days a week.

3.2   Part B: Gathering Online Schedule Data

In the existing process, shipping schedules that the service provider A could gather were limited to the offline data which were collected from advertisers but there were open shipping schedules that shipping companies who were not A's clients uploaded on the internet.

For this reason, a web crawler which could gather schedule data in table forms was utilized to expand the range of data collection onto websites. Online schedules collected by web crawler were automatically accumulated in the integrated DB with offline data. They were applied for a web schedule search service.

3.3 Part C: Advanced Web Service

The service provider A had provisioned a text-based shipping schedule search service through the website but data were not enough to charge a fee. In short, it was a perfunctory service.

Related to this issue, a new online service was launched with the extended (integrated) schedule DB. This service was outsourced to a SI (system integrator) and developed by Flash- one of the RIAs (rich internet applications) [15] for more interactive user interfaces. As a result, the advanced web service showed satisfactory improvements in terms of service quality and revenue diversity. Paid-up members can select any one between Flash and text modes.

3.4 Part D: Strengthened Offline Service

At the end of the publication process, the desktop publishing S/W was replaced from QuarkXpress to InDesign for securing the stability. Lastly, we added inland transport schedules connected with shipping schedules such as TCR (Trans-Chinese Railway), TSR (Trans-Siberian Railway) into the weekly magazine for strengthening the offline service. With the last development, our project was successfully completed that it took about 2 years from 2010.

Since then, practical problems were not found for the next two years except an additional requirement to increase the number of convertible file forms as advertisers increased.

4   Conclusion

The previous researches utilized shipping schedules for logistics decisions like ship operations, unit/facility dispatching, etc. but in our research, they were considered as a business model. In addition, the shipping schedule service is well-known on the logistics scene but not in the realm of research. Consequently, our research case itself has academic significance.

On the other hand, we carried out a project to improve the existing process of the service provider A from collecting schedule data to provisioning the schedule service. Through the development of the new file conversion system, we found that work time was reduced by 25% from four days to three days a week. And by replacing the existing DBMS and desktop publication system with new ones, the stability of the overall publication process was secured. Lastly, the additional utilization of online shipping schedules and inland transport schedules connected with shipping schedules conduced to high quality service and revenue diversification. These findings were confirmed by the A.

Though our artefacts were not fully elaborated due to industrial secrets, our research presented a real case which improved the existing shipping schedule service that it could be an academic reference for an industrial information service specific in the shipping and logistics industries.

In the following research, we will study a logistics cloud service as an alternative to converting heterogeneous shipping schedules through the common platform. The consequences will provide insights to overcome our limitation in terms of smooth information flow that we could not provide a solution available to change all schedule forms automatically because of highly complicated data forms.

References

1. Zhou, H., Benton Jr., W.C.: Supply Chain Practice and Information Sharing. Journal of Operations Management. **25**, 1348–1365 (2007)
2. Frankel, R., Li, X.: Characteristics of a Firm's Information Environment and the Information Asymmetry between Insiders and Outsiders. Journal of Accounting and Economics. **37**, 229–259 (2004)
3. McQuaid, R.W.: Job Search Success and Employability in Local Labor Markets. The Annalys of Regional Science. **40**, 407–421 (2006)
4. Indonesia Shipping Gazette, http://www.indoshippinggazette.com
5. Korea Shipping Gazette, http://www.ksg.co.kr
6. New Zealand Shipping Gazette, http://www.shipdata.co.nz
7. Microsoft Office, http://products.office.com
8. Macworld, http://www.macworld.com/article/1049078/quark7firstlook.html
9. Technied.Com, http://www.technied.com/tag/logo
10. 4D, http://www.4d.com
11. QuarkXpress, http://www.quark.com/Products/QuarkXPress
12. Choi, H.R., Kim, H.S., Park, B.J., Kang, M.H., Jung, J.U.: Development of an agent system to collect schedule information on the web for intermodal transportation network planning. In: 2007 Annual Conference on International Conference on Computer Engineering and Applications, pp. 366–371. World Scientific and Engineering Academy and Society, Wisconsin (2007)
13. Using a Macintosh for Desktop Publishing, http://desktoppub.about.com/cs/computers/i/mac_for_dtp_2.htm
14. The Apache POI Project, http://poi.apache.org
15. Lawton, G.: New Ways to Build Rich Internet Applications. Computer. **4**(8), 10–12 (2008)

Machine Learning

A Novel Algorithm for Finding Overlapping Communities in Networks Based on Label Propagation

Bingyu Liu[1(✉)], Cuirong Wang[1], Cong Wang[1], and Yiran Wang[2]

[1] College of Information Science and Engineering,
Northeastern University, Shenyang 110819, China
liuby78@163.com
[2] School of Economies and Business, Northeastern University at Qinhuangdao,
Hebei 066004, China

Abstract. Community discovery in Social network is one of the hot spots. In real networks, some nodes belong to several different communities. Overlapping community discovery has been more and more popular. Label propagation algorithm has been proven to be an effective method for complex network community discovery, this algorithm has the characteristics of simple and fast. For the poor stability problem of Label propagation algorithm, this article proposes a stable overlapping communities discovery method based on the label propagation algorithm: SALPA. At the beginning of the method, introduce the influence of nodes, which is used to measure the influence of nodes, select the most influential nodes as the core nodes, in the propagating stage, when there are more than one label with the same degree of membership, select the connectivity lager than the threshold. The method has been carried out in three real networks and two big synthetic networks. Compared with the classical algorithm, experiment results demonstrate the effectiveness, stability and computational speed of the method have been improved.

Keywords: Label propagation · Overlapping community · Complex networks · Community discovery

1 Introduction

Complex network has the properties of small-world, scale-free, community[1] is another major characteristic of complex networks. With the emergence of new types of social networks, community discovery has stimulated a great deal of interest, not only in the field of sociology, but also include the fields of cooperation networks, public opinion discovery and E-commerce Recommender Systems etc. The community has been found to play an important role in the real network, can be used to find social organizations with same hobby in social networks, to master the transmission of resources in network, the prediction of the change of the network topology etc.

Because in the real network, some nodes may simultaneously belong to many communities, such as music lovers may also love sports. The network has many overlapping interconnected communities, in recent years overlapping community discovery has been

© Springer International Publishing Switzerland 2015
Y. Tan et al. (Eds.): ICSI-CCI 2015, Part II, LNCS 9141, pp. 325–332, 2015.
DOI: 10.1007/978-3-319-20472-7_35

studied so much. The overlapping community discovery algorithm is mainly divided into three kinds, graph based partitioning, optimization and link partition. The main algorithm based on the partition algorithm of graph is CPM(Clique Percolation Method) proposed by Palla[2]. The basic idea of the algorithm is one node within the community more likely to form clique because of its high density; optimization algorithm partitions are based on the community metrics, such as density, modularity and conductivity. The OSLOM[3] algorithm is by adding or removing nodes within the community to strengthen community fitness function; the most famous is Baumes who proposed a two stage method in the literature[4], first divides the network into disjoint "seeds" community, then by adding or removing some of the neighboring nodes to achieve maximum community "density". Most of these algorithms are in need of a priori knowledge or time disadvantage of high complexity.

Because the community networks are very large, community discovery algorithm's speed become particularly important. The RAK[5] algorithm by far is one of the fastest algorithm, it is proposed by Raghavan, the algorithm not only has linear time complexity characteristics but also has the advantages of simple operation and without prior knowledge etc. But, like most community discovery algorithm, it can only be found not overlapping communities and not stable. One literature improves the RAK algorithm[6], proposed algorithm for mining overlapping community label propagation based (COPRA),the algorithm can dig out the overlapping community, but not on the stability of the community for effective treatment.

To solve the existing problems of traditional label propagation algorithm, the label propagation algorithm in this paper improved the initial stage, computing the node's influence, select the most influential nodes as the central nodes. In the label propagation process, assigned a unique label to center nodes. In order to reduce the complexity of the algorithm, all nodes are not calculated in the influence of the initialization phase, only calculate part of the nodes. In the label propagation stage, in view of the traditional label propagation algorithm is not stable, proposes a new label propagation algorithm, the algorithm in multi-dimensional measure of neighbor labels, select the largest as the node's label.

The rest of this paper is organized as follows. Section two introduces overlapping community discovery methods based on label propagation algorithm; the third section described the SALPA algorithm in detail, and the algorithm is analyzed; the fourth section algorithm is studied by experiments and compared with the classical algorithms; finally concludes the paper.

2   Related Works

The label propagation algorithm is a learning method of graph based machine. In 2007, Raghavan[5] first applied the label propagation algorithm to the community discovery.

The main idea of Raghavan's community discovery algorithm is to assign a label to each node, in iterative process, each node updates its own label, the labels appear the most times from the neighbor node updates for their new label. If many of the same number of most frequency labels exist, randomly select one as the new tag value. After several times of iterative, densely connected nodes converge to the same label. Finally, the node with the same label values belong to the same community.

Several label propagation algorithms proposed to improve Raghavan's algorithm. But these algorithms are only found the non-overlapping community. Then the label propagation algorithm is further improved, which is applied to overlapping community discovery. One of the first to use the label propagation algorithm to solve the problem of overlapping community detection algorithm is proposed by Steve, it is COPRA[6]. The COPRA algorithm allows nodes to have at most V tags, V as the algorithm parameters to input. The initial stage of the algorithm, given a unique label to each node, and set membership is 1. Then according to the neighbor node labels, calculated for each node may labeled and its membership, and standardization of membership. Finally, delete those be contained entirely in the community, and then the internal disconnected community split. Although the algorithm can solve the overlapping community detection problem, but the COPRA algorithm require input the number of overlapping communities each node can belong to, it is not suitable for large scale network. In a large network, some nodes may belong to 2 communities, and some nodes may belong to the 4, 5 different communities. Set a global upper bound for the number of overlapping communities one node can belong to is not suitable and there are still shortcomings such as not stable. Stability of some algorithm for label propagation algorithm is improved. But these algorithms have some limitations.

3   Community Discovery Algorithm

Our algorithm is divided into two parts; the first part is to find the most influential nodes. If the node's influence is less than a threshold, given the node's influence 0. The second part is to use the label propagation algorithm to detect overlapping communities.

In label propagation algorithm the sequence of nodes is also important, it affect the efficiency of label propagation algorithm. In the label initialization phase, we first select the most influential nodes. Next to choose label for every node, retain the node's membership greater than the given threshold. According to the six degree theory, we select N/6 nodes, whose influence is larger than other nodes, and N is the count of nodes in the community.

The importance of nodes usually defined as degree index, closeness and betweenness. Degree of a node is defined as the number of neighbors of the node, the local information of the degree can only characterize the node, and calculate the compactness and betweenness is higher complexity. Importance index of the network node is mainly based on local network attribute index, based on the network global attribute index and based on network location attribute index, which is based on global properties index, due to consider the global information of the network, the calculation time complexity is high, so it is not suitable for large scale network. In this algorithm, we don't need to accurately sort of community nodes. Network based on local property index although simple, but according to the evolution characteristics of Centola[7], the study found that simply select the degree of nodes as a measure index is not scientific. In 2010, Kitsak pointed out degree and betweenness cannot accurately describe a node's communication ability, and the use of K-kernel[8] decomposition method, the number of kernel of a node can better reflect the node's importance, but the algorithm is more suitable for single source communication network. Considering the characteristics of the above methods, the algorithm selects the partial attribute index and network location attribute based index as the measure of the importance of nodes.

Assume the network G= (V, E) composed of |V|=N nodes and |E|=M edges. Node i's communication capability is expressed as

$$P(i) = \frac{f_i}{\sqrt{\sum_{j=1}^{t} f_j^2}} + \frac{k_i + \max(k_j)}{\sqrt{(k_i + \max(k_j))^2}} \tag{1}$$

Where f_i represents the sum of the node's degree and its neighbor nodes' degree. f_j is the degree of node j, where j is the neighbor of node i, t is the count of node i's neighbor, k_i is the Ks value of node i. In order to improve the efficiency of the algorithm, select the most N/6 influential node temporarily as center node of each community and give the only label.

In order to be able to label overlapping nodes, using the same method as COPRA, use (C, b) on community nodes as labelled set, where C denotes the tag number, B is membership, and all node's label membership is 1. When more than one have the same membership, compare connectivity of the node to the tagged community, maintain the tag whose connectivity is greater than a certain threshold. Connection between node and communities is defined as the proportion of the neighbor in the community and all.

$$con(i,c) = \frac{|N(i,c)|}{|NC(i)|} \tag{2}$$

Where |N(i,c)| is the count of node i's neighbor who is in the community c, |NC(i)| is the number of communities that node's neighbor belong to. $con(i, c)$ is the connectivity of node i and Community c.

In the label propagation phase, when the new label membership degree is larger than the threshold, maintain multiple tags, the node is the overlapping node. In the SALPA algorithm, use synchronous update process, the end condition of iteration same as COPRA algorithm. In the initial stage of the algorithm, we choose the center node identifier as the old node's identifier, *labeln* is used to the new label storage iterative, initialized to null. Algorithm for calculating the results stored in the *labelres*. At the end stage of the algorithm, because the center points be extracted in the initial stages, the center point is not accurate, so get the community for further processing. Compared to many communities, identified, remove inclusion relation sub community.

4   Experiments and Results

In this paper, the experiment was conducted in comparison algorithm known community division of real networks and a large real network.

4.1 Experimental Data

In order to verify the validity and accuracy of the method, selecting three real networks and two different characteristics of the production of LFR artificial network as the experimental data. Three real networks are USA Zachary karate club network[9], one is Dolphins network[10], and another is PGP network. Zachary is American Karate Club network. Zachary is the most commonly used community found a test data set, the network consists of 34 nodes with representatives of members of the club, also has 78 edges represent links between the relationships between members of the club. Connections between nodes represent other occasions that two members often appear together in the club activities outside, was not at the club, they are friends. Dolphins' network is a dolphin social network, including 62 sections, 159 edges. PGP network is contains 10680 nodes and 24316 edges of the social network.

The next experiment was carried out in two different characteristics of the LFR artificial network, the two network are large-scale network (LG) and large community network (LC). Large scale network (LG) set N=6000, <k>=10, kmax=30, Cmin=10, Cmax=50. Community network (LC) set the Cmin=30, Cmax=100 and other parameters of large scale network agreement.

4.2 Evaluation Standard

For a real network, using overlapping community modularity as evaluation index of Q_{ov}[11] algorithm. But for artificial network, you can use the NMI (Normalized Mutual Information) the matching degree between the real community[12] to evaluate community and network built-in algorithm of finding.

$$Q_{ov} = \frac{1}{2m} \sum_{i,j} \frac{1}{O_i O_j} \left(A_{ij} - \frac{d_i d_j}{2m} \right) \delta(C_i, C_j) \tag{3}$$

Where O_i, O_j is the count of community that node i or node j belongs, d_i, d_j is the degree of node i and node j, when a node i and node j belong to the same community, $\delta(C_i, C_j)$ =1, otherwise zero.

NMI is usually used to measure the algorithm accuracy rate found in known community structure in the network community, defined as

$$I(A,B) = \frac{-2 \sum_{i=1}^{C_A} C_{ij} \sum_{j=1}^{C_B} C_{ij} \log \left(\frac{C_{ij} C}{C_i C_j} \right)}{\sum_{i=1}^{C_A} C_i \log \left(\frac{C_i}{C} \right) + \sum_{j=1}^{C_B} C_j \log \left(\frac{C_j}{C} \right)} \tag{4}$$

A and B are two different community group of network C, CA (CB) is the count of community in network A (B), C_{ij} is the count of node that in the community of node i of A and   at the same time the nodes in the community of node j in network B, C_i is the node count of the ith community in network A, C is the number of nodes in network C. NMI embodies the consistency of the two sets A and B, it is in range [0,1].

4.3   The Results of Experiments

4.3.1   The Efficiency of the Implementation of Algorithm
In order to test the efficiency of the implementation of SALPA algorithm, compared with the COPRA and CFinder algorithm in artificial network. In the experiment, the parameter V value of COPRA algorithm is 4, the threshold value of SALPA algorithm is 0.6. The experimental LG data using the same configuration, the N value from 1000 to 100000. The experimental results shown in Figure1, the SALPA algorithm is superior to COPRA and CFinder, the efficiency of the overall implementation of the two algorithms.

4.3.2   Test the Validity of the Algorithm
To further verify the effectiveness of the algorithm, some experiments were carried out in two artificial network respectively, after 50 independent replicate experiments, mean values of NMI. The experimental results as shown in Table 1.

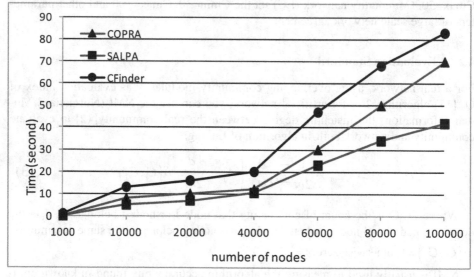

Fig. 1. Comparison of different scale efficiency of the implementation of the network

Table 1. SALPA algorithm implemented in the artificial network results

networks	Max NMI	AVG NMI
LG	1	0.9871
LC	1	0.9705

4.3.3   Comparison of Algorithm Implementation Results
During the experiment, we firstly selects three real networks to compared with COPRA algorithm. Parameter V and its results are the selection of the optimal results in COPRA algorithm. The comparison results are shown in Table 2, the community

structure of module SALPA algorithm to get the optimal ratio of COPRA algorithm to get the results of higher.

Table 2. The results of COPRA and SALPA executed in real networks

networks	COPRA		SALPA
	Best parameters	Q_{ov}	Q_{ov}
Zachary	3	0.44	0.72
Dolphins	4	0.705	0.74
PGP	11	0.7943	0.8014

5   Conclusion

In order to overcome the shortcomings of traditional label propagation algorithm, we propose a new stable community discovery method based on label propagation, without increasing the complexity of the algorithm, and the stability is enhanced. In the label propagation process, considering characteristics of neighbor nodes. Future work is extended to social network data with temporal relations, thus mining the user's behavior in the community, further discover the time behavior characteristics of users.

Acknowledgements. This paper was supported by National Natural Science Foundation of China (61300195), Natural Science Foundation of Hebei Province (F2014501078), The General Project of Liaoning Province Department of Education Science Research (L2013099), The Research Foundation of Northeastern University at Qinhuangdao (XNK201402, XNK201501) and The Doctor Found of Northeastern University at Qinhuangdao (XNB201317).

References

1. Fortunato, S.: Community detection in graphs. Physics Reports **486**(3), 75–174 (2010)
2. Palla, G., Derényi, I., Farkas, I., Vicsek, T.: Uncovering the overlapping community structure of complex networks in nature and society. Nature **435**(7043), 814–818 (2005)
3. Lancichinetti, A., Radicchi, F., Ramasco, J.J., Fortunato, S.: Finding statistically significant communities in networks. PloS one **6**(4), e18961 (2011)
4. Baumes, J., Goldberg, M., Magdon-Ismail, M.: Efficient Identification of Overlapping Communities. In: Kantor, P., Muresan, G., Roberts, F., Zeng, D.D., Wang, F.-Y., Chen, H., Merkle, R.C. (eds.) ISI 2005. LNCS, vol. 3495, pp. 27–36. Springer, Heidelberg (2005)
5. Raghavan, U.N., Albert, R., Kumara, S.: Near linear time algorithm to detect community structures in large-scale networks. Physical Review E **76**(3), 036106 (2007)
6. Gregory, S.: Finding overlapping communities in networks by label propagation. New Journal of Physics **12**(10), 103018 (2010)
7. Centola, D.: The spread of behavior in an online social network experiment. Science **329**(5996), 1194–1197 (2010)

8. Kitsak, M., Gallos, L.K., Havlin, S., Liljeros, F., Muchnik, L., Stanley, H.E., Makse, H.A.: Identification of influential spreaders in complex networks. Nature Physics **6**(11), 888–893 (2010)
9. Zachary, W.W.: An information flow model for conflict and fission in small groups. Journal of Anthropological Research, 452–473 (1977)
10. Lusseau, D., Schneider, K., Boisseau, O.J., Haase, P., Slooten, E., Dawson, S.M.: The bottlenose dolphin community of Doubtful Sound features a large proportion of long-lasting associations. Behavioral Ecology and Sociobiology **54**(4), 396–405 (2003)
11. Shen, H., Cheng, X., Cai, K., Hu, M.B.: Detect overlapping and hierarchical community structure in networks. Physica A: Statistical Mechanics and its Applications **388**(8), 1706–1712 (2009)
12. Lancichinetti, A., Fortunato, S., Kertész, J.: Detecting the overlapping and hierarchical community structure in complex networks. New Journal of Physics **11**(3), 033015 (2009)

Label Propagation for Question Classification in CQA

Jun Chen, Lei Su[✉], Yiyang Li, and Peng Shu

School of Information Engineering and Automation,
Kunming University of Science and Technology, Kunming 650051, China
s28341@hotmail.com

Abstract. Questions in Community question answering (CQA) consisting of some labeled questions and numerous unlabeled questions are so complex and irregular. Therefore, question classification in CQA has become the research hotspot in recent years. In this paper, we propose to classify the questions in CQA through the label propagation algorithm (LPA) based on the concept of graph, where nodes represent the labeled and unlabeled sample questions and edges represent the distance between the sample questions, through the node label propagation to realize question classification. Experiments on corpuses from "Baidu Knows", the accuracy in question classification through the LPA is not only higher than that through the KNN algorithm and SVM algorithm that have applied the labeled samples, but also higher than that through the SVM-based Bootstrapping algorithm that has utilized the labeled and unlabeled samples.

Keywords: Community question answering (CQA) · Question classification · Graph-based · Label propagation

1 Introduction

With the advent of Web2.0 era, numerous novel service modes based on web2.0 have merged on internet, such as forums, CQA [1] and SNS (social networking services). The traditional SE (search engine) utilizes keyword matching to return numerous web pages for the users to make a choice, which has also brought great convenience to the users. However as an open and interactive network platform, CQA is able to utilize the collective intelligence of web users to provide direct answers to the questions through user participation. In recent years, the mainstream CQA systems such as "Baidu Knows" and "Yahoo! Answers" have been developed quickly [2], where mass ask-answer data have been accumulated when people participate in the interaction and communication on CQA platform. At present, question classification on "Baidu Knows" has been involved with 23 major fields, including computer/network and life/fashion etc. to make question classification become one of the important parts to constitute the CQA system.

However, questions in CQA are rather complex and the classification hierarchy is not so apparent, then if question classification is conducted based on the concept of hierarchical classification proposed by Li et al. [3][4], the result might not be so favorable. Zhang et al. [5] proposed to classify the questions through the supervised learning method according to the n-gram features on the surface layer. Actually

© Springer International Publishing Switzerland 2015
Y. Tan et al. (Eds.): ICSI-CCI 2015, Part II, LNCS 9141, pp. 333–340, 2015.
DOI: 10.1007/978-3-319-20472-7_36

there's abundant ask-answer data in a CQA system, where most of the data is unlabeled. Therefore it becomes necessary to label all of the questions if questions are classified according to the supervised learning method. But it will cost lots of human and material resources. On account of this, it's advisable to take full use of those unlabeled questions during the question classification in CQA.

Currently, great progress has been made in question classification through the semi-supervised learning method [6], such as the self-training, Co-training and graph-based methods. The Tri-Training algorithm and Co-forest algorithm proposed by Zhou et al. [7][8] determine the labels on the unlabeled instances through the integration approach, while the SVM-based Bootstrapping algorithm proposed by Zhang et al. [9] utilizes the circular classification method to classify those unlabeled instances. It's a common point that all of the above algorithms are able to classify the labeled unlabeled samples into the labeled samples. But it's also a deficiency that all of them can just utilize the unlabeled samples locally without the consideration of the relevance and similarity between the unlabeled samples, and they cannot utilize fully the unlabeled samples from an overall perspective.

Hence this paper proposes a graph-based semi-supervised learning algorithm, which is also called as label propagation algorithm (LPA) [10]. First, extract the high-frequency words as the classification feature and acquire the eigenvector for question classification by words similarity algorithm based on Tongyici Cilin [11]. Then utilize the eigenvector of the question to compute the similarity between the questions. Finally classify the questions according to the LPA.

The second section in this paper introduces the feature extraction of questions in CQA, the third section introduces the question classification through the LPA, and the fourth section describes the experiment on the question classification in the life field on "Baidu Knows" and the comparative analysis on the result. The fifth section is the conclusion about our research.

2   Feature Extraction of Questions in CQA

Feature extraction is one of the critical processes in question classification. We choose bag of words as the feature to avoid the problem of severe data sparseness since there're numerous words in a text. However due to the fact that there're only a few words contained in a question, severe data sparseness would occur when bag of words is chosen as the feature. For this reason, it's necessary to solve the problem about data sparseness.

In this paper, feature space is constructed through the selection of some high-frequency words from multiple classifications. However since there're only a few words contained in a question and very few feature dimensions can be found in the feature space of a question, then it happens constantly that the characteristic value of the feature dimension is 0 for a feature vector in the feature space. This is actually the problem of data sparseness described as above. In order to solve this problem, this paper adopts words similarity algorithm based on Tongyici Cilin to compute the characteristic value of a question among the corresponding feature vectors. It turns out that this method is able to solve the problem of data sparseness efficiently.

However how to obtain the corresponding feature vector for every question? First, segment the words in a question and remove the stop words. Then match the segmented words in a question with the preset high-frequency words in terms of the similarity according to the words similarity algorithm based on Tongyici Cilin that is the "HIT-CIR Tongyici Cilin (Extended)". Since a word might have several meanings, then it might contain multiple semantic items. And word similarity is obtained by choosing the maximum similarity from multiple sense similarities. However when two semantic items are not in the same tree, the sense similarity can be computed according to Formula (1),

$$Sim(A, B) = f .$$ (1)

When two semantic items are in the same tree, the sense similarity can be computed according to Formula (2)

$$Sim_x(A, B) = X \cdot \cos(n \cdot \frac{\pi}{180}) \cdot (\frac{n-k+1}{n}) .$$ (2)

Where X represents a, b, c and d, $Sim_a(A,B)$ is the computational formula for the second tier of branch, $Sim_b(A,B)$ is the computational formula for the third tier of branch, $Sim_c(A,B)$ is the computational formula for the fourth tier of branch and $Sim_d(A,B)$ is the computational formula for the fifth tier of branch, n is the total number of the nodes in the layer of branches and k is the distance between two branches. Through artificial evaluation, the initial value for the number of layers can be set as $a=0.65$, $b=0.8$, $c=0.9$, $d=0.96$, $e=0.5$ and $f=0.1$.

Through the computation on the word similarity, the characteristic value of a question among the corresponding feature vectors can be obtained.

3   Question Classification in CQA Based on LPA

3.1   Establishment of LPA-Based Graph Model

Generally, all of the graph-based learning methods are established based on such a hypothesis that if two nodes are characterized with the same features, they tend to be classified into the same category. However the hypothesis for question classification is that: if the feature vector between two questions is similar, then they tend to be grouped into the same category. It reveals that the hypothesis for question classification coincides with that for the graph-based learning method. Hence we would utilize graph to construction a question classification model and then use massive unlabeled data for semi-supervised learning.

Assume that $(x_1,y_1)...(x_l,y_l)$ is the labeled data, $Y_L=\{y_1...y_l\} \in C$ is the classification label and C is the set of classification for those labeled data. Also assume that $(x_{l+1},y_{l+1})...(x_{l+u},y_{l+u})$ is the unlabeled data, $Y_U=\{y_{l+1}...y_{l+u}\}$ is unknown and the data set $X=\{x_1...x_{l+u}\} \in T$, where T is the set of feature vectors for a question. Then our purpose is to forecast Y_U, the category of the unlabeled data according to X and Y_L.

In order to integrate efficiently the information about the aforesaid labeled sample data and the unlabeled sample data in the learning process, our model defines a complete graph, $G=(V,E)$, where V, the set of nodes represents the various labeled samples and unlabeled samples in the dataset, while E, the edge to connect two random nodes, x_i and x_j can be computed according to Formula (3) to figure out the weight.

$$W_{ij} = \exp(-\frac{d_{ij}^2}{\sigma^2})=\exp(-\frac{\sum_{d=1}^{D}(x_i^d - x_j^d)^2}{\sigma^2}) .$$

(3)

Where d_{ij} represents the Euclidean distance between two random nodes and the weight, W_{ij} is subject to the parameter σ. In this paper, the value of the parameter σ is set to be 0.99.

As described above, we build a graph-based learning model.

3.2   Label Propagation Algorithm

After the graph strategy-based question classification model has been constructed, the nodes will contain both of the labeled samples and the unlabeled samples. However how to select an algorithm to transfer the tag information contained in the labeled sample data on the whole graph has played a critical role. Many researchers have proposed to dig the potential cluster structure information implied in the data through this graph-based semi-supervised machine learning algorithm to achieve the global consistency during the transfer of label information.

During the graph-based question classification in CQA, every node label will be transferred circularly to the nearby nodes through the weighted edges until global stability is achieved finally to forecast the label information for those unlabeled samples based on the LPA [1]. In this process of label propagation, the label of the labeled samples will remain unchanged. Then the greater the weight of the edge between the nodes is, the more easily the label information will be transferred between the nodes. Therefore the more similar the sample nodes are, the more likely that the nodes will have the same label.

In order to measure the possibility that a node label can be propagated to the other nodes through the edge, it's necessary to define a $(l+u) \times (l+u)$ probability transfer matrix, T according to Formula (4).

$$T_{ij} = P(j \rightarrow i) = \frac{w_{ij}}{\sum_{k=1}^{n} w_{kj}} .$$

(4)

Where T_{ij} is the probability of the transmission from Node x_j to Node x_i.

Also it's necessary to define a $(l+u) \times C$ label matrix of Y, where y_{ij} is the possibility that Node y_i should be allocated with Label x_j.

The specific steps for the application of LPA are provided as below:

Input: One piece of labeled data, u pieces of unlabeled data and Classification C.

Output: u classifications of unlabeled data.

a) Initialization: utilize Formula (3) to compute the edge weight matrix W_{ij} so as to get the similarity between data.

b) Obtain W_{ij} according to Step a) and then utilize Formula (4) to compute the probability of transmission from Node x_j to Node x_i.

c) Define a $(l+u) \times C$ label matrix, Y.

d) Compute the probability value for the label transferred from the surrounding nodes to this node according to the transmission probability and then update the probability distribution of the label matrix, Y.

e) Set a limitation on the labeled data and repeat the operation of Step d) to update continuously the label matrix, Y until there's a convergence.

4   Experiment and Result Analysis

The experimental data is involved with totally 7000 instances about the questions in the field of life on "Baidu Knows". All of these instances that are manually labeled have been classified into 14 sub-classes under five major categories. The specific classification is shown in Table 1.

Table 1. Question classification system in the field of life on "Baidu Know"

Categorie	Subcategorie
Gourmet Cooking	Cooking Methods, The Hotel Restaurant, Alcoholic Drinks, Kitchenware
Beauty Sculpting	Makeup, Hair Care, Skincare, Skin Whitening, Medicated Acne, Body Shaping
Household Appliances	Household Appliances
Buy Car And Keep Car	Car Buying, Car Care
Parenting	Parenting

In the following experiments, we performed the sampling for 20 times on a group of data sets, which consist of two categories. The test set accounts for 10% of the whole dataset and the training set takes up 90% of the dataset. Finally we take the average value of the 20 results as the final result. In order to compare the classification effect through the supervised learning algorithm and the LPA, the training set is grouped into L, the labeled instance set and U, the unlabeled instance set according to the different proportions of the unlabeled instances. Through the supervised learning method, test can be conducted after the training of the labeled instance set, L alone. However through the graph-based semi-supervised learning method, test won't be conducted unless both of the labeled instance set L and the unlabeled instance set U are used in the training.

4.1   Comparison of Question with Different Unlabeled Proportions

The proportion of the unlabeled instances in the training set may influence the classification effect. In this experiment, four groups of data are adopted with each group consisting of two categories. The proportion of the unlabeled instances in each group of data is separately 20%, 40%, 60% and 80%. The detail information about these four groups of data is shown in Table 2.

Table 2. Experimental dataset for question with different unlabeled proportions

Data set	Feature dimension	Unlabeled proportion	Labeled data	Unlabeled data	Test data
Data1		80%	180	720	100
Data2	500	60%	360	540	100
Data3		40%	540	360	100
Data4		20%	720	180	100

The benchmark algorithms used in this experiment are the supervised learning-based KNN algorithm and the SVM algorithm, which have also been compared with the LPA in this paper. Table 3 gives the classification accuracy through different algorithms with different proportions of the unlabeled instances.

Table 3. Experimental result about different unlabeled proportions

Data set	Unlabeled proportion	KNN	SVM	LP
Data1	80%	79.6	69.9	83.4
Data2	60%	82.3	82.2	85.2
Data3	40%	83.3	81.9	85.7
Data4	20%	84.2	83.3	86.1

The experimental results in Table 3 reveal that with the decrease in the proportion of the unlabeled instances in the training set, the classification accuracy through KNN, SVM and LP has been improved as well due to the increase in the data size in L to acquire the initial model for more adequate training. In the case that there're different proportions of unlabeled instances, the classification accuracy through LP is higher than that through the KNN and even through the SVM. All of these prove that the question classification model established based on the labeled training instances will improve more efficiently the classification accuracy when the unlabeled instances are introduced into the training set. In this experiment, LPA, which has utilized the training classification model containing both of the labeled and unlabeled instances, performs better in the classification effect than the KNN algorithm and SVM algorithm that has adopted such a training classification model, where only labeled instances are contained.

4.2   Comparison of LPA with SVM-Based Bootstrapping Algorithm

In order to verify the influence of the relevance and similarity between the unlabeled instances on the classification result, this paper adopts the same four datasets in the previous experiment to compare the LPA with the SVM-based Bootstrapping algorithm. Table 4 shows the classification accuracy through the LPA and the SVM-based Bootstrapping algorithm in the case that the proportion of the unlabeled instances is different.

Table 4. Experimental result about LP and SVM-based Bootstapping

Data set	Unlabeled proportion	Bootstapping	LP
Data1	80%	80.9	83.4
Data2	60%	82.5	85.2
Data3	40%	84.3	85.7
Data4	20%	84.9	86.1

Table 4 reveals that with the decrease in the proportion of the unlabeled instances in the training set, that's to say, the increase in the data size of the labeled instances and the decrease in the data size of the unlabeled instances in the training set, LPA performs better in the classification accuracy than the SVM-based Bootstrapping algorithm. It proves that the classification accuracy will be improved more effectively when the relevance and similarity between the unlabeled instances are taken into account in a training question classification model. In this experiment, since LPA is able to take the relevance and similarity between the unlabeled instances into account, it's superior to the SVM-based Bootstrapping algorithm that doesn't take the relevance and similarity between the unlabeled instances into account from the perspective of the classification effect.

5   Conclusion

In this big CQA platform, user interaction has occurred increasingly frequently, more and more questions with more and more types have been raised by the users. If the supervised learning method is adopted, the effect becomes increasingly unfavorable when questions have been classified in such a training classification model that contains only the labeled samples. Currently since it's an easy work to collect numerous unlabeled samples, the semi-supervised learning method that can utilize the training classification model containing a few labeled samples and numerous unlabeled samples to classify questions has been the research focus. In this paper, questions have been classified through the graph-based LPA, which can not only utilize the unlabeled samples, but also consider the relevance and similarity between the unlabeled instances to achieve such a training model based on the concept of global consistency to realize question classification.

Acknowledgments. This work was supported by the National Natural Science Foundation of China (No.61365010), Yunnan Nature Science Foundation (2011FZ069), Yunnan Province Department of Education Foundation (2011Y387).

References

1. Qiu-dan, Z.Z.F.L.: Studies on Community Question Answering—A Survey. Computer Science **11**, 008 (2010)
2. Yan, X., Fan, S.: CQA-Oriented Coarse-Grained Question Classification Algorithm. Jisuanji Yingyong yu Ruanjian, **30**(1) (2013)
3. Roth, D., Small, K.: The role of semantic information in learning question classifiers. In: Proceedings of the Conference First International Joint Conference on Natural Language Processing (2004)
4. Xin, L., Dan, R.: Learning question classifier. In: Proceedings of the 19th International Conference on Computational Linguistics, Taipei, pp. 556–562 (2002)
5. Zhang, D., Lee, W.S.: Question classification using support vector machines. In: Proceedings of the 26th Annual International ACM SIGIR Conference on Research and Development in Informaion Retrieval, pp. 26–32. ACM (2003)
6. Zhu, X.: Semi-supervised learning literature survey (2005)
7. Zhou, Z.H., Li, M.: Tri-training: Exploiting unlabeled data using three classifiers. Knowledge and Data Engineering, IEEE Transactions on **17**(11), 1529–1541 (2005)
8. Li, M., Zhou, Z.H.: Improve computer-aided diagnosis with machine learning techniques using undiagnosed samples. IEEE Transactions on Systems, Man and Cybernetics, Part A: Systems and Humans **37**(6), 1088–1098 (2007)
9. Zhang, Z.: Weakly-supervised relation classification for information extraction. In: Proceedings of the Thirteenth ACM International Conference on Information and Knowledge Management, pp. 581–588. ACM (2004)
10. Zhu, X., Ghahramani, Z.: Learning from labeled and unlabeled data with label propagation. Technical Report CMU-CALD-02-107, Carnegie Mellon University (2002)
11. Jiu-le, T.I.A.N., Wei, Z.H.A.O.: Words similarity algorithm based on Tongyici Cilin in semantic web adaptive learning system. Journal of Jilin University (Information Science Edition) **28**(6), 602–608 (2010)
12. Hotho, A., Staab, S., Stumme, G.: Ontologies improve text document clustering. In: Third IEEE International Conference on Data Mining. ICDM 2003, pp. 541–544. IEEE (2003)

Semi-supervised Question Classification
Based on Ensemble Learning

Yiyang Li, Lei Su$^{(\boxtimes)}$, Jun Chen, and Liwei Yuan

School of Information Engineering and Automation,
Kunming University of Science and Technology, Kunming 650051, China
s28341@hotmail.com

Abstract. In the traditional task of question classification, a mass of labeled questions are required. However, it's very hard to obtain many labeled questions in the real world. Meanwhile, it is very easy to obtain vast unlabeled question samples. Therefore, how to utilize these unlabeled samples to improve the question classification accuracy has been the core question of the question classification. In this paper, a semi-supervised question classification method based on ensemble learning, semi-Bagging, is proposed. The method utilizes a handful of labeled question samples to train the classifier. And then the classifier use a large number of unlabeled question samples which have pseudo labels to train again. Finally, during the experiments on question samples of 15 classes extracted from the community question answering system, the method could effectively utilize a large number of unlabeled question samples and a few of labeled question samples to improve the question classification accuracy.

Keywords: Ensemble semi-supervised · Semi-supervised learning · Question classification · Semantic extension

1 Introduction

CQA (Community Question Answering) is an emerging model of knowledge sharing, it has the characteristics of interactivity and openness [1]. To the questions of different classes, it has different quality of the answers and different degree of active of users [2]. Therefore, in a CQA, the question classification plays an important role, and classification accuracy directly affects the quality of the CQA. Question classification is roughly divided into two, one is rule-based and the other is statistical-based. The rule-based classification algorithm is simple and its classification accuracy is high, however, it needs human being to built rule base which is toilsome and not agile [3]. Therefore, the statistical-based method takes hold. Xin Li [4] put forward the idea of hierarchical classification and uses SNoW [11] classifier to classify 6 big classes and 50 small classes, and the classification accuracy is 91% and 84.2% respectively. However, the above question classification methods are supervised learning classifications which need a large number of labeled question samples; however, it is time-consuming and resource-consuming. Meanwhile, it is very relaxed to achieve a mass

© Springer International Publishing Switzerland 2015
Y. Tan et al. (Eds.): ICSI-CCI 2015, Part II, LNCS 9141, pp. 341–348, 2015.
DOI: 10.1007/978-3-319-20472-7_37

of unlabeled question samples. Therefore, the semi-supervised question classification algorithms which utilize a few of labeled question samples and a large number of unlabeled question samples have becoming the problems needed to be study.

There have been a lot of semi-supervised learning classification methods in machine learning field. Co-training put forward by Blum and Mitchell [5] uses two classifiers, it trains two classifiers and each one expands training space for another in semi-supervised process. Zhou proposed Tri-training [6] which uses three classifiers, and after trains three classifiers use two of them to expand training space for the third classifier in the semi-supervised process. Zhou also put forward Co-forest [7] algorithm, the method is a ensemble semi-supervised classification algorithm, which utilizes 6 classifiers and adopts majority voting mechanism to expand training space of every classifier.

In this pager, a semi-supervised question classification algorithm Semi-Bagging is proposed which uses Bagging [8] to combine classifiers. To train Bagging firstly, then it expands training space for itself. In semi-supervised process, data editing technology based on KNN is adopted, accordingly reduce the error markers in the expanded training set so that to improve the accuracy of classification. During the experiments on 15 classes, the semi-supervised question classification method Semi-Bagging can effectively use a large number of unlabeled question samples to improve question classification performance.

The rest of this paper is organized as follows. Section 2 presents feature extraction methods of Chinese question; Section 3 presents the semi-supervised question classification method Semi-Bagging based on ensemble; Section 4 reports the question classification experiments of 15 classes on life with analysis and contrast.

2　　Chinese Question Feature Extraction

This paper selects word bag as the question classification feature model, in order to achieve the feature vector of every question, to make word segmentation for every question, then use TFIDF to proceed feature extraction. However, there is difference between question classification and general text classification, a question only can separate about ten words even less. So, there are many 0 in the feature space, i.e. sparse matrix. In order to solve the problem of sparse matrix, this paper introduces word semantic extension calculation method based on Tongyici Cilin [12] proposed by Liu and Wang [10].

Word semantic extension consists of the word similarity and the word relevance. At first proceed word similarity calculation, adopt word similarity calculation algorithm based on encoding distance in dictionary. Word similarity calculation as shown in formula (1)

$$Sim(w_1, w_2) = d \cdot (\frac{n-k+1}{n}) \cdot \cos(n \cdot \frac{\pi}{180}) .\tag{1}$$

Where the $Sim(w_1,w_2)$ is semantic similarity($0<Sim<1$); d is a coefficient, it's up to the branch layer of the corresponding code of two words; n is the number of panel point in branch layer; k is the distance of two branch point.

Then to proceed word semantic relevance calculation, utilizing semantic relationships and statistical methods between two words of Tongyici Cilin [12] to calculate relevance. Word similarity calculation as shown in formula (2)

$$rel(w_1,w_2) = \frac{count(w_1,w_2)}{min(count(w_1),count(w_2))} . \tag{2}$$

Where $count(w_1,w_2)$ is the number of occurrences of w_1 and w_2 at the same time, $count(w_1)$ and $count(w_2)$ is the number of occurrences of $w1$ and w_2 individually, $min(count(w_1), count(w_2))$ is the minimum value of individual occurrences of w_1 and w_2.

3   Semi-supervised Question Classification Based on Ensemble Learning

The main idea is that, given a weak classifier and a training set L, let the classifier train several round, the training set L_i of every round consist of n training samples which are randomly selected from L, and the initial training sample could appear many times also can not appear in the training set of some one round. Achieving a estimate algorithm list $h_1,...,h_m$ after training, the final estimate algorithm H adopts the majority of the voting mechanism to classify [8].

In this paper, an ensemble semi-supervised classification method Semi-Bagging based on Bagging is used to classify questions. Let U denote the unlabeled question sample set, and L denote the labeled question sample set. Combining classifiers by Bagging technology, an ensemble classifier is obtained. The ensemble classifier is initially trained uses the labeled question samples L, afterwards the classifier label every sample of unlabeled question samples U, i.e. give them pseudo label, then a new training set L' is obtained by combining the new labeled samples and L. Finally, Bagging re-train with the training set L' that a new classifier Semi-Bagging is produced.

However, in the process of Bagging labeled the unlabeled samples, there may be many error labels. In order to avoid the case that error pseudo labels affect classification result, a simple data editing technology is adopted. The technology wipe off some error pseudo labels to guarantee that classification result less or not be affected by error labels. The data editing technology based on KNN which uses KNN to structure simple classifier. Only just the estimate h_1 of Bagging is equal to the estimate h_2 of KNN allow Bagging to label the corresponding unlabeled sample in the semi-supervised process. Otherwise, Bagging cannot label the unlabeled sample and don't add the unlabeled sample to L.

The pseudo-code of Semi-Bagging is presented in table1. The function buildclassifier(L) is used to generate classifier. The classifiers of Bagging and KNN are combined to expand the training set L after they are generated. Then the new classifier Semi-Bagging is generated utilize expanded data set L and function buildclassifier(L).

Table 1. Pseudo-code describing the semi-Bagging algorithm

Algorithm: Semi-Bagging
    Input: the labeled set L, the unlabeled set U
   Process: Bagging ← Bagging.buildclassifier(L)
         Knn ← Knn.buildclassifier(L)
         for every $x \in U$ **do**
           if $h_1(x) = h_2(x)$
           then $L ← L \cup \{x, h_1(x)\}$
         end of for
         Semi-Bagging ← Bagging.buildclassifier(L)
   Output: h(x) ← $\underset{y \in \text{label}}{\text{argmax}} \sum_{i:h_i(x)=y} 1$

---

4   Experiments and Result Analysis

4.1   Experiment Data

The data is Chinese question about life, there are 12000 samples labeled by human including 15 classes. The detailed information of the two sets of data is shown as table2.

The experiment divide 12000 questions into testing set T_1 and training set T_2 according to the proportion by 10-folds cross validation, the ratio of T_1 is 10% and ratio of T_2 is 90%, i.e. the 1200 of 12000 questions is the testing set and the rest is training set. The T_2 is divided into labeled sample set L and unlabeled sample set U in order to compare the effect of supervised learning and semi-supervised learning, for example, 10% of T_2 is the semi-supervised sample set and the rest of T_2 is the supervised sample set. The experiment deal with 100 times like this, then calculate the average of the 100 times.

Table 2. The experiment data set of different feature extraction method

Data Set	The Number of Train Data	The Number of Test Data	The Classes of Data Set
12000	10800	1200	costume jewelry, facial toning, gourmet cooking, buying a property, furniture decoration, home appliance, health, car-buying subsidy, transportation, shopping, life common sense, marriage, named fortunetelling, etiquette and parenting etc.

4.2   Feature Extraction Methods

The different feature extraction methods have different effects on machine learning. In this paper, a lexical semantic extension method based on Tongyici Cilin is used to extract feature, the TFIDF is also adopted to compare in order to validate the effect of the method in Chinese question classification. And the feature vector structure feature space of 200 dimension.

In this experiment, two sets of data including 15 classes, D1 and D2, are used. The two sets of data divide T_2 into unlabeled sample set U and labeled sample set L according to the ratio. The value of radio of U is 90% and the value of radio of L is 10%. The detailed information of the two sets of data is shown as table3.

Table 3. The experiment data set of different feature extraction method

Data Set	Feature Extraction Method	Unlabeled Data Set	The Number of Labeled Data	The Number of UnLabeled Data	The Number of Test Data
D1	TFIDF	90%	1080	9720	1200
D2	Semantic Extension				

In order to compare to the effect of different feature extraction methods in the question classification, the supervised learning and the semi-supervised learning adopt same classifier J48. Supervised learning adopts a single classifier and ensemble classifier based on Bagging and semi-supervised learning adopts ensemble semi-supervised classifier based on Bagging. The classification accuracy of the classifier is shown as the table4. The number of classifiers of ensemble supervised learning and the ensemble semi-supervised learning is 3 and supervised learning training only in the labeled sample set L.

Table 4. Different semantic extension method by J48 classifier

Data Set	Feature Extraction Method	Supervised Learning		Semi-Supervised Learning
		J48	Bagging	Semi-Bagging
D1	TFIDF	41.42%	43.21%	43.01%
D2	Semantic Extension	78.01%	80.85%	81.91%

The information of table4 shows that, the classification accuracy of supervised learning and semi-supervised learning are lower if TFIDF is adopted to extract feature. On the contrary, the classification accuracies of supervised learning and semi-supervised learning are higher.

And the feature extraction method only just utilize high-frequency words can't achieve ideal classification effect. And it hardly effectively uses unlabeled data. The feature extraction method semantic extension achieves better effect, which states that the feature extraction method is helpful to improve accuracy of question classification in the very great degree.

4.3   Compare to Semi-bagging and Bagging

In order to verity the effect of semi-supervised learning classification algorithm, different radios of unlabeled data are adopted, then compare to the ensemble supervised learning method Bagging.

6 sets of data are used in this experiment, including 15 classes. They are D1, D2, D3, D4, D5 and D6. These 6 sets of data by the different radios of T_2 contain 10800 questions. And the feature vector structure feature space of 200 dimension. Their feature extraction methods are TFIDF and semantic extension. The detailed information of the 6 sets of data is shown as table5.

Table 5. The experiment data of Semi-Bagging and Bagging

Data Set	Feature Extraction Method	UnLabeled Radio	Labeled Data	UnLabeled Data	Test Data
D1	TFIDF	95%	540	10260	1200
D2	Semantic Extension				
D3	TFIDF	90%	1080	9720	1200
D4	Semantic Extension				
D5	TFIDF	80%	2160	8640	1200
D6	Semantic Extension				

The classifier J48graft and J48 are used in this experiment, and the number of ensemble is 3. The classification result of Semi-Bagging and Bagging is shown as table6.

Table 6. The classification accuracy of Semi-Bagging and Bagging

Unlabeled Radio	Classifier	TFIDF		Semantic Extension	
		Bagging	Semi-Bagging	Bagging	Semi-Bagging
95%	J48graft	42.15%	41.93%	77.17%	78.33%
	J48	41.67%	41.50%	77.70%	79.22%
90%	J48graft	43.52%	43.37%	80.39%	81.14%
	J48	43.21%	43.01%	80.85%	81.91%
80%	J48graft	44.26%	44.26%	82.62%	82.96%
	J48	44.15%	43.92%	83.05%	83.54%

Can be seen from table6, if feature extraction method TFIDF is used, the classification accuracy of semi-supervised learning method Semi-Bagging is slightly higher than supervised learning method Bagging's, and lower sometimes. However, the classification accuracy of semi-supervised learning method Semi-Bagging is much higher than supervised learning method Bagging's based on the feature extraction method semantic extension.

Furthermore, if the two classifiers are used the classification accuracies of semi-supervised learning method Semi-Bagging are higher than supervised learning method Bagging's based on the feature extraction method semantic extension. For example, the classification accuracy of ensemble semi-supervised method is higher 1.52% than the ensemble supervised method which combines 3 J48 classifiers, during the case that the radio of unlabeled examples is 95%. Table6 also shows that different radios of unlabeled examples generate different results. The higher the radio of unlabeled examples is, the greater the classification accuracy of Bagging and the worse the effect of semi-supervised learning based on the feature extraction method semantic extension will be.

Therefore, this experiment verifies that the feature extraction method is effective and the ensemble semi-supervised classification algorithm based on Bagging is an efficient method in the case of different classifiers and different radios of unlabeled samples.

5   Conclusion

In a CQA, question classification is the core component. The experiment verify that the semi-supervised classification algorithm based on ensemble can combine a few of labeled question samples and vast unlabeled question samples to perfect the classification performance.

Acknowledgment. This work was supported by the National Natural Science Foundation of China (No.61365010), Yunnan Nature Science Foundation (2011FZ069), Yunnan Province Department of Education Foundation (2011Y387).

References

1. Zhongfeng, Z., Qiudan, L.: Studies on Community Question Answering-A survey. Computer Science **37**(1), 19–20 (2010)
2. Liu, Y, Agichtein, E.: On the evolution of the yahoo! answers QA community. In: The ACM SIGIR International Conference on Research and Development in Information Retrieval. Singapore, pp. 737–738 (2008)
3. Sichun, Y., Chao, G., Feng, Q.: The Question Feature Model of The Basic Features and Word Bag BindinFusion. Journal of Chinese **26**(5), 46–52 (2012)
4. Zhang, D., Lee, W.S.: Question classification using support vector machines. In: Proceedings of the 26th Annual International ACM SIGIR Conference on Research and Development in Information Retrieval. Toronto, Canada, pp. 26–32 (2003)
5. Blum, A., Mitchell, T.: Combining labeled and unlabeled data with co-training. In: Proceedings of the 11th Annual Conference on Computational Learning Theory, Wisconsin, MI, pp. 92–100 (1998)
6. Zhou, Z.-H., Li, M.: Tri-training: Exploiting unlabeled data using three classifiers. IEEE Transactions on Knowledge and Data Engineering **17**(11), 1529–1541 (2005)

7. Li, M., Zhou, Z.-H.: Improve computer-aided diagnosis with machine learning techniques using undiagnosed samples. IEEE Transactions on Systems, Man and Cybernetics-Part A **37**(6), 1088–1098 (2007)
8. Breiman, L.: Bagging Predictors. Machine Learning **24**(2), 123–140 (1996)
9. Nigam, K., Ghani, R.: Analyzing the effectiveness and applicability of co-training. In: Proceedings of the 9th International Conference on Information and Knowledge Management, McLean, VA, pp. 86–93 (2000)
10. Riu-yang, L., Liang-fang, W.: Keyword Extraction Algorithm Combining Semantic Extension Degree and Lexical Chain. Journal of Computer **40**(12), 265–266 (2013)
11. Xun-hua, X., Ji-cheng, W.: The Multi-Classification Algorithm of SVM. Microelectronics and Computer **21**(10), 149–152 (2004)
12. Jiu-le, T., Wei, Z.: Word Similarity Calculation Method based on Tongyici Cilin. Journal of Jilin University (Information science edition) **28**(6), 603–604 (2010)

Semi-supervised Community Detection Framework Based on Non-negative Factorization Using Individual Labels

Zhaoxian Wang[1], Wenjun Wang[1], Guixiang Xue[2], Pengfei Jiao[1], and Xuewei Li[1(✉)]

[1] Tianjin Key Laboratory of Cognitive Computing and Application,
School of Computer Science and Technology, Tianjin University, Tianjin 300072, China
{zhaoxian_wang,wjwang,pjiao,lixuewei}@tju.edu.cn
[2] School of Computer Science and Software, Hebei University of Technology,
Tianjin 300130, China

Abstract. Community structure is one of the most significant properties of complex networks and is a foundational concept in exploring and analyzing networks. Researchers have concentrated partially on the topology information for community detection before, ignoring the prior information of the complex networks. However, background information can be obtained from the domain knowledge in many applications in advance. Especially, the labels of some nodes are already known, which indicates that a point exactly belongs to a specific category or does not belong to a certain one. Then, how to encode these individual labels into community detection becomes a challenging and interesting problem. In this paper, we present a semi-supervised framework based on non-negative matrix factorization, which can effectively incorporate the individual labels into the process of community detection. Promising experimental results on synthetic and real networks are provided to improve the accuracy of community detection.

Keywords: Community detection · Semi-supervised framework · Non-negative Matrix Factorization (NMF) · Individual label

1 Introduction

Many systems take the form of networks, such as social and biological networks. An important property of the network is community structure which is first proposed by Girvan and Newman [1]. Community is a subgraph in which the vertices are more tightly connected with each other than with the vertices outside the subgraphs [2]. The nodes in the same community have similar features. Detecting the community can help us understand and analyze the network more deeply.

In the past few years, a large number of methods have been proposed to detect communities in the complex networks, including GN algorithm proposed by Girvan and Newman [3], modularity-based methods [4], stochastic blockmodels [5] and so on. Most of these approaches only take the topology information into consideration,

© Springer International Publishing Switzerland 2015
Y. Tan et al. (Eds.): ICSI-CCI 2015, Part II, LNCS 9141, pp. 349–359, 2015.
DOI: 10.1007/978-3-319-20472-7_38

little considering the background information. However, in the real world, some prior information can be learned from the network, which should be useful for us to identify the community structure.

Recently, many semi-supervised community detection algorithms have been proposed [6]-[9]. Ma el al proposed a semi-supervised method based on symmetric non-negative matrix, which incorporates the pairwise constraints into the adjacency matrix for finding the community structure [6]. A semi-supervised method based on the spin-glass model from statistical physics can integrate the prior information in forms of individual labels and pairwise constrains into community detection proposed by Eaton and Mansbach [7].Zhang [8] studied a semi-supervised learning framework which encodes pairwise constraints by modifying the adjacency matrix of network, which can also be regarded as de-noising the consensus matrix of community structures. Later, Zhang [9] added a logical inference step to utilize the must-link and cannot-link constraints fully. These algorithms use the prior information by transferring and modifying the adjacency matrix directly. After reconstructing the adjacency matrix, the semi-supervised problem is transformed into an unsupervised one [10].

Will the important nodes in the priors affect the result of community detection? We first extract the individual labels randomly, and later select the nodes in prior information concerning its importance. The centrality of nodes in networks can be assessed by degree, betweenness, closeness, eigenvector and so on [11-12].

In this paper, we propose a semi-supervised framework for community detection based on the NMF. One contribution of our framework is that it constructs a matrix by the positive and negative labels to more fully utilize the prior information. Another contribution is that we research the effect of important nodes in the priors on the community detection. This framework is applied to the artificial and real networks. The experimental results show that the framework can significantly improve the detection performance.

The remainder of this paper is organized as follows. Section 2 includes the review of basic problem formulation and notations used in our framework. In Section 3, we describe our semi-supervised community detection framework in detail. Experimental results on artificial and real-world networks are given in Section 4. Finally, a conclusion is presented in Section 5.

2   Problem Formulation and Notations

We first give the notation of network which will be used throughout the paper. A network can be modeled as a graph G= (V, E), where V is the node set and E is the edge set. The network structure is defined by a N×N adjacency matrix A. N is the number of nodes. If there is an edge between node i and j, we set the element A_{ij} to 1, otherwise to 0. We assume G is an undirected and unweighted graph. Self-connections are not allowed.

NMF was first introduced by Lee and Seung as a method for study the substructure of data matrix [13]. It was defined as the factorization of a non-negative matrix A into the multiplication of two other non-negative matrices U and V, where A is a $N×N$

matrix, U and V are $N \times K$ matrices, where K is the target number of communities to be detected in the network. In other words, NMF was aimed at mining the Euclidian distance between A and UV^T. The community structure can be inferred from V: node i belongs to the community k if V_{ik} is the largest element in the i-th row. We use the next objective (loss) function to quantify the quality of the factorization result. This function is based on the square loss function [14], which is equivalent to the square of the Frobenius norm of the difference between two matrices and is presented as follows.

$$L_{LSE}(A, UV^T) = ||A - UV^T||_F^2.$$ (1)

Lee and Seung [15] presented iterative updating algorithms to minimize to the objective function L_{LSE} as follows.

$$(U)_{ij} \leftarrow (U)_{ij} \frac{(AV)_{ij}}{(UV^TV)_{ij}}, (V)_{ij} \leftarrow (V)_{ij} \frac{(A^TU)_{ij}}{(VU^TU)_{ij}}.$$ (2)

The prior information contains individual labels and pairwise constraints and we use the former one in this paper. There are positive and negative labels in the individual labels. If a node does belong to a community, we call this positive label (PL), while if a node does not belong to a community, we regard it as the negative label (NL). The matrix O of size $N \times K$ is constructed from the background information, where N is the number of nodes and K is the target number of communities in the complex network. For any node i, we define i-th row of O as follows.

1. If node i has the PL, and i belongs to the j-th community, then

$$O_{ik} = \begin{cases} 1, & \text{if } k = j \\ 0, & \text{if } k \neq j \end{cases}, \text{ for k=1, 2,...K.}$$ (3)

2. If node i has the NL, and i does not belongs to the j-th community, then

$$O_{ik} = \begin{cases} 0, & \text{if } k = j \\ \frac{1}{K-1}, & \text{if } k \neq j \end{cases}, \text{ for k=1, 2,...K.}$$ (4)

3. If node i has no priors, then

$$O_{ik} = \frac{1}{K}, \text{ for k=1, 2,...K.}$$ (5)

In this paper, we use the normalized mutual information (NMI) to evaluate the performance of our framework on detecting the community structure [16]. This value can be formulated as follows. In Eq.6, R is the real community result and B is the found community result. In general, the larger the value of NMI, the better the partition of the network will be.

$$\text{NMI}(R, B) = \frac{-2 \sum_{i=1}^{cR} \sum_{j=1}^{cB} N_{ij} \log\left(\frac{N_{ij}N}{N_{i.}N_{.j}}\right)}{\sum_{i=1}^{cR} N_{i.} \log\left(\frac{N_{i.}}{N}\right) + \sum_{j=1}^{cB} N_{.j} \log\left(\frac{N_{.j}}{N}\right)}.$$ (6)

3   The Semi-supervised Community Detection Framework Based on NMF Using Individual Label

In this section, based on the individual labels discussed above, we first present the semi-supervised community detection framework which incorporates the prior information into the NMF objective function. Then we will see how the important nodes in prior information affect the result of community detection.

3.1   Description of Our Framework Based on NMF with Individual Labels

In this section, we propose the semi-supervised framework based on NMF which can make use of the individual labels to improve the performance of community detection. NMF can factorize a non-negative matrix A into the multiplication of two other non-negative matrices U and V, where A is an N×N matrix, both U and V are N×K matrices. We can infer the community structure in the network from V. In the i-th row, it is easy to known, if V_{ik} is the largest element, then node i belongs to the community k. If we have known that node i belongs to the community k, then we can enhance the value of V_{ik}, however, if that node i does not belong to the community k, then the value of V_{ik} will be punished.

To use the individual labels to improve the result, we denote the new representation of V where the matrix O summarized from the individual labels are used as a multiplication factor in Eq.7. In this paper, the sign of operation $\otimes$ indicates the dot product.

$$d(O, V) = O \otimes V. \tag{7}$$

For NMF there is an interesting fact that the estimates are always scale invariant. For example, we add a multiplication factor c to U and the other factor $\frac{1}{c}$ to V to get different U and V. The product UV^T will not change. Although there is no explicit control for NMF, standard NMF tends to estimate sparse components. The factorized matrices are obtained through minimizing an objective function defined in Eq.8.

$$\min_{U \geq 0, V \geq 0} ||A - UV^T||^2 + \lambda_s \sum_{k=1}^{K} ||V_k||_1. \tag{8}$$

In the formula (5), the parameter $\lambda_s \geq 0$. Adding penalties to NMF is a common strategy since they not only improve the interpretability, but also improve numerical stability of the estimation by making the NMF optimization less under constrained. The assessment algorithm for the penalized NMF is studied in many papers. The main iteration rule in our work is presented as follows. And the parameter $\ddot{e}_s$ is set to 1 in the experiments.

The algorithm of Sparse NMF with individual labels is described.

```
program Inflation
  const λs = 1;
  begin
    construct O;
    initialize {U, V}, positive random matrices;
    repeat
```
$$\text{set } (U)_{ij} \leftarrow (U)_{ij} \frac{(AV)_{ij}}{(UV^TV)_{ij}};$$

$$\text{set } (V)_{ij} \leftarrow (V)_{ij} \frac{(A^TU)_{ij}}{(VU^TU)_{ij}+\lambda_s};$$
```
    V=O⊗V;
    normalization of U,V;
    until convergence
end
```

3.2   The Evaluation of the Important Nodes

In the research of the social network, many methods have such a hypothesis, namely the importance of node is equivalent to its connection with other nodes, which makes the node significant. The basic idea of these methods is the importance of difference between different nodes in the network is obtained by some useful information, such as the degree of node, the shortest path, the weight of nodes and edges.

The proposed indexes of important nodes mainly can be divided into centrality and prestige. Measurement methods mainly include the node degree, betweenness, closeness, eigenvector and so on. In this paper, we use the degree and betweenness of nodes. A brief introduction about these two methods will be presented in the following.

The degree of nodes refers to the number of edges connected to the node in the network. The size of degree can reflect the importance of nodes to a certain extent. The larger the degree of node, the more important the node may be, because it may be located in the center of the network.

Betweenness was first put forward for measuring the individual's social status in the study of social network in 1977 by Freeman [17]. The betweenness of node u refers to all the shortest paths in the network through the node u. We define the set of shortest path between nodes i and j as s_{ij}, and the betweenness formula of u after normalization is presented as follows.

$$B_u = \sum_{i,j} \frac{\sum_{l \in s_{ij}} \delta_l^u}{|s_{ij}|}. \tag{9}$$

The symbol $\sum_{l \in s_{ij}} \delta_l^u$ is the number of shortest paths through node u.

The size of betweenness can reflect the importance of nodes in a way. The larger betweenness of the node, the more important the node is. The betweenness is useful for us to find the important nodes with large flow.

According to the importance of nodes, we select some prior information on purpose in the following experiments. The detail description about experiments will be presented in section 4.

4   Experiment and Discussion

In this section, we design a set of experiments, whose data set are LFR artificial networks [18] and real-world networks including Amazon's network of political books [19], the network of blogs about US politics [20] and adjacency network of common adjectives and nouns in the novel David Copperfield by Charles Dickens [21]. The normalized mutual information (NMI) is used to evaluate the performance of detecting communities with our framework, which is discussed above. The closer to 1 the NMI, the better the partition of the network will be.

4.1   Artificial Networks

In this subsection, the experiments include evaluating the performance of the framework with different percentage of priors and measuring the ability of the framework to detect communities with different important nodes in the background information.

The LFR benchmark network is an artificial network for community detection, which is claimed to possess some basic statistical properties found in real networks. The generator of LFR allows us to specify the number of nodes (N), average degree (k), maximum degree (maxk), exponent for the degree sequence (t1), exponent for the community size distribution (t2), minimum for the community sizes (minc), maximum for the community sizes (maxc) and the mixing parameter (u). In LFR, both community size and degree distributions are power laws, where vertices and communities are generated by sampling. With the increase of u, the structure of network becomes vague, and the detection of communities becomes more challenging.

In this paper, we set the number of nodes to 1000, the minimum community size to 10 and 20, the maximum community size to five times the minimum community size, the average degree to 20, the exponent of the vertex degree and community size to -2 and -1, respectively, and the mixing parameter to different values 0.7 and 0.8.

The percentage of the labeled nodes in the network is an important factor in the experiments. To fully use the individual labels, we construct the matrix O and incorporate it into the updating process of NMF, The average performance of our framework based on different percentage of the used priors of half positive labels and half negative labels is displayed in Fig. 1. There is a positive correlation between NMI and the used priors. There are abnormal points in the first row and the second column picture, where the value of NMI decreases when the used priors is 1%. The NMI of the standard NMF is the NMI where the used priors are 0. Compared with standard NMF, the NMIs of our framework are higher. Obviously, when u becomes 0.8, the result of community detection by our framework turns to be weak. There is a question that the value of NMI is below to 1 when the used priors are 100% in Fig.1. In Fig.1, the priors are half positive labels and half negative labels, so when the used priors are 100%, there are some fuzzy labels in the prior information, so the result of community detection is not exactly the same as the real one.

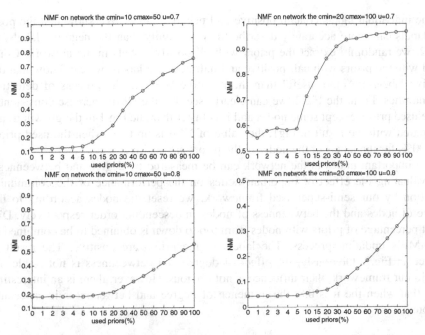

Fig. 1. Performance of our framework in the terms of NMI as a function of the percentage of priors with half positive and negative labels added on LFR networks

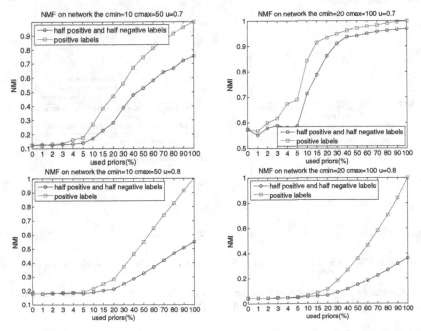

Fig. 2. Performance of our framework in the terms of NMI as a function of the percentage of priors with positive labels added on LFR networks

The individual labels contain positive and negative labels. We know that the positive labels can more accurately describe the community than the negative labels. In Fig. 2, we randomly extract the priors with all positive labels in the network. Compared with the priors with half positive and half negative labels, we can know that the positive labels are more useful than the negative labels in the process of detecting communities. From the Fig.2, we can clearly see that the NMIs increase consistently as the used priors except some nodes and it is faster than the Fig 1 in the growth trend. Compared with the result in Fig1, the value of NMI is up to 1, when the used priors are 100%, for the labels of the nodes in the priors are positive.

The important nodes in the network can be measured by degree and betweenness. To evaluating the effect of important nodes on the performance of the community detection by our semi-supervised framework, we reset the nodes according to the degree of nodes and the betweenness of nodes in descending order respectively. Different percentage of priors with nodes from top to down is obtained to be combined to the NMF's updating process. The labels in the priors are positive. The result is showed in Fig.3. Obviously, the effect of degree and betweenness is not stable. At least in our framework, their influence is not obvious. However, there is an interesting thing that when the u is 0.8, the influence of degree and betweenness is especially obvious.

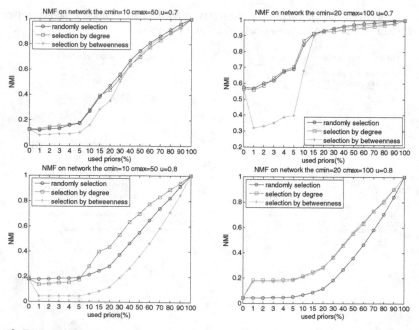

Fig. 3. Performance of our framework in the terms of NMI as a function of the percentage of priors with top degree and betweenness added on LFR networks

In summary, incorporating the individual labels into the process of NMF updating process can effectively improve the performance of the community detection in the

LFR network, especially the positive labels. However, the effect of important nodes is not obvious.

4.2   Real-World Networks

In this subsection, we test our framework with real-world networks, Amazon's network of political books, the network of blogs about US politics and the adjacency network of common adjectives and nouns in the novel David Copperfield by Charles Dickens. Firstly, we will give the data description, and then the experiments' performance will be presented.

The network of Amazon's political books contains 105 books on US politics and 441 edges. The nodes are books sold by bookseller Amazon, which have been manually given labels as "liberal", "neutral", or "conservative". Edges represent co-purchasing of books. The network of blogs about US politics consists of 1490 nodes and is treated as an undirected network in this paper. The nodes in the network are divided into "liberal" and "conservation" according to the content in the blogs, which represent the blogs and the edges in the network represent that a URL presented on the page of one blog references another political blog. If there is reference relationship between two blogs, the edge between blogs forms. There are 112 vertices and 850 edges in the adjacency network of common adjectives and nouns in the novel David Copperfield by Charles Dickens where the vertices represent common adjectives and nouns and the edges connect any two words that appear adjacent to one another at any point in the book.

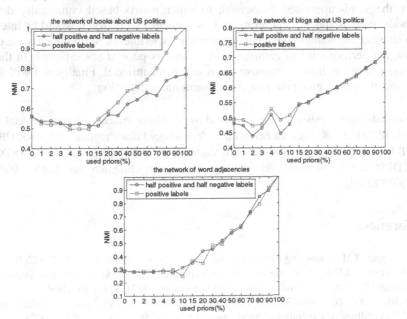

Fig. 4. Performance of our framework in the terms of NMI as a function of the percentage of priors with top degree and betweenness added on the real-world networks

Applying our proposed framework to these real-world networks, the result of the community detection is shown in Fig.4. The performance of our framework on the real-world networks is consistent with that on the LFR network. However, the result of the network of blogs about US politics is abnormal and when the used priors is set 100%, the NMI is less than 1. In Fig.4b and Fig.4c the two lines overlap. There are two communities in these two real networks, which is known previously, based on the construction of O discussed above, we can find that if node i does not belong to one community, it must belong to the other one, then there is no difference between the positive labels and negative labels because we can construct the same O at node i. Further, the percentage of priors is important to the result of community detection.

5   Conclusions

In this paper, a semi-supervised community detection method based on NMF with individual labels is proposed. Unlike previous works which transfer the individual labels into the adjacency matrix, we formulate it into the objective function and incorporate it into the process of NMF updating. As can be seen from the extensive experiments on both artificial and real networks that using the individual labels can significantly improve performance, especially in the situation where the individual labels are positive. Moreover, we extract some important nodes with large degree and betweenness into the priors and the effect of these nodes on the community detection is not obvious.

A number of improvements of our framework may be possible. Firstly, we hope to apply the semi-supervised framework to other matrix-based community detection methods, such as spectral clustering and its variants. Secondly, it would be interesting to investigate the abnormal phenomenon in the experiments. With the increase of used priors, the performance of community detection is poor at some points. In this case the research about how to improve the result is meaningful. Finally, we will investigate how the priors guide the process of community detection.

Acknowledgments. This work was supported by the Major Project of National Social Science Fund (14ZDB153), the National Science and Technology Pillar Program (2013BAK02B06 and 2015BAL05B02), Tianjin Science and Technology Pillar Program (13ZCZDGX01099, 13ZCDZSF02700), National Science and Technology Program for Public Well-being (2012GS120302).

References

1. Strogatz, S.H.: Exploring complex networks. J. Nature **410**(6825), 268–276 (2001)
2. Newman, M.E.J.: Detecting community structure in networks. J. The European Physical Journal B-Condensed Matter and Complex Systems **38**(2), 321–330 (2004)
3. Girvan, M., Newman, M.E.J.: Community structure in social and biological networks. J Proceedings of the National Academy of Sciences **99**(12), 7821–7826 (2002)
4. Newman, M.E.J.: Modularity and community structure in networks. J Proceedings of the National Academy of Sciences **103**(23), 8577–8582 (2006)

5. Karrer, B., Newman, M.E.J.: Stochastic blockmodels and community structure in networks. J. Physical Review E **83**(1), 016107 (2011)
6. Ma, X., Gao, L., Yong, X., et al.: Semi-supervised clustering algorithm for community structure detection in complex networks. J Physica A: Statistical Mechanics and its Applications **389**(1), 187–197 (2010)
7. Eaton, E., Mansbach, R.: A Spin-Glass Model for Semi-Supervised Community Detection. AAAI (2012)
8. Zhang, Z.Y.: Community structure detection in complex networks with partial background information. J. EPL (Euro physics Letters) **101**(4), 48005 (2013)
9. Zhang, Z.Y., Sun, K.D., Wang, S.Q.: Enhanced community structure detection in complex networks with partial background information. J. Scientific reports (2013)
10. Yang, L., Cao, X., Jin, D., et al.: A Unified Semi-Supervised Community Detection Framework Using Latent Space Graph Regularization. J (2014)
11. Nan, H., Wen-Yan, G.: Evaluate nodes importance in the network using data field theory. In: International Conference on Convergence Information Technology, pp. 1225–1234. IEEE (2007)
12. Freeman, L.C.: Centrality in social networks conceptual clarification. J. Social networks **1**(3), 215–239 (1979)
13. Lee, D.D., Seung, H.S.: Learning the parts of objects by non-negative matrix factorization. J. Nature **401**(6755), 788–791 (1999)
14. Wang, R.S., Zhang, S., Wang, Y., et al.: Clustering complex networks and biological networks by non-negative matrix factorization with various similarity measures. J. Neurocomputing **72**(1), 134–141 (2008)
15. Lee, D.D., Seung, H.S.: Algorithms for non-negative matrix factorization. Advances in neural information processing systems, 556–562 (2001)
16. Zhong, S., Ghosh, J.: Generative model-based document clustering: a comparative study. J. Knowledge and Information Systems **8**(3), 374–384 (2005)
17. Freeman, L.C.: A set of measures of centrality based on betweeness. J. Sociometry, 35–41 (1977)
18. Lancichinetti, A., Fortunato, S., Radicchi, F.: Benchmark graphs for testing community detection algorithms. J. Physical review E **78**(4), 046110 (2008)
19. Newman, M.E.J.: Modularity and community structure in networks. J Proceedings of the National Academy of Sciences **103**(23), 8577–8582 (2006)
20. Adamic, L.A., Glance, N.: The political blogosphere and the 2004 US election: divided they blog. In: Proceedings of the 3rd international workshop on Link discovery, pp. 36–43. ACM (2005)
21. Newman, M.E.J.: Finding community structure in networks using the eigenvectors of matrices. J. Physical review E **74**(3), 036104 (2006)

Machine Learning Interpretation of Conventional Well Logs in Crystalline Rocks

Ahmed Amara Konaté[1(✉)], Heping Pan[1(✉)], Muhammad Adnan Khalid[2],
Gang Li[1], Jie Huai Yang[1], Chengxiang Deng[1], and Sinan Fang[1]

[1] Institute of Geophysics and Geomatics, China University of Geosciences, Wuhan, China
konate77@yahoo.fr, panpinge@163.com
[2] Institute of Automation, Chinese Academy of Sciences, Beijing, China

Abstract. The identification of lithologies is a crucial task in continental scientific drilling research. In fact, in complex geological situations such as crystalline rocks, more complex nonlinear functional behaviors exist in well log interpretation/classification purposes; thus posing challenges in accurate identification of lithology using geophysical log data in the context of crystalline rocks. The aim of this work is to explore the capability of k-nearest neighbors classifier and to demonstrate its performance in comparison with other classifiers in the context of crystalline rocks. The results show that best classifier was neural network followed by support vector machine and k-nearest neighbors. These intelligence machine learning methods appear to be promising in recognizing lithology and can be a very useful tool to facilitate the task of geophysicists allowing them to quickly get the nature of all the geological units during exploration phase.

Keywords: k-nearest neighbors · Support vector machine · Artificial neural network · Machine learning · Pattern recognition · Conventional well log · Geophysical exploration

1 Introduction

Geophysical well logging (GWL) involves the monitoring and evaluation of drilled rock from the earth's crust. The measurements generally fall into three groups namely; nuclear, electrical and acoustic. GWL provides continuous records on the composition and structural features of the penetrated rock. In this context, lithology and rock properties can be estimated. GWL is largely used for reservoir evaluation, and lithological identification in sedimentary rocks. As a result, log responses in sedimentary rocks are well known; even though, this is not the case for crystal-line rocks ([2]). Compared with sedimentary rocks, crystalline rocks are more di-verse, with more complex compositions, textures, and structures, ensuing in various challenges in their lithological identification and prediction. Although most petroleum downhole logging technique maybe utilized in crystalline rocks; however, interpretation are necessary because of environmental differences.

© Springer International Publishing Switzerland 2015
Y. Tan et al. (Eds.): ICSI-CCI 2015, Part II, LNCS 9141, pp. 360–370, 2015.
DOI: 10.1007/978-3-319-20472-7_39

Statistical cross-plotting have been used on geophysical log data classification in the context of crystalline rocks. Examples could be found in Pechnig et al.([11]) ; Niu et al. ([9]); Luo and Pan ([7]) where the various authors revealed the application of cross plotting in crystalline rocks. While cross-plotting can be utilized to explore relationships between two logs, nevertheless it is unable to clearly reveal with understanding the relationships that may be in the whole data ([6]). Another problem in this area is that the cross-plots approach of log interpretation knowledge through the geophysical log specialist has proven to be both time-consuming and error-prone ([1], [13]). In other words, converting geophysical log data into lithological units continue to be difficult and not automatically solvable using conventional techniques ([19]). This reality has been an important motivation for using machine learning techniques.

Automating well log interpretation provides two immediate benefits: firstly, it makes it possible to process large amounts of logs rapidly, thus supporting the creation of large knowledge bases. Secondly, automation produces uniform results. Since the early days of the introduction of computers to geosciences, algorithms combined with geophysicist reasoning have made a significant contribution to the field of geophysics (e.g.,[20],[21],[22],[23]). The most important of these are machine learning methods. Machine learning methods have been recently propose for conventional well log data classification in the context of crystalline rocks (e.g.,[8],[10],[15]). Nonetheless this study propose k-nearest neighbors (k-NN) as an alternative to classify conventional well log data. It is pertinent to state that k-NN has been applied successfully to various logging data problems (e.g.,[1],[5]) Nevertheless, so far this has not yet been widely used in the context of crystalline rocks. Especially within the study area of this research. Therefore, the aim of this work is to explore the capability of k-NN classifier and to demonstrate its performance compare to other classifiers such as artificial neural network (ANN) and support vector machine (SVM).

This study highlights an additional contribution of machine learning methods as the best possible way of utilizing conventional well logs to make decision automatically in terms of lithology recognition task in structurally complex, crystalline metamorphic terrain. The results show that these intelligence machine learning methods appear to be promising in recognizing lithology and can be a very useful tool to facilitate the task of geophysicists allowing them to quickly get the nature of all the geological units during exploration phase in realistic crystalline environments.

2   Data

Scientific holes are drilled and cored precisely in order to be able to determine the local geological situation of the drilled area. It provides unique and vital data and samples for a very wide range of geological problems. The 5158m deep main borehole of the Chinese continental scientific drilling project (CCSD-MH) penetrated five main lithological units: orthogneiss, paragrneiss, eclogite, amphibolite and ultramafic rocks in the Dabie-Sulu ultrahigh pressure metamorphic (UHPM) terrane in eastern part of China. The Dabie-Sulu UHPM belt is one of the largest belt in the world ([24]). A description of the CCSD project, its geological and geophysical logging implications may be found in Ji and Xu [16].

The experimentation consists of recognizing the lithology from conventional well logs using machine learning methods. Experiments were performed with 38937 samples from the interval depth 100-5000 m of CCSD-MH. For each sample there are 10 inputs variables (=log curve). There are: CNL (Compensated Neutron), DEN (Compensated Bulk Density), PE (photoelectric absorption capture cross section), GR (gamma ray), K (potassium content), KTH (Potassium Plus Thorium), TH (Thorium), U (Uranium content), RD (resistivity of laterolog deep)and RSFL (Spherically Focused Resistivity). These log curves have been proved to be tied directly to mineralogical changes in crystalline metamorphic rocks ([10]).Table 1 shows the description of the log curves used in this study.

Table 1. The description of the log curves used in this study

NO	Logs	Unit	Tool type Description
1	RD	ohm-m	electrical resistivity
2	RSFL	ohm-m	electrical resistivity
3	DEN	g/cm^3	Gamma
4	CNL	p.u.	Neutron
5	PE	b/e	Gamma
6	GR	API	Gamma
7	K	%	Spectral Gamma Ray
8	KTH	API	Spectral Gamma Ray
9	TH	ppm	Spectral Gamma Ray
10	U	ppm	Spectral Gamma Ray

Machine learning methods (classifiers) such as k-NN, SVM, ANN being supervised, the output was obtained according to the core description that is, orthogneiss, paragneiss, eclogite, amphibolite and ultramafic rocks. Each classifier has certain parameters the values of which need to be fixed appropriately for controlling undertraining and overtraining hence generalization. That is, number of neighbor k for k-NN, penalty C and radial basis function kernel spread σ for SVM and hidden nodes NHN for ANN. So with this in mind, the dataset was divided in to two subdata D1 (27773 samples) and D2 (11164 samples). D1 subdata was used for finding the optimal parameters through 3-fold cross validation ([12]) in which the data was separated in to 3 different sets. Then, the model is trained using 2 of these sets, and its performance is tested on the third. This is repeated 3 times so that each set in turn is utilized as the test set. Therefore, the performance for the given classifier is finally achieved as the aver-age performance on the test sets. Table 2 shows the best performance of cross validation. After this, each classifier was tested on D2 which has never been used during the optimization process. The key objective of learning a classifier is to build model with good generalization capability. Generalization is a central issue for the development of mathematical and statistical models. It refers to the ability of a model to accurately represent the underlying data generation process, rather than the noise features of the training data ([26]). Beside, to make sure that each variable is treated similarly in the model, the dataset were scaled in [0, 1].

Table 2. The best performance of cross validation

Classifier	Parameters	Cross validation accuracy (%)
k-NN	k= 47	71.20
SVM	C=7; σ =0.4	74.60
BPNN(One hidden layer)	NHN= 26	87.32

3   Methods

3.1   k-NN

In this section, a brief introduction of k-NN that is needed for the definition of their functionality is presented. A more comprehensive presentation can be found in Fukunaga ([4]) and Cheriet et al. ([3]). In classification problems holding moderately high dimensional input space without priori statistical knowledge of objects being classified, the k-NN method offers simplicity and relatively high convergence speed. k-NN classifier finds the k nearest neighbours of the sample x, and then, through a voting process, it classifies x in the class which has most representatives among those k nearest neighbours. Action of k-NN classifier is given in Fig 1.

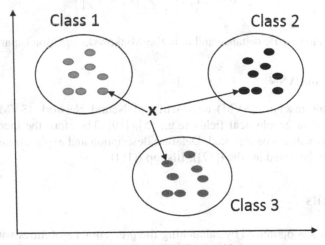

Fig. 1. Action of k-NN classifier

When the number of samples in training dataset is large, this algorithm shows to be very efficient, for reducing the misclassification error. However, the classification times becomes longer. Another benefit of the k-NN as compared to other supervised learning such as SVM, decision tree, and ANN is that it can simply deal with problems in which the class size is higher ([14]). Generally, k-NN classification accuracy mainly is influenced by k value and the type of distance metrics used for computing nearest distance. Therefore, in this study we have determined the optimal k value and distance metric using resampling technique on D1. After evaluating these best possible values of k and distance metric, a *k*-NN classifier was tested for metamorphic rocks prediction on the independent subdata D2. The distance metrics which were experimented in this study were Euclidean, City Block and Minkowski distance.

Let assume that p as the number of variables in the dataset, distance between the points x_i and x_j is defined as a follows:

Euclidean distance

$$d\left(x_i, x_j\right) = \sqrt{\sum_{k1}^{p} \left(x_{ik} - x_{jk}\right)^2} \tag{1}$$

City block distance

$$d\left(x_i, x_j\right) = \sum_{k=1}^{p} \left|x_{ik} - x_{jk}\right| \tag{2}$$

Minkowski distance

$$d\left(x_i, x_j\right) = \left(\sum_{k=1}^{p} \left|x_{ik} - x_{jk}\right|\right)^{\frac{1}{m}} \tag{3}$$

Where *d* is the metric distance and *m* in the Minkowski equation is parameter

3.2   SVM and ANN

Support vector machine (SVM) and Artificial Neural Network (SVM) have been widely applied in geophysical fields (e.g., [5], [19]).Therefore the theory aspect of these methods will not be repeated. Detailed description and explanations about SVM and ANN may be found in Abe ([17]); Bishop ([18]).

4   Results

The choice of k is optimized by calculating the prediction capabilities with different k values using 3 fold cross validation technique on D1. *k* value from 1 to 47 were evaluated. Fig 2 shows the performance of k value vs. cross-validation accuracy. From Fig 2, it was found that in general the accuracy rate increase with the increase of *k* value. This is because larger value of *k* reduce the effect of noise on classification. However, this can be time consuming in context of a large number of samples as we remarked in this study.   Moreover, a large amount of memory was required to run the *k*-NN classifier.

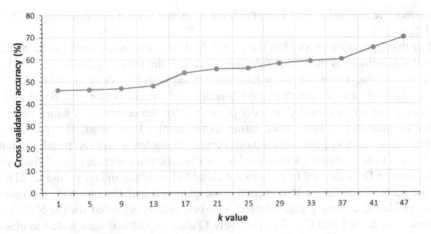

Fig. 2. Performance of k value vs. cross-validation accuracy

The highest accuracy was reached when $k = 47$ using Euclidean distance. A correct classification rate of 71.37% were obtained on D2. Figure 3 shows the performance of k-NN into each category of metamorphic rocks using D2.

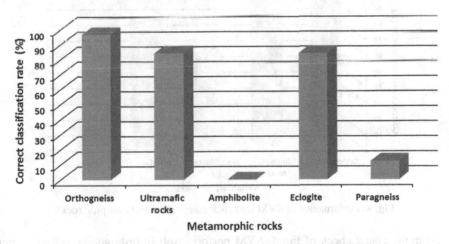

Fig. 3. Performance of k-NN into each category of metamorphic rocks

From Fig 3, k-NN show poor outcome in amphibolite and paragneiss rocks recognition. Because it exhibits a classification rate of 0% for amphibolite and 12.17% for paragneiss rocks. In contrast, under the circumstances of orthogneiss, eclogite and ultramafic rocks recognition k-NN classification approaches demonstrated high performance. A classification rate of 96.85%, 84.11% and 84.37% were obtained respectively. In summary, the results have clearly shown the potential of k-NN for solving lithology identification problems in crystalline rocks. That is, the k-NN classifier performs well in the case of orthogneiss, eclogite and ultramafic rocks recognition. However, it is important to highlight that k-NN classifier is an expensive method in terms of memory requirements and computational demands during classification of CCSD-MH data.

For the application of SVM to the CCSD-MH data, the radial basis function (RBF) kernel was utilized as the kernel function, since it is the most popularly used in classification purpose. There are two parameters related with the SVM: σ and C. The first one is the RBF spread, which plays a fundamental role in performing SVM and the last one is the upper bound C, penalty parameter for the error terms, which has to be added in order to take into consideration those samples that cannot be separated. For optimal classification performance both kernel parameter, σ and the penalty parameter, C have to be optimized with regard to the classification accuracy. So, 3 fold cross-validation on D1 was used to explore the suitable kernel parameter σ, and C. Mainly, all the pairs of (σ, C) for RBF kernel were explored and the one with the optimal cross-validation accuracy was selected. The best results achieved for the SVM parameters were σ= 0.4 and C = 7 respectively (Table 2). D2 was then tested to classify metamorphic rocks. A correct classification rate of SVM on D2 was 74.77%. Fig 4 shows the performance of SVM into each category of metamorphic rocks from D2.

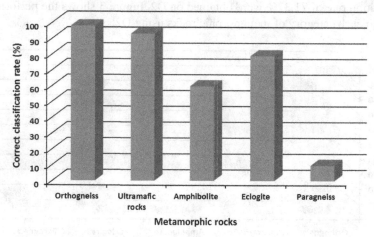

Fig. 4. Performance of SVM into each category of metamorphic rocks

From the visual check of Fig 4, SVM perform well in orthogneiss, eclogites, amphibolite and ultramafic rocks recognition. The SVM exhibits a classification rate of 97.76% for orthogneiss, 78.70 % for eclogite, and 59. 54 % for amphibolite and 92.82% for ultramafic rocks. SVM shows poor performance in paragneiss rocks identification and classification that is SVM shows 9.59 % of correct recognition rate. In summary, it can be stated that, the experimentation on CCSD-MH data show that SVM can be used as a classifier, in order to provide satisfactory recognition of metamorphic rocks.

For the application of ANN, a feed forward back-propagation neural network (BPNN) with a single hidden layer and 5 output nodes was used. Geophysical log set are used as input of BPNN to train the classifier. There are many algorithms available to train the BPNN ([18]). In this study, scaled conjugate gradient algorithm developed by (Moller [25]) was used.

One factor that determines the performance of the ANN is its size (number of hidden nodes). It plays a critical role in performance of BPNN. Hence, inappropriate selection of this parameter may cause over-fitting or under-fitting problem. Therefore, an optimal number of hidden nodes (NHN) for the classifier that can accurately classify metamorphic rocks needs to be determined. In this study, 3-fold cross-validation was applied on D1 to investigate the appropriate NHN. BPN's with their number of hidden nodes between from 4 to 50 were evaluated to obtain the best architecture. Each of the BPN's was trained for up to 10000 training epochs with the stopping condition of 0.000001 applied. The best test result among all the BPNN architectures and all the training epochs examined were selected. Fig 5. shows the best classification performance using 3-fold cross-validation.

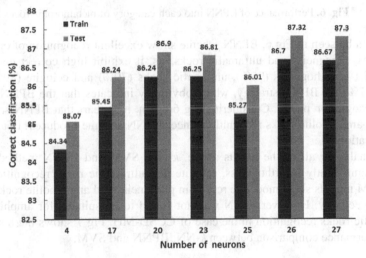

Fig. 5. The best Performance of BPNN using 3 cross- validation

From Fig 5, in terms of correct classification accuracy model (training) BPNN [10-26-5] scheme provides best results; since BPNN [10-26-5] scheme show 86.70%, highest correct classification rate. Also in terms of correct classification accuracy (testing), BPNN [10-26-5] shows better generalization predictiveness on unseen data ability; because, it exhibits 87.32% highest value of correct classification rate. The aim of ANN is to generalize effectively. Generalization gives us a more convincing estimate of the validity of the ANN. Therefore, generalization is taken as the analytical accuracy using BPNN in this study. So, the best model was achieved with the number of neuron, NHN= 26. After successful training and testing using 3 fold cross validation, the BPNN [10-26-5] scheme were also tested on D2. A correct classification rate of 82.21% were obtained on D2. Fig 6. displays the test on D2; and visually illustrate the correct classification rates for individual metamorphic rock type's values, thus providing a visual exhibition of accuracy.

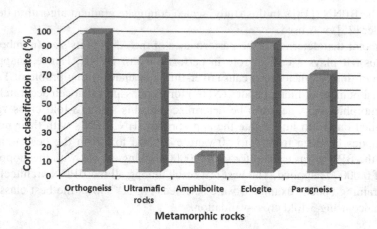

Fig. 6. Performance of BPNN into each category of metamorphic rocks

As can be seen in Fig 6, BPNN scheme show excellent recognition of orthogneiss, ecologies, paragneiss and ultramafic rocks, as, it exhibit high correct classification rate. That is orthogneiss (95%), ultramafic rocks (79%) and eclogite (89%), paragneiss (67%) for BPNN strategy, which obviously indicates that the BPNN has these rocks recognition power. Considering Fig 6 again, it appears that BPNN scheme suffered in amphibolite rocks recognition since, BPNN scheme produces11 % of correct classification rates.

By analyzing all of the results above, k-NN, SVM and BPNN methods can improve significantly in orthogneiss, eclogite and ultramafic rocks recognition. BPNN and SVM models show moderate results in paragneiss and amphibolite rocks recognition respectively. However, k-NN was not found to be suitable for amphibolite and ultramafic rocks recognition in the case of CCSD-MH. Fig 7 shows the histogram of the performance comparison between k-NN, BPNN and SVM.

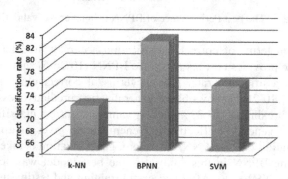

Fig. 7. Performance comparison between k-NN, BPNN and SVM

From Fig 7, it can observed that the different classifiers based on CCSD-MH data can be effective tools for the geophysical log data classification purpose. However, the best results were achieved by BPNN, followed by SVM and then k-NN.

This because BPNN showed 82.21% followed by SVM (74.77%) and k-NN (71.37%). This performance of BPNN has also been confirmed by Pan et al. ([10]). Based on results obtained, it can fairly be stated that with large amount of sample data, BPNN is able to give satisfactory results thus achieving generalization. This has further confirm the assertion by many researchers that to achieve generalization more samples should be considered, so that the misclassification rate of training and testing can be reduced drastically.

5   Conclusion

This study has investigated the performance of supervised classifier such as BPNN, k-NN and SVM in geophysical log data classification in the context of crystalline metamorphic rocks. From the experimental results in this study, the consensuses of all methods are helpful for the lithology classification using geophysical well logs from crystalline rocks. Nonetheless, BPNN was the best among all the classification algorithms investigated. It is important to note that k-NN classifier was an expensive method in terms of memory requirements and computational demands during classification of CCSD-MH data. The present study was done using a single deep well. However, it gets more interesting in applying these intelligence machine learning methods to other geologically complex zone of interest. We look forward to see more of this scenario in future research works.

References

1. Amirgaliev, E., Isabaev, Z., Iskakov, S., Kuchin, Y., et al.: Recognition of rocks at uranium deposits by using a few methods of machine Learning. Soft computing in machine learning. Advances in intelligent systems and computing **273**, 33–40 (2014)
2. Bartetzko, A., Delius, H., Pechnig, R.: Effect of compositional and structural variations on log responses of igneous and metamorphic rocks. I: mafic rocks. In: Harvey, P.K., Brewer, T.S., Pezard, P.A., Petrov, V.A. (eds.) Petrophysical Properties of Crystalline Rocks, pp. 255–278. Geological Society, London, Special Publications (2005)
3. Cheriet, M., Kharma, N., Liu, C.L., Suen, C.Y.: Character recognition systems: a guide for students and practioners. Published by John Wiley & Sons Inc, Hoboken (2007)
4. Fukunaga, K.: Introduction to Statistical Pattern Recognition. Academic Press Professional Inc, San Diego (1990)
5. Gelfort, R.: On classification of logging data. Ph.D. thesis, Clausthal University of Technology, Germany (2006)
6. Kassenaar, J.D.C.: An application of principal components analysis to borehole geophysical data. In: Proceedings of the Fourth International Symposium on Borehole Geophysics for Minerals, Geotechnical and Groundwater Applications, Toronto, Ontario, pp. 211–218 (1991)
7. Luo, M., Pan, H.P.: Well Logging Responses of UHP Metamorphic Rocks from CCSD Main Hole in Sulu Terrane, Eastern Central China. Journal of Earth Science **21**(3), 347–357 (2010)

8. Maiti, S., Tiwari, R.K.: A hybrid Monte Carlo method based artificial neural networks approach for rock boundaries identification: a case study from KTB Borehole. Pure Appl Geophysics **166**, 2059–2090 (2009)

9. Niu, X.Y., Pan, H.P., Wang, W.X., Zhu, L.F., Xu, D.H.: Geophysical Well Logging in Main Hole (0–2 000 m) of Chinese Continental Scientific Drilling. Acta petrologica Sinica **20**(1), 109–118 (2004)

10. Pan, H.P., Luo, M., Zhao, Y.G.: Identification of metamorphic rocks in the CCSD main hole. In: IEEE Sixth International Conference on Natural Computation (ICNC), pp. 4049–4051 (2010)

11. Pechnig, R., Delius, H., Bartetzko, A.: Effect of compositional variations on log responses of igneous and metamorphic rocks. II: acid and intermediate rocks. In: Harvey, P.K., Brewer, T.S., Pezard, P.A., Petrov, V.A. (eds.) Petrophysical Properties of Crystalline Rocks, pp. 279–300. Geological Society, London, Special Publications (2005)

12. Stone, M.: Cross-Validatory Choice and Assessment of Statistical Predictions. Journal of the Royal Statistical Society. **36**(2), 111–147 (1974)

13. Saggaf, M.M., Nebrija, E.L.: Estimation of Lithologies and Depositional Facies from Wire-Line. Logs AAPG Bulletin. **84**, 1633–1646 (2000)

14. Yazdani, A., Ebrahimi, T., Hoffmann, U.: Classification of EEG signals using dempster shafer theory and a K-nearest neighbor classifier. In: Proc of the 4th int IEEE EMBS Conf on Neural Engineering, pp. 327–30 (2009)

15. Bosch, D., Ledo, J., Queralt, P.: Fuzzy Logic Determination of Lithologies from Well Log Data: Application to the KTB Project Data set. Surveys in Geophysics **34**(4), 413–439 (2009)

16. Ji, S.C., Xu, Z.Q.: Drilling deep into the ultrahigh pressure (UHP) metamorphic terrane. Tectonophysics **475**, 201–203 (2009)

17. Abe, S.: Support Vector Machines for Pattern Classification, Advances in Pattern Recognition, 2nd edn. Springer (2010). doi:10.1007/978-1-84996-098-4_7

18. Bishop, C.M.: Neural Networks for Pattern Recognition. Oxford Press (1995)

19. Ehret, B.: Pattern recognition of geophysical Data. Geoderma, 111–125 (2010)

20. Sandham, W., Leggett, M. (eds.): Geophysical Applications of Artificial Neural Networks and Fuzzy Logic. Series: Modern Approaches in Geophysics (2003)

21. Li, Y., Bian, Z., Yan, P., Chang, T.: Pattern recognition in geophysical signal processing and Interpretation. Handbook of Pattern Recognition and Computer Vision, 511–539 (1993)

22. Aminzadeh, F. (ed): Handbook of Geophysical Exploration: Section I. Seismic Exploration, 20, Pattern Recognition & Image Processing, Geophysical Press, London (1987)

23. Palaz, I., Sengupta, S.K. (eds.) Automated Pattern Analysis in Petroleum Exploration. Springer New York (1992)

24. Xu, Z.Q., Wang, Q., Tang, Z., Chen, F.: Fabric kinematics of the ultrahigh-pressure metamorphic rocks from the main borehole of the Chinese Continental Scientific Drilling Project: Implications for continental subduction and exhumation. Tectonophysics **475**, 235–250 (2009)

25. Moller, M.F.: A scaled conjugate gradient algorithm for fast supervised learning. Neural Networks. **6**, 525–533 (1993)

26. May, R.J., Maier, H.R., Dandy, G.C.: Data splitting for artificial neural networks using SOM- based stratified sampling. Neural Networks **23**(2), 283–294 (2010)

Pre-Scaling Anisotropic Orthogonal Procrustes Analysis Based on Gradient Descent over Matrix Manifold

Peng Zhang[✉], Zhou Sun, Chunbo Fan, and Yi Ding

Data Center, National Disaster Reduction Center of China, Beijing, People's Republic of China
{zhangpeng,sunzhou,fanchunbo,dingyi}@ndrcc.gov.cn

Abstract. This paper proposes a pre-scaling extension of the Orthogonal Procrustes Analysis (OPA), where anisotropic scaling occurs before rigid motion. We propose an efficient algorithm to solve this problem based on gradient descent method over matrix manifold. We show that the proposed algorithm is monotonically convergent and provide an acceleration procedure. Its performance is validated through a series of numerical simulations.

Keywords: Procrustes analysis · Gradient descent · Stiefel manifold · Pre-scaling

1 Introduction

Orthogonal Procrustes analysis (OPA), named by Hurley and Cattel [1], is an important technique to compare the similarity between two sets of observations under rigid motion. It is originally designed to assess the embedding learned by MDS [2,3,4] and has been successfully applied to shape analysis [5,6] and embedding quality assessment for manifold learning [7].

Ordinary OPA only admits a global scaling factor. However, anisotropic scaling, that is, separate scaling along each coordinate axis, is a common scene in computer science and engineering [10,11]. To extend the application range of OPA, anisotropic orthogonal Procrusets analysis (AOPA) [10,11,12] is proposed to address such issue and has successful applications to computer vision [8,9] and manifold learning [13].

In this paper, we focus on the pre-scaling case of AOPA, that is, anisotropic scaling occurs before rigid motion. In literature, two approaches, namely, block relaxation (BR) [10] and majorization principle (MP) [12], are proposed to solve the pre-scaling AOPA problem. Different from these works, in this paper we describe an alternative solution to this issue using gradient descent over matrix manifold, based on our previous work [13]. Compared with BR and MP, the proposed algorithm has straightforward geometric intuition and much higher accuracy. Compared with our early work [13], which is mainly designed for embedding quality assessment, we reformulate that into the context of AOPA and provide a proof of algorithm convergence. As to traditional gradient descent methods in Euclidean space, we propose an extension to non-Euclidean domain, where the feasible region is a Riemannian submanifold of matrices. Besides, a new acceleration strategy is proposed for faster convergence.

© Springer International Publishing Switzerland 2015
Y. Tan et al. (Eds.): ICSI-CCI 2015, Part II, LNCS 9141, pp. 371–379, 2015.
DOI: 10.1007/978-3-319-20472-7_40

Experimental results from numerical simulations validate the effectiveness of the proposed method.

The rest parts of this paper are organized as follows. Section 2 reviews related works on pre-scaling anisotropic orthogonal Procrustes analysis in literature. Section 3 describes the proposed gradient descent method. Section 4 demonstrates experimental results and Section 5 concludes this paper.

2   Related Works

In this section, we briefly review two existing approaches for pre-scaling anisotropic orthogonal Procrustes analysis (PAOPA), namely, block relaxation and majorization principle. We first give main notations of symbols used in this paper for convenience of presentation. Then we formulate the PAOPA problem and state how the aforementioned two methods work.

Table 1. Main notations

Symbols	Explanations
x_i	n-dimensional observation, $i = 1,2,...,N$.
X	n by N matrix with x_i its i-th column.
y_i	m-dimensional observation, $i = 1,2,...,N$
Y	m by N matrix with y_i its i-th column
P	n by m rotation matrix, $P^T P = I_m$.
I_m	Identity matrix of order m.
D	m by m diagonal scaling matrix with non-zero diagonal entries.

2.1   Problem Formulation

Given two sets of observations $\{x_i\}$ and $\{y_i\}$, lying in R^n and R^m respectively, we assume that they are both centered at the origin, that is, their means are removed. Then the orthogonal Procrustes analysis aims to find a rotation which optimally transforms $\{y_i\}$ to $\{x_i\}$. This is equivalent to the following optimization problem

$$\min_P \quad \|X - PY\|_F$$
$$\text{s.t.} \quad P^T P = I_m \qquad (1)$$

Pre-scaling anisotropic orthogonal Procrustes analysis (PAOPA) further assumes that there exists separate scalings along each dimension of $\{y_i\}$, and it tries to recover such anisotropic scaling together with the optimal rotation. This can be formulated by modifying (1) into

$$\min_{P,D} \quad \|X - PDY\|_F$$
$$\text{s.t.} \quad P^T P = I_m \quad , \qquad (2)$$
$$D_{ii} \neq 0$$

where D_{ii} is the i-th diagonal entry of D.

2.2   Block Relaxation

Gower and Dijksterhuis [10] proposed a block relaxation (BR) algorithm to address optimization problem (2), which alternatively computes P and D. For fixed P, the global optimal D is computed as

$$D_{ii} = (B^T P)_{ii} \Big/ (YY^T)_{ii} \, , \tag{3}$$

where $B = XY^T$. Now fix D, the optimal P is given by the orthogonal polar factor of BD. Let USV^T be the singular value decomposition of BD, we have

$$P = UV^T \, . \tag{4}$$

The iteration process terminates until P converges.

2.3   Majorization Principle

Dosse and Berge [11,12] proposed a majorization principle algorithm based on a series of theoretical analysis. Given an initial guess of P, namely P_0, the update of P is defined as the orthogonal polar factor of $B\mathrm{diag}(P_0^T B)$, where the diag operator transforms a square matrix into a diagonal one by keeping its diagonal entries. The iteration continues by updating P_0 with previously computed P and terminates until P converges.

3   Pre-Scaling Anisotropic Orthogonal Procrustes Analysis Based on Gradient Descent

3.1   Gradient Descent on the Stiefel Manifold

In this section, we present a new solution to the pre-scaling anisotropic orthogonal Procrustes analysis problem based on gradient descent method over matrix manifold, named as PAOPA-GD for short. The proposed approach consists of two steps, which are computations of the optimal anisotropic scaling matrix D and rotation matrix P, respectively.

Similar to the BR method, we start by fixing P to derive an explicit solution to D. Formally, the objective function in (2) can be expanded as

$$f(P, D) = \sum_i A_{ii} D_{ii}^2 - 2 \sum_i B_{ii} D_{ii} + \mathrm{tr}(XX^T) \, , \tag{5}$$

where $B = XY^T$, $A = YY^T$, and A_{ii} is its i-th diagonal entry. This is a convex function over D_{ii} and admits a global optimum, which is given in (3).

Substituting (3) into (5), the objective function now reads

$$f(P) = -\mathrm{tr}\big((P^T M) \odot (P^T M)\big) + \mathrm{tr}(XX^T) \, , \tag{6}$$

where $M = XY^T \text{diag}(YY^T)$ and $\odot$ stands for the Hadamard product over matrices, that is, entry-wise multiplication. Then the optimization problem (2) can be transformed into

$$\max_P \quad g(P) = tr\big((P^T M) \odot (P^T M)\big)$$
$$\text{s.t.} \quad\quad\quad P^T P = I_m \tag{7}$$

The orthogonal constraint over P implies that P lies on the Stiefel manifold $St(n, m)$, which is formed by all n by m orthogonal matrices and a Riemannian submanifold embedded in $R^{n \times m}$. A straightforward strategy to solve (7) is to use gradient descent over the matrix manifold $St(n, m)$. Since the feasible region is a Riemannian submanifold rather than trivial Euclidean space, at each P, we need to project the gradient in the ambient $R^{n \times m}$ to its tangential space. Then P is updated along such projection in the tangential space. After that, the updated P is retracted back onto $St(n, m)$.

First, we derive the gradient of $g(P)$ in $R^{n \times m}$, denoted by ∇g. By matrix caculus [15] and algebraic deduction, the differentiation of g is given by $Dg = 2\text{vec}(M\text{diag}(P^T M))$, where the vec operator reformulates a $n \times m$ matrix into a nm dimensional vector by stacking its columns one underneath other. Then the gradient is simply

$$\nabla g = 2\,M\text{diag}(P^T M) \,. \tag{8}$$

By properties of $St(n, m)$ [14], the projection of ∇g onto the tangential space at P, denoted by $\nabla_T g$, is computed as follows

$$\begin{aligned} \nabla_T g &= \nabla g - \frac{P(P^T \nabla g + (\nabla g)^T P)}{2} \\ &= 2\,M\text{diag}(P^T M) - PP^T M\text{diag}(P^T M) - P\text{diag}(P^T M)M^T P \end{aligned} \tag{9}$$

At each iteration, P is updated along $\nabla_T g$ with given step length s and then retracted back onto $St(n, m)$. Let $QR = P + s\nabla_T g$ be the QR decomposition of the update on the tangential space, the aforementioned retraction is simply given by Q. The iteration terminates until $\|\nabla_T g\|_F$ is smaller than some given threshold.

3.2 Convergence Analysis

In this subsection, we prove the convergence of the gradient descent approach given in Section 3.1, which is presented in Lemma 1. For clarity of presentation, we use upper-script with brackets for counts of iterations.

Lemma 1. *For any initial $P^{(0)} \notin \text{Null}(M)$, the iteration of P using gradient descent converges monotonously.*

Proof: By expanding the objective function in (7) and using Cauchy-Schwarz inequality, we have

$$g(P) = \sum_i P_i^T M_i M_i^T P_i$$
$$\leq \sum_i \|P_i\|^2 \|M_i\|^2 , \qquad (10)$$
$$= \sum_i \|M_i\|^2$$

where P_i and M_i are the i-th columns of P and M, respectively. Here we use the fact that P is orthogonal, hence P_i has unit norm. The above deduction shows that $g(P)$ is upper-bounded.

Note that the initial $P^{(0)} \notin \text{Null}(M)$, then $\nabla_T g(P^{(0)})$ does not vanish and $g(P^{(i)}) \leq g(P^{(i+1)})$ as P is updated along the gradient, which is the descent direction of increasing $g(P)$. This implies that $\{g(P^{(i)})\}$ is a non-decreasing and upper-bounded series, which converges to some constant. ∎

3.3 Acceleration of Iteration

Since the update of P involves QR-decomposition, which does not admit an explicit formulation, then traditional line search strategies cannot be applied to find an optimal step length. Instead, we propose a forward-backward approach to accelerate the iteration process, which consists of two steps.

- **Forward Step.** In the i-th iteration, if $\text{tr}((\nabla_T g(P^{(i)}))^T g(P^{(i-1)})) < 0$, then decrease the step length s to αs with $0 < \alpha < 1$.
- **Backward Step.** In the i-th iteration, if $\text{tr}((\nabla_T g(P^{(i)}))^T g(P^{(i-1)})) \geq 0$, then increase the step length s to $(1 + \beta)s$ with $0 < \beta < 1$.

Such strategy aims to pull back the update if $P^{(i)}$ "goes ahead" the optima and push the update otherwise. In this paper, we set $\alpha = \beta = 0.5$.

3.4 Computational Complexity

To end this section, we briefly analyze the computational complexity of the proposed PAOPA-GD algorithm. The computation of gradient in (9) only involves matrix multiplication, which costs $O(mn)$ time. The retraction back onto the manifold involves a QR decomposition, which costs $O(n^2 m)$ time for the gradient in $R^{n \times m}$. All together, the overall complexity of PAOPA-GD is $O(kn^2 m)$, where k is the number of iterations.

4 Experimental Results

In this section, we conduct three numerical simulations to validate the effectiveness of the proposed PAOPA-GD algorithm. We show that PAOPA-GD can efficiently recover optimal rotation and scaling matrices for pre-scaling anisotropic orthogonal Procrustes problem, and its accuracy is much higher than the block relaxation (BR) and majorization principle (MP) approaches. Besides, we demonstrate the application of PAOPA-GD to embedding quality assessment for manifold learning. In all experiments, we use GD to denote the proposed gradient descent based approach. The stopping criterion for all three algorithms is 10^{-4} and the initial step length for GD is 0.01.

In the first simulation, we compare the performances among GD, BR, and MP for randomly generated data sets under various dimensions. For each dimension from 10 to 100 with step 10, we randomly generate an orthogonal matrix P, an anisotropic scaling matrix D, and 200 observations from a uniform distribution to form Y. The target observation matrix X is computed as PDY. Then X and Y are used as inputs for GD, BR, and MP to find optimal P^* and D^*. The approximation error is defined as

$$e = \|X - P^*D^*Y\|_F . \tag{11}$$

This process is repeated 50 times for each dimension. The mean error together with standard deviation are recorded and shown in Fig. 1. From Fig. 1, we can clear see that the mean error of GD is much lower than BR and MP and as low as 10^{-6}. The near-zero error of GD also suggests that it has successfully find optimal P^* and D^* to match Y to X.

The second simulation has similar settings to the first one. The only difference is that the dimension of Y is only a half of that of X. Experimental results are illustrated in Fig. 2, which again demonstrates the superiority of GD over BR and MP on accuracy.

In the last simulation, we apply GD to embedding quality assessment for manifold learning, which is a popular technique for nonlinear dimensionality reduction in recent decades. Manifold learning methods can effectively recover meaningful low-dimension embedding from high-dimensional samples, which reside on a low- dimensional manifold embedded in the ambient Euclidean space. However, most manifold learning methods output a normalized embedding, that is, each dimension of the embedding is scaled separately. In such case, standard orthogonal Procrustes analysis

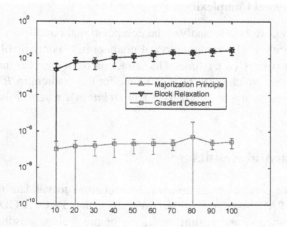

Fig. 1. Simulation results of Experiment 1. X and Y have the same dimensions. Approximation error versus dimensionality of observations is illustrated. Blue upper triangle: MP. Black lower triangle: BR. Red square: GD.

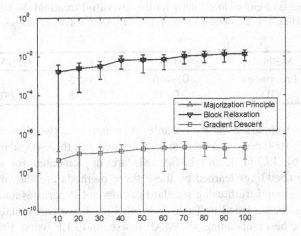

Fig. 2. Simulation results of Experiment 2. X and Y have different dimensions. Approximation error versus dimensionality of observations is illustrated. Blue upper triangle: MP. Black lower triangle: BR. Red square: GD.

would fail to assess the quality of learned embedding even providing the ground truth, while GD can be applied to this issue since anisotropic scaling caused by normalization can be recovered.

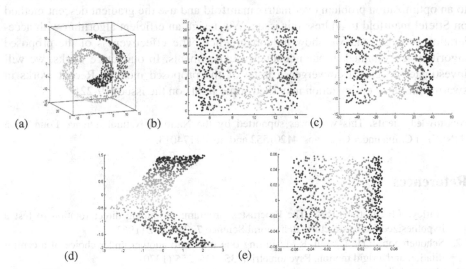

Fig. 3. Learning results of ISOMAP, LLE, and LTSA on the SwissRoll manifold. (a) The manifold. (b) Ground truth of global coordinates. (c) ISOMAP. (d) LLE. (e) LTSA.

Table 2. Assessments of embedding quality for the SwissRoll manifold. Matching errors from the learned embedding to the ground truth are presented. Low values correspond to embeddings of good quality.

Methods	ISOMAP	LLE	LTSA
Procrustes	0.6446	0.4955	0.1292
GD	0.0234	0.6598	0.0032

We use the benchmark SwissRoll manifold, which is a two-dimensional surface embedded in R^3. 1000 points are randomly generated and three popular methods, that are, ISOMAP[16], LLE[17], and LTSA[18], are implemented for dimensionality reduction. The embeddings learned by these three methods are shown in Fig. 3 and compared to the ground truth using standard orthogonal Procrustes analysis and GD. Experimental results are listed in Table 2. From Fig. 3, we can visually inspect that LTSA outputs the best embedding, ISOMAP worse, and LLE worst. From the matching errors in Table 2, we can see that GD correctly assesses the quality of the three embeddings, while standard orthogonal Procrustes analysis misjudged ISOMAP and LLE.

5   Conclusion

In this paper, we propose an alternative solution to pre-scaling anisotropic orthogonal Procrustes analysis (PAOPA). Based on our previous work, we reformulate PAOPA to an optimization problem over matrix manifold and use the gradient descent method on Stiefel manifold to address this issue. We design an efficient algorithm with acceleration strategy, and we show its convergence. The effectiveness of the proposed algorithm is examined through conducted experiments. In our future works, we will investigate the global convergence issue of the proposed method. Recent works in swarm and fuzzy optimization may shed some lights on the issue [19-22].

Acknowledgments. This work was supported by the National Natural Science Foundation (NNSF) of China under Grant nos. 41201552 and no. 41174013.

References

1. Hurley, J.R., Cattell, R.B.: The Procrustes program: producing direct rotation to test a hypothesized factor structure. Behavioral Science **7**, 258–262 (1962)
2. Schonemann, P.H., Carroll, R.M.: Fitting one matrix to another under choice of a central dilation and a rigid motion. Psychometrika **35**, 245–255 (1970)
3. Gower, J.C.: Statistical methods of comparing different multivariate analyses of the same data. In: Mathematics in the Archeological and Historical Sciences, pp. 138–149. University Press (1971)
4. Ten Berge, J.M.F.: Orthogonal Procrustes rotation for two or more matrices. Psychometrika **42**, 267–276 (1977)

5. Goodall, C.: Procrustes methods in the statistical analysis of shape. Journal of the Royal Statistical Society, Series B **53**, 285–339 (1991)
6. Dryden, I., Mardia, K.: Statistical shape analysis. John Wiley and Sons (1998)
7. Goldberg, Y., Ritov, Y.: Local procrustes for manifold embedding: a measure of embedding quality and embedding algorithms. Machine Learning **77**(1), 1–25 (2009)
8. Garro, V., Crosilla, F., Fusiello, A.: Solving the PnP problem with anisotropic orthogonal Procrustes analysis. In: Proceedings of the 2012 Second Joint 3DIM/3DPVT Conference: 3D Imaging, Modeling, Processing, Visualization & Transmission, 262–269 (2012)
9. Chen, E.C., McLeod, A.J., Jayarathne, U.L., Peter, T.M.: Solving for free-hand and real-time 3d ultrasound calibration with anisotropic orthogonal Procrustes analysis. In: Proceedings of SPIE, vol. 9036, 90361Z-1-7 (2014)
10. Gower, J.C.: Dijksterhuis, G.B.: Procrustes problems. Oxford University Press (2004)
11. Dosse, M.B.: Extension of Generalized Procrustes Analysis. Agrostat, Rennes, pp. 1–16 (2004)
12. Dosse, M.B., Berge, J.T.: Anisotropic orthogonal Procrustes Analysis. Journal of Classification **27**, 111–128 (2010)
13. Zhang, P., Ren, Y., Zhang, B.: A new embedding quality assessment method for manifold learning. Neurocomputing **97**, 251–266 (2012)
14. Absil, P.A., Mahony, R., Sepulchre, R.: Optimization Algorithms on Matrix Manifolds. Princeton University Press, Princeton, NJ, USA (2007)
15. Magnus, J.R., Neudecker, H.: Matrix differential calculus with applications in statistics and econometrics. 2nd edn. John Wiley and Sons (1999)
16. Tenenbaum, J.B., Silva, V., Langford, J.C.: A global geometric framework for nonlinear dimensionality reduction. Science **290**(5500), 2319–2323 (2000)
17. Roweis, S.T., Saul, L.K.: Nonlinear dimensionality reduction by locally linear embedding. Science **290**(5500), 2323–2326 (2000)
18. Zhang, Z., Zha, H.: Principal manifolds and nonlinear dimensionality reduction via tangent space alignment. SIAM Journal on Scientific Computing **26**(1), 313–338 (2005)
19. Valdez, F., Melin, P., Castillo, O.: An improved evolutionary method with fuzzy logic for combining Particle Swarm Optimization and Genetic Algorithms. Applied Soft Computing **11**(2), 2625–2632 (2010)
20. Precup, R.-E., David, R.-C., Petriu, E.M., Preitl, S., Paul, A.S.: Gravitational Search Algorithm-Based Tuning of Fuzzy Control Systems with a Reduced Parametric Sensitivity. In: Gaspar-Cunha, A., Takahashi, R., Schaefer, G., Costa, L. (eds.) Soft Computing in Industrial Applications. AISC, vol. 96, pp. 141–150. Springer, Heidelberg (2011)
21. Wu, Z., Chow, T., Cheng, S., Shi, Y.: Contour gradient optimization. International Journal of Swarm Intelligence Research **4**(2), 1–28 (2013)
22. El-Hefnawy, N.: Solving bi-level problems using modified particle swarm optimization algorithm. International Journal of Artificial Intelligence **12**(2), 88–101 (2014)

Life Record: A Smartphone-Based Daily Activity Monitoring System

Pei-Ching Yang[1(✉)], Shiang-Chiuan Su[1], I-Lin Wu[1], and Jung-Hsien Chiang[1,2]

[1] Department of Computer Science and Information Engineering,
National Cheng Kung University, Tainan, Taiwan
{yang.peiching,smatch,perado,jchiang}@iir.csie.ncku.edu.tw
[2] Institute of Medical Informatics National Cheng Kung University Tainan,
Tainan, Taiwan

Abstract. In this paper, we propose a two-layered classification approach to effectively recognize the physical activities while the smartphone is placed at any four common positions on the body. Then we implement a Life Record app on smartphone that automatically classifies physical activities and records them as the personal life logs. For assisting users in comprehending their daily activities, the system also provides the visualization interface that shows the brief descriptions of their life logs.

We demonstrate that the system possesses less limitation to monitor daily activities that the users are not restricted to carry their smartphones in specific positions. Another major benefit of our system is to provide a complete overview of personal activities, which enhances the self-awareness of physical activity in our daily life through an intuitive visualization interface. Furthermore, analysis of life logs can also be applied in specific services or recommendation applications in the future.

Keywords: Activity monitoring · Life record · Physical inactivity · Pattern recognition · Data visualization

1 Introduction

Physical inactivity is one of the most important modifiable risk factors, and it causes unhealthy life habits such that most people spend their leisure time involved in sedentary pursuits. Due to this lack of physical activity, more and more people have become overweight (body mass index ≥ 25 kg/m^2) and even have become obese (body mass index ≥ 30 kg/m^2). Globally, in 2005, it was estimated that over 1 billion people were overweight, including 805 million women, and that over 300 million people were obese. By 2015, it has been estimated that over 1.5 billion people will be overweight [1]. Overweight and obesity, moreover, will cause increasing risk for various chronic diseases, such as diabetes, cardiovascular diseases, hypertension and cancer.

© Springer International Publishing Switzerland 2015
Y. Tan et al. (Eds.): ICSI-CCI 2015, Part II, LNCS 9141, pp. 380–385, 2015.
DOI: 10.1007/978-3-319-20472-7_41

Several commercially available devices, such as Fitbit One [2] and Fitbit Flex [3], Bodymedia [4], and Jawbone UP [5], have embedded these wearable sensors to provide daily activity monitoring, to compute caloric expenditure, and even trace sleeping status during the night. Although these can provide daily activity monitoring or specific exercise training, they need to be worn in a specific body position and are not popular for the general public.

In addition to hardware, smartphones also provide a developmental platform on which developers can easily design applications. Several studies [6-9] have proposed human activity recognition systems based on smartphones to facilitate long-term daily activity monitoring.

In this research, we use Android smartphones to collect daily activity data, so this system is applicable to individuals who own Android smartphones. Users can observe activity habits through their life logs from the visualizations on the smartphone application.

2   Related Work

Human Activity Recognition research mostly uses observation of human actions to obtain an understanding of types of activities that they perform within a specific time interval. Typical activities under consideration vary from mechanical process such as activities of daily living (ADL) to socio spatial processes like meetings. To recognize the activities that occur in daily living, wearable sensors have been used to acquire the signals from different body positions intended to detect movements. Accelerometers have been the most commonly used device to recognize human activity during high performance in physical activity recognition (PAR) research. So far, almost all studies of PAR differ according to the type and number of activities identified and by the location, type and number of accelerometers used.

Wu et al. [40] proposed a system called SensCare, which was a semi-automatic lifelog summarization system for elderly care. SensCare fuses heterogeneous sensor information and automatically segments and recognizes user's daily activities in a hierarchical way. It combines unsupervised activity segmentation and activity recognition to segment an activity with the specific time period related to its occurrence. GPS data is fused with the activity segmentation to predict high-level daily activities.

Other studies referenced in [9-13] were all aimed at providing ubiquitous recognition systems used in long-term health care monitoring. In short-term supervised monitoring situations, large numbers of body-fixed sensors can be used to allow the collection of greater quantities of information, leading to very accurate assessments of movement; however, in long-term, unsupervised monitoring environments, subject compliance is essential if the system is to be used [14].

Human life models are useful in a variety of applications, such as the detection of abnormal behavior. They can also be used to analyze correlations between the regularity of workers' behavior and their levels of stress [15].

3   Material and Method

In this research, we develop a smartphone-based Daily activity monitoring system for measurement of daily activities in human life. We design a two-layered activity classification scheme to recognize the state of activity and the type of activity from the sensor signals of the smartphone. We implement the visualization of personal activity record and quantify the activities in daily life by the duration and the calories consumed from them.

3.1   System Architecture

The development process consists of two phases: (1) we first construct the two-layered activity classification scheme, which contains the data processing and model building. We collected the data from an accelerometer and an orientation sensor and extract the feature set as the input of the classification model. Then, the subset of features is selected to train the two-layered classification model. After the model building, it is tested with both the laboratory dataset and the real dataset; (2) we utilize the model to develop the Life Record app on smartphone for recording physical activities in users' daily life. We also implement the visualization methods on the app to provide an overview of personal daily activity. Finally, we analyze the activity patterns and the relationships from users' life logs for discovering the model of their daily life.

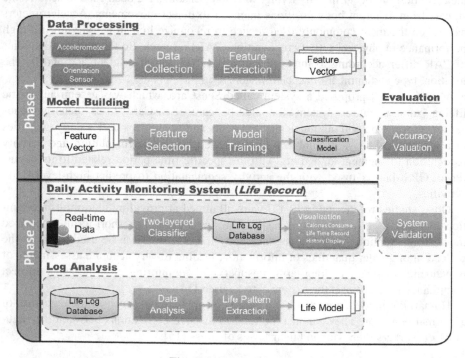

Fig. 1. System architecture

3.2   Two-Layered Activity Classification

We build a two-layered hierarchical activity classifier for recognizing the state and class of activity based on the sensor in smartphone.

Figure 3.4 shows the flow for building the two-layered classification model. First, a feature selection approach was used to pick out a subset of relevant features for the purpose of building a robust learning model before we built each classification model in the two-layered classifier. Then, the classification models were trained individually with the corresponding selected feature set. Finally, the classifier was constructed using the models for each layer, and it could be implemented for the Life Record app.

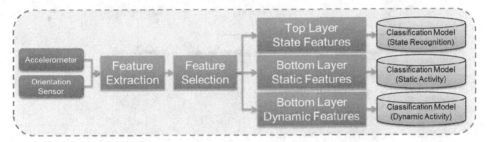

Fig. 2. Procedure for classification model building

3.3   Life Record App

We performed an evaluation of iHOPE. We invited twenty medical professionals and ten graduate school students to try out the system on a daily basis for two weeks, and they are asked to fill out the questionnaire for feedback after two weeks of practice. The participants all used the smartphone in a regular basis in their daily lives.

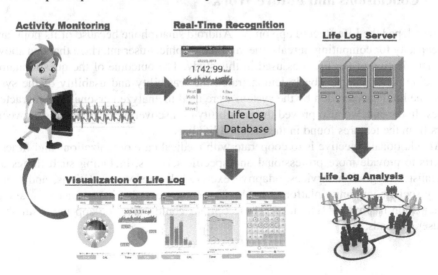

Fig. 3. Life Record scenario

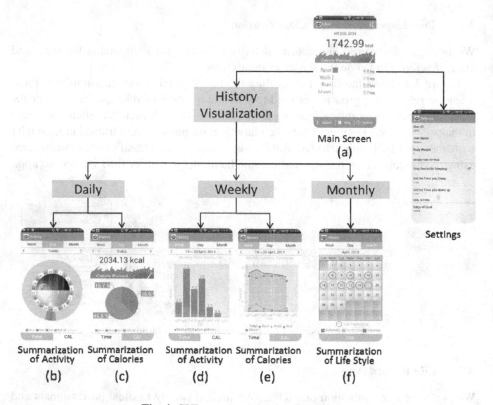

Fig. 4. GUI screens navigation flowcharts

4   Conclusions and Future Work

We developed the Life Record app on the Android smartphone because of its popularity, capacity for computing in real-time, and the graphical user interface that can show the visualization of the life logs used in this study. The outcome of the questionnaire providing participant feedback demonstrated the feasibility and usability of the system. The life logs recorded by the system were used to analyze personal life characteristics. In the end, we also proved the possibility of discovering relationships between users from the features found in the personal life logs.

An additional objective is to cooperate with medical care organizations and fitness experts to provide more professional and specific services, including such things as specialists' suggestion services, adaptive exercise plan recommendations, and combining social community platforms to share or compare with friends for encouraging participation in physical activities. These options could make the app more suitable for user demands.

References

1. W. H. Organization, Preventing chronic diseases: a vital investment: WHO global report. World Health Organization, Geneva (2005)
2. Fitbit One. http://www.fitbit.com/one
3. Fitbit Flex. http://www.fitbit.com/flex
4. Bodymedia. http://www.bodymedia.com/
5. Jawbone UP. https://jawbone.com/up
6. Bicocchi, N., Mamei, M., Zambonelli, F.: Detecting activities from body-worn accelerometers via instance-based algorithms. Pervasive and Mobile Computing 6(4), 482–495 (2010)
7. Luo, Z.C.: The development of activity recognition system using the smartphone with accelerometer. Institute of Medical Informatics, National Cheng Kung University, Master (2010)
8. Wu, P., Peng, H.-K., Zhu, J., Zhang, Y.: SensCare: Semi-automatic Activity Summarization System for Elderly Care. In: Zhang, J.Y., Wilkiewicz, J., Nahapetian, A. (eds.) MobiCASE 2011. LNICST, vol. 95, pp. 1–19. Springer, Heidelberg (2012)
9. Zhao, Z., et al.: Cross-people mobile-phone based activity recognition. In: IJCAI, vol. 11 (2011)
10. Lee, Y.-S., Cho, S.-B.: Activity Recognition Using Hierarchical Hidden Markov Models on a Smartphone with 3D Accelerometer. In: Corchado, E., Kurzyński, M., Woźniak, M. (eds.) HAIS 2011, Part I. LNCS, vol. 6678, pp. 460–467. Springer, Heidelberg (2011)
11. Longstaff, B., Reddy, S., Estrin, D.: Improving activity classification for health applications on mobile devices using active and semi-supervised learning. In: 2010 4th International Conference on Pervasive Computing Technologies for Healthcare (PervasiveHealth). IEEE (2010)
12. Martin, E., et al.: Enhancing context awareness with activity recognition and radio fingerprinting. In: 2011 Fifth IEEE International Conference on Semantic Computing (ICSC). IEEE (2011)
13. Nickel, C., et al.: Using hidden markov models for accelerometer-based biometric gait recognition. In: 2011 IEEE 7th International Colloquium on Signal Processing and its Applications (CSPA). IEEE (2011)
14. Mathie, M.J., et al.: Accelerometry: providing an integrated, practical method for long-term, ambulatory monitoring of human movement. Physiological Measurement 25(2), R1 (2004)
15. Okada, S., et al.: Analysis of the correlation between the regularity of work behavior and stress indices based on longitudinal behavioral data. In: Proceedings of the 14th ACM International Conference on Multimodal Interaction. ACM (2012)

Blind Source Separation

Blind Source Separation and Wavelet Packet Based Novel Harmonic Retrieval Algorithm

Fasong Wang[1(✉)], Zhongyong Wang[1], Rui Li[2], and Linrang Zhang[3]

[1] School of Information Engineering, Zhengzhou University, Zhengzhou 450001, China
fasongwang@126.com
[2] School of Sciences, Henan University of Technology, Zhengzhou, China
[3] National Key Laboratory for Radar Signal Processing, Xidian University, Xi'an, China

Abstract. In this paper, a novel one dimensional harmonic retrieval (HR) algorithm is proposed, which can be applied in additive colored Gaussian or non-Gaussian noise when the frequencies of the harmonic signals are closely spaced in frequency domain. Resorting to the blind source separation (BSS) based harmonic retrieval model, the main algorithm is developed mainly using the wavelet packet (WP) decomposition approach, where the criterion is formed as the cumulant based approximation of the mutual information (MI) for the selection of optimal sub-band of WP decomposition with the least-dependent components between the same nodes. Simulation results show that the proposed algorithm can retrieve the harmonic source signals and yield good performance.

Keywords: Blind source separation (BSS) · Harmonic retrieval (HR) · Wavelet packet (WP) · Mutual information (MI)

1 Introduction

The one- and multi-dimensional harmonic retrieval (HR) problem arises in various areas of physical systems including communications, geophysics and radar signal processing, which has been an active research area during the past few decades [1-4]. HR in additive colored noise is one of important research topics in which the number of harmonics and their frequencies often need to be estimated from noisy data accurately, especially when the frequencies of the harmonic signals are very close in frequency domain and corrupted with additive colored Gaussian or non-Gaussian noise.

The realization of HR in additive noise includes traditional approach and modern methods. Traditional approach has been achieved by the Fourier transformation (FT), which is stability but the resolution of the solution is low. In order to increase the resolution of the spectral estimation, some modern methods have been developed, but all these methods have low signal-to-noise ratio (SNR).

In addition, most HR approaches either assumed white noise and utilized correlation based methods or assumed colored Gaussian noise and employed higher-order statistic based methods [2-3]. The former class methods generally assume that either the noise is white or its covariance matrix is given. Even though there has been some work addressing colored noise, typically, however, the noise is colored and a priori

© Springer International Publishing Switzerland 2015
Y. Tan et al. (Eds.): ICSI-CCI 2015, Part II, LNCS 9141, pp. 389–396, 2015.
DOI: 10.1007/978-3-319-20472-7_42

knowledge (or an estimate) of the noise covariance matrix is unavailable. By exploiting the fact that higher-than-second-order cumulants are zero for Gaussian processes, several researchers have demonstrated that cumulant-based HR methods can suppress the effect of colored Gaussian noise [2-3]. In all these methods, it is common assumed that the additive noise has Gaussian distribution or non-Gaussian distribution but the model of the noise must be restricted.

The resolution of the classical power spectrum estimators for HR is of order $1/T$, where T is the effective window length. In contrast, the resolution of the periodogram is $1/Q$, where Q is the data length, and since $T \ll Q$, the periodogram exhibits higher resolution than the classical power spectrum estimators. But, the choice of the window function dictates the resolution-variance tradeoff. When the frequencies of harmonic signals are closely spaced in frequency domain, the problem can't get ideal identification results using the conventional HR methods.

In this paper, we don't presume the distribution, color and model of the additive noise except that it is stationary. The proposed HR algorithm is realized by using the wavelet packet (WP) decomposition approach and the popular used method called blind source separation (BSS), in addition, we just utilize a single channel mixture of the harmonic sources.

2   Basic Model

In general, the discrete noiseless one-dimensional real harmonic signals which contain P sinusoids are modeled as follows:

$$s[k] = \sum_{k=1}^{P} a_k \cos[\omega_k k + \varphi_k] . \tag{1}$$

where P is the number of the harmonic signals, a_k and ω_k are the unknown constants called amplitudes and frequencies, and $\omega_i \neq \omega_j$ for $i \neq j$, amplitudes are assumed positive and $0 < \omega_i < \pi, i = 1, 2, \cdots, P$. Additionally, the phases φ_k are i.i.d. random variables uniformly distributed over $(-\pi, \pi]$. Due to the presence of noise, one observes a noise contaminated version of $s[k]$, namely

$$x[k] = s[k] + n[k] . \tag{2}$$

where $n[k]$ is the additive noise which is stationary and statistical independent to harmonic signals. The problem of interest is to estimate harmonic number P and their frequencies ω_i using just the noisy observations $x[k], t = 1, 2, \cdots, T$.

Let us denote the N source signals by the vector $s[k] = (s_1[k], \cdots, s_N[k])^T$, and the observed signals by $x[k] = (x_1[k], \cdots, x_M[k])^T$. Now the mixing process of BSS can be expressed as

$$x[k] = As[k] + n[k] . \tag{3}$$

where the matrix $\mathbf{A} = [a_{ij}] \in \mathbf{R}^{M \times N}$ collects the mixing coefficients. No particular assumptions on the mixing coefficients are been made. For technical simplicity, we shall also assume that all the signals have zero mean, but this is no restriction since it simply means that the signals have been centered. $\mathbf{n}[k] = (n_1[k], \cdots, n_M[k])^{\mathrm{T}}$ is a vector of additive noise. The problem of BSS is now to estimate both the source signals $s_j[k], j = 1, \cdots, N$ and the mixing matrix $\mathbf{A}$ based on observations of the $x_i[k], i = 1, \cdots, M$ alone [5-7].

Following the procedure of [8], we will build the basic real harmonic BSS model using just the one channel observed signal $x[k], t = 1, 2, \cdots, T$ producing by equation (2). Utilizing equation (1) and equation (2), for arbitrary integer $l(1 \le l \le T)$, we get

$$\mathbf{x}[k] = \mathbf{As}[k] + \mathbf{n}[k] .\tag{4}$$

where

$$s[k] = \begin{pmatrix} a_1 \cos[\omega_1 k + \varphi_1] \\ a_2 \cos[\omega_2 k + \varphi_2] \\ \vdots \\ a_P \cos[\omega_P k + \varphi_P] \end{pmatrix}^{\mathrm{T}} .\tag{5}$$

$$\mathbf{A} = \begin{pmatrix} 1 & 1 & \cdots & 1 \\ \cos \omega_1 & \cos \omega_2 & \cdots & \cos \omega_P \\ \vdots & \vdots & \ddots & \vdots \\ \cos(P-1)\omega_1 & \cos(P-1)\omega_2 & \cdots & \cos(P-1)\omega_P \end{pmatrix} .\tag{6}$$

$$\mathbf{n}[k] = \frac{1}{2} \left\{ \begin{pmatrix} n[k] \\ n[k+1] \\ \vdots \\ n[k+P-1] \end{pmatrix} + \begin{pmatrix} n[k] \\ n[k-1] \\ \vdots \\ n[k-P+1] \end{pmatrix} \right\} .\tag{7}$$

The covariance of the noise $\mathbf{n}[k]$ can be obtained as:

$$E\{\mathbf{n}[k]\mathbf{n}^{\mathrm{T}}[k]\} = \frac{1}{2} \left\{ \begin{pmatrix} r_n[0] & \cdots & r_n[P-1] \\ r_n[1] & \cdots & r_n[P-2] \\ \vdots & \ddots & \vdots \\ r_n[P-1] & \cdots & r_n[0] \end{pmatrix} + \begin{pmatrix} r_n[0] & \cdots & r_n[P-1] \\ r_n[1] & \cdots & r_n[P] \\ \vdots & \ddots & \vdots \\ r_n[P-1] & \cdots & r_n[2P-2] \end{pmatrix} \right\} .\tag{8}$$

where $r_n[\cdot]$ is the correlation function of the channel noise signals.

At the same time, we can get $|\mathbf{A}| \ne 0$, $\forall i \ne k, i, k = 1, 2, \cdots, P$, and it is very interesting that model (4) is a typical even-determined BSS model. Moreover, from equation (6), we notice that if we can identify the frequency of the harmonic signals successfully, the mixing matrix $\mathbf{A}$ is then obtained directly.

3   Proposed Algorithm

It is assumed that source signals are non-Gaussian and statistically independent when one utilizes ICA algorithms to solve the BSS problem (3). However, the independence property of source signals may not hold in some real-world situations. Among many extensions of the basic ICA models, sub-band decomposition ICA (SDICA) [9] assumes that each source signal is represented as the sum of some independent subcomponents and dependent subcomponents.

As described above, BSS based HR problem (4) accords with the SDICA model well in appearance. Now, we will give the separation criteria using the idea of SDICA based on WP decomposition. In SDICA model, source signals can be represented as:

$$s_i[k] = s_{i,1}[k] + s_{i,2}[k] + \cdots + s_{i,L}[k] . \tag{9}$$

where $s_{i,j}[k]$, $i = 1, \cdots, N$, $j = 1, \cdots, L$ are sub-band subcomponents. And the subcomponents are mutually independent for only a certain set of j. The observations are generated from the sources s_i according to equation (4). Here we assume that the number of sources is equal to that of the observations and that the observations are zero mean. Similar to ICA, the goal of SDICA is to find the separation matrix $\mathbf{W} \triangleq \mathbf{A}^{-1}$, which estimates the original sources

$$\mathbf{y}[k] = \mathbf{W}\mathbf{x}[k] . \tag{10}$$

where $\mathbf{y}[k] = \hat{\mathbf{s}}[k]$ and $\mathbf{y}(t) \in \mathbf{R}^N$. We shall assume that for certain set of j, sub-bands in equation (9) are least dependent or possibly independent [9]. Under presented assumptions, the standard linear ICA algorithms can be applied to the selected set of j sub-bands in order to learn demixing matrix $\mathbf{W}$ as follows:

$$\mathbf{y}_j[k] = \mathbf{W}\mathbf{x}_j[k] . \tag{11}$$

So, the problems to be resolved can be described as two separated problems: i) the preprocessing transform should be used to obtain sub-band representation of the original wideband BSS based HR problem (4); ii) the criteria should be constructed to select the set with least dependent sub-bands.

In order to solve the BSS based HR problem (4), one can use any linear operator on $\mathbf{s}[k]$ which will extract a set of optimal sub-bands. WP was introduced by Coifman et al. [10] as a generalization of wavelet in the sense that instead of dividing only the approximation space, as in the standard orthogonal wavelet transform, the detail spaces are also divided. Each source and noise signal can be expressed in terms of its decomposition coefficients as:

$$s_{ji}^q[k] = \sum_\alpha c_{ji\alpha}^q \varphi_{q\alpha}[k] , \quad n_{ji}^q[k] = \sum_\alpha e_{ji\alpha}^q \varphi_{q\alpha}[k] . \tag{12}$$

where the indexes q, j, i, α represent the scale level, the sub-band index, the source index and the shift index, respectively, herein $j = 1, 2, \cdots, 2^q$. $\varphi_{q\alpha}[k]$ is the chosen wavelet and $c_{ji\alpha}^q$, $e_{ji\alpha}^q$ are the corresponding decomposition coefficients. If we choose the same representation space as for the source signals, each component of the observed signals $\mathbf{x}$ can be written as

$$x_{jm}^q[k] = \sum_\alpha d_{jm\alpha}^q \varphi_{j\alpha}[k] . \tag{13}$$

where m is the observed signal index. Let vectors $\mathbf{c}_{j\alpha}^q = [c_{j1\alpha}^q, c_{j2\alpha}^q, \cdots, c_{jN\alpha}^q]^\mathrm{T}$, $\mathbf{e}_{j\alpha}^q = [e_{j1\alpha}^q, e_{j2\alpha}^q, \cdots, e_{jN\alpha}^q]^\mathrm{T}$ and $\mathbf{d}_{j\alpha}^q = [d_{j1\alpha}^q, d_{j2\alpha}^q, \cdots, d_{jM\alpha}^q]^\mathrm{T}$ be constructed from the l-th coefficients of the sources, noises and mixtures, respectively. From equation (12) and (13) using the orthogonally property of the functions $\varphi_{q\alpha}[k]$, one obtains

$$\mathbf{d}_j^q = \mathbf{A}\mathbf{c}_j^q + \mathbf{e}_j^q . \tag{14}$$

Then, the estimation of the mixing matrix is performed using the decomposition coefficients $\mathbf{d}_j^q = [d_{j1}^q, \cdots, d_{jM}^q]^\mathrm{T}$ of the mixtures corresponding a special shift index α. Also, when the noise is present, equation (14) becomes approximately,

$$\mathbf{d}_j^q \doteq \mathbf{A}\mathbf{c}_j^q . \tag{15}$$

From equation (4) and (14) we can see that the relation between decomposition coefficients of the mixtures and the sources is exactly the same as in the original domain of signals. From equation (12), (13) and (15), we obtain:

$$\mathbf{x}_j^q = \mathbf{A}\mathbf{s}_j^q . \tag{16}$$

The sub-bands selection with most independent components $\mathbf{s}_j$ can be done by using the measure MI between the same nodes in the WP trees. Under weak correlation and weak non-Gaussian assumptions, it has been shown in [11] that MI can be approximated via small cumulant approximation of the Kullback-Leibler (KL) divergence as

$$I_j^q(x_{j1}^q, x_{j2}^q, \cdots, x_{jN}^q) \doteq \frac{1}{4} \sum_{\substack{0 \le n < l \le N \\ n \ne l}} c_{nl}^2 + \frac{1}{2} \sum_{r \ge 3} \frac{1}{r!} \sum_{i_1 i_2 \cdots i_r = 1}^N c_{i_1 i_2 \cdots i_r}^2 . \tag{17}$$

where $c_{i_1 i_2 \cdots i_r}^2$ denotes square of the related r-th order cross-cumulant and $i_1 i_2 \cdots i_r$ denotes the partition of indices such that they are not all identical and c_{nl} denotes cross-cumulant or second-order cross-cumulant. But in order to estimate the MI in equation (17), one must estimate the joint probability density function (PDF) of the mixtures which is a difficult task in practice. So we will use another method to estimate MI as proposed in [5-7], and $I_j^q(x_{j1}^q, x_{j2}^q, \cdots x_{jN}^q)$ was approximated by the sum of pairwise independences. Approximation of the joint MI by the sum of pairwise MI is commonly used in the ICA community in order to simplify computational complexity of the linear instantaneous ICA algorithms. Then, the approximation of MI by the sum of pair-wise MI is described as follows [11]:

$$
\begin{aligned}
I_j^q(x_{j1}^q, x_{j2}^q, \cdots, x_{jN}^q) &\doteq \frac{1}{4} \sum_{\substack{0 \le n < l \le N \\ n \ne l}} \mathrm{cum}^2(x_{jn}^q, x_{jl}^q) \\
&+ \frac{1}{12} \sum_{\substack{0 \le n < l \le N \\ n \ne l}} (\mathrm{cum}^2(x_{jn}^q, x_{jn}^q, x_{jl}^q) + \mathrm{cum}^2(x_{jn}^q, x_{jl}^q, x_{jl}^q)) \\
&+ \frac{1}{48} \sum_{\substack{0 \le n < l \le N \\ n \ne l}} \Big(\mathrm{cum}^2(x_{jn}^q, x_{jn}^q, x_{jn}^q, x_{jl}^q) + \mathrm{cum}^2(x_{jn}^q, x_{jn}^q, x_{jl}^q, x_{jl}^q) \\
&+ \mathrm{cum}^2(x_{jn}^q, x_{jl}^q, x_{jl}^q, x_{jl}^q)\Big) .
\end{aligned}
\tag{18}
$$

where $\mathrm{cum}(\bullet)$ denotes second, third or fourth-order cross-cumulants.

After the analysis of the model and approximation of the MI estimator $I_j^q(x_{j1}^q, x_{j2}^q, \cdots x_{jN}^q)$, we now build the separation criteria using WP decomposition method based on the built BSS HR model (4).

Once the sub-band is selected, we obtain either estimation of the inverse of the separation matrix $\mathbf{W}$ or estimation of the basis matrix $\mathbf{A}$ by applying standard ICA algorithms on equation (14). Reconstructed source harmonic signals $\hat{s}$ are obtained by the estimated mixing matrix as model (5).

4   Simulation Results

In order to confirm the validity and performance of the proposed one-dimensional HR algorithm-WP-BSS-HR, simulations using Matlab are given below with four source signals which have different waveforms and contaminated with additive noise. The example demonstrates the comparisons of the proposed WP-BSS-HR algorithm with the classic HR MUSIC algorithm [1, 2] and fastICA algorithm [5] directly at different SNR levels. The four source signals are generated as follows:

$$s_1[k] = 0.3\cos[2\pi k N_1 / 1024 + 1.1]; \quad s_2[k] = 0.7\cos[2\pi k N_2 / 1024 + 1.8];$$

$$s_3[k] = 1.2\cos[2\pi k N_3 / 1024 + 1.1]; \quad s_4[k] = 0.5\cos[2\pi k N_4 / 1024 + 0.5].$$

where $N_1 = 50$, $N_2 = 170$, $N_3 = 290$, $N_4 = 410$. The source signals were contaminated with additive moving average (MA) (2) noise. The mixed harmonic signals in noise is generated by the BSS based HR models.

Simulation results over 200 independent trials are given in Table 1 using three methods referred above. Obviously, the WP-BSS-HR algorithm outperformed the fastICA algorithm and MUSIC algorithm at different SNR levels, as expected, the proposed algorithm is robust even in low SNR.

Table 1. Right frequency detection rates (if in one experiment the frequency emerges more than one time, it should be recorded as only one time) of the proposed WP-BSS-HR algorithm after 200 simulation experiments at different SNR levels. The corresponding results of fastICA algorithm and MUSIC algorithm are also given.

		source 1	source 2	source 3	source 4
	WP-BSS-HR	100%	100%	100%	100%
10dB	*fastICA*	80%	89%	79.5%	82%
	MUSIC	100%	100%	100%	100%
	WP-BSS-HR	100%	100%	100%	100%
5dB	*fastICA*	85%	82%	73.5%	69%
	MUSIC	100%	100%	100%	100%
	WP-BSS-HR	100%	100%	100%	100%
0dB	*fastICA*	75%	71.5%	58%	67%
	MUSIC	95%	95.5%	94%	95.5%
	WP-BSS-HR	100%	100%	100%	100%
-5dB	*fastICA*	65.5%	51%	49%	38.5%
	MUSIC	69.5%	65%	67%	70%
	WP-BSS-HR	100%	100%	100%	100%
-10dB	*fastICA*	50%	41%	41.5%	33%
	MUSIC	61%	56.5%	53.5%	45.5%
	WP-BSS-HR	83%	88.5%	82%	89%
-15dB	*fastICA*	38%	27.5%	43%	42%
	MUSIC	50.5%	41.5%	39%	44%
	WP-BSS-HR	75%	80%	71%	74.5%
-20dB	*fastICA*	39%	38.5%	33%	36%
	MUSIC	40%	31.5%	37%	35.5%

5   Conclusions

HR in additive colored Gaussian or non-Gaussian noise, especially when the frequencies of the harmonic signals are closely spaced in frequency domain, is a frequently encountered problem in many signal processing problems. In this paper, we developed a BSS and WP decomposition based algorithm from linear mixtures of harmonic signals using only one observed channel signal. The algorithm can be seen as a new avenue to treat HR problem. There are two conclusions, firstly, based on the BSS based HR model, we proposed the HR algorithm called WP-BSS-HR using the WP decomposition method. Secondly, we gave some extensive simulations, the simulation results show that the proposed WP-BSS-HR algorithm could separate or extract the harmonic source signals and yield good performance.

Acknowledgments. This research is financially supported by the National Natural Science Foundation of China (No.61401401, 61172086, 61402421, U1204607), the China Postdoctoral Science Foundation (No.2014M561998) and the young teachers special Research Foundation Project of Zhengzhou University (No. 1411318029).

References

1. Sherman, P.J.: Three techniques for harmonic retrieval in unknown colored noise. Mech. Syst. Signal Pr. **5**(3), 183–197 (1991)
2. Swami, A., Mendel, J.M.: Cumulant-based approach to the harmonic retrieval and related problems. IEEE Trans. Signal Proces. **39**(5), 1099–1109 (1991)
3. Zhang, X.D., Liang, Y.C., Li, Y.D.: A hybrid approach to harmonic retrieval in non-Gaussian noise. IEEE Trans. Inform. Theory **40**(4), 1220–1226 (1994)
4. Yang, S.Y., Li, H.W., Jiang, T.: Detecting the number of 2-D harmonics in multiplicative and additive noise using enhanced matrix. Digit. Signal Process **22**(3), 246–252 (2012)
5. Hyvarinen, A., Karhunen, J., Oja, E.: Independent Component Analysis. Wiley-Interscience, New York (2001)
6. Cichocki, A., Amari, S.: Adaptive blind signal and image processing: learning algorithms and applications. Wiley-Interscience, New York (2002)
7. Comon, P., Jutten, C.: Handbook of blind source separation: independent component analysis and applications. Elsevier, Oxford (2010)
8. Wang, F.S., Zhang, L.R., Li, R.: Harmonic Retrieval by Period Blind Source Extraction Method: Model and Algorithm. Digit. Signal Process **22**(5), 569–585 (2012)
9. Zhang, K., Chan, L.W.: An adaptive method for subband decomposition ICA. Neural Comput. **18**(1), 191–223 (2006)
10. Coifman, R., Wickerhauser, M.: Entropy-based algorithms for best basis selection. IEEE Trans. Inform. Theory **38**(3), 713–718 (1992)
11. Cardoso, J.F.: Dependence, correlation and gaussianity in independent component analysis. J. Mach. Learn. Res. **4**, 1177–1203 (2003)

Underdetermined Blind Speech Signal Separation Method Based on the Improved Shortest Path Method

Shuping Lv$^{(\boxtimes)}$, Chuci Liu, Cheng Zhang, and Jieao Wen

College of Automation, Harbin Engineering University, 150000, Harbin, China
lvshuping@hrbeu.edu.cn,
{8287123,279542151,413768469}@qq.com

Abstract. Underdetermined blind speech signal separation is a widespread case of blind source separation. Underdetermined blind signal separation is the case that the number of sensors is less than the number of sources. A two-step method is used to solve the underdetermined blind speech signal separation, but it has some shortcomings that are weak noise defense, bad signal sparsity, et al. This paper focuses on the study of sparse signals. Then we proposed a smoothing method to improve the shortest path method. The improved shortest path method can remove the false peaks potential function and enhance the accuracy of the mixing matrix estimation. At last, we gained the accurate estimation of the sources by the improved shortest path method.

Keywords: Blind signal separation · Mixing matrix · Potential function · Underdetermined system · Sparse signals

1   Introduction

Blind speech signal separation can be widely used in robot control [1], speech recognition [2], digital communication [3], et al. The determined blind speech signal separation is nearly perfect, but the undetermined speech signal separation only has a few methods [4-6], among which Bofill firstly proposed a two-step method [7]. The method derives a series of new algorithms. The first step is to estimate the mixing matrix through the linear clustering characteristic of the sparse signal and the second step is to estimate source through the mixing matrix and signal sparse constraints. If the mixing matrix isn't estimated accurately, we cannot gain accurate estimated sources. The more accurate the mixing matrix is estimated, the more accurate the sources are. Because the estimation of the mixing matrix is related to the sparsity of the speech signal [8], the sparsity of the speech signal must be studied in the paper.

2   The Mathematical Models and the Analysis of Signal Sparsity

There are two kinds of mathematical models for the underdetermined system. One contains noise, the other doesn't contain noise. The model used in this article is the former, whose math equation is as follows.

© Springer International Publishing Switzerland 2015
Y. Tan et al. (Eds.): ICSI-CCI 2015, Part II, LNCS 9141, pp. 397–404, 2015.
DOI: 10.1007/978-3-319-20472-7_43

$$x(t) = As(t) + N(t), \quad t = 1, 2, \ldots T \tag{1}$$

T is the signal length. The two vectors: $x(t) = [x_1(t), x_2(t) \ldots x_M(t)]^T$ and $s(t) = [s_1(t), s_2(t) \ldots s_N(t)]^T$ are composed of M observation signals and N sources respectively, N>M. $N(t) = [N_1(t), N_2(t) \ldots N_N(t)]^T$ is the noise signal vector and A is the mixing matrix.

Sparse signal is the basis of underdetermined blind source separation [9]. At the time t, the equation (1) can be rewritten as

$$x(t) = a_i s_i(t) + N_i(t) \tag{2}$$

$a_i = [a_{i1}, a_{i2} \ldots a_{iN}]^T$ is the i-th column of A. In the absence of noise, the relationship between the observed signals $x_j(t)$ and $s_i(t)$ as the equation (2) can be inferred to $\dfrac{x_i(t)}{s_j(t)} = a_{ji}$.

$$\frac{x_i(t)}{x_k(t)} = \frac{x_i(t)}{s_j(t)} \bigg/ \frac{x_k(t)}{s_j(t)} = \frac{a_{ji}}{a_{ki}} \tag{3}$$

The equation above shows that if the observed signals are sparse, points of the observed signals distribute on a straight line, the slope is determined by $\dfrac{a_{ji}}{a_{ki}}$.

$s = [s_1, s_2, s_3, s_4]^T$ consists of four original speech signals which is shown in Figure 1. Each mixing channel signal goes on Fourier transform and single wavelet transform, at last we get the two transform's scatter plot respectively.

$$A = \begin{bmatrix} 6.29 & 7.90 & 2.91 & -0.99 \\ 3.48 & 7.13 & -1.89 & -1.19 \end{bmatrix}$$

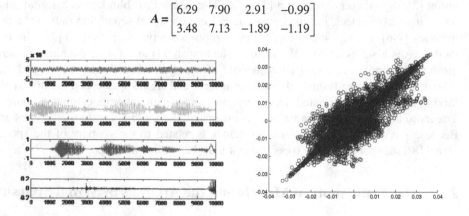

Fig. 1. The original speech signals waveform   **Fig. 2.** The Domain sparse scatter plot

$x = As$ is the mixing process, Figure 2 is a two mixed speech signals time domain scatter plot, Figure 3 is the scatter lot of mixed signal made by short-time Fourier transform, Figure 4 is the scatterplot of mixed signal made by wavelet transform.

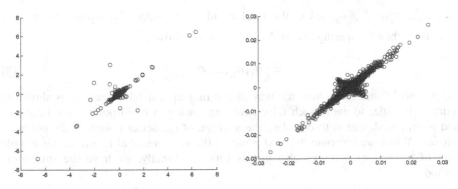

Fig. 3. Frequency Domain scatter plot   **Fig. 4.** Single wavelet transform scatter plot

As is shown in the Figure 2,3 and 4 the sparsity of the time-domain speech signal is the poorest and the transformed signals has better sparsity and better   linear clustering.

3   Estimating the Mixing Matrix Based on the Potential Function

Bofill [11] has been proposed theory to estimate   the mixing matrix A .When mixture space is a plane and directions can be parameterized using the angle θ in polar coordinates.

$$\begin{cases} l_t = \sqrt{\left(x_1^t\right)^2 + \left(x_2^t\right)^2} \\ \theta_t = \tan^{-1}\left(x_2^t / x_1^t\right) \end{cases} \tag{4}$$

l_t and θ_t are the radius and angle respectively, α is the angle between an arbitrary direction and θ_t . A basis function ϕ around x^t is defined as follows.

$$\phi(\alpha) = \begin{cases} 1 - \dfrac{\alpha}{\pi/4} & when |\alpha| < \pi/4 \\ 0 & others \end{cases} \tag{5}$$

A global potential function ϕ was defined over the absolute angle θ as follows.

$$\phi(\theta, \lambda) = \sum_t l_t \phi\left(\lambda(\theta - \theta_t)\right) \tag{6}$$

Bofill used a parameter λ to adjust the desired angular width or resolution of the local contributions and used a weight l_t to put more emphasis on the more reliable data. From his work we can conclude that each peak of the potential function is a column vector of the mixing matrix. We can get the mixing matrix column vector.

$$a_i = \left[\cos(\theta_i), \sin(\theta_i) \right]^T \tag{7}$$

$i = 1, 2, ..., npeak$, $npeak$ is the total number of peaks, θ_i represents the i-th peak angle, then the mixing matrix A can be got as follows.

$$\tilde{A} = \left[a_1, a_2, ..., a_{npeak} \right] \tag{8}$$

The potential function through the way above may appear false peak, as is shown in figure 5. In order to avoid such false peaks appearing, a multi-point smoothing method is proposed, that is to calculate the average of the series points of the potential function. When we compute the next point of the new potential function, the sample point of the former potential also moves forward. Finally, we have the smoothing method.

$$b(j) = (a_{j-n} + a_{j-n+1} + ... + a_j ... + a_{j+n-1} + a_{j+n})/2n \tag{9}$$

After the potential function is smoothed, to some extent the false peaks will be restrained, Figure 6 is figure5's 50-point smoothing.

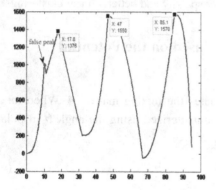

Fig. 5. The potential function

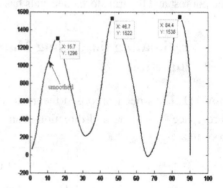

Fig. 6. The smoothed potential function

As is shown in figure 6 there is no false peaks, so this method can improves the accuracy for the column vector estimation of A and source estimation. From Figure 6,

we can get A 's estimation $\tilde{A}$.

$$\tilde{A} = \begin{bmatrix} \cos(15.7°) & \cos(46.7°) & \cos(84.4°) \\ \sin(15.7°) & \sin(46.7°) & \sin(84.4°) \end{bmatrix}$$

4   The Improved Shortest Path Method

When we know the given mixing matrix A and the observed signal x, the original signal is assumed to be sparse, we can't have the only estimation of the sources. We can maximize a posteriori likelihood method to estimate the source signals. Maximizing a posteriori likelihood problem can be converted into a linear programming problem [11]:

$$min \sum_{t=1}^{T} \sum_{j=1}^{N} |s_j(t)|, \quad As(t) = x(t) \tag{10}$$

In the transform domain, the equation becomes:

$$min \sum_{t=1}^{T} \sum_{j=1}^{N} |\tilde{s}_j(t)|, \quad A\tilde{s}(t) = \tilde{x}(t) \tag{11}$$

The shortest path method is a common way to achieve the estimated signals[11]. The shortest path method is described in equation (11), we find the shortest path from the origin o to $x(t)$ is the vector a and b. The sketch map is shown as figure 7.

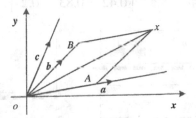

Fig. 7. The sketch map of the shortest path method

$A_r = [a,b]^T$ is the estimated mixing matrix A's any second-order sub-matrix. $s_r(t)$ is component and $x(t)$ is in both directions along the a and b.

$$\begin{cases} s_r = A_r^{-1} x(t) \\ s_j(t) = 0, j \neq a, b \end{cases} \tag{12}$$

This method is computationally intensive, the situation may also occur irreversible. After we remove the signals whose energy is relatively small or weak, the classification approaches can reduce the amount of calculation and improves the classification results. In order to achieve better separation effect, setting $r = 0.1 * \max(l_t)$, $l_t < r$ would make the values of these points zero [12].

Based on the principle of frequency masking method, we remove the points around vectors a_i and these points must satisfy the condition as follows.

$$\frac{1}{6}(\theta_{a2} - \theta_{a1}) \leq \theta \leq \frac{1}{6}(\theta_{a3} - \theta_{a2}) \tag{13}$$

According to the principle of time-frequency masking, we have

$$\begin{cases} s_a(t) = x_i(t)/a_{ii} \\ s_j(t) = 0, j \neq a \end{cases} \quad (14)$$

From the above equation, the inverse operation is avoided so that the amount of computation can be reduced. And then we calculate the remaining points through the method above, we can obtain the estimated signals.

5   Blind Sparse Speech Signals Separation Simulation

The sources used in the simulation are from the ICA Institute at Helsinki University of Technology. The sampling frequency is 8 kHz and the sampling points are 50000.

The simulation process uses three original speech signals shown in Figure 8, it is mixed by a 2×3 random matrix A.

$$A = \begin{bmatrix} 0.85 & -0.12 & -0.9 \\ 0.42 & 0.83 & 0.2 \end{bmatrix}$$

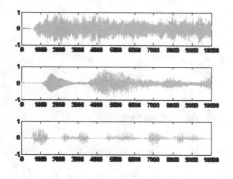

Fig. 8. The speech sources waveform

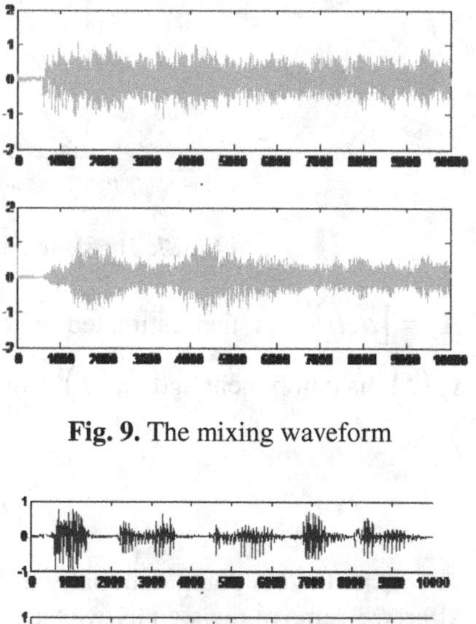

Fig. 9. The mixing waveform

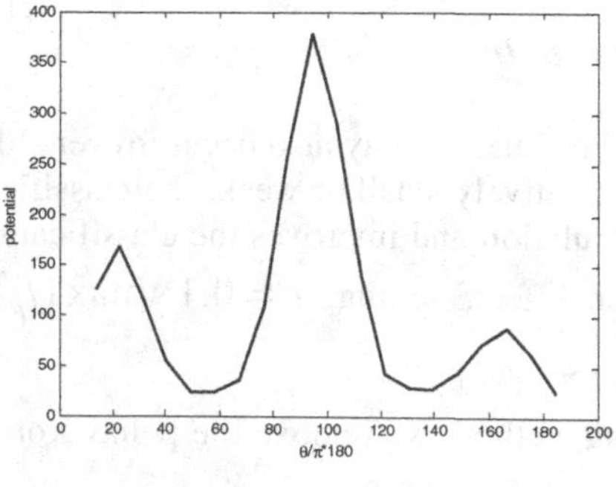

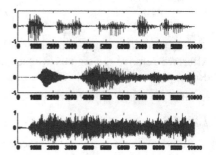

Fig. 10. The potential function

Fig. 11. Estimate of the original signal waveform

The observation signals x is obtained through the mixing process $x = As$, whose waveforms are shown in Figure 9.

Figure 9 shows two channel signals through wavelet transform, the potential function is shown in figure 10.

Through the potential function in figure 10, the mixing matrix A can be got.

$$\tilde{A} = \begin{bmatrix} \cos(20°) & \cos(91°) & \cos(172°) \\ \sin(20°) & \sin(91°) & \sin(172°) \end{bmatrix}$$

$$\tilde{A} = \begin{bmatrix} 0.94 & -0.02 & -0.99 \\ 0.34 & 1 & 0.14 \end{bmatrix}, A = \begin{bmatrix} 0.85 & -0.12 & -0.9 \\ 0.42 & 0.83 & 0.2 \end{bmatrix}. \text{ We fined the two}$$

matrices are very close. We get the estimation of the sources through the improved shortest path method and the result is shown in the figure 11.

Figure8 is compared with figure 11 and we fined the estimated waveform and the sources are very close.

Table 1. Results of Bofill's method in reconstruction ratio of the separation   dB

The number of signal sources	y_1	y_2	y_3	y_4	y_5
2	27.7	24.7			
3	25.8	22.1	19.4		
4	21.7	19.4	15.7	16.6	
5	12.4	12.7	10.4	11.7	9.8

Table 2. Improved method in reconstruction ratio of the separation results   dB

The number of signal sources	y_1	y_2	y_3	y_4	y_5
2	34.9	32.6			
3	29.2	28.4	24.6		
4	25.3	24.7	20.7	21.2	
5	20.4	19.5	15.7	15.8	13.5

Judging from the value of the reconstruction ratio, we find in the same number of the source, the separation results of this paper's method is larger than the separation results of Bofill. This paper describes that our method has a greater improvement than the method of Bofill. With the number of signal source increasing, the reconstruction radio gradually decreases, because with the increase in the number of source signals, signal sparsity declines. In the case of too many sources, this method cannot achieve mixed signal separated.

6   Conclusions

In this paper, we studied the speech signal's sparsity and improved the shortest path method. The signals through STFT and wavelet transform have stronger sparsity.

The improved shortest path method smoothes the potential function and more accurate mixing matrix can be got. At last, we get better separation result than Bofill's method. Our method has better robustness and anti-interference ability. This method uses only two observed signals. As is known to all the less observed signals are, the more sources are separated. The method is more practical and it does not require much sensor acquisition speech signals in practice.

Acknowledgments. The authors are grateful to College of Automation Harbin Engineering University; at the same time, thanks MA student adviser for technical help; at last, thanks our families for their understanding and support.

References

1. Tamal, Y., Sasakl, Y., Kagami, S., et al.: Three ring microphone array for 3D sound localization and separation for mobile robot audition. In: Proceedings of the IEEE Conference on Intelligent Robots and Systems, pp. 1956–1963. IEEE Press, USA (2005)
2. Ehlers, F., Schuster, H.G.: Blind Separation of Convolutive Mixtures and an Application in Automatic Speech Recognition in a Noisy Environment. J. Gnal Rong Ranaon on, 08–12 (1997)
3. Communication signal blind reconnaissance technology based on probability density estimation blind sources separation. J. (2006)
4. Weihong, F., Aili, L., Lifen, M., et al.: Underdetermined blind separation based on the estimated parameters of the potential function method. J. Systems Engineering and Electronics, 19–23 (2014)
5. Weihong, F., Lu, W., Lifen, M.: An improved potential function underdetermined blind source separation. J. Xi'an University of Electronic Science and Technology, 11–16 (2014)
6. Guopeng, W., Yulin, L., Yinggung, L.: Research sparsity of underdetermined blind source separation method based on speech. J. Computer Applications, 06–08 (2009)
7. Bofill, P.: Underdetermined blind separation of delayed sound sources in the frequency domain. J. Neurocomputing, 27–41 (2003)
8. El-shabrawy, O.:Underdetermined Blind Source Separation based on Fuzzy C-Means and Semi-Nonnegative Matrix Factorization . J. Damietta University Researches, 01–04 (2014)
9. Lee, T-W., Lewicki, M.S., Gipolami, M., et al.: Blind source separation of more sources than mixtures using overcomplete representations .J. Signal Processing Letters. 87–90 (1999)
10. Araki, S., Sawada, H., Mukal, R., et al.: Underdetermined blind sparse source separation for arbitrarily arranged multiple sensors. J. Signal Processing, 33–47 (2007)
11. Bofill, P. Zibulevsky, M.:Underdetermined blind source separation using sparse representations. J. Signal Processing, 53–62 (2001)
12. Ning, Z., Xiuyu, X., Qingchao, S., et al.: Based on an improved separation of the shortest path method underdetermined blind speech. J. Sichuan University Natural Science, 09–14 (2012)

Improved Blind Recovery Algorithm
for Underdetermined Mixtures by Compressed Sensing

Fasong Wang[1(✉)], Rui Li[2], Zhongyong Wang[1], and Dong Li[3]

[1] School of Information Engineering, Zhengzhou University, Zhengzhou 450001, China
fasongwang@126.com
[2] School of Sciences, Henan University of Technology, Zhengzhou, China
[3] Key Laboratory of Aerocraft Tracking Telemetering & Command and Communication,
Ministry of Education, Chongqing University, Chongqing 400044, China

Abstract. Linear underdetermined blind source separation (UBSS) is a useful but difficult problem for its illness settings. In this paper, based on compressed sensing (CS) theory, the model inherent connections between UBSS and CS is analysed on the basis of sparsity of the source signals. The mathematical model of underdetermined blind source recovery by CS is built. In order to build the estimated measurement matrix, the source number and mixing matrix are estimated using the refined clustering procedure based on unsupervised robust C prototypes (URCP) method, the measurement matrix and the measurement equation are obtained according to the proposed combined underdetermined blind source recovery model. Then, the proposed blind compressed recovery (BCR) algorithm is derived based on the signal sparse compressive sampling matching pursuit (SSCoSaMP) scheme, which realizes the reconstruction of the underdetermined sparse source signals efficiently. Simulations are provided to show the effectiveness of the proposed method using artificial data.

Keywords: Underdetermined blind source separation (UBSS) · Compressed sensing (CS) · Sparse representation (SR) · Compressive sampling matching pursuit (CoSaMP) · Unsupervised robust C prototypes (URCP)

1 Introduction

Linear underdetermined blind source separation (UBSS) problem is a common issue in practice [1]-[6]. In this case, the inverse of mixing matrix does not exist and, consequently, a solution for source estimation should also be found even if the mixing matrix has been estimated, which makes the conventional independent component analysis (ICA) based BSS algorithm cannot separate or extract all the potential source signals successfully. To dispose of linear UBSS problem, some a priori information should be resorted, a powerful framework for solving linear UBSS is to exploit the sparsity of source signals in a given signal representation dictionary, which is called sparse component analysis (SCA) [6].

There are mainly two kinds of methods to treat UBSS using SCA, the first one is to estimate the mixing matrix and source signals simultaneously [7], and the second one

© Springer International Publishing Switzerland 2015
Y. Tan et al. (Eds.): ICSI-CCI 2015, Part II, LNCS 9141, pp. 405–412, 2015.
DOI: 10.1007/978-3-319-20472-7_44

is the so called two-stage method, where the mixing matrix is estimated in the first stage assuming that the source signals can be sparse represented in some basis or dictionaries and then the source signals are recovered in the consequent stage [8].

The two-stage approach is an attractive and intuitive method that has attracted much more attentions compared with the first method. Moreover, the overwhelming majority natural source signals have the inherent sparsity properties. Sparse representation (SR) of natural signals on a specific set of basis or dictionary has emerged as one of the leading concepts in a wide range of signal processing applications owing to the new sampling theory, compressed sensing (CS) [9]-[12], which provides an alternative to the Nyquist-Shannon sampling theory.

In this paper, following the two-stage strategy, we will combine the model of linear UBSS with CS, according to analyzing the model inherent connections between them, the mathematical model of underdetermined blind sparse signal reconstruction is constructed. Essentially, the linear UBSS is a form of compressive sampling, so we consider the linear UBSS problem directly from the compressed mixtures obtained from CS measurements, that is, instead of solving the UBSS problem in high dimensional data domain, but in a low dimensional measurement domain. But how to design the measurement matrix is not trivial. On the other hand, the source separation process is somehow equivalent to the sparse signal recovery problem confronted in CS and the efficient signal sparse compressive sampling matching pursuit (SSCoSaMP) [13] algorithm will be applied to the underdetermined blind sparse signal reconstruction model to realize the source signals recovery even if the sparse dictionary is truly redundant and not orthogonal anymore.

In this paper, at first, the source number and mixing matrix will be estimated using the unsupervised robust C prototypes (URCP) [14], the measurement matrix and the measurement equation can be obtained according to the proposed combined mathematical model of underdetermined blind sparse signal reconstruction. Then, the proposed blind compressed recovery (BCR) algorithm is derived based on the SSCoSaMP method [13], which realizes the reconstruction of the underdetermined sparse source signals. Simulations of artificial data and real-world data are provided to show the effectiveness of the proposed BCR algorithm.

2   Combined Mathematical Model

BSS is the process of separating a set of unknown original source signals from the observed mixtures without any knowledge about the mixing process or the source signals [15]. The linear BSS problem can be described as:

$$\mathbf{X} = \mathbf{AS} + \mathbf{N} .\tag{1}$$

where $\mathbf{X} = \left[\mathbf{x}_1, \mathbf{x}_2, \cdots, \mathbf{x}_I\right]^{\mathrm{T}} \in \mathbb{R}^{I \times T}$ are the observed signals, $\mathbf{S} = \left[\mathbf{s}_1, \mathbf{s}_2, \cdots, \mathbf{s}_J\right]^{\mathrm{T}} \in \mathbb{R}^{J \times T}$ are the unknown latent source signals. $\mathbf{A} \in \mathbb{R}^{I \times J}$ is the unknown mixing matrix with full row rank and $I \leq J$. $\mathbf{N}$ is the additive noise. In this paper, we consider the noiseless case, that is $\mathbf{N} = \mathbf{0}$.

The main objective of BSS is to estimate the mixing matrix $\mathbf{A}$ and the source signals $\mathbf{S}$. In general, the performance of the conventional BSS methods is poor to UBSS problem and SCA methods are used alternative [1]–[8]. In addition, generally, most signals are not sparse in the original domain, but they may be sparse in another linear transformed domain.

Taking M linear measurements of a signal $\mathbf{v} \in \mathbb{R}^N$ corresponds to applying a measurement matrix $\mathbf{M} \in \mathbb{R}^{M \times N}$,

$$\mathbf{u} = \mathbf{M} \mathbf{v} \ . \tag{2}$$

where the vector $\mathbf{u} \in \mathbb{R}^M$ is the measurement vector. The main interest is in the vastly undersampled case $M \ll N$. Without further information, it is, of course, impossible to recover $\mathbf{v}$ from $\mathbf{u}$ since the linear system (2) is highly underdetermined. However, if the vector $\mathbf{v}$ is sparse, then the situation dramatically changes. This leads to finding the sparsest solution to the l_0-minimization of the underdetermined problem:

$$\min_{\mathbf{v} \in \mathbb{R}^{N \times 1}} \|\mathbf{v}\|_{l_0} \quad \text{s.t.} \quad \mathbf{u} = \mathbf{M} \mathbf{v} \ . \tag{3}$$

Unfortunately, this combinatorial minimization problem is NP-hard in general and computationally intractable [17]. If signal $\mathbf{v}$ is not sparse in the original analysis domain, from the theory of harmonic analysis, $\mathbf{v}$ can be expressed as:

$$\mathbf{v} = \sum_{i=1}^{N} \varphi_i \alpha_i = \Phi \alpha \ . \tag{4}$$

which implies $\alpha_i = \langle \mathbf{s}, \varphi_i \rangle = \varphi_i^{\mathrm{T}} \mathbf{s}$. In practice, the signal $\mathbf{v}$ is not measured directly, alternatively, we measure their linear projections:

$$\mathbf{u} = \mathbf{M} \mathbf{v} = \mathbf{M} \Phi \alpha = \Theta \alpha \ . \tag{5}$$

using an $M \times N$ measurement matrix $\mathbf{M}$, the matrix $\Theta = \mathbf{M} \Phi \in \mathbb{R}^{M \times N}$ is called sensing matrix. $\mathbf{u}$ can be treated as the measurements of sparse signal α projection on the sensing matrix Θ. It is shown that if the matrix $\mathbf{M}$ satisfies the restricted isometry property (RIP) on K-sparse vectors, then we can use $M = \mathrm{O}(K \log(N / K))$ measurements and perform the l_1-minimization by linear programming instead of the l_0-minimization [18], that is:

$$\min_{\alpha \in \mathbb{R}^{N \times 1}} \|\alpha\|_{l_1} \quad \text{s.t.} \quad \mathbf{y} = \mathbf{M} \Phi \alpha = \Theta \alpha \ . \tag{6}$$

There are approximately three practical methods to solving sparse reconstruction problems (3) and (6) in the literatures: 1) Convex optimization algorithms; 2) Combinatorial algorithms; 3) Greedy algorithms, which proved to give equivalent solutions to Eq.(3) with high probability such as orthogonal matching pursuit (OMP) [19] and compressive sampling matching pursuit (CoSaMP) [20] algorithms. The greedy algorithm is in some sense a good compromise between those extremes concerning computational complexity and the required number of measurements.

Considering the general BSS model (1), rewrite source signals in matrix form as $\mathbf{S}$, similarity, the observed signals are $\mathbf{X}$, the relationship between source signals and observed signals can be described as:

$$\left[\mathbf{x}_1, \mathbf{x}_2, \cdots, \mathbf{x}_I\right]^{\mathrm{T}} = \mathbf{A}\left[\mathbf{s}_1, \mathbf{s}_2, \cdots, \mathbf{s}_J\right]^{\mathrm{T}} . \tag{7}$$

Assuming that the data length is T, we can get:

$$\breve{\mathbf{s}} = [s_1(1), \cdots, s_1(T), s_2(1), \cdots, s_2(T), \cdots, s_I(1), \cdots, s_I(T)]^{\mathrm{T}} . \tag{8}$$

$$\breve{\mathbf{y}} = [x_1(1), \cdots, x_1(T), x_2(1), \cdots, x_2(T), \cdots, x_J(1), \cdots, x_J(T)]^{\mathrm{T}} . \tag{9}$$

where $\breve{\mathbf{s}} \in \mathbb{R}^{IT}$ and $\breve{\mathbf{y}} \in \mathbb{R}^{JT}$ respectively. According to Eq.(8) and Eq.(9), Eq.(7) can be described as:

$$\breve{\mathbf{y}} = \mathbf{M}\breve{\mathbf{s}} . \tag{10}$$

where $\mathbf{M} = \begin{pmatrix} \mathbf{M}_{11} & \mathbf{M}_{12} & \cdots & \mathbf{M}_{1I} \\ \vdots & \vdots & \ddots & \vdots \\ \mathbf{M}_{J1} & \mathbf{M}_{J2} & \cdots & \mathbf{M}_{JI} \end{pmatrix} \in \mathbb{R}^{JT \times IT}$ and $\mathbf{M}_{ij} = \mathrm{diag}\left(a_{ij}\right) \in \mathbb{R}^{T \times T}$.

So, Eq. (10) is the measurement equation of CS, $\breve{\mathbf{y}}$ is the measurement vector, $\mathbf{M} = [\mathbf{M}_{ij}] \in \mathbb{R}^{JT \times IT}$ is the measurement block matrix and $\breve{\mathbf{s}}$ is the source signals. If the source signals are sparse under some sparse dictionary, that is, $\breve{\mathbf{s}} = \sum_{i=1}^{I} \varphi_i \alpha_i = \Phi\alpha$, then Eq.(10) is the typical CS model. If the mixing matrix $\mathbf{A}$ has been estimated, we can get the estimated measurement matrix $\hat{\mathbf{M}}$, that is: $\mathbf{M} = \hat{\mathbf{M}}$. The mathematical model for the linear underdetermined blind source recovery using CS can be summarized as:

Given observed signals $\breve{\mathbf{y}}$ and estimated measurement matrix $\hat{\mathbf{M}}$, seeking sparse source signals $\breve{\mathbf{s}}$ which satisfies $\breve{\mathbf{y}} = \hat{\mathbf{M}}\breve{\mathbf{s}}$, where $\breve{\mathbf{s}}$ can be decomposed as $\breve{\mathbf{s}} = \Phi\alpha$ and Φ is the proper sparse dictionary, α are the K-sparse signals.

In the next two sections, the measurement matrix is first estimated, then the source signals will be reconstructed using the efficient CS sparse signal recovery method.

3   Measurement Matrix Construction

One of the main tasks of UBSS problem is the identification of line orientations vectors from the observed mixed signals, so the sparsity is a basic requirement for good estimation of the source number and mixing matrix. In this case a possible solution is to look for a linear sparse transformation $\mathcal{T}[\cdot]$ such that the new representation of the data is sparse. In this paper, we suppose that we have an optimal approach to get the

SR of the source signals, and then the sparsity can be utilized to estimate the source number and mixing matrix from the mixtures.

Before estimating the mixing matrix, one must know or accurately estimate the number of source signals. The BSS problem with an unknown number of source signals is an important practical issue that is usually skipped in many papers by assuming that the number of source signals is known or equal to the number of mixtures. In practice, however, such an assumption does not often hold. In this section we will exploit the refined clustering procedure based on **URCP algorithm [21]**. The proposed procedure can automatically estimate the number of source signals and mixing matrix simultaneously efficient.

Note that the values of threshold parameters ρ_d and ρ_T should be selected properly according to the practical applications in advance. As stated in [21], for example, if the source signals are not sparse enough or the environment is noisy, a small value should be selected for ρ_d while a large value for ρ_T.

4   SSCoSaMP Based BCR Algorithm

In this paper, we consider to use a greedy type method called SSCoSaMP, which considers the situation where the sparse dictionary is truly redundant or overcomplete by utilizing the D-RIP, a condition on the sensing matrix similar to the well-known RIP on measurement matrix of CS. In contrast to prior work, the SSCoSaMP algorithm is "signal-focused", that is, they are focused on recovering the source signal rather than its sparse dictionary coefficients [13].

After the work done in section 2 and section 3, we can obtain the estimation of mixing matrix, then according to Eq.(10), the mathematical model of underdetermined blind sparse signal reconstruction using CS can be derived too. Next, based on the obtained measurement signal $\mathbf{y}$ and measurement matrix $\hat{\mathbf{M}}$.

Note that, the approaches searching for near-optimal supports $\mathcal{S}_D(\tilde{\mathbf{s}}, K)$ are not arbitrary [13], [22], for cases where the nonzero elements of $\boldsymbol{\alpha}$ are scattered well, the OMP and ℓ_1-minimization algorithm can be chosen as the optimal $\mathcal{S}_D(\tilde{\mathbf{s}}, K)$ solvers under the SSCoSaMP framework. On the contrary, when the nonzero elements of $\boldsymbol{\alpha}$ are clustered, the CoSaMP based $\mathcal{S}_D(\tilde{\mathbf{s}}, K)$ solvers and corresponding SSCoSaMP algorithm perform well.

5   Simulations

The proposed BCR algorithm includes two parts: estimating source number and mixing matrix simultaneously to form the measurement matrix of CS and reconstructing the original sparse source signals. So, we will consider these two aspects in the simulations. We will firstly give an experiment to demonstrate the efficiency of the proposed mixing matrix estimation method and the sparse source signals reconstruction method to artificial data, and then a comparison simulation is demonstrated to show the performance of the proposed algorithm to conventional UBSS algorithm. In all the two experiments, we set $\rho_d = 0.5$ and $\rho_T = 0.1$, we also assume that the source signals are sparse, so the dictionary $\mathbf{D} = \mathbf{I}$.

In order to show the ability of the proposed algorithm in identifying the source number and estimating the mixing matrix, the following numerical experiment is carried out in underdetermined cases. The data length of the four artificial sparse source signals is 1024. All source signals have different *spike* numbers which are 12, 15, 16 and 20, where the *spike* numbers means the numbers of non-zero elements for different source signals. The *spike* locations are chosen in random. The mixing matrix $\mathbf{A}$ is set as $\mathbf{A} = \begin{bmatrix} 1.0 & 1.0 & 1.0 & 1.0 \\ 0.6 & 1.3 & 1.9 & 0.8 \end{bmatrix}$. The mixed signals are generated using the mixing matrix $\mathbf{A}$. To check how well the mixing matrix can be estimated, we use the normalized mean square error (NMSE) in dB as a performance index [16].

Using the proposed **URCP** based refined clustering algorithm, we can obtain the estimation of source number and the mixing matrix $\hat{\mathbf{A}}$ with best performance. Because of the ambiguity of the estimated mixing matrix and the true one, before calculate the estimation error, we must adjust the sign and order of the columns of $\hat{\mathbf{A}}$ to make it corresponds to $\mathbf{A}$. After these procedures, the estimated mixing matrix can be obtained as $\hat{\mathbf{A}} = \begin{bmatrix} 1.0021 & 1.0058 & 0.9934 & 1.0131 \\ 0.6001 & 1.2944 & 1.8990 & 0.7954 \end{bmatrix}$. After 100 simulations, the average result of the performance index of $\mathbf{A}$ is $\mathrm{NMSE}(\mathbf{A}) = -45.2605$ dB.

For comparison, we use the single-source-points (SSPs) algorithm proposed in [16], using the same mixing matrix $\mathbf{A}$ as above, similarly, after 100 simulations, the average result of the performance index of $\mathbf{A}$ is $\mathrm{NMSE}(\mathbf{A})_{SSP} = -40.5433$ dB. It should be note that the SSPs algorithm must know the accurate source number before execution. From the experiment results, we can conclude that the proposed BCR algorithm for mixing matrix and source number estimation can works well.

In order to testify the performance of the proposed BCR algorithm, a comparison of the proposed BCR algorithm with conventional UBSS algorithm is presented, we demonstrate an experiment that separate 4 sparse source signals from 2 mixtures. The data length of the source signals is 1024. All source signals have different spike numbers which are 15, 25, 35 and 45. The compared conventional UBSS algorithms are chosen as the popular time-frequency ratio of mixtures (TIFROM) algorithm [23], [24] and the degenerate unmixing estimation technique (DUET) algorithm [25].

In order to compute the $\mathcal{S}_{\mathbf{D}}(\tilde{\mathbf{s}}, K)$, in this experiment we also use OMP as the optimal solver to execute the proposed SSCoSaMP algorithm to reconstruct the sparse source signals. The experiment settings of TIFROM and DUET algorithm are tuned to the best performance. We repeat each of the experiment 100 times and calculate the average performance. Table 1 gives the average simulation results. In Table 1, the reconstruction performance index S/N is the signal-to-noise ratio of the error between source signals $s_i(t)$ and their estimates $\hat{s}_i(t)$, which is calculated to measure the accuracy of the estimations of the source signals and defined as [26]. As shown in Table 1, the proposed BCR algorithm can successfully reconstruct the sparse source signals and has better performance than TIFROM and DUET algorithms.

Table 1. The comparison results of the proposed BCR algorithm with TIFROM algorithm and DUET algorithm of four sparse source signals: S/N(dB)

Algorithm	S/N (dB)			
	source 1	source 2	source 3	source 4
BCR	61.52	59.68	57.31	55.44
TIFROM	18.25	19.33	17.21	16.67
DUET	16.32	17.14	15.80	14.60

6   Conclusions

In this paper, we build the underdetermined blind source recovery mathematical model using CS. We develop the source number and mixing matrix estimation method based on refined clustering procedure using URCP **algorithm**. Then, the measurement matrix is constructed. Consequently, the proposed blind compressed reconstruction algorithm is derived based on the SSCoSaMP method to realize the reconstruction of the underdetermined sparse source signals. Simulations show the effectiveness of the proposed method for artificial data, some comparisons with the state-of-the-art algorithms are also provided.

Acknowledgments. This research is financially supported by the National Natural Science Foundation of China (No.61401401, 61172086, 61402421, U1204607), the China Postdoctoral Science Foundation (No.2014M561998) and the young teachers special Research Foundation Project of Zhengzhou University (No. 1411318029).

References

1. Plumbley, M.D., Blumensath, T., Daudet, L., Gribonval, R., Davies, M.E.: Sparse representations in audio music: from coding to source separation. Proc. of IEEE **98**(6), 995–1005 (2010)
2. Liu, B., Reju, V.G., Khong, A.W.H.: A linear source recovery method for underdetermined mixtures of uncorrelated AR-model signals without sparseness. IEEE Trans. Signal Process. **62**(19), 4947–4958 (2014)
3. Gu, F.L., Zhang, H., Wang, W.W., Zhu, D.S.: PARAFAC-based blind identification of underdetermined mixtures using Gaussian mixture model. Circuits Syst. Signal Process. **33**(6), 1841–1857 (2014)
4. Li, Y., Cichocki, A., Amari, S.: Analysis of sparse representation and blind source separation. Neural Comput. **16**(6), 1193–1234 (2004)
5. Aissa-El-Bey, A., Linh-Trung, N., Abed-Meraim, K., Belouchrani, A., Grenier, Y.: Underdetermined blind separation of nondisjoint sources in the time-frequency domain. IEEE Trans. Signal Process. **55**(3), 897–907 (2007)
6. Georgiev, P.G., Theis, F., Cichocki, A.: Sparse component analysis and blind source separation of underdetermined mixtures. IEEE Trans. Neural Netw. **16**(4), 992–996 (2005)
7. He, Z., Xie, S., Zhang, L., Cichocki, A.: A note on Lewicki-Sejnowski gradient for learning overcomplete representation. Neural Comput. **20**(3), 636–643 (2008)

8. Yang, Z.Y., Tan, B.H., Zhou, G.X., Zhang, J.L.: Source number estimation and separation algorithms of underdetermined blind separation. Sci. China Ser. F **51**(10), 1623–1632 (2008)

9. Eldar, Y.C., Kutyniok, G.: Compressed sensing: theory and applications. Cambridge University Press, USA (2012)

10. Candes, E., Romberg, J., Tao, T.: Robust uncertainty principles: exact signal reconstruction from highly incomplete frequency information. IEEE Trans. Inform. Theory **52**(2), 489–509 (2006)

11. Donoho, D.L.: Compressed sensing. IEEE Trans. Inform. Theory **52**(4), 1289–1306 (2006)

12. Baraniuk, R.G.: Compressive sensing. IEEE Signal Proc. Mag. **24**(4), 118–120 (2007)

13. Davenport, M.A., Needell, D., Wakin, M.B.: Signal space CoSaMP for sparse recovery with redundant dictionaries. IEEE Trans. Inform. Theory **59**(10), 6820–6829 (2013)

14. Frigui, H., Krishnapuram, R.: A robust algorithm for automatic extraction of an unknown number of clusters from noisy data. Pattern Recognit. Lett. **17**(6), 1223–1232 (1996)

15. Comon, P., Jutten, C.: Handbook of blind source separation: independent component analysis and applications. Elsevier, Amsterdam (2010)

16. Reju, V.G., Koh, S.N., Soon, I.Y.: An algorithm for mixing matrix estimation in instantaneous blind source separation. Signal Process. **89**(9), 1762–1773 (2009)

17. Duarte, M.F., Eldar, Y.C.: Structured compressed sensing: from theory to applications. IEEE Trans. Signal Process. **59**(9), 4053–4085 (2011)

18. Candes, E., Tao, T.: Decoding by linear programming. IEEE Trans. Inform. Theory **51**(12), 4203–4215 (2005)

19. Tropp, J., Gilbert, A.: Signal recovery from random measurements via orthogonal matching pursuit. IEEE Trans. Inform. Theory **53**(12), 4655–4666 (2007)

20. Needell, D., Tropp, J.A.: CoSaMP: Iterative signal recovery from incomplete and inaccurate samples. Appl. Comput. Harmon. A **26**(3), 301–321 (2009)

21. Lv, Q., Zhang, X.D.: A unified method for blind separation of sparse sources with unknown source number. IEEE Signal Process. Let. **13**(1), 49–51 (2006)

22. Giryes, R., Needell, D.: Near oracle performance and block analysis of signal space greedy methods, submitted for publication. arXiv: 1402. 2601 (2014)

23. Abrard, F., Deville, Y.: A time-frequency blind signal separation method applicable to underdetermined mixtures of dependent sources. Signal Process. **85**(7), 1389–1403 (2005)

24. Li, R., Wang, F.S.: Blind dependent sources separation method using wavelet. Int. J. Comput. Appli. Tech. **41**(3), 296–302 (2011)

25. Yilmaz, O., Rickard, S.: Blind separation of speech mixtures via time-frequency masking. IEEE Trans. Signal Process. **52**(7), 1830–1847 (2004)

26. Bofill, P., Zibulevsky, M.: Underdetermined blind source separation using sparse representations. Signal Process. **81**(11), 2353–2362 (2001)

Swarm Interaction Behavior

Established Routine of Swarm Monitoring Systems Functioning

Alexey Ivutin[1]([⊠]), Eugene Larkin[2], and Vladislav Kotov[2]

[1] Department of Computer Technology, Tula State University, Tula 300012, Russia
alexey.ivutin@gmail.com, elarkin@mail.ru, vkotov@list.ru
[2] Department of Robotics and Industry Mechanization,
Tula State University, Tula 300012, Russia

Abstract. Questions of simulation of established conditions of swarm automatic monitoring systems, which include number of assemblies and blocks been operated according to a local cyclic program are considered. Sample cyclic program, which provides switches on/off board equipment for monitoring is formed. Simplification of sample cyclic program is obtained. With use of semi-Markov matrix, which circumscribe sample cyclic program, densities of times of active and passive states of equipment is obtained. Formulae for evaluation of probabilities of active and passive states of pre-determined quantity from common number of swarm monitoring units are received.

Keywords: Swarm monitoring system · Unit of swarm monitoring system · Cyclic program · Semi-markov process · Ergodic process · Switches on/off · Active state · Passive state · External observer

1 Introduction

At present swarm automatic monitoring systems are widely used for observation of scenes in areas of ecology and technogenic disasters, military conflicts, intense building of large objects etc. For ensuring of most full coverage of observed territory by fields of vision of monitoring units there are used complexes of units, both stationary and mobile, such as mobile robots, marine robots, unmanned aerial vehicles etc. Control of complexes is done from point of management, to where data from all sensors is transmitted [1].

Distinctive feature of such systems is that every monitoring unit operates autonomously in accordance with a program, which is embedded to its onboard computer. Program operates on cyclic algorithm, One of algorithm's command transfers onboard sensors in active state, and other command switches off sensors until the next transferring of it in active state [2].

During operation of swarm monitoring systems emerging two problems. First of it is the problem of covering with sensor fields of view the proper square in current time. The second problem consists in limitation of capacity of communication channel through which information from monitoring units arrives to operator observer. Due to the fact, the task of evaluation number of active units in current time is quite actual.

2 Mathematical Model

For solving the problem of evaluation of state of swarm monitoring system one can assume that group includes k units of hardware. Unit with number k operates in accordance with k-th cyclic program and every state of k-th cyclic program is linked with state of k-th unit. For external observer switches to adjacent states of cyclic program have a stochastic nature. Time of presence in states is accidental too. So in common, operation of a swarm monitoring system may be described with the next formalism:

$$M = \{M_1, \ldots, M_k, \ldots, M_K\}, \tag{1}$$

where M_k - is the model of operation of k-th unit of swarm.

Natural formalism for description of k-th unit of group is semi-Markov sub-process [3, 4] of the form

$$M_k = \{A_k, \mathbf{R}_k, \mathbf{h}_k(t)\}, \tag{2}$$

where $A_k = \{a_{1(k)}, \ldots, a_{j(k)}, \ldots, a_{J(k)}\}$ - is a set of states of k-th cyclic program; $\mathbf{R}_k = [R_{j(k),n(k)}]$ - is the adjacency matrix having the size $J(k) \times J(k)$, which describes a structure of k-th cyclic program; $\mathbf{h}_k(t)$ - semi-Markov matrix having the size $J(k) \times J(k)$ which specifies stochastic and temporal parameters of k-th cyclic program implementation;

$$R_{j(k),n(k)} = \begin{cases} 1, & \text{if from the state } a_{j(k)} \text{ one can to switch to the state } a_{n(k)}; \\ 0, & \text{if from the state } a_{j(k)} \text{ one cannot to switch to the state } a_{n(k)}; \end{cases}$$

$$\mathbf{h}(t) = \mathbf{p} \otimes \mathbf{f}(t) = [p_{j(k),n(k)} \cdot f_{j(k),n(k)}(t)]; \tag{3}$$

$\mathbf{p}_k = [p_{j(k),n(k)}]$ — is the stochastic matrix, which defines probabilities of switches from the state $a_{j(k)}$ to the state $a_{n(k)}$; $\mathbf{f}_k(t) = [f_{j(k),n(k)}(t)]$ - is the matrix of densities of time of presence of k-th semi-Markov sub-process in state $a_{j(k)}$, if after that it switches to the state $a_{j(k)}$.

On elements of the adjacency matrix, the stochastic matrix and the matrix of time densities next restrictions may be imposed:

$$p_{j(k),n(k)} = 0, \text{ if } r_{j(k),n(k)} = 0;$$

$$\sum_{n(k)=1}^{J(k)} p_{j(k),n(k)} = 1 \text{ for all } 1 \leqslant j(k) \leqslant J(k); \tag{4}$$

$$f_{j(k)n(k)}(t) = 0 \text{ if } t \leqslant 0.$$

In such a way, taking into account accepted restrictions all K semi-Markov processes, included in set (1) are ergodic ones, and structures of it are characterized by strongly connected graph.

Control of switching on monitoring hardware is defined with one of states of cyclic program. Without loss of generality, one can accept that in all cyclic programs state $a_{1(k)}$ switches on the hardware, and $a_{m(k)}$ switches of it. This assumption leads to the next additional restrictions laid on semi-Markov processes M_k (fig. 1)

in sets A_k –

$$A_k = \{a_{1(k)}\} \cup \{a_{2(k)}, \ldots, a_{j(k)}, \ldots, a_{m(k)-1}\} \cup \{a_{m(k)}\} \cup$$
$$\{a_{m(k)+1}, \ldots, a_{i(k)}, \ldots, a_{J(k)}\};$$

in the adjacency matrix $\mathbf{R}_k$ and semi-Markov matrix $\mathbf{h}_k(t)$ –

$R_{j(k),n(k)} = 0$ and $h_{j(k),n(k)}() = 0$, if $1 \leqslant j(k) \leqslant m(k) - 1$ and $n(k) = 1$,

or if $j(k) = 1$ and $m(k) \leqslant n(k) \leqslant J(k)$,

or if $2 \leqslant j(k) \leqslant m(k) - 1$ and $m(k) + 1 \leqslant n(k) \leqslant J(k)$,

or if $j(k) = m(k)$ and $1 \leqslant n(k) \leqslant m(k) - 1$,

or if $m(k) + 1 \leqslant j(k) \leqslant J(k)$ and $2 \leqslant n(k) \leqslant m(k) - 1$,

or if $m(k) \leqslant j(k) \leqslant J(k)$ and $n(k) = m(k)$.

A rest elements of adjacency matrix $\mathbf{R}_k$ and semi-Markov matrix $\mathbf{h}_k(t)$ take such a value as to ensure an ergodicity of semi-Markov process

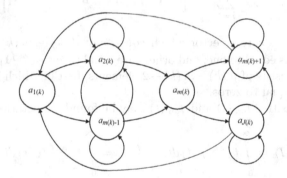

Fig. 1. A graph of control of switching on/off of monitoring hardware

Additional restriction is determined to the fact that in correctly designed cyclic algorithm a switching on the hardware must follow a switching off, and a switching off must follow a switching on. Twice in row hardware mast not be switched on or off.

3   Evaluation of Time Intervals

Let us evaluate time intervals between both switching on - switching off and switching off - switching on [5,6].

For evaluation of on/off time interval let us exclude from matrices $\mathbf{R}_k$ and $\mathbf{h}_k(t)$ columns and rows with numbers from $[m(k)+2]$-th till $J(k)$-th. As a result squire matrices ${}^{on}\mathbf{R}_k$ and ${}^{on}\mathbf{h}_k(t)$ of size $m(k) \times m(k)$ are obtained.

A time of hardware switching on is defined as time density of semi-Markov process ${}^{on}\mathbf{h}_k(t)$ achievement of state $m(k)$ from the state $1(k)$ with use of dependence

$$
{}^{on}f_k(t) = \sum_{r=1}^{\infty} \Im^{-1}\left[{}^{on}\mathbf{I}_1\left[\Im\left[{}^{on}\mathbf{h}_k(t)\right]\right]^r {}^{on}\mathbf{I}_{m(k)}\right], \tag{5}
$$

where ${}^{on}\mathbf{I}_1$ - is the row vector, which consists of $m(k)$ elements, first of which is equal to one and others are equal to zeros; ${}^{oon}\mathbf{I}_{m(k)}$ - is the column vector which consists of $m(k)$ elements, last of which is equal to one and others are equal to zeros; $\Im[\ldots]\,\Im^{-1}[\ldots]$ - direct and inverse Fourier transforms respectively.

For evaluation of off/on time interval let us exclude from matrices $\mathbf{R}_k$ and $\mathbf{h}_k(t)$ columns and rows with numbers from second till $[m(k)+2]$-th. As a result squire matrices ${}^{on}\mathbf{R}_k$ and ${}^{on}\mathbf{h}_k(t)$ of size $[J(k)-m(k)+2] \times [J(k)-m(k)+2]$ are obtained.

A time of hardware switching off is defined as time density of semi-Markov process ${}^{off}\mathbf{h}_k(t)$ achievement of first state $m(k)$ from the state $m(k)$ with use of dependence

$$
{}^{off}f_k(t) = \sum_{r=1}^{\infty} \Im^{-1}\left[{}^{off}\mathbf{I}_{m(k)}\left[\Im\left[{}^{off}\mathbf{h}_k(t)\right]\right]^r {}^{off}\mathbf{I}_1\right], \tag{6}
$$

where ${}^{off}\mathbf{I}_{m(k)}$ - is a row vector which consists of $J(k)-m(k)+2$ elements, second of which is equal to one, and others are equal to zeros; ${}^{off}\mathbf{I}_1$ - is a column-vector, which consists of $J(k)-m(k)+2$ elements, first of which is equal to one and others are equal to zeros.

For time densities $f_{on}(t)$ and $f_{on}(t)$ may be found expectations

$$
{}^{on}T_k = \int_0^{\infty} t \cdot {}^{on}f_k(t)\,dt; \quad {}^{off}T_k = \int_0^{\infty} t \cdot {}^{off}f_k(t)\,dt. \tag{7}
$$

4   Simplified Semi-Markov Process

In such a way common semi-Markov process (1), which describes states of swarm gets especially simple form (fig. 2) [7]:

$$
M_k' = \left\{\{b_{1(k)},\, b_{2(k)}\},\, \begin{pmatrix} 0 & 1 \\ 1 & 0 \end{pmatrix},\, \begin{bmatrix} 0 & {}^{on}f_k(t) \\ {}^{off}f_k(t) & 0 \end{bmatrix}\right\}; 1 \leqslant k \leqslant K,
$$

where $b_{1(k)}$ - is the state, which complies to switched on k-th unit; $b_{2(k)}$ - is the state, which complies to switched off k-th unit.

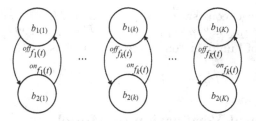

Fig. 2. The simplified form of semi-Markov process (1)

In established routine for external observer probabilities of "on" and "off" states of hardware in current time are equal to

$$^{on}p_k = \frac{^{on}T_k}{^{on}T_k + ^{off}T_k}; \quad ^{off}p_k = \frac{^{off}T_k}{^{on}T_k + ^{off}T_k}. \tag{8}$$

Let us define probabilities of l from K "on" states of swarm monitoring system $(0 \leqslant l \leqslant K)$. Let us produce set N_K K-digit binary natural numbers, k-th digit of which, σ_k, is fasten to k-th unit and can get two meanings:

$$\sigma_k = \begin{cases} 0, & \text{if } k\text{-th unit is in "on" state;} \\ 1, & \text{if } k\text{-th unit is in "off" state.} \end{cases} \tag{9}$$

Let us select from the set N_K a subset $N_K^l \subset N_K$ binary K-digit numbers, which have l units and $K - l$ zeros:

$$N_K^l = \{n_1, \ldots, n_{c(K,l)}, \ldots, n_{C(K,l)}\}, \tag{10}$$

where $C[K,l] = \frac{K!}{l! \cdot (K-l)!}$ - is quantity of K-digit numbers with l ones and $K - l$ zeros, which is equal to l-th binomial coefficient; $c(K,l)$ - is number of $n_{c(K,l)}$'s in set N_K^l;

$$n_{c(K,l)} = \left\langle \sigma_1^{c(K,l)}, \ldots, \sigma_k^{c(K,l)}, \ldots, \sigma_K^{c(K,l)} \right\rangle. \tag{11}$$

Let us define the function $P\left(\sigma_k^{c(K,l)}\right)$, which take the form

$$P\left(\sigma_k^{c(K,l)}\right) = \begin{cases} ^{on}p_k, & \text{if } \sigma_k^{c(K,l)} = 1; \\ ^{off}p_k, & \text{if } \sigma_k^{c(K,l)} = 0. \end{cases} \tag{12}$$

With (12), probability of "on" state l units from K is defined as

$$^{on}P_K^l = \sum_{c(J,l)=1}^{C(K,l)} \prod_{k=1}^{K} P\left(\sigma_k^{c(K,l)}\right), \tag{13}$$

where $\prod_{k=1}^{K} P\left(\sigma_k^{c(K,l)}\right)$ - probability of "on" state of k-th units of swarm, which corresponds to number $c(K,l)$ from all set of combinations.

In case of homogeneous swarm, $M_1 = \ldots = M_k \ldots = M_K$, and $^{on}p_1 = \ldots = {}^{on}p_k \ldots = {}^{on}p_K = {}^{on}p^{off}p_1 = \ldots = {}^{off}p_k \ldots = {}^{off}p_K = {}^{off}p$. In this case probabilities of l units in "on" state is defined as

$$^{on}P_K^l = C\,(K, l)\,(^{on}p)^l \cdot \left(^{off}p\right)^{K-l}. \tag{14}$$

5   Experimental Verification of Model

For verification of proposed method direct computer experiment was executed with use the Monte-Carlo method. Homogeneous swarm monitoring system model under verification includes 100 units of hardware. Distribution densities of stay of every unit in one of possible states were described by exponential laws with parameters λ and μ.

In the program k-th unit was described by the next data: B_k - current state ($B_k = 1$ - on state, $B_k = 0$ - off state); λ and μ - intensities of streams of on and off: events, correspondingly; T_k - random time of stay in current state. Besides global current time T, lately supervision time t and period of supervision Δt are defined.

Computer experiment was carried out as follows:

1. For all units of swarm initial state of B_k ($1 \leq k \leq K$) is accepted equal to 0, global current time T and lately supervision time t are established as $T = 0, t = 0$.
   Besides, the period Δt of supervision of swarm is established too.
2. For all units of swarm computation of random time of stay in current state is carried out as

$$T_k\,(B_k) = \begin{cases} \frac{-ln(1-\xi)}{\lambda}, & \text{if } B_k = 0; \\ \frac{-ln(1-\xi)}{\mu}, & \text{if } B_k = 1. \end{cases} \tag{15}$$

   where ξ - data is obtained from RAND-program (evenly distributed in the interval $0 \leq \xi \leq 1$).
3. The index k of unit with smallest time of stay in current state, $k^* = \arg\min_{1 \leq k \leq K} \{T_k\}$, is defined.
4. For all units correction of time of stay in current state is executed:

$$T_k = T_k - T_{k^*}.$$

5. The global current time T is increased on the value T_{k^*}.
6. For k^*-th unit current state B_{k^*} is changed on the inverse one and generation of the next random time interval of stay of k-th unit in current state, $T_k\,(B_k)$, in accordance with (15), is carried out.
7. If $(T - t) < \Delta t$, then transition to 10, otherwise transition to 8.
8. In accordance with the current distribution of states of swarm units, quantity of units, being in state 0, is defined. Defined number is added to the statistics of supervision of swarm.
9. Lately supervision time t is increased on Δt.

10. If global current time T is less than the end time of finish of experiment, then transition to 3, otherwise end of computer experiment.

On fig. 3 histograms of supervision of swarm for different ratios λ/μ are shown. Parameters of computer experiment were the next: $K = 1000$, $\Delta t = 10$, $T_e nd = 100000$. Histograms rather precisely correspond to theoretical distributions (are shown by continuous line) defined by (14).

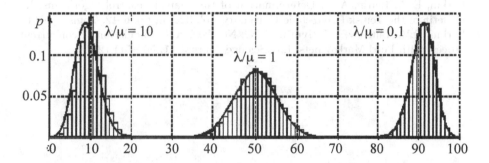

Fig. 3. Histograms of distribution of devices' number included in the swarm

6   Conclusion

In such a way mathematical apparatus for analytical description of programmable controlled objects, which include set of subsystems, every of which is functioned on its own algorithm. States of subsystems are formed the semi-Markov sub-processes, each state of which simulates the interpretation of the program operator. It is possible to determine with accuracy to time density distribution the time intervals between states. Timing and probabilistic characteristics of the states of swarm monitoring systems were obtained in a general form. They are essential to the achievement of quality control parameters.

Further research in this area may be directed to the development of simple engineering techniques for estimation of numerical characteristics of timing intervals, for instance by imitational simulation of polling process, or by the development of a time-compiler, which in parallel with the translation of the control program from high-level language will evaluate numerical characteristics of density of distributions between selected algorithm's operators.

References

1. Beni, G., Wang, J.: Swarm intelligence in cellular robotic systems. In: Sandini, G., Aebischer, P. (eds.) Robots and Biological Systems: Towards a New Bionics?. NATO ASI Series, vol. 102, pp. 703–712. Springer, Heidelberg (1993)
2. Larkin, E.V., Ivutin, A.N.: Estimation of latency in embedded real-time systems. In: 3-rd Mediterranean Conference on Embedded Computing (MECO-2014), pp. 236–239. IEEE (2014)

3. Howard, R.A.: Dynamic Probabilistic Systems. Semi-Markov and Decision Processes, vol. II. Dover Publications, NY (2007)
4. Ivutin, A.N., Larkin, E.V.: Generalized semi-Markov model of digital devices control algorithm. In: Bulletin of the Tula State University, vol. 1, pp. 221–228. TSU, Tula (2013)
5. Ivutin, A.N., Larkin, E.V.: Time and probability characteristics of a transaction in a digital control system. In: Bulletin of the Tula State University, vol. 1, pp. 252–258. TSU, Tula (2013)
6. Larkin, E.V., Ivutin, A.N.: Determination of time gaps in control algorithms. In: Bulletin of the Tomsk Polytechnic University, vol. 324(5), pp. 6–12, Tomsk (2014)
7. Larkin, E.V., Lutskov, Y.I., Ivutin, A.N., Novikov, A.S.: Simulation of concurrent process with Petri-Markov nets. Life Science Journal 11, 506–511 (2014)

A Gamified Online Forum Inspiring Group Intelligence Distillation for Policy Making

Shueh-Cheng Hu[1] and I-Ching Chen[2(✉)]

[1] Department of Computer Science and Communication Engineering,
Providence University, Taichung City, Taiwan, ROC
shuehcheng@gmail.com
[2] Department of Information Management,
Chung Chou University of Science Technology, Changhua County, Taiwan, ROC
jine@dragon.ccut.edu.tw

Abstract. Public affair administrators around the globe show increasing interests in seeking the public's opinions and insights into critical issues through Internet. The public's collective intelligence could be distilled via the deliberative processes in various online forums. To successfully serve as a source of group intelligence, an online forum needs active participation and contributions from its visitors. However, just like common information systems, most online forum systems were designed without consideration of motivating users. This article presents the gamification process of an open-sourced online forum system. The gamified JForum embeds several game mechanisms motivating visitors to participate and contribute more actively, which facilitates forming a resource of productive consensus.

Keywords: Public policy making · Group intelligence · Online forum · Gamification

1 Introduction

Today, people widely use the Internet as media, commercial channels, and social platforms. Besides, increasing public affair administrators rely on the Internet to offer public services due to merits such as enhanced efficiency and transparency [1, 2]. Among many administrative tasks, policy making is very critical since the qualities of executed policies directly influence citizens' satisfaction toward administration organizations. Toward that end, many progressive administrators tend to seek diverse opinions and insightful thoughts from the public on various online forums before making the corresponding policies [3]. That's because online forums are able to broaden the source of innovative ideas and insightful thoughts from a large-scaled public engagement. Furthermore, gradually-formed consensus to certain extent represents group intelligence that could be distilled through time-consuming deliberative processes at online forums. Obviously, in order to serve as a productive arena of distilling group intelligence for policy making, an online forum needs intensive engagement as well as constructive contributions from its participants.

© Springer International Publishing Switzerland 2015
Y. Tan et al. (Eds.): ICSI-CCI 2015, Part II, LNCS 9141, pp. 423–430, 2015.
DOI: 10.1007/978-3-319-20472-7_46

However, many online forums gradually withered due to decreasing participation and contributions. Unfortunately, the presence of these gradually withered online forums will be further deteriorated by many search engines that favor Web sites with up-to-date and intensively-referred contents [4]. Even worse, the lack of participation, contribution, and deteriorated search engine visibility will form a vicious circle. Accordingly, motivating participants is a critical task for successfully operating an effective online forum.

There are many options for people who need an online forum system, and many of the options even come with free licenses. However, these online forum systems were designed by people who applied conventional software development procedure and principles, which results in a software system with complete and correct functions, but their users might use these functions inactively. Obviously, that kind of systems cannot become sources of productive group intelligence for policy makers. In light of the significance of an animating online forum system and the lack of relevant studies, this work tried to renovate an open-sourced online forum system: JForum [5] through gamifying it.

2   Prior Studies Review

Most information systems were designed without consideration of motivating users because the traditional design philosophy only takes functionality and accountability into account while ignore the role that an user's motive plays in an individual's overall productivity. As a result, the so-called well-designed information systems offer complete functions enabling users to accomplish their assigned tasks correctly, but did not equip any mechanisms to motivate users to perform tasks more actively or even enthusiastically.

The routine tasks bore people; prior study indicated that employee's working quality will degrade if they experience boredom [6]. In highly informationized environments, people heavily rely on various information systems to complete their routine tasks. Consequently, designers need to take inspiring user's motive into account while they are developing an information system. To create a more animating working environment in the age of informationization, gamification of information systems rationally emerges as a popular approach.

Gamification refers to planting game mechanisms into a non-game environment such as information system [7, 8]. The original idea is blending users' engagement and addiction that could be found while they are playing various games into their working environments. Its purpose is to strengthen users' motivation; i.e., make users perform tasks with more fun, stronger motive, and deeper engagement. Once each individual is well motivated, the overall performance of an organization could be improved consequently.

With wide recognition of its effects, gamification has been applied by enterprises to animate their employees; i.e., users of their information systems. According to a report from Gartner, over 70% of the global 2000 enterprises have used gamification to renovate their information system before 2015 [9]. Among many successful cases,

Starbucks gamified one supply chain management system by ranking suppliers in a leaderboard according to their on-time delivery records. The operations of the gamified system urged suppliers in the supply chain try their best to fulfill orders on time, thus earned better ranking in the leaderboard, which meant better efficiency and administration from outsides' perspectives. The benign consequence brought to Starbucks by this gamification work were higher percentage of on-time delivery of supplies, lower logistics cost, and more profits [10]. Delta airlines successfully used a smartphone APP including a number of game mechanisms to enhance its public recognition, customer loyalty, and revenue [11]. Not only in traditional manufacturing and service industries, gamification also was adopted by software developers, who integrated gamification components into software development process, and the preliminary results indicated that improved quality of software and corresponding documents [12].

3   Gamification Analysis and Design

Including the mentioned cases, the success of many prior experiences [13-16] collectively point out an important fact: successful gamification of an information system does not necessarily rely on complex game mechanisms. By contrast, the key factors are identifying the functional parts that need to motivate users, and then embed proper game mechanisms to increase users' motivation and engagement.

The JForum, an online forum system was selected to gamify due to its openness and popularity. The gamification implies renovating an existing information system rather than creating a new one from scratch, thus openness is critical. Generally speaking, a gamification process comprises the following key activities: understanding the users, setting missions, identifying motivations, and embedding effective game mechanisms accordingly [17], and this section delineates the particular process.

3.1   Target Users and Missions

Rationally, target users of an online forum share a common profile: they are willing to acquire knowledge during the course of interacting with peers, so they likely to be socializers, explorers, and achievers according to the Bartle's player type categorization [18]. Socializers enjoy the interactions with peers in the forum, explorers are happy to find new information that they did not know before, and achievers can feel satisfaction by observing peers responded or recognized the helpful information that they provided.

Obviously, reasonable target business outcomes of a gamified online forum include more active participation and more productive contributions from users. Consequently, the missions of a gamification work should be to encourage people to join the forum, participate the discussions, and contribute (raise issues or provide information and feedback) more valuable contents.

3.2   Motivational Drivers Analysis

With clear missions, then we need to identify factors that are able to drive people perform what we intend them to do. A number of motivational drivers [19] that suits with target users are described as follows:

1. Collecting: For a long time, people enjoy collecting of either physical or abstract things such as coins or number of friends on social networks, which are meaningful to the collectors, in terms of value, security, or social status. Furthermore, once a collection starts, people tend to complete it, so if a collection could be infinite, the collecting activities will keep going.
2. Connecting: Connecting with people, especially those who share common interests or characteristics with us, makes our life enjoyable. This explains the foundation of various associations, clubs, fellowships, etc. Although being a member of a forum itself is making connections with other people, users still have motivation to expand the connections with people outside of the forum.
3. Achievement: People get great satisfaction from achievement, which usually means successfully dealing with challenges. The positive psychological feedback makes us be willing to rise to the same challenge repeatedly; even we know that we probably fail sometimes.
4. Feedback: The feedback means acknowledgement, recognition, or just response to initiators' actions or messages. Feedback enhances the sense of being noticed, so not receiving feedback is extremely demotivating to anyone. Providing feedback is very important to encourage continuous participation and contributions in a forum.
5. Autonomy: Just like average people, members of a forum do not want the contents quality or the forum atmosphere shift to a situation, which they dislike to see but unable to restrain it. Therefore, if there is a trigger available, they tend to take necessary actions when they encounter any potential leading to these situations, such as offensive or inappropriate contents.
6. Fear of punishment: Members in a forum, just like most members in a society, tend to avoid speech and behaviors leading to punishment, this tendency gradually develops social norms or the corresponding regulations in written. Unlike other motivational drivers; this factor prevents members from doing somethings.

After setting missions and identifying motivation drivers, the next step is to embed proper game mechanisms that are able to motivate the users.

3.3   Game Mechanisms Selection

A game mechanism refers to a component with which users interact during the process of playing games. Besides its visible part displaying on the user interface, a mechanism also includes a set of rules that govern how this mechanism works. To realize the mentioned missions, the following 5 game mechanisms were selected to embed into the JForum. Each mechanism's characteristics and the motivational drivers it offers were described as follows:

1. Points: The most popular mechanism in various games and have been widely applied in commercial contexts to reward customers' loyalty. Points motivate users due to humans' intrinsic desire to collect things such as money, stamps, antiques, etc. Besides, awarding points to users for her/his participation and contributions is a rational approach to recognizing their activities in the forum. In this work, points will be awarded to encourage certain types of actions including login, raise new topic, post new message, reply a message; as well as will be deducted to discourage other types of actions such as posting inappropriate contents, abusive reporting, etc.
2. Leaderboard: This mechanism ranks users according to their achievements within a specific context, which usually be representable in the form of points. It forms a competitive atmosphere, which encourages those who dislike following behind peers to engage the forum more actively.
3. Badges: To award members' accumulating a certain amount of points, specific types of badges will be granted. Usually, there are multiple types of badges honoring achievements in different levels of difficulty, or with different types of works. In the latter case, collecting badges will motivate some members.
4. Facebook likes: To use the plugins API of the Facebook, members' messages could be exposed on the largest social network. This allows members to make connections with other friends who are not in the same forum, as well as receive feedbacks of relevance from friends on Facebook.
5. Report of inappropriate contents: This mechanism echoes members' motivations in two facets; one is autonomy and another is fear of punishment. The former one drives members to control the quality of contents or the atmosphere of the forum through suppressing inappropriate contents. The latter one holds back members from posting contents that are evidently not suitable in a particular forum. To avoid abusive reporting, the reported contents will be sent to administrators for judging whether they are really inappropriate or not.

4   System Implementations

The JForum is a Java servlet application being able to run on Apache Tomcat. Its design generally complies with the model-view-controller (MVC) architectural pattern [17], it has the 3-tier structure. Accordingly, planting game mechanisms needs to deal with components in different tiers.

To minimize the interrelationship (coupling) between the original parts in JForum and the newly parts due to the gamification, the works were conducted with a loosely-coupled style. The figure 1 illustrates the major components that need to be added and updated for gamifying the JForum, and major components are described as follows.

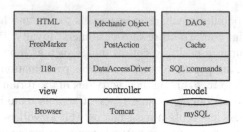

Fig. 1. Architecture of a Gamified JForum

4.1   User Interface

The JForum uses the FreeMarker package [20] as a tool to separate Web pages design (view) and business logic programming (controller) works. The FreeMarker engine generates textual contents based on a template that contains HTML and FTL tags for dynamic Web contents, as well as a data model specifying data sources. Consequently, to embed a new game mechanism that will bring some new messages and graphics dynamically, it is necessary to update both the template and the data model of correspondence; the template decides the visual effect and format of the new components, and the data model tells where the displayed data come from dynamically when a particular page being accessed.

In addition, the JForum uses the I18 internationalization package for global users. For that reason, all new textual messages for embedding game mechanisms need to be added into the property file listing messages for a particular language (locale).

4.2   Gaming Objects and Rules

When thinking in the object-oriented way, each game mechanism obviously needs an object for holding its attributes and defining how it works. The PostAction object handles all actions for managing topics and posts, such as creating new topics or posting new messages, replying, deleting messages, etc. In other words, it deals with most major actions that the gamification work should focus on. So that, it needs linkage with some new game mechanisms. For example, when a user post a new message, the action of awarding points will be initiated by this object.

The DataAccessDriver abstract class defines the interface for linking the game mechanism and the persisting of the added game objects. Thus, all game mechanisms except the Facebook connection need to rely on one of its realization. Generally, the attributes of game objects will be fetched or stored via one of the corresponding concrete classes.

4.3   Persistence Model

Taking flexibility of persistent storage into account, data access object (DAO) design pattern was applied to separate the objects and its underlying storage mechanism. Thus, the object codes do not need to be changed due to switching to different data sources or

APIs. DAOs provide abstraction and encapsulate all details for accessing the data source, which might be relational database, cloud storage, LDAP, mainframe file systems, etc. In the JForum system, each object was further divided into two layers of DAO for more flexibility, one is entity specific DAO, another is the generic DAO.

Cache mechanism in the JForum speeds up the object access operations by storing copies of object contents in Java virtual machine. To make contents of the persistent game mechanisms cacheable, addition of all persistent mechanisms needs the corresponding update in the cache mechanism. Besides the properties of game mechanism objects, many gaming rules could also be encoded and then persisted in the database for more flexibility and maintainability.

5   Conclusions

The quality of public policies influences the public's well-being; accordingly, it is not only rational but wiser to take group intelligence into account before making public policies. In the age of Internet, online forums enable large-scaled and efficient deliberation of opinions and thoughts regarding public issues, but the pre-condition is enthusiastic engagement and active contributions from participants.

Borrowing the attractive features of various games, gamification was applied to improve users' experiences and engagement in non-game contexts. The embedding of game mechanisms into an online forum system makes the system more engaging to participants, who will feel more animating while they are interacting with peers and contributing productive thoughts.

This article describes the work of gamifying a JForum system. Besides, the present work shows the feasibility of gamifying an open-sourced online forum system by embedding 5 popular mechanisms into its original code base. The preliminary trials and interviews with users indicated that the most significant impact brought by the present gamification on users is that they can observe the aspiring and competitive atmosphere, which was shaped by the badges and leaderboard and to some extent encouraged them to play more active roles in the information exchanging platform. The Facebook connections enable users to spread forum contents to their own social networks. The mechanic of reporting inappropriate contents enable users to collectively maintain the quality of forum, which is important to the sustainable operation of a forum. By contrast, rewarding points made little difference due to the lack of redeeming mechanism enabling users to consume what they earned.

The works being worthy of further investigations include the quantitative analysis of participants' perceptions or satisfaction toward the embedded game mechanisms, and the evaluation of performance and productivity influence after the gamification being deployed, particular the influences of messages on the made policies.

Acknowledgements. This research work has being funded by the grant from the Ministry of Science and Technology, Taiwan, ROC, under Grant No. MOST 103-2221-E-126-014. We deeply appreciate their financial support and encouragement.

References

1. Noveck, B.S.S.: Wiki government: how technology can make government better, democracy stronger, and citizens more powerful. Brookings Institution Press (2009)
2. Lathrop, D., Ruma, L.: Open government: Collaboration, transparency, and participation in practice. O'Reilly Media, Inc. (2010)
3. Chadwick, A.: Internet politics: States, citizens, and new communication technologies. Oxford University Press, New York (2006)
4. Killoran, J.B.: How to use search engine optimization techniques to increase website visibility. IEEE Transactions on Professional Communication 56(1), 50–66 (2013)
5. JForum, T.: jforum is a powerful and robust discussion board system implemented in Java. http://jforum.net/
6. Watt, J., Hargis, M.: Boredom Proneness: Its Relationship with Subjective Underemployment, Perceived Organizational Support, and Job Performance. Journal of Business & Psychology 25(1), 163–174 (2010)
7. Aparicio, A.F., et al.: Analysis and application of gamification, p. 17
8. Deterding, S., et al.: From game design elements to gamefulness: defining "gamification". In: Proceedings of the 15th International Academic MindTrek Conference: Envisioning Future Media Environments, Tampere, Finland, pp. 9–15 (2011)
9. Pettey, C., van der Meulen, R.: Gartner Predicts Over 70 Percent of Global 2000 Organisations Will Have at Least One Gamified Application by 2014. Gartner, Inc., Barcelona (2011)
10. Sheely, E.: Case Study: How Starbucks Improved Supply Chain Efficiency with Gamification. http://www.gamification.co/2013/11/18/fostercooperationgamfication/
11. Hendricks, J.: Case Study: Delta's Nonstop NYC Game That Got 190K Interactions in 6 Weeks. http://www.gamification.co/2013/11/20/delta-nonstop-nyc/
12. Dubois, D.J., Tamburrelli, G.: Understanding gamification mechanisms for software development. In: Proceedings of the 2013 9th Joint Meeting on Foundations of Software Engineering, Saint Petersburg, Russia, pp. 659–662 (2013)
13. Rodrigues, L.F., Costa, C.J., Oliveira, A.: The adoption of gamification in e-banking. In: Proceedings of the 2013 International Conference on Information Systems and Design of Communication, Lisboa, Portugal, pp. 47–55 (2013)
14. Krishna, D.: Application of Online Gamification to New Hire Onboarding, pp. 153–156
15. Korn, O.: Industrial playgrounds: how gamification helps to enrich work for elderly or impaired persons in production. In: Proceedings of the 4th ACM SIGCHI Symposium on Engineering Interactive Computing Systems, Copenhagen, Denmark, pp. 313–316 (2012)
16. Gnauk, B., Dannecker, L., Hahmann, M.: Leveraging gamification in demand dispatch systems. In: Proceedings of the 2012 Joint EDBT/ICDT Workshops, Berlin, Germany, pp. 103–110 (2012)
17. Zichermann, G., Cunningham, C.: Gamification by Design: Implementing Game Mechanics in Web and Mobile Apps. O'Reilly Media (2011)
18. Bartle, R.: Hearts, clubs, diamonds, spades: Players who suit MUDs. Journal of MUD Research 1(1), 19 (1996)
19. Kumar, J.M., Herger, M.: Gamification at Work: Designing Engaging Business Software. The Interaction Design Foundation, Aarhus (2013)
20. Geer, B., Bayer, M., Revusky, J.: The FreeMarker template engine (2004)

The Web as an Infrastructure for Knowledge Management: Lessons Learnt

Aurélien Bénel[1]([✉]) and L'Hédi Zaher[2]

[1] ICD/Tech-CICO, Troyes University of Technology, Troyes, France
aurelien.benel@utt.fr
[2] IM Développement, Tunis, Tunisia
hedi.zaher@imdev.tn

Abstract. Research works, whether they aim at building a 'Semantic Web' or a 'Social Semantic Web', consider as a prerequisite that the ideal architecture for managing knowledge would be the Web. Indeed, one can only admire how the CERN internal hypertext scaled out to a world wide level never seen before for this kind of applications. However, current knowledge structures and related algorithms cause new kind of architectural issues. About these issues faced by both communities, we would like to bring out three lessons learnt, three steps in setting up a scalable infrastructure. We will focus on a typical case of knowledge management but with a higher than usual volume of data. Starting with SPARQL, a commonly used Semantic Web technology, we will see the benefits of the REST architecture and the MapReduce design pattern.

1 Introduction

This paper deals with Web services design for knowledge management. Although HTTP scaled out remarkably for the World Wide Web, scaling out knowledge management using web services is still an open issue.

Weirdly enough, in the 'Semantic Web' program [1], the World Wide Web Consortium focused more on formats and languages than on the use of its own protocol. Meanwhile, through trials and errors, we developed an experience in the design of an infrastructure for a 'Social Semantic Web' [3], [4]. Even if this approach has been developed as an opposite to the 'Semantic Web', we think that the lessons learnt in setting up a scalable infrastructure could benefit both approaches.

We will focus on a typical case of knowledge management but with a higher than usual volume of data. Then, we will bring out three steps in designing an infrastructure. Starting with SPARQL [2], a commonly used Semantic Web technology, we will see the benefits from the REST architecture and the *MapReduce* design pattern.

2 Requirements

We will illustrate our experience feedback with *i-Semantec*, a project related to knowledge capitalization, management and reuse in large industrial companies.

© Springer International Publishing Switzerland 2015
Y. Tan et al. (Eds.): ICSI-CCI 2015, Part II, LNCS 9141, pp. 431–438, 2015.
DOI: 10.1007/978-3-319-20472-7_47

In these firms, data and documents from various stakeholders and relating to the lifecycle of the product are already managed by integrated systems named 'PLM' (for 'Product Lifecycle Management'). But there are mainly two problems with these integrated systems. First, because the integration is based on the formalization of the main business processes, it often ignores the specificity of each profession. And then each profession tends to implement its own databases and documents, which escape capitalization. Second, in a time of market instability, the information system should be more adaptable to the continuous change of processes and partners networks needed to meet customers' requirements.

Since this project only serves as an example we will not write more about functional and technical requirements. We will only focus on the feature list we had to implement:

- browsing a technical data warehouse which models may vary depending on the project;
- enhancing these data with freely defined attributes.

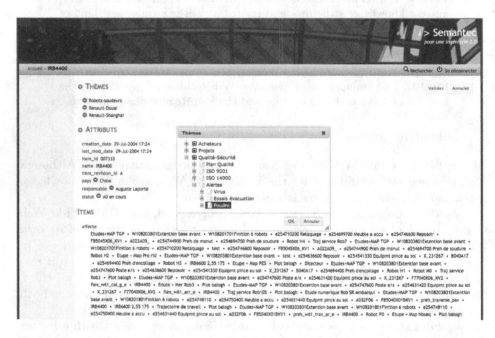

Fig. 1. Enriching the technical data of a robot in *i-Semantec* (*Agorae* screenshot) : Digital workspace for topic maps cooperative building ([5])

Items had to be browsable:

- by class of items,
- by (used) attribute and value,
- from another item through a composition relationship, a sequence in a production line, etc.

3   First Step: Semantic Web Technologies

In the *i-Semantec* project, our partners were specialists in product data management. They extracted data from two industrial projects as triples and stored them in a RDF data warehouse system [6]. This system implemented the HTTP biding for SPARQL [7]. Each project had been provided with its data model as a schema in RDFS. We then tried to build the features described earlier on this typical Semantic Web infrastructure. As we will see, we experienced serious and blocking issues.

3.1   Browsing Items

The first issue encountered in implementing the Web service for items browsing was about performance. As shown in Tab. 1, response times which were quite acceptable for 1k triples, scaled very badly with 1 million triples. Moreover, the only mechanism for getting better response times on successive identical queries was the in-memory cache of the database, which is very dependent on free memory.

Table 1. Comparison of response time to SPARQL queries (seconds) on databases containing respectively 834 and 931,338 RDF triples. Tests are done with RAP 0.9.4 by Ch. Bizer and MySQL on Linux with a Xen virtual machine equivalent to an AMD Athlon 64 X2 with 2 Gb memory.

	approx. 1k items	approx. 1G items
Item details	0.5	1
Listing items types	0.5	0.5
Listing items for a type	0.2	40
Listing used attributes	0.4	30
Listing values for an attribute	0.1	3
Listing items for an attribute value		5

3.2   Enriching Items

Triples model, aka Entity-attribute-value model (EAV), is a well known model which can combine multiple schemas and accept user-defined attributes.

However, in its RDF/XML implementation, attributes are XML elements. As XML elements, they should be defined at design time in a schema. This makes it more difficult to let users define attributes on the fly, contrary to JSON for instance, where an attribute (key) is just a string [8].

Moreover, SPARQL Update, the extension to the SPARQL query language that provides the ability to add, update, and delete RDF triples was not implemented at the time of the project. The five-year lag between the normalization of queries (2008) and updates [10] (2013) is probably indicative of the difference in priorities between the Semantic Web and the Social Semantic Web.

Therefore, we had to use our own knowledge management Web service for user-defined enrichments, integrated with a read-only connector used to query the RDF data warehouse through our own protocol.

Weirdly enough, in an opportunistic and serendipitous manner, the fact that it was a read-only access and a non-updatable data warehouse solves the first issue: payloads computed from the RDF data could be cached once and for all on disk.

In other words, to meet the requirements we had to 'hack' the Semantic Web technologies, which appeared more a burden than a help for this project.

4   Second Step: REST Architecture

From an architectural perspective, one notable difference between our protocol [9] and SPARQL was that our protocol was 'RESTful'.

REST is an architectural style for 'distributed hypermedia systems', introduced by one of the author of HTTP in his thesis [11]. It aims at generalizing the scalable design of the original Web to the complex Web applications and Web services of nowadays. As its complete name suggests ('representational state transfer'), the main idea of REST is that in a protocol like HTTP, a payload should always be a state of a resource representation. This implies that neither resource representation nor resource identifiers should refer to actions. Instead, one should use the corresponding methods defined in the protocol (e.g. GET, POST, PUT, DELETE, HEAD, etc. in HTTP). To qualify these methods, the HTTP specification introduces two important notions: 'safe' and 'idempotent'.

> "In particular, the convention has been established that the GET and HEAD methods SHOULD NOT have the significance of taking an action other than retrieval. These methods ought to be considered 'safe'." [12].

Being *safe*, requests using GET (accordingly to the specifications) will be cacheable, preemptively loadable, or even usable in a repeatable history. However to be cached by a server, a client, a proxy or a reverse proxy, changing POST with GET (like SPARQL does) will not be enough. The server has to implement a cache invalidation mechanism based on an update timestamp or on a content hash. Moreover, a resource representation should be cached only if there are chances that it will be retrieved again. Therefore a URL should map to an identifiable object rather than to a common query. This also makes integration easier, since using a different data management system would not require to emulate a complete query language but just one query.

> "Methods can also have the property of 'idempotence' in that (aside from error or expiration issues) the side-effects of $N > 0$ identical requests is the same as for a single request. The methods GET, HEAD, PUT and DELETE share this property." [12].

Being *idempotent*, requests using PUT or DELETE (accordingly to the specifications) will be resilient to client retries on timeouts. Contrary to the updates done with POST by SPARUL [10], an update with PUT will never cause multiple creations due, for example, to an excessive load of the server.

At this stage, we have what most bloggers call a REST service [13]:

- one updatable resource by object (with computed links to other objects),
- one 'super-resource' by class (in order to list instances).

However, we will see in the next section that this design is not ideal for cache management and for complex actions performance, and we will introduce a more efficient RESTful design.

5   Third Step: *MapReduce* Design Pattern

In the previous section, we described an approach in which every object is mapped to an HTTP resource. The main issue of this naive way to follow the REST architecture is the difficulty to implement a server-side cache. Indeed, if the resource representation includes data from other resources (typically for reverse links), it becomes quite complex to determine from the history if a resource representation has changed without computing it again.

A second issue in this approach is that the more objects you need to load, the more requests you have to send. For example, loading the data of a prolific user of our software (to run a data visualization algorithm [14] required 25 000 requests with this architecture. Because of each request latency, the overall time needed to load these data is 31 minutes! To reduce latency to a constant time, the number of requests needed for a complex operation should be constant. Therefore, on large datasets, a bulk of objects should be mapped to a single resource. Then a new problem would arise: the gain of caching a resource representation will be very low since it would change much more frequently. Instead, one should cache partial results needed for generating this representation. This can be fairly complex, but luckily, the *MapReduce* design pattern addresses this problem.

MapReduce [15] is a design pattern for processing large data sets. In a *MapReduce* framework, developers only have to implement (see Fig. 2):

- a *map* function that "generates [for each chunk of data] a set of intermediate key/value pairs",
- and a *reduce* function that "merges all intermediate values associated with the same intermediate key".

Then the framework handles the cache of partial results and the scaling of the algorithm over different processes and computers.

One should note that processing data with *MapReduce* is drastically different from doing it with a relational database. In a relational approach, the index depends on the data model, and is designed to optimize any queries that could be defined on it. In a *MapReduce* approach, the index depends on the map functions. Algorithms must be adapted to use intermediate values sorting wisely

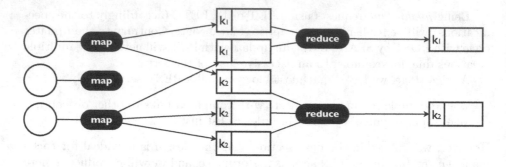

Fig. 2. *MapReduce*: data flow

(similarly to what was done with IBM/BULL card sorters). Compared with the relational approach, *MapReduce* is like computing the results of a query for every possible parameter found in the data. What could be seen as a burden is usually an optimization, since only partial results affected by an update are computed again.

As observed in a database system like *CouchDB* [16], using *MapReduce* has three important impacts on a REST service interface.

First, different kinds of resources are needed: the objects that are updatable and the *views* that are selections on the results of applying *map* and (optionally) *reduce* functions on these objects. In other words, contrary to what we saw in the previous section, computed values are not displayed in updatable resources anymore but in different resources that are read-only.

Second, owing to the implementation of the *MapReduce* framework (partial results cache and index), getting a broader *view* is incredibly faster than getting the same data with a bunch of narrower ones. Therefore, the granularity of resources depends on their type: *views* tend to be far more coarse-grained than objects (whose granularity corresponds to what is usually updated at the same time).

Third, because *MapReduce* aims at distributing computing, creating an object identifier should be done in a distributable way. As many peer-to-peer software, CouchDB uses 'universal unique identifiers' (UUID). But because a UUID [17] corresponds to a 'uniform resource name' (URN) rather than to a 'uniform resource location' (URL), we need to reinterpret the browsability principle of REST services [18] in the light of the original chapter thesis about REST [11] which stated that the use of URNs instead of URLs could "improve the longevity of resources references". Untying a resource reference from a location has also interesting effects on services integration, since clients can aggregate the descriptions of a resource that are scattered over different services that do not necessarily 'know' each other. This can be particularly handy to meet the functional requirements given earlier, since data coming from the existing database system can be provided through a read-only *adapter* service, while every community (or person) can model and store its (his) 'viewpoint' on its (his) own service.

6   Conclusion

The three steps we experienced in designing a scalable Web service infrastructure for knowledge management could be summed up by focusing on what an URL maps to:

1. a query,
2. an object with computed attributes,
3. an object or a view.

As we saw in this paper, this progression has drastic effects on caching, distribution and integration.

If our proposition is still novel, it is probably because very few in the 'Semantic Web' community have had interest in enterprise technologies like REST [19] or *MapReduce* [20]. As we did with use models, we hope that performance issues will be considered by the community in order to meet real user needs.

References

1. Berners-Lee, T., Hendler, J., Lassila, O.: The Semantic Web. Scientific American, May 17, 2001
2. Prud'hommeaux, E., Seaborne, A.: SPARQL Query Language for RDF. Recommendation. W3C (2008)
3. Zacklad, M., Cahier, J.-P., Pétard, X.: Du web cognitivement sémantique au web socio sémantique: Exigences représentationnelles de la coopération. Web sémantique et Sciences humaines et sociales (2003)
4. Bénel, A., Zhou, C., Cahier, J.-P.: Beyond web 2.0... and beyond the semantic web. In: Randall, D., Salembier, P. (eds) From CSCW to Web 2.0: European Developments in Collaborative Design. Computer Supported Cooperative Work, pp. 155–171. Springer (2010)
5. Zaher, L., Cahier, J.P., Zacklad, M.: The Agoræ/Hypertopic approach. In: Harzallah, M., Charlet, J., Aussenac-Gilles, N. (eds) Workshop on Indexing and Knowledge in Human Sciences (IKHS). Actes de la semaine de la connaissance, vol. 3, pp. 66–70 (2006)
6. Sriti, M.-F., Eynard, B., Boutinaud, Ph., Matta, N., Zacklad, M.: Towards a semantic-based platform to improve knowledge management in collaborative product development. In: Proceedings of the Thirteenth International Product Development Management Conference (2007)
7. Clark, K.G., Feigenbaum, L., Torres, E.: SPARQL Protocol for RDF. Recommendation. W3C (2008)
8. Crockford, D.: The application/json Media Type for JavaScript Object Notation (JSON). RFC 4627. IETF (2006)
9. Zhou, C., Bénel, A., Lejeune, C.: Towards a standard protocol for community-driven organizations of knowledge. In: Proceedings of the Thirteenth International Conference on Concurrent Engineering. Frontiers in Artificial Intelligence and Appl., vol. 143, pp. 438–449. IOS Press (2006)
10. Gearon, P., Passant, A., Polleres, A.: SPARQL 1.1 Update. Recommendation. W3C (2013)

11. Fielding, R.T.: Architectural Styles and the Design of Network-based Software Architectures. PhD thesis. University of California (2000)
12. Fielding, R.T., Gettys, J., Mogul, J., Frystyk, H., Masinter, L., Leach, P., Berners-Lee, T.: Hypertext Transfer Protocol - HTTP/1.1. RFC 2616. IETF (1999)
13. Tilkov, S.: A brief introduction to REST. InfoQueue, December 10, 2007
14. Zhou, C., Bénel, A.: From the crowd to communities: new interfaces for social tagging. In: Proceedings of the Eighth International Conference on the Design of Cooperative Systems, pp. 242–250 (2008)
15. Dean, J., Ghemawat, S.: MapReduce: simplified data processing on large clusters. In: Proceedings of the Sixth Symposium on Operating System Design and Implementation (2004)
16. Anderson, J.C., Lehnardt, J., Slater, N.: CouchDB: The Definitive Guide. O'Reilly (2010)
17. Leach, P., Mealling, M., Salz, R.: A Universally Unique IDentifier (UUID) URN Namespace. RFC 4122. IETF (2005)
18. Fielding, R.T.: REST APIs must be hypertext-driven. Blog post, October 20, 2008
19. Ogbuji, C.: SPARQL 1.1 Uniform HTTP Protocol for Managing RDF Graphs. Working draft. W3C (2010)
20. Oren, E., Delbru, R., Catasta, M., Cyganiak, R., Stenzhorn, H., Tummarello, G.: Sindice. com: A document-oriented lookup index for open linked data. International Journal of Metadata, Semantics and Ontologies 3(1), 37–52. Inderscience (2008)

Using Extensible Metadata Definitions to Create a Vendor-Independent SIEM System

Kai-Oliver Detken[1(✉)], Dirk Scheuermann[2], and Bastian Hellmann[3]

[1] DECOIT GmbH, Fahrenheitstraße 9 28359, Bremen, Germany
detken@decoit.de
[2] Fraunhofer Institute for Secure Information Technology, Rheinstrasse 75 64295,
Darmstadt, Germany
dirk.scheuermann@sit.fraunhofer.de
[3] University of Applied Sciences and Arts of Hanover, Ricklinger Stadtweg 120 30459,
Hanover, Germany
bastian.hellmann@hs-hannover.de

Abstract. The threat of cyber-attacks grows up, as one can see by several negative security news and reports [8]. Today there are many security components (e.g. anti-virus-system, firewall, and IDS) available to protect enterprise networks; unfortunately, they work independently from each other — isolated. But many attacks can only be recognized if logs and events of different security components are combined and correlated with each other. Existing specifications of the Trusted Computing Group (TCG) already provide a standardized protocol for metadata collection and exchange named IF-MAP. This protocol is very useful for network security applications and for the correlation of different metadata in one common database. That circumstance again is very suitable for Security Information and Event Management (SIEM) systems. In this paper we present a SIEM architecture developed during a research project called SIMU. Additionally, we introduce a new kind of metadata that can be helpful for domains that are not covered by the existing TCG specifications. Therefore, a metadata model with unique data types has been designed for higher flexibility. For the realization two different extensions are discussed in this paper: a new *feature model* or an additional *service identifier*.

Keywords: Security Information and Event Management (SIEM) · Anomaly detection · IF-MAP · Metadata schema · Trusted computing · Feature model

1 Introduction

Security Information and Event Management (SIEM) systems are seen as an important security component of company networks and IT infrastructures. These systems allow to consolidate and to evaluate messages and alerts of individual components of an IT system. At the same time messages of specialized security systems (firewall-logs, VPN gateways etc.) can be taken into account. However, practice showed that these SIEM systems are extremely complex and only operable with large personnel effort. Many times SIEM systems are installed but neglected in continuing operation.

© Springer International Publishing Switzerland 2015
Y. Tan et al. (Eds.): ICSI-CCI 2015, Part II, LNCS 9141, pp. 439–453, 2015.
DOI: 10.1007/978-3-319-20472-7_48

SIEM systems are typically only suitable for the use in huge enterprise environments, mainly because of the following reasons:

a. Deficient scalability to small and medium-sized networks.
b. High costs for installation and maintenance because new components (collectors) of IT infrastructure have to be installed, configured and maintained.
c. High costs for the operation due to the necessity of extensive expert knowledge for the policy and rule definition as well as for the correct analysis and the right interpretation of the output of SIEM systems.

Therefore, the main goal of the SIMU project is the development of a system, similar to SIEM, which significantly improves IT security in a corporate network without making great effort. In addition to its simple integration into IT infrastructures of SME and its easy traceability of relevant events and processes in the network, it is to be realized without great effort of configuration, operation and maintenance. On the functional level SIMU works like common SIEM systems, which means it monitors processes and events within the corporate network and automatically initiates proactive real-time measures to improve security. [1]

The remainder of this paper is organized as follows: Section 2 gives a short definition of SIEM systems, followed by the overall architecture of the SIMU research project in section 3, including a short introduction into the IF-MAP specification of the Trusted Computing Group. Section 4 describes the general requirements on extensible metadata models and the process of publishing them in the context of already given specifications, worked out within the ESUKOM project [9]. The specific and abstract metadata defined in the ESUKOM research project, implementing the requirements from section 4 as well as their mapping to IF-MAP is then shown in section 5. In section 6 we wind up our findings by creating and publishing own metadata definitions and by explaining how they allow creating an open-source based SIEM system as in SIMU.

2   A Definition of SIEM Systems

The acronyms SEM, SIM, and SIEM are often used in the same context, although correctly the term SIEM is a combination of the other two. The first area provides long-term storage, analysis and reporting of log data and is known as Security Information Management (SIM). The second area deals with real-time monitoring, correlation of events, notifications and console views and is commonly known as Security Event Management (SEM) [3]. Both areas can be combined differently to set-up a SIEM system.

SIEM technology provides in detail real-time analysis of security alerts, which have been generated by network hardware and applications. SIEM can be used as software, appliances or managed services, and is also applied to log security data and generate reports for compliance purposes. The objective of SIEM is to help companies respond faster to attacks and organize mountains of log data.

The term Security Information and Event Management (SIEM) has been published by Mark Nicolett and Amrit Williams of Gartner in 2005[4] and describes the product capabilities of gathering, analyzing and presenting information from network and security devices. Further features are identity and access management applications, vulnerability management and policy compliance tools, operating system, database and application logs, and external threat data. A key focus is to monitor and help manage user and service privileges, directory services and other system configuration changes, as well as providing log auditing and review and incident response.

A complete SIEM system consists of different modules (e.g. event correlation, anomaly detection, identity mapping). These modules are responsible for the "intelligence" of a SIEM system, determining the complexity of events to be detected by the system.

3   The SIMU Research Project

The architecture of the SIMU research project, as shown in Figure 1, uses the IF-MAP protocol as central mechanism to exchange metadata between different security components. Besides the used and adapted open source tools, two German vendor solutions from NCP (VPN) and macmon (NAC) have been integrated.

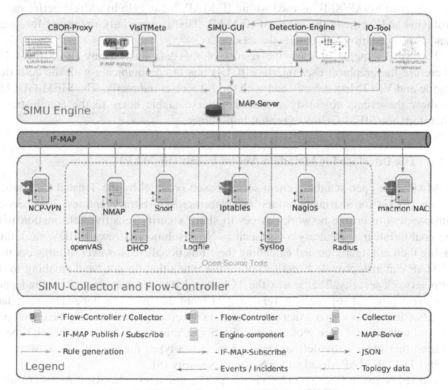

Fig. 1. SIEM architecture of the project SIMU

The architecture is divided into two layers, which are connected by the IF-MAP protocol:

a.   The *SIMU collector* and *flow-controller layer*, where the components are re-
     sponsible for data collection and enforcement.
b.   The *SIMU engine*, which includes the components for central data and know-
     ledge storage, the data correlation, aggregation, and visualization of data, as
     well as interfaces to other protocols.

The SIMU engine is the processing and presenting component of the architecture.
It includes the MAP server as the central communication point for information
exchange, the VisITMeta (a software and research project aiming at storing and visualizing IF-MAP graph data [10]) component for data storage and metadata graph
visualization, the correlation engine for situation detection and policy checking and
the SIEM-GUI – a graphical user interface which presents the analysis results and
according incidents to the administrator in an understandable manner. The IO-Tool
(Interconnected-asset Ontology) [7] works on ontology basis and extends the database
with further asset information of the network infrastructure to enable a correlation
later on. The CBOR (Concise Binary Object Representation) protocol (RFC-7049) is
a data exchange protocol, which focuses on multiple design goals including small
code and message size, and extensibility. CBOR is used in this architecture as an
alternative to SOAP/XML as used within IF-MAP. It can help to address performance
problems and facilitate the usage of IF-MAP. That is especially important for small
bandwidth scenarios with mobile devices.

The SIMU engine presents the results of the data analysis by the SIEM-GUI.
Therefore the graphical user interface (GUI) has to communicate with the detection
engine and VisITMeta directly and with the IO toolset indirectly. The SIEM-GUI has
to show the events obviously and send understandable notes to the administrator.
Therefore, the SIEM-GUI is of central importance.

3.1   The Interface for Metadata Access Points (IF-MAP)

IF-MAP is an open standard, client-server based protocol by the Trusted Computing
Group (TCG) for sharing arbitrary metadata across arbitrary entities. Its intended
purpose was to enable network devices to share security sensitive information with
the goal to integrate arbitrary tools (such as NAC solutions, firewalls, IDS, etc.), thus
easing their configuration and extending their functionality. However, it turns out that
IF-MAP can also provide benefit to other use cases that do not have anything to do
with network security. That is why the TCG decided to separate the use case independent base protocol [6] (current version 2.2) from the use case dependent metadata
specifications. This ensures that new metadata specifications can easily be developed
without touching the base protocol specification. Currently, there is one official specification that specifically defines standard metadata types for the field of network security: TNC IF-MAP Metadata for Network Security [5].

Trusted Network Connect (TNC) is the TCG approach for Network Access Control
(NAC) solutions. TNC is the reference architecture for NAC that defines the necessary entities and the interfaces through which they are communicating in an interoperable way. IF-MAP is part of the TNC framework.

The base specification defines the two roles – client and server – and three different operations: publish, search and subscribe to distribute and access information. In addition, the basic data model is defined, consisting of identifiers (entities) and metadata, which can be attached either to the identifiers directly or connect two identifiers as a kind of relationship, called link. Thereby, an undirected information graph originates. Both metadata and identifiers provide specified instruments to use them for arbitrary domains.

IF-MAP can provide the following benefits:

a. Integration of existing security systems by a standardized, interoperable network interface
b. Avoidance of isolated data silos within a network infrastructure
c. Extended functionality of existing security tools (e.g. automatic responses on detected intrusions, identity-based configuration of packet filters)
d. Vendor independence

3.2   Collector and Flow Controller Components

Flow controller and collectors are typical security components and services in a network infrastructure. They collect information or manage the network behavior. Several clients have been adopted to support integration into an IF-MAP environment, e.g. providing the IF-MAP data model with their information or using IF-MAP information for decisions. The following IF-MAP client collectors for the architecture of SIMU have been implemented (as shown in Fig. 1):

a. **DHCP collector:** extracts metadata of actual IP leases from the lease file.
b. **RADIUS collector:** delivers metadata regarding user logins and the user itself (groups, authority).
c. **Syslog collector:** delivers metadata regarding the status of arbitrary syslog clients - hosts and services (e.g. CPU load or false logins).
d. **Nagios collector:** publishes extracted metadata from Nagios regarding the status of hosts and services (i.e. availability of the network).
e. **Icinga REST collector:** works as an alternative for Nagios and has the same functionality.
f. **Snort collector:** translates Snort alerts in IF-MAP metadata.
g. **NMAP:** Network Mapper integration into the IF-MAP environment. Detects devices, server, services, etc.
h. **OpenVAS:** publication of the scan results obtained by the vulnerability scanner. Allows for periodic or regularly triggered scans and can automatically respond to requests for investigation.
i. **Log-file collector:** generic collector for analysis of arbitrary log-files and translation of the log information in IF-MAP metadata.
j. **Android collector:** delivers metadata regarding the status and behavior of an Android smartphone (e.g. firmware, kernel, build number, traffic on different network interfaces, CPU load).

k. **LDAP collector:** manages a connection to the directory service and delivers according IF-MAP metadata.

Furthermore the architecture includes the following flow-controller components, which partly have also collector functionality:

a. **iptables flow-controller:** makes automated set-up of firewall rules possible as a reaction to special metadata events and publishes these metadata also as enforcement reports for the MAP server.

b. **macmon NAC flow-controller:** This network access control system from macmon publishes different information of connected endpoint-devices, especially authorization information of well-known end-devices, the location in the network (e.g. physical port, WLAN-AP), and further device characteristics (e.g. operating system, open ports).

c. **NCP-VPN flow-controller:** This virtual private network solution of NCP can deliver several relevant metadata to the server, such as authorization, IP address, data throughput, and connection time of a user. Enforcement is possible on the VPN layer.

d. **OpenVPN flow-controller:** SSL-based alternative VPN solution with similar features.

All these components combined present the sensors of a SIEM system and collect data from the network. The key feature of a SIEM system is to correlate these data efficiently and usefully to find out which event is an anomaly and which is not. Therefore, it is necessary to analyze a data basis of the same format. IF-MAP can handle this with its extensible metadata definition.

4   Requirements and Strategies for Metadata Definition

Metadata plays an important role for securing network applications. The TCG already established specifications providing large amounts of standardized metadata and identifiers useful for network security. However, the existing standard data schemes often do not provide appropriate types for all data objects relevant for a certain application like a desired SIEM system. The design of new applications often requires the definition of additional and domain-specific metadata. The German research project ESUKOM [9] addressed the problem of real-time security for enterprise networks. The general approach was to establish a metadata model allowing the discovery of anomalies and unwanted situations in a network by consolidation of metadata. With the ESUKOM project and its covered use cases and prototype developments as an example, we want to outline the process of building use-case specific data models and looking for appropriate data types and possible enlargements of the existing TCG specifications.

4.1   Scientific and Technical Goals of ESUKOM

The ESUKOM project aimed to develop a real-time security solution for enterprise networks that is based on the correlation of metadata. The ESUKOM approach focused on the integration of available and widely deployed security measures based upon the Trusted Computing Group's IF-MAP specification. The idea was to operate on a common data pool that represents the current status of an enterprise network. Currently deployed security measures were integrated and able to share information as needed across this common data pool. This enables the ESUKOM solution to realize real-time security measures. All data shared across the common pool were formulated according to a well-defined data model.

In order to achieve this goal, the following tasks were accomplished:

    a.   Implementation of IF-MAP software components
    b.   Development of an advanced metadata model
    c.   Development of correlation algorithms
    d.   Integration of deployed security tools

For this paper, especially the metadata model approach is interesting. The IF-MAP specification currently defines a model for metadata that specifically targets use cases in the area of network security. The metadata model of IF-MAP has been extended and refined for ESUKOM to get the possibility to add new types of metadata as well as to improve drawbacks of the current metadata model.

4.2   Problem Description

Within the SIMU research project, the main goal was to design and develop a real-time security SIEM system for networks of small and medium-sized enterprises (SME). The mobility of modern endpoints such as smartphones, tablets and notebooks and their corresponding threats to the overall network security were especially considered while developing the system.

Several user scenarios and use cases have been defined within the project ESUKOM before. For a detailed data model specification and a final development of a client-server application, a generic scenario has been developed for which the following relevant key features appeared:

    a.   **Anomaly Detection:** recognition of illegal system states by detection of abnormal behavior. Such behavior could be excessive use of network resources or unusual usage, and must be considered within contextual information like time or location.
    b.   **Mobile Device Awareness:** recognition of devices as mobile devices and application of corresponding policies. Smartphone specific policies could, for example, allow to ensure the non-use of sensors.

c.   **Location-based Services:** providing services depending on detected position of devices. For example, a device could be allowed to access data only when it is present at a specific location, and is denied access when outside this location.

d.   **Detection of critical attacks on vulnerable components:** When known attacks are detected within the network on a component that has a known vulnerability in one of its provided services, it must be recognized instantly and appropriate countermeasures need to be taken.

e.   **Real-time Enforcement:** If abnormal behavior or malicious endpoints are detected within the network, an immediate reaction must occur. Therefore information about the detection has to be made available to enforcement components as soon as they appear.

The security solution to address all key features uses metadata, i.e. information gathered about the network – like the participating components and their capabilities – and the actions that occur within. These metadata can be generated by different components of the network itself, which allows using already existing components of today's networks, like DHCP servers, flow controllers or intrusion detection systems (IDS). Thus, the specific view of each component on traffic and endpoints in the network can be used. The idea within the project was to gather all this separated information in such a way, that the information could be used by any other component. Therefore, the information itself had to be gathered and stored in a uniform way, both regarding the exchange over the network and the data model itself.

4.3   Requirements for the Data Model

The following requirements for a data model suitable for the key features were found:

a.   **Integration of arbitrary metadata:** Metadata from multiple different domains must be used within the new data model. As different components have a specific view onto a network, the data model has to be flexible and non-restrictive in terms of what values can be expressed.

b.   **Technology independence:** The model itself has to be independent of any concrete technology. An implementation of the data model, i.e. a mapping to a concrete technology, will then have to ensure that all needed components to solve the key features can exchange the data in a platform independent way.

c.   **Allowing enlargements:** To allow the use of our model in future use cases and scenarios, the model itself has to be extensible. Thus, the definition of data has to be done in a flexible way.

d.   **Covering all intended use cases:** All previously identified concepts and key features need to be represented by the model. The model has to be able to include all metadata that is needed to solve the key features.

This abstract data model allows defining all metadata that are needed to implement all key features within the ESUKOM project.

4.4   Strategies for Additional Data Definition

To embed the functionality of the data model developed within the ESUKOM project into the already existing specifications by the TCG, several strategies can be followed:

1. **Additional vendor specific data:** A simple strategy is to use already existing functionality for extension within a specification. Although this would leave the specification as it is, it will not be appropriate for future standard applications, as they would also have to use the ESUKOM specific – and thus not standard- definitions. Interoperability with non-ESUKOM components would be a problem.

2. **Enlargement of specification:** The original specification can be extended by the types and attributes of the data model. This process usually takes a long time, as changing a specification involves multiple rounds of review by the corresponding working group, and has the disadvantage that it cannot always be done (e.g. new data definitions are too use-case specific).

3. **Define a standardized way to enhance metadata specifications:** A third strategy and kind of compromise of the two previous suggestions is to encourage the specification work group to adjust their own policies for enlargement so that additions like the ones of the ESUKOM project can be easily added to the specification. That way third-party vendors can also use the then more powerful metadata definitions of other research projects, companies etc.

The requirements for the ESUKOM metadata model must now be realized in a concrete metadata model, suitable for all use cases and scenarios. Afterwards, this metadata model has to be mapped onto an existing technology that defines both a data model, which is flexible enough to adapt the requirements of the ESUKOM data model, as well as a communication protocol and architecture to gather and exchange the metadata.

5   Data Model for Non-Proprietary SIEM Systems

In the last sections, we have identified the major properties of the IF-MAP protocol and general requirements for data models as well as possible strategies. The ESUKOM project now took the approach to design an IF-MAP based data model based on these aspects. In this section, the ESUKOM data model will be presented and it is exemplified how the model can be mapped onto entities within the IF-MAP scope.

5.1   General Data Model

The general scope of the model is to cover the following aspects:

a. **Specification of data objects:** Identifiers shall be defined for instances publishing and subscribing data as well as formats for the exchanged data (metadata).

b. **Specification of anomalies:** Abnormal system states shall be represented by combinations of certain data values.

c. **Specification of policies:** Actions shall be taken if certain system states, including anomalies, are detected.

First of all, the following basic data components are defined to cover the intended data objects. Later on, they will be mapped onto identifiers and metadata in compliance with [5] and [6]:

a. **Feature (F)**: A feature represents an elementary unit of the metadata model containing a measured value inside the application.

b. **Category (C)**: A category represents a collection of features belonging to the same group. Nested structures are possible, i.e. a category may also represent a collection of subcategories.

c. **Context Parameter (Ctx-P)**: Context parameters provide additional information connected with a feature describing the closer context, in particular information about time, location and connected devices.

On the way to the detection of anomalies, the following components are defined to represent certain system states:

a. **Context (Ctx)**: A context is a Boolean combination of Ctx-Ps as mentioned above. Independent of certain features, it just gives out Boolean information that a certain set of Ctx-Ps values fulfill certain conditions.

b. **Signature (Sig)**: A signature represents a pattern describing a certain system state. It consists of a set of feature instances with corresponding values. Optionally, a signature instance may also contain a set of context instances (see above).

c. **Anomaly (A)**: An anomaly represents a system state deviating from normal system's behavior. It is composed of so called hints, each one describing a feature deviating from its expected value range. As well as signatures, anomalies may also optionally contain sets of context instances. The major difference to signatures consists in the fact that anomalies are not directly visible by exact feature values but must be determined with a more complex analysis.

After specification of data objects and system states, the next step is the definition of policies how to react to certain system states, in particular anomalies, by certain action. In our model, a policy (P) consists of a collection of rules (R) whereby a rule is represented by a statement of the form "if condition do action". A condition is represented by a Boolean combination of contexts, signatures and anomalies, and an action consists of a creation, deletion or modification of features. Figure 2 shows the overview of the components of the ESUKOM model.

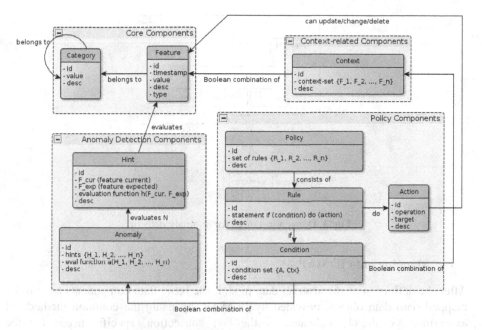

Fig. 2. ESUKOM model as a whole

5.2   Identification of Domain Instances

Before looking for a formal description of the data model components described above, ESUKOM first looks for concrete data objects needed for the key features of the project listed in section 4.2. This resulted in sets of so called domain instances including categories, features, signatures, policies etc. with assigned values. Figure 3 shows the domain instances used to represent the key feature needed to detect suspicious login attempts (many false attempts in a short time or nearly simultaneously login attempts of the same user at different locations) or abnormal network traffic. The grey rectangles define the categories with their logical structure: the user with a corresponding login history and the collection of dataflow parameters. All yellow rectangles represent the features. For all other key features within ESUKOM, different sets of domain instances were defined. Some of them work in conjunction with domain instances of other key features. As an example, domain instances that define a smartphone with its operating system version, its apps, the apps permissions, and so on, can also be used to detect anomalies in the behavior of a device.

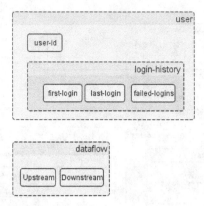

Fig. 3. Domain instances for anomaly detection

5.3   Mapping onto IF-MAP

After the definition of the abstract data model, the basic data components have to be mapped onto data objects provided by IF-MAP, by using the common methods of extension. As already indicated in the last subsection, specific needs for the ESUKOM key features shall be considered:

a.   **Mapping of categories (C):** In our data model, categories work as structuring elements and do not contain any metered values. Therefore, they are mapped onto identifiers according to [6]. An identifier of type *other* is used, whereby the *other-type-definition* attribute is set to the type of the category. For handling the hierarchy of categories, a new metadata type *subcategory-of* without any further contents or attributes is defined. Publishing this metadata type as a link between two categories indicates their subcategory relation.

b.   **Mapping of context parameters (Ctx-P):** Context parameters represent information directly applied to feature data. Therefore, they are mapped onto XML attributes. For this purpose, a new attribute group *contextParameters* is defined that covers the attribute values intended to be used for the context information (time, location and connected devices).

c.   **Mapping of features (F):** It is pretty obvious that features are represented by metadata according to [5]. These metadata are published by identifiers representing categories, and they contain attribute groups representing context parameters. However, the exact approach how to map features onto metadata types is not as straightforward as the mapping concept for categories and context parameters.

First of all, it seems to be reasonable to search for appropriate standard metadata and to only define new vendor specific data if none of the data types already specified within [5] fits to the data objects specified for the key features. However, this approach appeared as problematic for the following reasons, in particular with respect to the concrete domain instances identified for our key features:

a.   A lot of new data types, currently not provided by [5] are needed for our key features. The few cases with existing usable standard data types appeared as exceptional cases. Sometimes, data types generally fitting to the feature exist but not with right structure (either with missing or superfluous components).

b.   According to TCG policies, standard metadata types must not be equipped with additional attribute groups, and existing attribute groups must not be enlarged with additional attributes or replaced by other attribute groups, although this would be compliant with the present XML syntax of the speci-fication. As far as enlargements are foreseen, they are to be done exclusively by the TCG. This condition provides a major problem for the mapping of context parameters onto attributes as foreseen by our model.

Therefore, the decision was to specify a new (vendor specific) metadata type fea-ture and to use this unique data type for all features identified within the ESUKOM features. The name of the feature is contained within an id attribute of the data type, and the values are classified by a type attribute that can be quantitative (concrete me-tered value), qualitative (enumeration) or arbitrary (any string). The metadata value itself is represented by a value attribute. Additionally, it can be enriched by context parameters like time or location as described above. Listing 1 shows the XML struc-ture of the new feature type.

```xml
<xs:element name="feature">
  <xs:complexType>
    <xs:sequence>
      <xs:element name="id" type="xs:string">
      </xs:element>
      <xs:element name="type">
        <xs:simpleType>
          <xs:restriction base="xs:string">
            <xs:enumeration value="quantitive"/>
            <xs:enumeration value="qualified"/>
            <xs:enumeration value="arbitrary"/>
          </xs:restriction>
        </xs:simpleType>
      </xs:element>
      <xs:element name="value" type="xs:string">
      </xs:element>
    </xs:sequence>
    <xs:attributeGroup ref="ifmap:multiValueMetadataAttributes"/>
    <xs:attributeGroup ref="contextParameters"/>
  </xs:complexType>
</xs:element>
```

Listing 1. XML definition of „Feature"

By this way, our model gets very flexible, and no general changes on the abstract model or the definition of new data types are needed if new necessary features for new applications like new SIEM systems are detected. This new metadata type is then to be published by identifiers representing categories as described before.

The strategies of the ESUKOM project regarding the data model design were also continued within the follow-up project SIMU. Here another extension was developed, where a new identifier called service was introduced to describe services running on a device. Together with new link-metadata and new identifiers for vulnerabilities and implementations, a sub-graph for a service within a network and its concrete implementation on a device (e.g. measured by nmap) as well as its detected vulnerabilities (e.g. measured by OpenVAS) can be described. This can then be used to correlate detected attacks that aim at a specific CVE with actually running services that are vulnerable to this exact CVE.

6   Conclusions

SIEM systems are complex solutions and consist of different modules, security components, and interfaces. With the use of SIEM systems there are much installation and service efforts associated. That is the reason why in small and medium-sized enterprises (SME) these kinds of systems are still not represented. The developed system in SIMU focuses on SME scenarios and includes an easier implementation strategy than other SIEM systems today. By the use of the protocol IF-MAP it is possible to receive the log-information from different security components on the same data format. Thereby, it is possible to collect these data in a common database and correlate this information base to find out anomalies or vulnerability.

After the identification of use cases and corresponding data objects, several strategies for an enlargement of the present IF-MAP standard metadata were discussed. Finally, it was decided to uniquely use the newly defined data type feature, which gives a high flexibility in case of further enlargements for other use cases.

Despite the good practical results, the ESUKOM data model also shows up some weaknesses when strictly used this way for any future applications. One weakness is provided by the fact that the new feature type represents a non-standard data type whereas some client developers do not need additional metadata beyond the existing specifications and therefore are not designed for usage of any other ones. But in summary, it can be recommended to partially use the ESUKOM model by always preferring standard metadata types but always having the option to use the feature data type as well as the new context attributes together with any metadata type.

Furthermore, the grouping of features into categories should be possible, but not mandatory. As a final result, a proposition for a TCG-driven enhancement of the IF-MAP specifications can be given by adopting the new feature type as a standard type. Furthermore, the newly developed attribute group contextParameters should be adopted as a new attribute group type optionally available for all standard types in order to have a unique format for time and location information as well as information about connected devices. In addition, a new identifier type should be defined for the optional categories. A realization of this proposition would always give client developers the option to make use of the ESUKOM data model within the IF-MAP standards, but its usage would not be mandatory.

This *feature model* approach makes it possible to create a flexible data model if needed. The data model of the SIMU project is based on the work of ESUKOM, although no immediate need for the *feature type* occurred. It is recommended to also include the service *identifier* (specified within the SIMU project) into future versions of the TCG specifications, given that it is as basic as the other standard identifier types, such as *ip-address* or *device*, when describing the state of a network with IF-MAP. As a final conclusion, it is recommended to use the existing specification description of the TCG regarding interoperability with other IF-MAP components. If an extension is needed it would be useful to integrate this definition also into the standard specification. If that is not possible the solution of ESUKOM can be used. But that includes interoperability lacks between different IF-MAP components and will work only in proprietary environment.

Acknowledgements. The authors give thanks to the German Ministry of Education and Research (BMBF) [2] for the financial support as well as to all other partners involved into the research projects ESUKOM [9] and SIMU [1] for their great collaboration. The projects consist of the industrial partners DECOIT GmbH, NCP engineering GmbH, macmon secure GmbH, and the research partners Fraunhofer SIT, and University of Applied Sciences and Arts of Hanover.

References

1. SIMU project website. http://www.simu-project.de
2. Federal Ministry of Education and Research. http://www.bmbf.de/en/index.php
3. Jamil, A.: The difference between SEM, SIM and SIEM, July 29, 2009
4. Williams, A.: The Future of SIEM – The market will begin to diverge, January 1, 2007
5. TCG: TNC IF-MAP Metadata for Network Security. Trusted Network Connect, Specification Version 1.1, Revision 8, Trusted Computing Group (2012)
6. TCG: TNC IF-MAP Binding for SOAP. Trusted Network Connect, Specification Version 2.2, Revision 9, Trusted Computing Group (2014)
7. Birkholz, H., Sieverdingbeck, I., Sohr, K., Bormann, C.: IO: an interconnected asset ontology in support of risk management processes. In: IEEE Seventh International Conference on Availability, Reliability and Security, pp. 534–541 (2012)
8. Shahd, M., Fliehe, M.: Fast ein Drittel der Unternehmen verzeichnen Cyberangriffe. BITKOM news release from 11th of March 2014, CeBIT, Hanover (2014)
9. ESUKOM project website. http://www.esukom.de
10. Ahlers, V., Heine, F., Hellmann, B., Kleiner, C., Renners, L., Rossow, T., Steuerwald, R.: Replicable security monitoring: visualizing time-variant graphs of network metadata. In: Joint Proceedings of the Fourth International Workshop on Euler Diagrams (ED 2014) and the First International Workshop on Graph Visualization in Practice (GViP 2014) Co-located with Diagrams 2014, Number 1244 in CEUR Workshop Proceedings, pp. 32–41 (2014)

Parameters and System Optimization

Parameter Estimation of Chaotic Systems Using Fireworks Algorithm

Hao Li[1,2]([✉]), Peng Bai[1], Jun-Jie Xue[1], Jie Zhu[1], and Hui Zhang[2]

[1] Air Force Engineering University, Xi'an 710051, China
snk.poison@163.com
[2] Department of Intelligence, Air Force Early-Warning Academy,
Wuhan 430019, China

Abstract. Chaotic system is a nonlinear deterministic system, and parameter identification for the chaotic system is an important issue in nonlinear science, such as secure communication, etc. By setting up an appropriate objective function, the parameter identification can be converted into a multi-dimensional optimization problem which can be solved by evolutionary algorithms. Emerging as an evolutionary algorithm, Fireworks Algorithm (FWA) has shown its good computational performance and robustness. In order to expand the application of FWA, several types of FWA are applied to estimate the parameters for two typical chaotic systems in which three parameters are totally unknown, simulation results show most of FWAs can have better estimation precision and robustness, and FWA is a new effective parameter identification method for the chaotic systems.

Introduction

Swarm intelligence is a complex, highly interactive process. Based on swarm intelligence, researchers have proposed a variety of optimization algorithms. Such as particle swarm optimization (PSO) algorithm, ant colony optimization (ACO) algorithm, differential evolution (DE) algorithm, artificial bee colony algorithm (ABC), fish school search algorithm (FSS) and wolf pack algorithm (WPA), etc. Swarm intelligence algorithm is a stochastic search method, with the ability to respond flexibly to changes in the internal and external, that is, when their failure or certain individuals change the search criteria, it is possible to escape from self-organization and find better value. Because of these unique advantages, swarm intelligence algorithms successfully solve many practical problems, including robot control, unmanned vehicles, predicting social behavior, strengthening the communication network and so on.

Inspired by fireworks explosion at night, conventional fireworks algorithm (FWA) was developed in 2010 [1]. This algorithm is quite effective in finding global optimal value. As a firework explodes, a shower of sparks will be shown in the adjacent area. Those sparks will explode again and generate other shows of sparks in a smaller area. Gradually, the sparks will search the whole solution space in a fine structure and focus on a small place to find the optimal solution.

© Springer International Publishing Switzerland 2015
Y. Tan et al. (Eds.): ICSI-CCI 2015, Part II, LNCS 9141, pp. 457–467, 2015.
DOI: 10.1007/978-3-319-20472-7_49

The Enhanced Fireworks Algorithm (EFWA)[2] is an improved version of the FWA. In the EFWA, many operators in the conventional FWA are improved or corrected. By defining different calculation method of explosion amplitude, the Dynamic Search Fireworks Algorithm (dynFWA) [3] and the Adaptive Fireworks Algorithm (AFWA) [4] are also proposed to improve the optimization performance.

Conventional fireworks algorithm and its variants are capable of dealing with optimization problems. Many researches used these algorithms in a variety of applications. Janecek et al. applied fireworks algorithm to non-negative matrix factorization (NMF) [5]. Gao et al. applied fireworks algorithm to digital filters design [6]. He et al. used fireworks algorithm for spam detection [7]. Du (Du, 2013) solved nonlinear equations with fireworks algorithm and compared it with ABC algorithm. Similarly, in order to expand the application of FWA, we use several types of FWA for the parameter estimation of two typical chaotic systems in chaotic control and synchronization. Experiments show that the method has better adaptability, reliability and high precision. It proves to be a successful approach in the parameter estimation for the chaotic systems.

Estimation Framework of Chaotic System

Motivation

Chaos embodies characteristics including complexity, internal randomness, initial value sensitivity, irregular order and so on [8]. In scientific research and engineering application, chaotic characteristics are presented in various systems and therefore chaotic system's control and synchronization have become an important research field and also widely applied in fields such as secure communication, medicine, biography and chemistry [9]. Many nonlinear system control methods such as adaptive control [10], active control [11], PI control [12] and secure communication [13] can be used in chaotic system's control and synchronization. However, if parameters of the chaotic system are unknown, all of these methods above are not applicable anymore. In engineering practice, there exist some unknown parameters caused by chaotic system's complexity, which makes some parameters hard to measure. Consequently, the parameter estimation of the chaotic system becomes a key issue in chaotic control and synchronization.

Principle

Essentially, the parameter estimation of the chaotic system is an optimization problem for multi-dimensional complexity function. Swarm intelligence has been widely applied in resolving various kinds of problems. In reference [14], genetic algorithm (GA) was used for a single parameter estimation of Lorenz system, achieving better results but slower convergence. The single parameter's estimation performance of ant colony algorithm for chaotic system has been studied in both noisy and non-noisy backgrounds in reference [15], but as its text is mentioned, ant colony algorithm for chaotic parameter estimation also has slow convergence problem. On the basis of opposition-based learning and harmony search algorithm, a hybrid biogeography-based optimization (HBBO) is proposed in reference [16] with better results of both Lorenz system and Rossler system, In addition, adaptive spatial contraction bee colony algorithm [17], chaotic invasive weed optimization [18], cuckoo adaptive search, simulated annealing hybrid algorithm [19], oppositional seeker optimization algorithm [20] and other algorithms are also used for the parameter estimation of chaotic systems.

However, according to the famous "No Free Lunch Theorems" that was proposed by Wolpert and Macready in 1997, there can be no solution to all kinds of problems. Therefore, a different parameter estimation method should be developed aiming at a specific problem.

Generally speaking, an n-dimensional chaotic system could be expressed in the following formula:

$$\dot{X} = F(X, X_0, \theta) \tag{1}$$

Where $X = (x_1, x_2, \ldots, x_n)^T \in R^n$ denotes n-dimensional state variable of original system, X_0 denotes system's initial state and $\theta = (\theta_1, \theta_2, \ldots, \theta_m)^T \in R^m$ is the unknown parameters vector, in the context that system structure is known, estimation system can be expressed as:

$$\dot{Y} = F(Y, X_0, \tilde{\theta}) \tag{2}$$

Where $Y = (y_1, y_2, \ldots, y_n)^T \in R^n$ denotes n-dimensional state variable of estimation system, and $\tilde{\theta} = (\tilde{\theta}_1, \tilde{\theta}_2, \ldots, \tilde{\theta}_m)^T \in R^m$ denotes estimation values of parameters.

Accordingly, the principle of chaotic system's parameter estimation is depicted in Fig. 1.

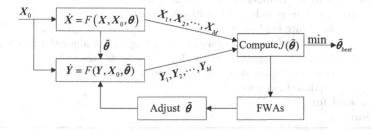

Fig. 1. The Principle of Chaotic Parameter Estimation

Fitness Function Selection

According to the principles described before, the parameters estimation for the chaotic system is transformed into an optimization problem as follows:

$$\min J(\tilde{\theta}) = \frac{1}{M} \sum_{k=1}^{M} \|X_k - Y_k\|^2 \tag{3}$$

In the expression above, M denotes sequence length of state variables in parameter estimation; X_k and Y_k denote state variables at kth moment of the real system and estimation system respectively. Obviously, the parameter estimation of the chaotic system is equivalent to a multi-dimensional continuous optimization problem which needs to search for the optimum of decision variable $\tilde{\theta}$ to acquire minimum value of fitness function $J(\tilde{\theta})$.

Due to the dynamic behavior of unstable chaotic system, the parameters are not easy to obtain. In addition, for the fitness function, there are multiple Variables, multiple local optimal, the traditional optimization method is very easy to capture in local minima, but difficult to achieve global optimal parameters.

In this paper, several types of FWA are applied to estimate the parameters for two typical chaotic system, and compared with other typical algorithm, the FWAs for the parameter identification of chaotic systems has very important significance. It is an important application for fireworks algorithm in the field of chaos synchronization and control.

Parameter Estimation Algorithm

Therefore, utilizing FWA for the chaotic system's parameter estimation, we need to set the parameters to a firework, and the corresponding fireworks with minimum fitness is the best parameter of chaotic system, and the specific process of the parameter estimation is showed as follows:

Parameter Estimation Algorithm

1: Initialize the location of N fireworks
2: Calculate the fitness of each firework $(J(\tilde{\theta}))$
   and record the firework with minimum fitness
3: **for** $i = 1$ to the number of Monte Carlo **do**
4:    **for** $j = 1$ to population iterations **do**
5:       Generate explosion sparks
6:       Generate Gaussian sparks (According to the algorithm needs)
7:       Evaluate the quality of the locations by fitness
         (Include fireworks, explosion sparks and Gaussian sparks)
8:       Update fireworks
9:    **end for**
10: **end for**

Simulations and Analysis

Lorenz Chaotic System

Presented by Lorenz in 1963, it can describe several different physical systems such as disk dynamos, laser devices and several problems related to convection. Dynamic equation of Lorenz system is expressed as follows:

$$\begin{cases} \dot{x} = a(y - x) \\ \dot{y} = bx - xz - y \\ \dot{z} = xy - cz \end{cases} \tag{4}$$

In which x, y and z represent state variables of system, $a = 10, b = 28$ and $c = 8/3$ are real parameter values. Using FWA to estimate a, b and c in Lorenz. The initial range of estimated parameters is $9 \leq a \leq 11, 20 \leq b \leq 30$ and $2 \leq c \leq 3$. For the FWA, the parameters are set as those: Iterations is 100 times, $N = 5, M_e = 50, a = 0.04, b = 0.8, \hat{A} = 40$ and $M_g = 5$. Table 3 are statistical optimums, mean and worst values of the parameter estimation found by the FWA, the EFWA, the dynFWA, the AFWA, the GA, the PSO and the BBO over 20 independent runs:

Fig. 2 and Fig. 3 are the average fitness evolution curve of FWA and convergence curves of parameter estimation after 20 times' independent running.

According to Table 2, Fig. 2 and Fig. 3, generally speaking, compared with other 3 algorithm, FWA and its variants can get a better parameter estimation

Table 1. Results of parameter estimation for Lorenz system

Fitness	Algorithm	a	b	c	J
Optimums	GA [14]	10.0671	27.9221	2.6635	4.3107
	PSO [21]	9.9953	28.0071	2.6670	0.0486
	BBO [16]	10.0068	27.9968	2.6667	2.36×10^{-5}
	FWA	10.0019	28.0011	2.6703	1.09×10^{-3}
	EFWA	10.0000	28.0000	2.6667	9.41×10^{-9}
	dynFWA	9.9977	28.0030	2.6663	2.25×10^{-5}
	AFWA	10.0000	28.0000	2.6667	4.02×10^{-9}
Averages	GA [14]	10.1398	27.7427	2.6486	943.7629
	PSO [21]	10.0184	27.9934	2.6663	4.1828
	BBO [16]	10.0183	27.9913	2.6671	0.0033
	FWA	9.9796	28.0001	2.6652	2.14×10^{-2}
	EFWA	10.0000	28.0000	2.6667	2.95×10^{-7}
	dynFWA	9.9989	28.0017	2.6671	1.67×10^{-3}
	AFWA	10.0002	28.0000	2.6667	1.30×10^{-4}
Worst values	GA [14]	10.9290	26.1276	2.5621	6461.4801
	PSO [21]	10.6082	27.7044	2.6572	39.4060
	BBO [16]	9.9440	28.0360	2.6509	0.0289
	FWA	9.8497	28.0715	2.6641	5.40×10^{-2}
	EFWA	10.0006	27.9991	2.6667	1.61×10^{-6}
	dynFWA	10.0730	27.9326	2.6741	1.35×10^{-2}
	AFWA	10.0203	27.9806	2.6681	1.05×10^{-3}

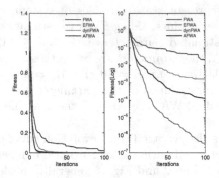

Fig. 2. Average fitness evolution curve for Lorenz system

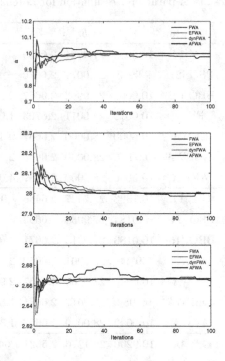

Fig. 3. Average parameter estimation curve of FWA for Lorenz system

accuracy, specifically EFWA and AFWA can be achieved 10^{-9}. In these types of fireworks algorithms, AFWA has the fastest fitness convergence rate, EFWA has the best fitness optimums and evolution curve.

Rossler Chaotic System

As a simplified model of a chemical reaction system, Rossler system's expression showed as below is also a very famous function in nonlinear dynamics:

$$\begin{cases} \dot{x} = -y - z \\ \dot{y} = ay + x \\ \dot{z} = b + z(x - c) \end{cases} \tag{5}$$

Rossler parameters' real values are set as $a = 0.2, b = 0.4, c = 5.7$. Search range of estimation parameters is $[0, 10]$. For the FWA, the parameters are set as those: Iterations is 100 times, $N = 5, M_e = 10, a = 0.04, b = 0.8, \hat{A} = 40$ and $M_g = 5$ (in total, 20 fireworks/sparks). 5 times of independent parameter estimation of Rossler system has been processed with FWA and simulation results are compared with results of HBBO in reference [8] and DE introduced in reference [22] in Table 4.

The average fitness evolution curve of FWA and convergence curves of parameter estimation are drawn in Figure 4 and Figure 5 respectively. This shows that FWAs have a fast convergence speed as well as outstanding performance in global optimum searching according to Figure 4 and Figure 5. Simulation results outline both the solving precision and the computation stability of FWAs.

Analysis

According to the experimental results, the feasibility of the parameter estimation by FWA and its variants have been proved. As can be seen from the results, in these types of fireworks algorithms, EFWA, dynFWA and AFWA have achieved better results. For Lorenz chaotic system, EFWA can reach optimal fitness accuracy at 10^{-9} in the case of a small number of iterations, and 10^{-11} for Rossler chaotic system, demonstrated its powerful search capability for the chaotic system. Fireworks and explosion sparks in FWA are similar to the coarse and fine in two-parameter control system. By changing the position of the fireworks and

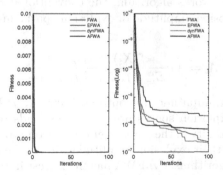

Fig. 4. Average fitness evolution curve for Rossler system

Table 2. Results of parameter estimation for Rossler system

Fitness	Algorithm	a	b	c	J
	DE [22]	0.2000	0.3530	5.6710	0.0090
Optimums	HBBO [8]	0.2000	0.4001	5.7000	2.02×10^{-10}
	FWA	0.1996	0.3978	5.6544	3.75×10^{-8}
	EFWA	0.2000	0.4001	5.7023	4.09×10^{-11}
	dynFWA	0.1998	0.3945	5.6226	4.88×10^{-8}
	AFWA	0.1999	0.3935	5.5969	7.19×10^{-8}
	DE [22]	0.1999	0.3728	5.7994	0.0248
Averages	HBBO [8]	0.1995	0.3910	5.6639	3.37×10^{-5}
	FWA	0.1991	0.4243	6.0550	2.07×10^{-6}
	EFWA	0.2000	0.4023	5.7373	2.37×10^{-7}
	dynFWA	0.2006	0.4015	5.7229	2.25×10^{-7}
	AFWA	0.2001	0.4050	5.7793	6.85×10^{-7}
	DE [22]	0.1852	0.2000	5.5839	0.884
Worst values	HBBO [8]	0.1978	0.3575	5.5296	1.60×10^{-4}
	FWA	0.1960	0.4540	6.4914	5.34×10^{-6}
	EFWA	0.2004	0.4197	6.0053	5.80×10^{-7}
	dynFWA	0.2020	0.4167	5.9494	5.50×10^{-7}
	AFWA	0.2007	0.4320	6.2059	1.52×10^{-6}

explosion sparks, algorithm will strike some optimal balance between search depth and breadth. At the same time, elitist in population iteration can speed up the convergence and ensure the continuous optimization, and Gaussian sparks provide the possibilities of jumping local optima.

In these types of fireworks algorithm, the conventional FWA has relatively poor results. The explosion amplitude of explosion sparks in conventional FWA is expressed as follow:

$$A_i = \hat{A} \cdot \frac{f(X_i) - y_{min} + \varepsilon}{\Sigma_{i=1}^{N}(f(X_i) - y_{min}) + \varepsilon} \tag{6}$$

A_i is the explosion amplitude, for optimal firework, $f(X_i) = y_{min}$, so the amplitude A_i of its explosion sparks are zero. This leads to its poor capacity optimization and unstability in the parameter estimation process, but for EFWA, dynFWA and AFWA, their explosion amplitude will not be zero.

In these types of algorithms, dynFWA has the fastest initial convergence rate due to its dynamic explosion amplitude, but with the iteration, dynFWA does not necessarily achieve the best results, which will also happened to AFWA.

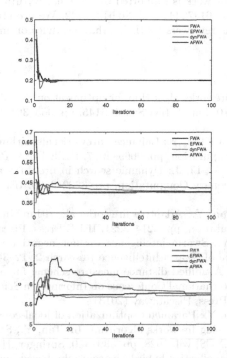

Fig. 5. Average parameter estimation curve of FWA for Rossler system

Instead, the EFWA, with its poorer performance than the other two algorithms in the standard function test, has obtained the best results for the chaotic system. In comparison with EFWA, AFWA and dynFWA, as the iteration, the explosion amplitude of three algorithms are also changing, they all have adaptive amplitude and elitist characteristics. Therefore, faced with the actually complex chaotic system, the improvement effect of the explosion amplitude for dynFWA and AFWA may not always be able to play an positive role. Nevertheless, EFWA, FWA and dynFWA are still better than other algorithms in the table above.

Conclusion

In this paper, the parameter estimation of the chaotic system is transformed into a class of multi-dimensional parameter optimization problem. In order to expand the application of FWA which has emerged recently, FWA was introduced, validated and finally applied to estimate the unknown parameters of the chaotic systems. With Lorenz and Rossler chaotic system, the simulation results show that the FWA can have better effect than other algorithms, it is a new and effective chaotic system parameter estimation method.

Acknowledgment. This work is supported by National Natural Science Foundation of China under Grant No.61472441 and No.61472442. We are grateful to Prof. Ying TAN and the anonymous reviewers for their valuable advice on improving the paper.

References

1. Tan, Y., Zhu, Y.: Fireworks algorithm for optimization. In: Tan, Y., Shi, Y., Tan, K.C. (eds.) ICSI 2010, Part I. LNCS, vol. 6145, pp. 355–364. Springer, Heidelberg (2010)
2. Zheng, S., Janecek, A., Tan, Y.: Enhanced fireworks algorithm. In: IEEE Congress on Evolutionary Computation, pp. 2069–2077. IEEE Press, Piscataway (2013)
3. Zheng, S., Janecek, A., Li, J.: Dynamic search in fireworks algorithm. In: IEEE Congress on Evolutionary Computation, pp. 3222–3229. IEEE Press, Piscataway (2014)
4. Li, J., Zheng, S., Tan, Y.: Adaptive fireworks algorithm. In: IEEE Congress on Evolutionary Computation, pp. 3214–3221. IEEE Press, Piscataway (2014)
5. Janecek, A., Tan, Y.: Swarm intelligence for non-negative matrix factorization. International Journal of Swarm Intelligence Research **2**, 12–34 (2011)
6. Zheng, S., Tan, Y.: A unified distance measure scheme for orientation coding in identification. In: International Conference on Information Science and Technology, pp. 979–985. IEEE Press, Piscataway (2013)
7. He, W., Mi, G., Tan, Y.: Parameter optimization of local-concentration model for spam detection by using fireworks algorithm. In: Tan, Y., Shi, Y., Mo, H. (eds.) ICSI 2013, Part I. LNCS, vol. 7928, pp. 439–450. Springer, Heidelberg (2013)
8. Wang, L., Xu, Y.: An effective hybird biogeography-based optimization algorithm for parameter estimation of chaotic sysytems. Expert Syst. Appl. **38**, 15103–15109 (2011)
9. Liu, L., Zhang, J., Xu, G., Liang, L., Wang, M.: A chaotic secure communication method based on chaos systems partial series parameter estimation. Acta Phys. Sin. **63**, 010501 (2014)
10. Hegazi, A., Agiza, H., Dessoky, M.: Adaptive Synchronization for Rossler and Chua's Circuit Systems. International Journal of Bifurcation and Chaos **12**, 1579–1597 (2002)
11. Huang, L., Feng, R., Wang, M.: Synchronization of chaotic systems via nonlinear control. Physic Letters A **320**, 271–275 (2004)
12. Cheng, D., Huang, C., Cheng, S., Yan, J.: Synchronization of optical chaos in vertical-cavity surface-emitting lasers via optimal PI controller. Expert Systems with Applications **36**, 6854–6858 (2009)
13. Liu, Y., Wallace, K.: Modified dynamic minimization algorithm for parameter estimation of chaotic sysytem from a time series. Nonlinear Dyn. **66**, 213–229 (2011)
14. Dai, D., Ma, X., Li, F., You, Y.: An approach of parameter estimation for a chaotic system based on genetic algorithm. Acta Phys. Sin. **51**, 2459–2462 (2002)
15. Li, L., Peng, H., Yang, Y., Wang, X.: Parameter estimation for Lorenz chaotic system based on chaotic ant swarm algorithm. Acta Phys. Sin. **56**, 51–55 (2007)
16. Lin, J., Xu, L.: Parameter estimation for chaotic systems based on hybrid biogeography-based optimization. Acta Phys. Sin. **62**, 030505 (2013)
17. Gao, F., Fei, F., Xu, Q., Deng, Y., Qi, Y., Balasingham, I.: A novel artifical bee colony algorithm with space contraction for unknow parameters identification and time-delays of chaotic sysytems. Applied Mathematics and Computation **219**, 552–568 (2012)

18. Ahmadi, M., Mojallali, H.: Chaotic invasive weed optimization algorithm with application to parameter estimation of chaotic systems. Chaos, Solitons & Fractals **45**, 1108–1120 (2012)
19. Sheng, Z., Wang, J., Zhou, S., Zhou, B.: Parameter estimation for chaotic systems using a hybrid adaptive cuckoo search with simulated annealing algorithm. CHAOS **24**, 013133 (2014)
20. Lin, J., Chen, C.: Parameter estimation of chaotic systems by an oppositional seeker optimization algorithm. Nonlinear Dyn. **76**, 509–517 (2014)
21. He, Q., Wang, L., Liu, B.: Parameter estimation for chaotic systems by particle swarm optimization. Chaos, Solitons & Fractals **34**, 654–661 (2007)
22. Peng, B., Liu, B., Zhang, F., Wang, L.: Differential evolution algorithm-based parameter estimation for chaotic systems. Chaos, Solitons & Fractals **39**, 2110–2118 (2009)

Research and Implementation of Parameters Optimization Simulation Environment for Hydrological Models

Jiuyuan Huo[1,2 (✉)] and Yaonan Zhang[2]

[1] School of Electronic and Information Engineering, Lanzhou Jiaotong University,
Lanzhou 730070, People's Republic of China
huojy@mail.lzjtu.cn
[2] Cold and Arid Regions Environmental and Engineering Research Institute,
Chinese Academy of Sciences, Lanzhou 730000, People's Republic of China
yaonan@lzb.ac.cn

Abstract. For the Hydrological models' complexity, more model parameters, and most parameters are unpredictable as well as errors of observation data in river basin will lead to lots of errors and uncertainties in Hydrological model parameter calibration, Hydrological modeling and forecasting. Therefore, we developed a unified integrated simulation environment of parameter estimation for Hydrological models in Heihe river basin to support the implementation the processes of simulation operation, parameter optimization, parameter evaluation, analysis and simulation results visualization of Hydrological models in Heihe river basin. By using long-term monitoring measured data, we validated the hydrologic models' parameters estimation integrated simulation environment in Xinanjiang Hydrological model. The results showed that research in this paper would reduce complex workload of data collection and program development, and significantly improve accuracy and efficiency for model simulation.

Keywords: Hydrological models · Parameter estimation · Optimization method · Simulation environment

1 Introduction

Parameter calibration is essential for simulation of Hydrological models. Before the Hydrological models were applied to simulate watershed Ecology-Hydrological processes, the values of these parameters of models must be first determined by means of parameter estimation methods [1, 2]. These simulation and research processes and steps are all involved in lots of expertise which will result in many difficulties in model simulation [3, 4]. Therefore, there is an urgent need for an integrated simulation environment for Hydrological model parameter optimization to provide a visual environment for Geo-Hydrological model of integrated operation and management. By interactively accessing to integrated environment of Hydrological model parameter estimation to solve problems and difficulties that researchers en-

© Springer International Publishing Switzerland 2015
Y. Tan et al. (Eds.): ICSI-CCI 2015, Part II, LNCS 9141, pp. 468–475, 2015.
DOI: 10.1007/978-3-319-20472-7_50

countered in their study of Hydrological model, it will improve the efficiency of the model operation and provide better information services for the scientific and technical personnel.

Because of the Heihe River Basin is a western inland river basin which has the most typical features of the landscape pattern, thus it has always been a typical research region for the scientists research in Meteorology, Ecology, Hydrology, water resources of cold and arid areas. Especially in the past 50 years, research institutions and local authorities in the Heihe River Basin have carried out a lot of researches and industrial projects, these long research works have accumulated a wealth of data, models and scientific achievements [5]. However, with the deterioration of ecological environment of the Heihe River Basin, water resource problems in Heihe have become a major limiting factor for regional development [6]. Many scientific personnel devote themselves into the modeling, simulation processes and integrated studies of watershed Ecology-Hydrological models in Heihe River Basin [7-9].

In the integrated modeling simulation of the Heihe River Basin, Cheng Guodong carried out a project of "Development and simulation environment of cross-integrated research models in Heihe River Basin", proposed a model suitable for the simulation of the Heihe River Basin, focused on solving the system coupling problems of atmosphere - vegetation - soil - permafrost - snow, and built the main frame of a decision support system based on USGS's MMS platform to couple of surface water models and groundwater models [10]. Nian Yanyun designed a Hydrological information system that suitable for integrating geospatial database, database observing system, and data sharing distribution system in Heihe River Watershed [11]. Nan Zhuotong has studied a preliminary application of integrated modeling environment in the Heihe River Basin. To achieve cross-platform feature of modeling environment and modules, the open source Qt/C++ language was adopted. The environment is mainly for Hydrology models and land surface process models in Heihe river [12]. Zhang Yaonan carried out lots research works in terms of research and demonstration application of modeling system and others. And a research support platform has been progressively developed to realize the functions of integrating data observation, data transfer, data management, data processing, model management, model building, model calculation, results visualization and collaborative work environment. This platform will effectively support integrated studies of the Ecological, Hydrological model of the Heihe River basin [13, 14].

In studies of integrated research environment for model parameters estimation, there are mainly SWAT-CUP [15-17], MODOPTIM [18]. SWAT-CUP is mainly for parameter estimation procedure of SWAT model, which introduced GLUE, ParaSol, SUFI -2, MCMC and PSO five different parameter estimation methods, and performed analysis tasks of sensitivity analysis, parameter estimation, parameter estimation and uncertainty analysis for the SWAT model. MODOPTIM is a generic optimization program for groundwater flow model calibration and groundwater management of MODFLOW model.

In summary, the research aspects of the Hydrological model parameter estimation have the following characteristics and trends: parameter estimation uncertainties were discussed fewer in the presence of researches, and lack of a complete parameter estimation and integrated assessment environmental. Therefore, it is unable to determine the validity and applicability of the parameter estimation methods, and could not meet the system needs of the different Hydrological model for parameter optimization method.

2   System Architecture

The integrated parameter estimation environment of Hydrological models is for the research staff that have certain of theoretical basis and practical experience to intuitive and easy to manage, use and run all kinds of Hydrological models optimization method, parameters and data provided by the integrated environment in GUI mode. This environment will help to achieve the visualization performance and flexible application simulation processes of Hydrological model to support the integration of models, data, optimize methods and the life-cycle throughout the Hydrological simulation processes of model runs and results analysis.

Parameter estimation integrated environment of Hydrological models has the characters of convenient operation, rich functions, and its the application process is closely meet the needs of researchers, and strive to complete functions in the simple logic way. Researchers can login the system, and implement data submission, model selection and setting, parameter estimation method selection and model simulation run configuration settings and other tasks by the client to run the submitted simulation task. After the calculations completed, the user can obtain a simulation output, and visualize the output data in graphical way. System workflow chart of the integrated parameter estimation environment for Hydrological models was shown in Figure 1.

3   Design of System Prototype

The prototype system of the integrated parameter estimation environment for Hydrological models was developed the core functionality of smallest system. The system uses C/S structure, and provides all kinds of model parameter optimization functions by way of graphical user interface (GUI). The main features of Hydrological models' parameter estimation integrated environment are:

(a) Management of Hydrological model: user can add, modify and delete Geo Hydrological models, and manage their metadata.
(b) Management of Hydrological model's parameters: user can add, modify and delete the parameters of the selected model, and manage metadata of the model's parameters.
(c) Management of the Hydrological model's output variable: user can add, modify and delete the output variables of the selected models, and manage metadata of the model's output variables.
(d) Management of model parameter optimization method: user can add, modify and delete Hydrological model's parameter optimization method, and manage metadata of optimization method's parameter.

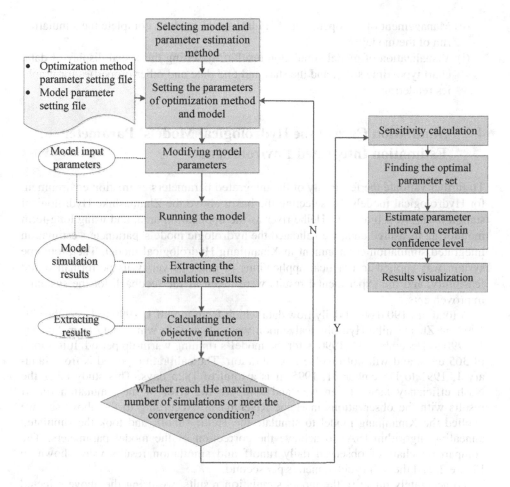

Fig. 1. System work flow of the integrated parameter estimation environment for Hydrological models

(e) Management of model parameter optimization method's parameters: user can add, modify and delete Hydrological model parameter optimization method's parameters, and manage metadata of optimization method's parameters.

(f) Management of the objective function: user can add, modify and delete the objective function of Hydrological models and manage metadata of the objective function.

(g) Management of operating parameters configuration of Hydrological model: users can add, modify and delete operating parameters required to run Hydrological models, such as choosing the appropriate model and the model parameters, setting the model's output variables, configuration parameters optimization method and its parameters, model's input data path and file name, model's output data path and file name, the start time and end time of the model's calibration and validation, and the model's running configuration.

(h) Management of the operation of Hydrological models: complete the simulation run of the model.

(i) Visualization of model simulation results: by setting the visual display of data, chart type, time scale, and the start and end date and other properties for graphics rendering.

4   Simulation Case of the Hydrological Models' Parameter Estimation Integrated Environment

To further validate the feasibility of the integrated parameters estimation environment for Hydrological models, by selecting the basin above the Zhamashike Hydrological station in the west branch of Heihe river as the application area, and using long-term monitoring measured data, we validated the hydrologic models' parameters estimation integrated simulation environment in Xinanjiang Hydrological model. The prototype system was applied in practical applications in order to verify its performance and feasibility, and the experimental results were taken as the feedback for the system's improvements.

A total of 2190 days of daily flow data which from month 1, 1990 to December 31, 1995, in Zhamashike Hydrological station was selected. In which data from January 1, 1990 to December, 31, 1990 is for the model's running warm-up period. It is a total of 365 days, and will not involve in calibration. The calibration period is from January 1, 1991 to December 31, 1995. It is a total of 1825 days. This study takes the Nash efficiency factor to analyze and compare the error of the simulation runoff results with the observations in Heihe River basin. According to the above set, we applied the Xinanjiang model to simulate the Heihe runoff, and took the simulated annealing algorithm (SA) to achieve the correction of the model parameters. The comparison chart of observed daily runoff and simulation results were shown in Figure 2, and the unit is cubic meters per second.

To accurately quantify the model simulation results, we using the above selected common NSE efficiency model evaluation factor to quantitative analyze the model's simulation results. The optimization parameters results of Xinanjiang model based SA algorithm were shown in Table 1, and the evaluation of runoff simulation results of the best five parameter sets were shown in Table 2. As it can be seen from the above runoff simulation results, the overall trend of daily runoff simulated values and observed values was consistent with each other, and the Nash-Suttclife model efficiency coefficient reached 0.75 or more. In the integrated parameter estimation environment for Hydrological models, the Xinanjiang model with optimized parameters can simulate the dynamic changes of runoff in Heihe River basin; the results meet the requirements, and have good geographical adaptability. But in the simulation comparison chart, it can also be seen that in the flood period of the wet years, here are still gaps and the oscillation phenomenon in the simulation of watershed peak flow in Xinanjiang model. Especially in June 1993 and June 1995, the simulation error is larger; the observed value is greater than the simulated value. It is demonstrated that there are still some uncertainties in runoff simulation and further in-depth analysis are needed to determine the sources of uncertainty.

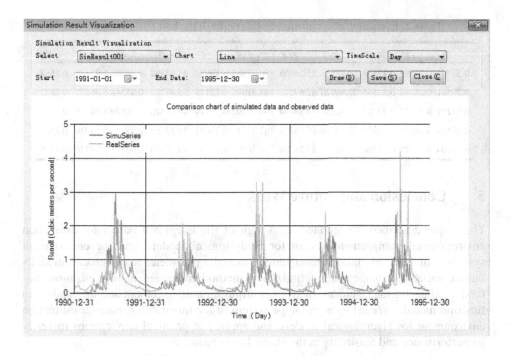

Fig. 2. Runoff simulated values and observed values in parameter validation period

Table 1. Xinanjiang model parameters optimization results based on SA algorithm

Parameter ID	Parameter Name	Unit	Parameter ranges	Parameter estimated value
0001	K	–	[0. 001, 1. 000]	0. 30707
0002	IMP	–	[0. 001, 0. 500]	0. 29885
0003	B	–	[0. 001, 1. 000]	0. 22557
0004	WUM	mm	[5. 000, 30. 000]	19. 48551
0005	WLM	mm	[50. 000, 100. 000]	85. 26665
0006	WDM	mm	[50. 000, 200. 000]	196. 77983
0007	C	–	[0. 001, 0. 300]	0. 19928
0008	FC	mm/h	[0. 001, 50. 000]	45. 04282
0009	KKG	–	[0. 001, 0. 990]	0. 98980
0010	Kr	–	[0. 001, 10. 000	5. 92115

Table 2. Evaluation of runoff simulation results of the best five parameter sets

K	IMP	B	WUM	WLM	WDM	C	FC	KKG	KR	NSE
0.30707	0.29885	0.22557	19.48551	85.26665	196.77983	0.19928	45.04282	0.98980	5.92115	**0.7554**
0.29561	0.28840	0.22565	22.86970	82.69571	183.51005	0.23178	39.63777	0.98848	5.95687	0.7532
0.31157	0.30223	0.23267	19.56256	84.84464	196.79392	0.19779	45.03510	0.98879	5.94170	0.7531
0.29588	0.28537	0.22304	22.66856	82.18327	182.77341	0.23095	39.32554	0.98868	5.93240	0.7525
0.31104	0.29927	0.21506	19.48723	84.78699	194.16962	0.19661	44.20319	0.98873	5.94736	0.7518

5   Conclusion and Future Work

This paper described the architecture design of the integrated parameter estimation environmental management system for Hydrological models, and the environment system is divided into model management, model's parameter management, model's output variable management, optimization method management, optimization method's parameter management, model's configuration management and the model run function module. We set up a prototype system of the integrated parameter estimation environment for Hydrological models, and applied in practical applications to verify its performance and feasibility in the Heihe River basin.

Acknowledgement. This work is supported by National Nature Science Foundation of China (Grant No. 61462058), China Postdoctoral Science Foundation funded project (2013M542398), and Science and Technology Program of Gansu Province (1308RJZA214).

References

1. James, W.: Simulation Modeling for Watershed Management. Springer, New York (2000)
2. Blasone, R.S., Madsen, H., Rosbjerg, D.: Uncertainty assessment of integrated distributed Hydrological models using GLUE with Markov chain Monte Carlo sampling. Journal of Hydrology **353**, 18–32 (2008)
3. Beven, K.J., Lamb, R., Quinn, P., Romanowicz, R., Freer, J.: TOPMODEL. In: Singh, V.P. (ed.) Computer Models of Watershed Hydrology. Water Resource Publications, Colorado (1995)
4. Carpenter, T.M., Georgakakos, K.P.: Impacts of parametric and radar rainfall uncertainty on the ensemble streamflow simulations of a distributed hydrologic model. Hydrology **298**, 27–60 (2004)
5. Cheng, G.D., Zhao, C.Y.: An Integrated Study of Ecological and Hydrological Processes in the Inland River Basin of the Arid Regions, China. Advances in Earth Science **23**, 1005–1010 (2008)
6. Lan, Y.C., Kang, E.S., Zhang, J.S., Hu, X.L.: Exploitation and Utilization Status of the Water Resources in the Heihe River Basin, Problems and Its Countermeasure. Journal of Arid Land Resources and Environment **17**, 34–39 (2003)
7. Zhao, W.Z., Cheng, G.D.: Frontier Issues and Experimental Observation on Ecohydrology. Advances in Earth Science **23**, 671–674 (2008)

8. Li, Q.S., Zhao, W.Z., Feng, Q.: Relationship between the dynamic change of water resources and the oasis development and evolution in the Heihe River Basin. Arid Land Geography **29**, 21–28 (2006)
9. Kang, E.S., Chen, R.S., Zhang, Z.H., Ji, X.B., Jin, B.W.: Some Scientific Problems Facing Researches on Hydrological Processes in an Inland River Basin. Advances in Earth Science **22**, 940–952 (2007)
10. Li, X., Cheng, G.D., Kang, E.S., Xu, Z.M., Nan, Z.T., Zhou, J., Han, X.J., Wang, S.G.: Digital Heihe River Basin 3: Model Integration. Advances in Earth Science **25**, 851–861 (2010)
11. Nian, Y.Y., Wu, L.Z.: A Review on Research and Practice for Hydrological Information System. Remote Sensing Technology and Application **28**, 391–398 (2013)
12. Nan, Z., Shu, L., Zhao, Y.: Integrated modeling environment and a preliminary application on the Heihe River Basin, China. Science China Technological Sciences **54**, 2145–2156 (2011)
13. Zhang, Y.N., Xiao, H.L., Li, X., Cheng, G.D.: E-Science Environment for Integrated Research of Inland River Basin. e-Science Technology & Application **2**, 52–64 (2019)
14. He, Z.F.: Simulation research of surface water and groundwater based on HOME modeling framework. PhD thesis, University of Chinese Academy of Sciences (2014)
15. Rouholahnejad, E., Abbaspour, K.C., Vejdani, M., Srinivasan, R., Schulin, R., Lehmann, A.: A parallelization framework for calibration of Hydrological models. Environmental Modelling & Software **1**, 28–36 (2012)
16. Arnold, J.G., Moriasi, D.N., Gassman, P.W., Abbaspour, K.C.: SWAT: Model Use, Calibration, and Validation. Transactions of the ASABE **55**, 1491–1508 (2012)
17. Betrie, G.D., Mohamed, Y.A., van Griensven, A., Srinivasan, R.: Sediment management modelling in the Blue Nile Basin using SWAT model. Hydrol. Earth Syst. Sci. **15**, 807–818 (2011)
18. Keith, J.H.: MODOPTIM: A General Optimization Program for Ground-Water Flow Model Calibration and Ground-Water Management with MODFLOW. Scientific Investigations Report, The Delano Max Wealth Institute (2012)

Automatic Binary De-obfuscation Based Compiler Optimization

Lin Wei[✉], Fei Jin-Long, and Cai Rui-Jie

State Key Laboratory of Mathematical Engineering and Advanced Computing,
Zhengzhou 450002, China
tiamo9880@gmail.com, feijinlong@126.com, wsxcrj@163.com

Abstract. Junk code increase manual analysis difficulty in reverse engineering, and seriously disturb the automatic analysis process of ant code obfuscating, so find a junk code removing method has a great significance in the field of reverse engineering. Based on this, aiming at the problem that the executable context-dependent junk code is difficult to remove automatically, this paper proposes a junk code removing method based on idle register slicing, through analyzing data dependencies between instructions, get all the idle register of all instruction in code block, slice the code block by idle registers respectively, remove all the junk code in the slice instruction. Experiments show that, this method can remove embedded executable junk code rapidly and accurately, improve the efficiency of reverse engineering.

Keywords: Junk code · Program slicing · Idle objects set

1   Introduction

At present, code obfuscation is widely used [1~4] in the anti-reverse analysis, Virtual machine protection, privacy data protection etc. software protection field. It can enhance the code quantity and complexity to prevent reverse analysis and feature extraction, which lead to new challenges of existing reversing analysis method. Code obfuscating means the code is transformed to another one. The original code P is transformed to object code P', both are identical logically, but P' is much less readable and understandable due to code complexity and code quantity. Junk code is one kind of code obfuscation.

Junk code, also called garbage code, is divided into non executable and executable [5]. Non executable one refers to a group of useless byte without executing actually which inserted into the original code. It's mainly with the opaque predicate jumps against static disassembling [6], but no effect on dynamic debugging and instruction tracing. The latter refers to a group of useless byte which executes but does not change the original logic; it's divided into context-dependent and context-free junk code.

L. Wei—Manuscript received November 20, 2014. This work was supported in part by the National Key Technology Support Program of China, (2012BAH47B01);National Natural Science Foundation of China (61309007).

Y. Tan et al. (Eds.): ICSI-CCI 2015, Part II, LNCS 9141, pp. 476–483, 2015.
DOI: 10.1007/978-3-319-20472-7_51

WU WM[7] proposed to use junk code to enhance the protection of virtual machine software , but the specific methods and algorithms about junk code inserting was not given; Mental Driller[8] proposed code morphing method and corresponding reduction method, but those methods are still based on pattern matching, not suitable for context-dependent junk code. Z0mbie[9] proposed the concept of an object set, gave the permutation conditions of exchanging two instructions, but didn't give the conditions of junk code elimination and inserting.

Aiming at the problem of context-dependent junk code is difficult to eliminate automatically. This paper proposes a junk code elimination method based on idle object slicing (JEIOS). By analyzing data dependence between instruction source data and destination data, we propose the concept of object set, define the idle object set, get all the idle object set of each instruction in code block, slice the code block by idle object respectively, and finally, eliminate all the junk code in the slice instruction.

2   Preparation Knowledge

As shown in Figure 1, code obfuscation can make the code understandable by invocating the instruction substitution recursively according to the requirement.

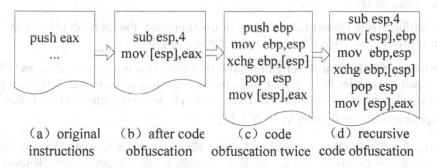

(a) original instructions   (b) after code obfuscation   (c) code obfuscation twice   (d) recursive code obfuscation

Fig. 1. Schematic diagram of code obfuscation

Code de-obfuscation can match instruction sequence according to pattern matching method with the help of disassemble engine, and then restore the original code recursively. But if obfuscated code inserts some executable junk code randomly, it will enhance the difficulty of restoring the original code greatly. It means when match instruction sequence with a predefined pattern, matching discrete instruction is needed, which will lead to massive false positives.

3   Junk Code Elimination Based on Idle Object

The basic idea of junk code elimination based on idle object slicing (JEIOS) is eliminating junk codes by analyzing data dependencies between instructions. Subsequent instructions will cover the semantics of junk code, so junk code can be executed and will not affect semantics. The key is to find out which instruction semantics are covered. Through the acquisition of each instruction idle object set, and then analyze the idle object set of all instructions in the entire code blocks by propagation algorithm, then according to the idle object clustering, appears idle blocks and non-idle blocks alternation. For an idle block, all the instruction semantics of which destination object set contain the object will be covered by the last one, so deleting these instructions will have achieved a purpose of junk code elimination.

Before the detailed description of JEIOS, we firstly introduce several basic concepts involved in JEIOS.

Definition 1(Object Set): Object set   is the set which instruction operation is affected and based on, including   all general registers, flag register, memory references, denoted as U. Taking the 32 bit X86 architectures as an example, U={EAX, EBX, ECX, EDX, ESI, EDI, ESP, EBP, flags, memory}.

Definition 2(Source Object Set): Source object set is the object set that one instruction INSK reads, denoted as S_K.

Definition 3(Destination Object Set): Destination object set is the object set that one instruction INS_K writes or affects, denoted as D_K.

Definition 4(Idle Object): For one instruction INS_K and one object P, if inserting any number of instructions between INS_{K-1} and INS_K which arbitrarily modify P and their D_K equals {P}, will not affect the following instructions execution, then P is called idle object for INS_K.

Definition 5(Idle Object Set): For one instruction INS_K, the set of all idle objects for this instruction is called idle object set, denoted as N_K, $N_K \subseteq U$.

Definition 6(Idle Object Set Propagation): Given two instructions INS_I and INS_J executed in tandem, the influence of N_J on N_I according to their data dependency is called idle object set propagation. Denote the propagating object set reckoned by propagation rules as F_I, then $N_I = N_I \bigcup F_I$.

Definition 7(Idle Object Scope) : If object K is idle object between INS_I and INS_J, then the scope between INS_I and INS_J is called Idle Object Scope, denoted as $S_K=[INS_I, \quad INS_J]$.

3.1   Overall Process

The overall process of idle object set propagation and program slicing is shown in Figure 4.

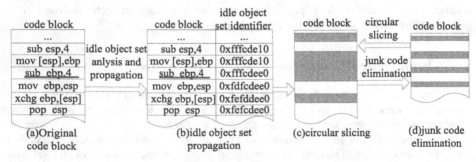

Fig. 2. Overall process

As shown in Figure 4, it including obtaining idle object set for single instruction, idle object set propagation, and program slicing based on idle object set.

3.2 Obtaining Idle Object Set for Single Instruction

Regulation 1: For one single instruction INS_K, idle object set N_K equals the intersection of the destination object set and a complementary set of source object set.

$$N_K = D_K \ \& \sim S_K \ . \tag{1}$$

$\sim S_K$ means the registers that INS_K does not refer to. But $\sim S_K$ is just idle registers for current instruction, which may be used in following instructions. $D_K \& \sim S_K$ represent the registers that are rewritten and not read, so those registers can be modified arbitrarily before this instruction. This rule is a relatively conservative rule, but $D_K \& \sim S_K$ must be idle objects which would be empty most of the time.

3.3 Idle Object Set Propagation

Regulation 2: Given the propagation direction which is from bottom to top, for two adjacent instructions INS_{K-1} and INS_K, so the elements in idle object sets N_K, which is not contained in the former instruction source object set S_{K-1}, can be propagated to former instruction idle object sets N_{K-1}.

$$N_{k-1} = N_{k-1} \bigcup (N_k - S_{k-1}) \ . \tag{2}$$

Take "ADD EAX ,4; MOV ECX ,8" for example, for the second instruction, ECX is idle object, and it is not contained in the former instruction source object set S_{K-1} {EAX}, so it is also a idle object for "ADD EAX ,4". We also can call this rule as the rule of "pass through".

Regulation 3: For two adjacent instructions INS_{K-1} and INS_K, if the elements in destination object sets D_{K-1} are all contained by the next instruction idle object set N_K, then the intersection of S_K and next instruction idle object sets N_K can be propagated to current instruction idle object sets N_{K-1}

$$if \ D_{k-1} \subseteq N_k, N_{k-1} = N_{k-1} \bigcup (N_k \bigcap S_{k-1}) \ . \tag{3}$$

If any element A of S_{K-1} is an idle object that means any modification of A won't change the following instruction semantics. But A have an impact on D_{K-1}, so any instruction which modify A will have an impact on two things: A and D_{K-1}. If A and D_{K-1} are both contained in idle object sets of next instruction, then A will be the idle object for this instruction.

Algorithm 1. Idle Object Set Propagation algorithm

INPUT: Instruction sequence list

OUTPUT: If success, return True; else return false

---

begin

1. **for** $(INS_K,$ K from 1 to N)

2. $INS_K.null\_set=XSET\_UNDEF|XSET\_MEM$//Initialize registers as non-idle object

3. **end**;

4. **for** $(INS_K,$ K from 1 to N)

5. $INS_K.null\_set \models INS_K.D_K \&(\sim INS_K.S_K)$// Initialize idle object set for instructions

7. **end**;

8. **for** $(INS_K,$ K from N-1 to 1)// propagate Idle Object Set from last instruction

9. $INS_K.null\_ set \models (INS_{K+1}.null\_set - INS_K.S_K)$ //rules 2

10.**if** $(INS_K.D_K \bigcap XSET\_MEM==NULL)$ && $(INS_K.D_K \subseteq INS_{K+1}.null\_set)$

11. $INS_K.null\_ set \models INS_K.S_K \& INS_{K+1}.null\_set$// rules 3

12. **end if**;

13. **end for**;

3.4 Junk Code Elimination Based on Idle Object Set

When all instructions idle object sets are gained, slice the code block by each element of the object set U, the process is shown in Figure 5:

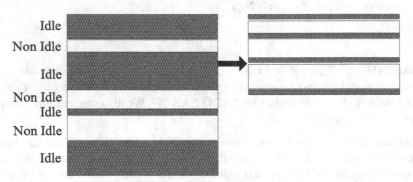

Fig. 3. Schematic diagram of the code block slicing

Hatched areas denote the instructions of which idle object set contain the EAX. Non-hatched region denote the instructions of which idle object set don't contain the EAX. EAX is idle object for each instruction in hatched areas, which means any modification on EAX value will not affect the code block semantics and the following instructions execution. So we can delete all instructions of which idle object set only contain EAX except the last one, because these instructions only affect EAX value and the last one cover the EAX value. The algorithm is as follows:

1) Initialize object set of each instruction, and call the above idle object set propagation algorithm to analyze the entire code block, obtain objective set of each instruction.

2) Slice the code block by each element RegSet[I] of object set U. That means through all instructions, determine whether RegSet[I] is included in the idle object set for each instruction. If true, add the instruction into instruction temp list TempList.

3) For each instruction in TempList, determine whether the idle object set only contain RegSet[I] or all elements in destination object set are marked as Deleted, if true, delete current instruction. If not, mark RegSet[I] in destination object set as Deleted.

4   Experimental Analysis

4.1   Validity Test

We take a part of the code in the first handle of the bubble sort algorithm protected by CodeVitualizer for example, to illustrate idle object set determining, idle object set propagation and program slicing based on idle object set.

XDE32 is an open source disassemble engine. It can analyze each instruction source object set and the destination object set denoted as a 32 bit DWORD. When an instruction source or destination object set including a register, the corresponding bit of the DWORD is set to 1. This paper also defines a DWORD REGUSED as registers used by an instruction. The REGUSED negated result means idle object set of this instruction.

1. mov eax, ecx	1. 0xfffbfff0
2. mov ebx, 0x3A	2. 0xfffb0fff
3. xor ebx, 0x98	3. 0xfffbffff
4. mov ecx, 0x98	4. 0xfffbff0f
5. add ecx, edx	5. 0xfffbffff
6. sub ebx, 0x16	6. 0xfffbffff
7. and eax, 0xff	7. 0xfffbffff
8. xor dword ptr ds:[edi+0x3e4], eax	8. 0xfffbffff
9. and ebx, ecx	9. 0xfffbffff
10. mov ebx, dword ptr ds:[edi+0x330]	10. 0xfffb0fff
11. mov ecx,0x10	11. 0xfffbff0f

(a) Idle Object Set determining of single instruction

1. mov eax, ecx	1. 0xfffb0ff0
2. mov ebx, 0x3A	2. 0xfffb0f0f
3. xor ebx, 0x98	3. 0xfffb0f0f
4. mov ecx, 0x98	4. 0xfffb0f0f
5. add ecx, edx	5. 0xfffb0f0f
6. sub ebx, 0x16	6. 0xfffb0f0f
7. and eax, 0xff	7. 0xfffb0f0f
8. xor dword ptr ds:[edi+0x3e4], eax	8. 0xfffb0f0f
9. and ebx, ecx	9. 0xfffb0f0f
10. mov ebx, dword ptr ds:[edi+0x330]	10. 0xfffb0f0f
11. mov ecx,0x10	11. 0xfffbff0f

(b) Idle Object Set Propagation

ebx is idle	1. mov eax, ecx	1. 0xfffb0ff0
	2. mov ebx, 0x3A	2. 0xfffb0f0f
	3. xor ebx, 0x98	3. 0xfffb0f0f
	4. mov ecx, 0x98	4. 0xfffb0f0f
	5. add ecx, edx	5. 0xfffb0f0f
	6. sub ebx, 0x16	6. 0xfffb0f0f
	7. and eax, 0xff	7. 0xfffb0f0f
	8. xor dword ptr ds:[edi+0x3e4], eax	8. 0xfffb0f0f
	9. and ebx, ecx	9. 0xfffb0f0f
ebx is not idle	10. mov ebx, dword ptr ds:[edi+0x330]	10. 0xfffb0f0f
	11. mov ecx,0x10	11. 0xfffbff0f

(c) program slicing results of EBX case

1. mov eax, ecx	1. 0xfffb0ff0
4. mov ecx, 0x98	4. 0xfffb0f0f
5. add ecx, edx	5. 0xfffb0f0f
7. and eax, 0xff	7. 0xfffb0f0f
8. xor dword ptr ds:[edi+0x3e4], eax	8. 0xfffb0f0f
10. mov ebx, dword ptr ds:[edi+0x330]	10. 0xfffb0f0f
11. mov ecx,0x10	11. 0xfffbff0f

(d) the results after JEIOS

Fig. 4. Junk code elimination process

1) Firstly, each instruction idle object set N_K is determined the by source object set S_K and destination object set D_K, the result is shown as figure 4(a), the left is Instruction sequence, and the right is the value of REGUSED.

2) Start idle object set propagation with the propagation rules in accordance with the instruction context. The result is shown as figure 4(b).

3) In accordance with the elements of the object set {EAX, EBX, ECX...} slice the program and cluster the instructions due to whether each instruction idle objects set include the element. Take EBX for example, the slicing result as shown as in figure 4(c). Idle objects set of instruction 1-10 include EBX, so they are idle block. The Idle object set of instruction 11 doesn't include EBX, so it is not idle block.

4) Instructions in idle block process junk code according to idle object set slicing algorithm. Take EBX for example, all the instructions which the destination object set is {EBX} are eliminated except the last one. The result is shown as figure 4(d).

4.2   Performance Test

Known from the analysis, the efficiency of idle object set propagation directly affects the performance of the whole system. Therefore this paper takes the instructions protected by virtual machine including massive junk code as experimental objects to test JEIOS processing capacity. We take all handles with original instruction "ADD EAX, EBX "protected by CodeVitualizer as the test object, 6022 instructions, and perform experiments with different numbers of Handle. The result is shown as figure 8:

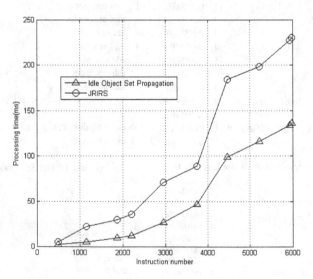

Fig. 5. Idle Object Set Propagation and JEIOS time overhead

As the result shows that:

1) JEIOS time overhead mainly used on idle object set propagation;

2) The instruction number has great influence on time overhead. So JEIOS is suitable for the code block analysis of which instruction number is below ten thousand.

3) Given hardware conditions, JEIOS has good data processing ability for the junk code, the time overhead of processing ten thousand conditions are below 0.8 seconds.

5 Conclusion

This paper presents the first effort that synergism idle object set and program slicing to solve junk code problem. The proposed solution, idle object sets, provoke interesting discussions from the assembly-level instruction data dependency analysis and open the door to a wide range of innovation opportunities. It not only can be used in automation junk code inserting or elimination, also can be used in instructions reordering. We expect to see a new generation of assembly-level instruction data dependency analysis method that is more effective, accurate, and easy to implement.

References

1. Su, Q.: Research and application of chaos opaque predicate in code obfuscation. Computer Science 40(6), 155–159 (2013)
2. Binh, N.T.: Hope: a framework for handling obfuscated polymorphic malware. In: Formal Methods, p. 26 (2014)
3. Mariano, C.: A family of experiments to assess the effectiveness and efficiency of source code obfuscation techniques. Empirical Software Engineering 19(4), 1040–1074 (2014)
4. Sebastian, S.: Covert computation: hiding code in code for obfuscation purposes. In: Proceedings of the 8th ACM SIGSAC Symposium on Information. Computer and Communications Security. ACM (2013)
5. Cai, Q.: Research and implementation of junk code obfuscating transform and encryption engine. Nanjing University of Posts and Telecommunications (2008)
6. Sun, G.Z., Cai, Q., Chen, D.W.: Study on encryption algorithm of sub-function junk code. Computer Engineering and Applications 45(3), 130–132 (2009)
7. Wu, W.M., Xu, W.F., Lin, Z.Y.: Software protection technique based on improved virtual machine. Computer Engineering & Science 36(4), 655–661 (2013)
8. Driller, M.: Metamorphism in practice or how I made metaphor and what I've learnt (2002). http://vx.netlux.org/lib/vmd01.html
9. z0mbie. Permutation conditions (2005). http://z0mbie.daemonlab.org/pcond.txt

Fast and High Quality Suggestive Contour Generation with L0 Gradient Minimization

Juncong Lin, Huiting Yang, Dazheng Wang, Guilin Li, Xing Gao$^{(\boxtimes)}$,
Qingqiang Wu, and Minghong Liao

Software School, Xiamen University, Xiamen 361005, Fujian,
People's Republic of China
{jclin,glli,gaoxing,liao}@xmu.edu.cn

Abstract. Line drawings are especially effective and natural in shape depiction. There are generally two ways to generate line drawings: object space methods and image space methods. Compared with object space methods, image space methods are much faster and independent of the 3D object, but easily affected by small noise in the rendered image. We suggest applying an edge-preserving L0 gradient smoothing step on the rendered image before line extraction. Experimental results show that our method can effectively alleviate unnecessary small scale lines, leading to results comparable to object space methods.

Keywords: Edge-preserving image smoothing · L0 gradient minimization · Contour generation

1   Introduction

Compared with realistic imagery, non-photorealistic rendering (NPR) techniques can be remarkably efficient at conveying shape and meaning with a minimum of visual distraction, and thus are vastly used in various area including illustrations, painterly rendering, and cartoon production. Line drawings are especially effective and natural in shape depiction. Lines like contours play an important role in shape recognition because they provide main cues for figure-to-ground distinction. There are generally two different strategies to generate contours from a 3D scene: Object space methods [4,12,13,27,29] compute the feature lines on the 3D objects directly; Image space methods [1,10,14,15,18,20,26] generate line drawing by applying certain image processing techniques on rendered images. Compared with object space methods, image space ones are much faster and independent of the 3D object. However, image space methods are limited by the precision of the images and easily affected by small noise presented in the rendered image. Thus, there are usually unnecessary small scale lines in the line drawing (see Fig.1(b)).

While the limitation of precision can be simply solved by increasing image resolution, those unnecessary small scale lines are rather difficult to alleviate. In this paper, we propose a L0 Gradient minimization based contour generation

© Springer International Publishing Switzerland 2015
Y. Tan et al. (Eds.): ICSI-CCI 2015, Part II, LNCS 9141, pp. 484–491, 2015.
DOI: 10.1007/978-3-319-20472-7_52

Fig. 1. With L0 gradient minimization based smoothing of the rendered image, our image space line drawing method can generated results comparable to object space methods: (a)original rendered image; (b) line drawing from (a); (c) smoothed image of (a); (d) line drawing from (c); (e) line drawing by object space method

method in the image space. To be specific, we apply a L0 gradient minimization process to smooth the rendered image while preserving edges before line extraction. Our method can effectively alleviate those small scale lines induced by noise in the rendered image, leading to results that are comparable to object space method (see Fig.1(d) and (e)).

2   Related Work

2.1   Edge-Preserving Image Filtering

Edge-preserving smoothing is frequently used in recent computational photography over the last decade. Traditional strategies decompose a given image into structure and detail by smoothing the image, simultaneously preserving or even enhancing image edges, and they differ from each other in how they define edges and how this prior information guides smoothing. Perona and Malik [16] presented an anisotropic diffusion model in which pixel-wise spatially-varying diffusivities are estimated from image gradients. These diffusivities prevent smoothing at image edges, and preserve important image structures while eliminating noise and fine details. In [25], Tumblin and Turk pioneered their use in tone mapping, with LCIS, a variant of anisotropic diffusion. However, anisotropic diffusion is rather time consuming and tends to oversharpen edges, and therefore is further investigated to address these limitations [22]. Another way is to use the bilateral filter which is popularized by Tomasi and Manduchi [24] and further improved by other researchers [6] mainly in efficiency. Due to its simplicity and effectiveness, bilateral filtering has been successfully applied to several computational photography applications [8,26]. Although bilateral filter works quite well for noise removal and detail extraction at a fine spatial scale, it is less appropriate to be extended for arbitrary scale. Some other researchers chose to adopt edge preserving regularization with either weighted least square [7] or total variation [19]. However, such kind of methods may influence contrast during smoothing as they penalize large gradient magnitudes too. Recently, Xu et al. [28] presented a complementary method to globally control how many non-zero gradients are based L0 gradient minimization , resulting into approximate prominent structure in a sparsity-control manner.

2.2 Image Space Contour Generation

We focus on relevant literature in image space contour generation. A simple way to generate a line drawing of a 3D scene would be to render the scene from the desired point of view, detect edges in the images, and display the edges. In [9], Gooch et al. introduced several simple algorithms on silhouette detection. The work of Raskar and Cohen [17] proposed a related work to allow variable-length line segments. However, the edges of a photograph do not typically correspond to the silhouette edges that we want to illustrate [21]. A better way is to extract the depth map, and apply an edge detector on it [3,5,20]. However, it cannot detect the boundaries between objects that are at the same depth, nor does it detect creases. It can be further improved by considering surface normal as well, to detect edges changing in surface orientation and incorporate these edges.

3 Background

Contours are lines where a surface turns away from the viewer and becomes invisible. From a mathematical view [2], contours of a smooth and closed surface S are the set of points that lie on this surface and satisfy:

$$\mathbf{n}(\mathbf{p}) \cdot \mathbf{v}(\mathbf{p}) = 0. \tag{1}$$

Here $\mathbf{c}$ is the view point, $\mathbf{p} \in S$ is a point on the surface, $\mathbf{n}(\mathbf{p})$ is the unit surface normal at $\mathbf{p}$, and $\mathbf{v}$ is the view vector: $\mathbf{v}(\mathbf{p}) = \mathbf{c} - \mathbf{p}$. DeCarlo et al. [4] noticed that the visual system are also sensitive to features where a surface bends sharply away from the viewer, yet remains visible, those that are almost contours and become contours in nearby views. They called these features *suggestive contours*, and formally defined them as the set of points on the surface at which its radial curvature κ_r is 0, and the directional derivative of κ_r in the direction of $\mathbf{w}$ (the projection of the view vector $\mathbf{v}$ on the tangent plane at $\mathbf{p}$) is positive:

$$D_w \kappa_r > 0. \tag{2}$$

Here, the directional derivative $D_w \kappa_r$ is defined as the differential of $\kappa_r(\mathbf{p})$ applied to $\mathbf{w}$, or $d\kappa_r(\mathbf{w})$.

Suggestive contour can be detected in the object space by finding zero crossing of κ_r exhaustively, filtering out those spurious zero crossings induced when the minimum principal curvature is close to zero and $\mathbf{w}$ roughly looks down its corresponding principal direction.

4 Image Space Suggestive Contour with L0 Gradient Minimization

Suggestive contours can also be generated in the image space. Note that the suggestive contour can also be defined as the set of minima of $\mathbf{n} \cdot \mathbf{v}$ in the direction of $\mathbf{w}$[4], thus we can firstly approximate $\mathbf{n} \cdot \mathbf{v}$ by rendering a smoothly shaded image with a diffuse light source placed at the camera origin and detect suggestive contours in this image as steep valleys in intensity, to find stable minima of $\mathbf{n} \cdot \mathbf{v}/\|\mathbf{v}\|$.

The key is to detect all the valleys effectively. There are various options [11,23], we can simply adopt one of them. However, noise and the high susceptibility of edge detector make high quality extraction difficult. We suggest stabilizing the extraction process with a prior suppression of low-amplitude details in the shaded image while preserving the main structures.

Inspired by [28], we also use a L0 gradient minimization scheme for the stabilization, which resolve issues arise in previous work due to its global nature.

4.1   L0 Gradient Minimization Based Stablization

Denoting the shaded image I, stablized image S, the gradient $\nabla S_p = (\partial_x S_p, \partial_y S_p)^T$ for each pixel p on S is computed as color difference between neighboring pixels along the x and y directions. The gradient measure is then defined as:

$$C(S) = \sharp\{p||\partial_x S_p| + |\partial_y S_p| \neq 0\}. \tag{3}$$

It count p whose magnitude $|\partial_x S_p| + |\partial_y S_p|$ is not zero. We can finally estimate by solving

$$\min_S \{\sum_p (S_p - I_p)^2 + \lambda \cdot C(S)\}. \tag{4}$$

The gradient magnitude $|\partial S_p|$ is defined as the sum of gradient magnitudes in rgb. The term $\sum(S - I)^2$ constrains image structure similarity.

The two terms in Equation (4) model the pixel-wise difference and global discontinuity statistically respectively, conventional discrete optimization method, such as gradient decent, are not appropriate. The basic idea is to expand the original terms by introducing auxiliary variables, and solve the problem with half quadratic splitting. To be specific, auxiliary variables h_p and v_p, corresponding to $\partial_x S_p$ and $\partial_y S_p$ are introduced, leading to a new objective function:

$$\min_{S,h,v} \{\sum_p (S_p - I_p)^2 + \lambda \cdot C(h,v) + \beta \cdot (\partial_x S_p - h_p)^2 + (\partial_y S_p - v_p)^2\}. \tag{5}$$

where $C(h,v) = \sharp\{p||h_p| + |v_p| \neq 0\}$ and β is an automatically adapting parameter to control the similarity between variables (h,v) and their corresponding gradients. The new objective function can be solved through alternatively minimizing (h,v) and S, with one set of the variables fixed in each pass:

- **Pass 1: computing S.** The S estimation stage corresponds to minimizing by omitting the terms not involving S in Equation 5

$$\{\sum_p (S_p - I_p)^2 + \beta \cdot (\partial_x S_p - h_p)^2 + (\partial_y S_p - v_p)^2\}. \tag{6}$$

– **Pass 2 : computing** (h, v) The objective function for (h, v) is

$$\min_{h,v}\{\sum_p ((\partial_x S_p - h_p)^2 + (\partial_y S_p - v_p)^2 + \frac{\lambda}{\beta} C(h, v)\}. \tag{7}$$

where $C(h, v)$ returns the number of non-zero elements in $|h_p| + |v_p|$. The energy can be spatially decomposed where each element h_p and v_p is estimated individually:

$$\sum_p \min_{h_p, v_p}\{\sum_p ((h_p - \partial_x S_p)^2 + (v_p - \partial_y S_p)^2 + \frac{\lambda}{\beta} H(|h_p| + |v_p|)\}. \tag{8}$$

and making the problem empirically solvable. $H(|h_p| + |v_p|)$ is a binary function returning 1 if $|h_p| + |v_p| \neq 0$ and 0 otherwise.

5   Experimental Results

We have implemented a prototype system to validate the feasibility of our approach. We ran various tests on a desktop PC with an Intel i5 CPU@2.70GHz, 8.00G RAM, GeForce GT 640M GPU@625MHz. The L0 gradient minimization

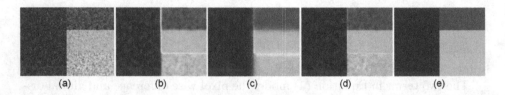

Fig. 2. Comparison among different image smoothing methods: (a)Noise image created by Farbman et al. [7]; (b)Gaussian filtering with $\sigma_s = 4$; (c)Bilateral filtering with $\sigma_s = 4, \sigma_r = 0.15$; (d)Weighted least square filtering with $\alpha = 1.2, \lambda = 0.25$; (e)L0 gradient minimization

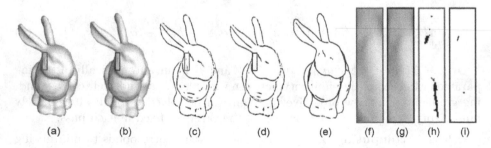

Fig. 3. Comparison of different line drawing method: (a)rendered image; (b) smoothed image; (c)line drawing from (a); (d) line drawing from (b); (e) line drawing in object space; (f)zoom in of (a); (g)zoom in of (b); (h)zoom in of (c); (i)zoom in of (d)

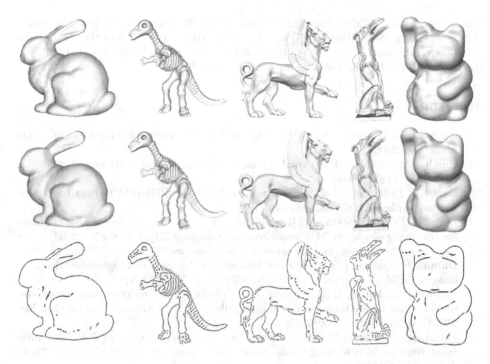

Fig. 4. More examples: First row shows the rendered images; Second row shows smoothed images; Third row shows the line drawing results

scheme can effectively small noise in image while preserving the main structures. To show that, we compare with several state-of-art methods in Fig.2. Fig.3 shows comparisons of the line drawing results of our method, direct image space method and object space method. We can clearly see from the Figure that the results generated by our method are more similar to the objet space method, with less small lines due to jittering in the shaded image. Fig.4 shows more results generated by our method.

6   Conclusions

In this paper, we present an image space line drawing method which can generate high quality line drawing results quickly. The key of our method is to smooth the rendered image with an edge-preserving Lo gradient minimization method before line extraction. Our scheme can effectively alleviate small scale lines induced by noise in the rendered image, leading to results that are comparable to object space method. Our method is independent of the complexity of the 3D object.

Acknowledgments. This work was supported by the National Natural Science Foundation of China (No.61202142), Joint Funds of the Ministry of Education of China and

China Mobile (No.MCM20130221), the National Key Technology R&D Program Foundation of China (Nos.2015BAH16F00/F02, 2015BAH34F00/F01) and the Fundamental Research Funds for the Central Universities (Nos. 20720140546, 2013121030),Key Technology R&D Program of Xiamen(No.3502Z20133011).

References

1. Buchanan, J.W., Sousa, M.C.: The edge buffer: a data structure for easy silhouette rendering (2000)
2. Cipolla, R., Giblin, P.: Visual Motion of Curves and Surfaces. Cambridge University Press (2000)
3. Curtis, C.J.: Loose and sketchy animation. In: ACM SIGGRAPH 1998 Electronic Art and Animation Catalog, p. 145 (1998)
4. DeCarlo, D., Finkelstein, A., Rusinkiewicz, S., Santella, A.: Suggestive contours for conveying shape. ACM Transacitons on Graphics **22**(3), 848–855 (2003)
5. Decaudin, P.: Cartoon-looking rendering of 3d-scenes. Tech. rep., INRIA (1996)
6. Durand, F., Dorsey, J.: Fast bilateral filtering for the display of high-dynamic-range images. ACM Transactions on Graphics **21**(3), 257–266 (2002)
7. Farbman, Z., Fattal, R., Lischinski, D., Szeliski, R.: Edge-preserving decompositions for multi-scale tone and detail manipulation. ACM Transactions on Graphics **27**(3), Article No. 67 (2008)
8. Fattal, R., Agrawala, M., Rusinkiewicz, S.: Multiscale shape and detail enhancement from multi-light image collections. ACM Transactions on Graphics **26**(3), Article No. 51 (2007)
9. Gooch, B., Sloan, P.P.J., Gooch, A., Shirley, P., Riesenfeld, R.: Interactive technical illustration. In: Proceedings of ACM Symposium on Interactive 3D Graphics, pp. 31–38 (1999)
10. Isenberg, T., Freudenberg, B., Halper, N., Schlechtweg, S., Strothotte, T.: A developer's guide to silhouette algorithms for polygonal models. IEEE Computer Graphics and Applications **23**(4), 28–37 (2003)
11. Iverson, L.A., Zucker, S.W.: Logical/linear operators for image curves. IEEE Transactions on Pattern Analysis and Machine Intelligence **17**(10), 982–996 (1995)
12. Judd, T., Durand, F., Adelson, E.: Apparent ridges for line drawing. ACM Transactions on Graphics **26**(3), Article No. 19 (2007)
13. Kolomenkin, M., Shimshoni, I., Tal, A.: Demarcating curves for shape illustration. ACM Transactions on Graphics **27**(5), Article No. 157 (2008)
14. Lee, Y., Markosian, L., Lee, S., Hughes, J.F.: Line drawings via abstracted shading. ACM Transacitons on Graphics **26**(3), Article No. 18 (2007)
15. Nienhaus, M., Doellner, J.: Edge-enhancement - an algorithm for real-time non-photorealistic rendering. Journal of WSCG **11**(2), 346–353 (2003)
16. Perona, P., Malik, J.: Scale-space and edge detection using anisotropic diffusion. IEEE Transactions on Pattern Analysis and Machine Intelligence **12**(7), 629–639 (1990)
17. Raskar, R., Cohen, M.: Image precision silhouette edges. In: Proceedings of Symposium on Interactive 3D Graphics, pp. 135–140 (1999)
18. Raskar, R., Tan, K.H., Feris, R., Yu, J., Turk, M.: Non-photorealistic camera: depth edge detection and stylized rendering using multi-flash imaging. ACM Transactions on Graphics **23**(3), 679–688 (2005)
19. Rudin, L.I., Stanley, J., Fatemi, E.: Nonlinear total variation based noise removal algorithms. Physica D: Nonlinear Phenomena **60**(1–4), 259–268 (1992)

20. Saito, T., Takahashi, T.: Comprehensible rendering of 3-d shapes. ACM SIGGRAPH Computer Graphics **24**(4), 197–206 (1990)
21. Sanocki, T., Bowyer, K.W., Heath, M.D., Sarkar, S.: Are real edges sufficient for object recognition? Journal of Experimental Psychology: Human Perception and Performance **24**(1), 340–349 (1998)
22. Scherzer, O., Weickert, J.: Relations between regularization and diffusion filtering. Journal of Mathematical Imaging and Vision **12**(1), 43–63 (2000)
23. Steger, C.: Subpixel-precise extraction of watersheds. In: Proceedings of the International Conference on Computer Vision, vol. 2, p. 884 (1999)
24. Tomasi, C., Manduchi, R.: Bilateral filtering for gray and color images. In: International Conference on Computer Vision, pp. 839–846 (1998)
25. Tumblin, J., Turk, G.: Lcis: a boundary hierarchy for detail-preserving contrast reduction. In: Proceedings of ACM SIGGRAPH, pp. 83–90 (1999)
26. Winnemoller, H., Olsen, S.C., Gooch, B.: Real-time video abstraction. ACM Transactions on Graphics **25**(3), 1221–1226 (2006)
27. Xie, X., He, Y., Tian, F., Seah, H.S., Gu, X., Qin, H.: An effective illustrative visualization framework based on photic extremum lines. IEEE Transactions on Visualization and Computer Graphics **13**(6), 1328–1335 (2007)
28. Xu, L., Lu, C., Xu, Y., Jia, J.: Image smoothing via l0 gradient minimization. ACM Transactions on Graphics **30**(6), Article No. 174 (2011)
29. Zhang, L., He, Y., Xia, J., Xie, X., Chen, W.: Realtime shape illustration using laplacian lines. IEEE Transactions on Visualization and Computer Graphics **17**(7), 993–1006 (2011)

Author Index

Printed in the United States
By Bookmasters

Printed in the United States
By Bookmasters